Lecture Notes in Artificial Intelligence 1404

Subseries of Lecture Notes in Computer Science
Edited by J. G. Carbonell and J. Siekmann

Lecture Notes in Computer Science

Edited by G. Goos, J. Hartmanis and J. van Leeuwen

Springer
Berlin
Heidelberg
New York
Barcelona
Budapest
Hong Kong
London
Milan
Paris
Santa Clara
Singapore
Tokyo

Christian Freksa   Christopher Habel
Karl F. Wender (Eds.)

Spatial Cognition

An Interdisciplinary Approach
to Representing and Processing
Spatial Knowledge

Springer

Series Editors
Jaime G. Carbonell, Carnegie Mellon University, Pittsburgh, PA, USA
Jörg Siekmann, University of Saarland, Saarbrücken, Germany

Volume Editors

Christian Freksa
Christopher Habel
Universität Hamburg, Fachbereich Informatik
Vogt-Kölln-Str. 30, D-22527 Hamburg, Germany
E-mail: {freksa, habel}@informatik.uni-hamburg.de

Karl F. Wender
Universität Trier, Fachbereich Psychologie
D-54286 Trier, Germany
E-mail: wender@cogpsy.Uni-Trier.de

Cataloging-in-Publication Data applied for

Die Deutsche Bibliothek - CIP-Einheitsaufnahme

Spatial cognition : an interdisciplinary approach to representing and
processing spatial kowledge / Christian Freksa ... (ed.). - Berlin ;
Heidelberg ; New York ; Barcelona ; Budapest ; Hong Kong ;
London ; Milan ; Paris ; Santa Clara ; Singapore ; Tokyo : Springer,
1998
  (Lecture notes in computer science ; Vol. 1404 : Lecture notes in
  artificial intelligence)
  ISBN 3-540-64603-5

CR Subject Classification (1991): I.2.4, I.2, J.2, J.4

ISSN 0302-9743
ISBN 3-540-64603-5 Springer-Verlag Berlin Heidelberg New York

Typesetting: Camera ready by author
SPIN 10637134   06/3142 – 5 4 3 2 1 0   Printed on acid-free paper

Preface

Research on spatial cognition is a rapidly evolving interdisciplinary enterprise for the study of spatial representations and cognitive spatial processes, be they real or abstract, human or machine. Spatial cognition brings together a variety of research methodologies: empirical investigations on human and animal orientation and navigation; studies of communicating spatial knowledge using language and graphical or other pictorial means; the development of formal models for representing and processing spatial knowledge; and computer implementations to solve spatial problems, to simulate human or animal orientation and navigation behavior, or to reproduce spatial communication patterns.

These approaches can interact in interesting and useful ways: Results from empirical studies call for formal explanations both of the underlying memory structures and of the processes operating upon them; we can develop and implement operational computer models obeying the relationships between objects and events described by the formal models; we can empirically test the computer models under a variety of conditions, and we can compare the results to the results from the human or animal experiments. A disagreement between these results can provide useful indications towards the refinement of the models.

The insight we gain in doing basic research on spatial cognition has a potential towards a great variety of applications. Without understanding human spatial cognition we will not be able to develop appropriate technology and interfaces for spatial information systems that communicate with humans by language and graphics in natural and efficient ways. Autonomous robots finding their ways in an unknown environment require abilities to infer locations of objects from incomplete and qualitative information from various sources and to follow imprecise instructions much like human beings. To use maps and other diagrams for the communication with computers we must understand how people generate and interpret them. To fully exploit the potential of virtual reality technology we must adapt its capabilities to human conceptions of space. To develop computers programmed by spatial structures rather than by sequential instructions we must more fully understand the relevant aspects of space.

In 1996, the Deutsche Forschungsgemeinschaft (DFG) established a priority program on spatial cognition to promote interdisciplinary research in this field. Fifteen research projects at thirteen research institutions across Germany cooperate in this program. In Fall 1997, a colloquium was held at the University of Trier. Fifteen projects from the priority program, two thematically related projects from other DFG programs, and five invited experts from other countries presented and discussed their work. After the discussions at the colloquium, the contributions were revised and underwent an anonymous reviewing and revision procedure. The resulting papers are collected in this volume.

The volume consists of 22 contributions and is structured into three sections: *Spatial knowledge acquisition and spatial memory, Formal and linguistic models,* and *Navigation in real and virtual worlds.* The first section consists of contributions describing empirical investigations and representations derived from such investigations; knowledge acquisition, memory organization, and spatial reference systems are addressed. The second section presents formal approaches to structuring spatial knowledge; the connection between language and spatial concepts and the formal organization of spatial concepts are addressed in this section. The third section brings together empirical and application-oriented views of navigation; the connections to robotics on one hand and to virtual reality on the other hand are addressed here.

We would like to thank all the authors for their careful work and for keeping our very tight deadlines. We thank our reviewers for their insightful and thorough comments which they prepared on a short notice. We thank Karin Schon for her superb editorial support and for motivating the authors to give this project top priority in their schedules. We thank the LNAI series editor Jörg Siekmann for proposing the publication of this volume in the Lecture Notes in Artificial Intelligence, and we thank Alfred Hofmann and Springer-Verlag for supporting our project and for helpful suggestions. Finally, we gratefully acknowledge the support by the Deutsche Forschungsgemeinschaft and we thank the reviewers and the administrator of the DFG priority program on spatial cognition for valuable advice.

April 1998   Christian Freksa

Christopher Habel

Karl F. Wender

Contents

Spatial Knowledge Acquisition and Spatial Memory

Formal and Linguistic Models

Navigation in Real and Virtual Worlds

Allocentric and Egocentric Spatial Representations: Definitions, Distinctions, and Interconnections

Roberta L. Klatzky[1]

Carnegie Mellon University, Professor and Head, Department of Psychology,
5000 Forbes Avenue, Pittsburgh, Pennsylvania, 15213, USA
klatzky@cmu.edu

Abstract. Although the literatures on human spatial cognition and animal navigation often make distinctions between egocentric and allocentric (also called exocentric or geocentric) representations, the terms have not generally been well defined. This chapter begins by making formal distinctions between three kinds of representations: allocentric locational, egocentric locational, and allocentric heading representations. These distinctions are made in the context of whole-body navigation (as contrasted, e.g., with manipulation). They are made on the basis of primitive parameters specified by each representation, and the representational distinctions are further supported by work on brain mechanisms used for animal navigation. From the assumptions about primitives, further inferences are made as to the kind of information each representation potentially makes available. Empirical studies of how well people compute primitive and derived spatial parameters are briefly reviewed. Finally, the chapter addresses what representations humans may use for processing spatial information during physical and imagined movement, and work on imagined updating of spatial position is used to constrain the connectivity among representations.

1 Reference Frames and Spatial Representations

Put simply, a reference frame is a means of representing the locations of entities in space. An entire chapter could be devoted to frames of reference, and several excellent ones have been (see, e.g., Berthoz, 1991; Brewer & Pears, 1993; Levinson, 1996; Soechting, Tong & Flanders, 1996). Nor is the literature deficient in discussions of allocentric and egocentric representations. In fact, the contrast between those two

[1] Chapter prepared for conference on Raumkognition, Trier, Germany, September 1997. The author acknowledges support of Grant 9740 from the National Eye Institute for the study of Klatzky et al. (in press). This work has benefited from extensive discussions of spatial cognition with Jack Loomis.

terms abounds in discussions of spatial perception, spatial cognition, and spatially directed action. Not everyone treats these terms equivalently, however. Allocentric is sometimes used synonymously with "exocentric" or "geocentric." To Paillard (1971) the geocentric reference frame was gravity-based and subsumed both allocentric and egocentric representations. Pick (1988) has pointed out that although allocentric implies reference to another human, the term has assumed more general use.

Exceptions notwithstanding, there is general understanding that in an egocentric reference frame, locations are represented with respect to the particular perspective of a perceiver, whereas an allocentric reference frame locates points within a framework external to the holder of the representation and independent of his or her position. While the general distinction between allocentric and egocentric representations is commonly made, it is far less common to see a specific proposal for what is represented in each.

2   Some Basic Definitions

In this section, I define critical parameters that are conveyed by spatial representations. It is important to note that the parameters are being defined independently of the nature of the representation that conveys them. For example, the egocentric bearing of a point in space will be defined as an angle that is measured with respect to an object, ego, within the space. The egocentric bearing is a particular numerical value that could be derived from an egocentric or allocentric representation, given relevant input information. It is the parameter definitions, rather than processes that derive the parameters, that are specified in this section.

2.1   Spatial Parameters

The parameters of a spatial representation are values that can be assigned to individual points (e.g., location of one point) or multiple points (e.g., distance between two points). *Primitive* parameters are those that the spatial representation conveys directly for all entities that are included in the representation. *Derived* parameters are those that can be computed from primitives, possibly in several computational steps. A *locational* representation is one that has primitives conveying the locations of points in space. A *heading* representation is one that has primitives conveying the heading of objects in space.

2.2   Points and Objects

A "point" refers to a spatial location for which the values of the primitive parameters in a locational representation are known. An "object" comprises multiple points that are organized into a coherent entity.

2.3 Axis of Orientation of an Object

The axis of orientation of an object is a line between points on the object that defines a canonical direction in space. Not all objects have an axis of orientation; for example, an object that is radially symmetrical has none. The axis of orientation of a person within a space is aligned with the sagittal plane. One can differentiate between the axis of orientation of the head vs. the body, but for most of the present paper, that distinction is irrelevant.

2.4 Heading of an Object

An object's heading in space is the angle between the object's axis of orientation and some reference direction external to the object (see Figure 1). The heading of a moving object can be differentiated from its *course*, or direction of travel as defined over the past few locations that were occupied. Because the reference direction is external to the object (a heading that was defined relative to its own axis of orientation would always be zero), heading will sometimes be referred to as *allocentric heading*.

2.5 Bearing Between Two Points

Like heading, bearing is defined with respect to a reference direction. The bearing from point A to point B is the angle between the reference direction and a line from A to B. If the reference direction is aligned with the axis of orientation of an "ego" (i.e., an oriented organism in the space), the bearing from A to B will be called *ego-oriented*. If any other reference direction is used, the bearing from A to B will be called *allocentric*. The *egocentric bearing* of a point, B, is equivalent to a bearing from ego to B, using ego's axis of orientation as the reference direction. Thus the egocentric bearing is a special case of the ego-oriented bearing, in which ego's location is the source point. The egocentric bearing of B is numerically (but not conceptually) equivalent to the difference between B's allocentric bearing from ego and ego's allocentric heading, when both are defined with respect to a common reference direction. (See Figure 1.)

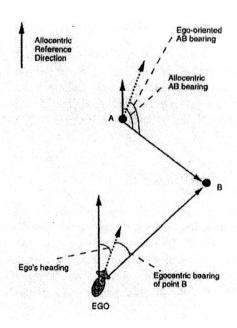

Figure 1. Illustration of basic terms introduced in text

2.6 Distance Between Two Points

The term distance is used here as it is commonly defined, i.e., as a metric relation between points corresponding to their separation in space (typically in this paper, Euclidean). It is sometimes useful to differentiate between egocentric and nonegocentric distance. The *egocentric distance* of some point P is the distance from ego to P; the distance between two points other than ego is called a *nonegocentric distance*.

3 Core Assumptions

This paper stipulates a core set of assumptions about representations of allocentric location, egocentric location, and allocentric heading, as follows.

i) Allocentric and egocentric locational representations convey the layout of points in space by means of an internal equivalent of a coordinate system (which may be distorted or incomplete).

ii) The primitives of allocentric and egocentric locational representations differ. The locational information provided by an allocentric representation is referred to space external to the perceiver; the information provided by an egocentric representation is referred to an ego with a definable axis of orientation. Specifically, the allocentric representation conveys the positions of points in the internal equivalent of Cartesian or Polar coordinates. The egocentric representation makes use of a special polar coordinate system in which the origin is at ego and the reference axis is ego's axis of orientation; it conveys the location of a point by egocentric distance and the egocentric bearing.

iii) In addition to the two types of locational representation, there is also an allocentric heading representation, which defines the angle of ego's axis of orientation relative to an external reference direction.

iv) What can in principle be computed from the primitives in the representations differs, as will be discussed further below. Point-to-point bearings are not stably defined with respect to the egocentric locational representation but can stably be computed from the allocentric one. To compute the heading of a unitary object other than the ego requires that the object be treated as multiple points (or point plus axis of orientation).

v) Connectivity among sensory systems and the different representations allows representations to be updated from sensory input. Sensory signals of changes in heading are input to the allocentric heading representation and from there are input to a locational representation (egocentric or allocentric) for purposes of updating. Signals of translatory changes of position are directly input into the egocentric and/or allocentric locational representation for purposes of updating.

vi) Connectivity between cognitive processes and the representations also allows representations to be updated from imagery -- apparently with limitations, as described below.

4 Primitives of Allocentric and Egocentric Representations

I will deal with primitives of representations in the context of a plane; generalization to three-dimensional space would be straightforward. For convenience, I will use a polar coordinate system to describe locations in the allocentric representation. The issue of Cartesian vs. polar coordinates was discussed by Gallistel (1990) in the context of updating during dead-reckoning navigation. He pointed out that computing within Cartesian coordinates is more stable, because it avoids feedback loops that compound error in updating. For present purposes, however, this issue is not critical, as the formalisms are interchangeable.

The key assumption made here is that different information is primitive in a navigator's allocentric and egocentric locational representations, as shown in Figure 2. An allocentric locational representation has an origin and reference direction. A point P in the allocentric representation has coordinates (d_o, β), where d_o is the distance from the origin and β is the bearing from the origin, defined with respect to the reference direction. A point in the egocentric locational representation has coordinates (d_e, μ) defined with respect to the ego, where d_e is the egocentric distance of the point and μ is its egocentric bearing. Note that the egocentric bearing is not defined relative to the navigator's heading (i.e., relative to an external reference direction); instead, egocentric bearing is defined with respect to an intrinsic axis of orientation that is imposed by the navigator's physical configuration.

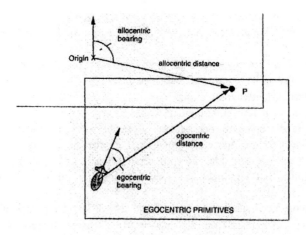

Figure 2. Egocentric and allocentric primitives

An important aspect of this formulation is that neither the egocentric nor allocentric locational representation directly conveys the headings of objects (including the navigator). Heading is not represented as a primitive in these systems, because they represent the locations of single points, which have no orientation. In contrast, object orientations -- which are needed in order to determine heading -- are defined with respect to the object's contours, comprising multiple points. An object has a heading in space only insofar as an axis of orientation can be defined, which requires that the object be represented as comprising at least two points, each with its own location. Although heading is not a primitive in the locational representation, it is a primitive in the allocentric heading representation, which conveys information solely about the heading of the navigator. The headings of other objects are derived properties rather than primitives.

5   Computation of Derived Properties from Allocentric and Egocentric Representations

In this section, I consider what derived properties could theoretically be derived from the various representations by a computational process. As was noted by Loomis et al. (1993), what can be computed in theory does not necessarily coincide with what actually is computable given the human architecture. It is unlikely, to say the least, that people have internalized computational algorithms equivalent to trigonometric axioms like the law of cosines or law of sines. Important spatial processes of humans reside in their ability to construct percept-like images of space that can be scanned, rotated, or otherwise transformed (see Finke, 1989, for review). The limitations on what can be computed are actually more interesting than the set of computable data, since the locational representations are rich enough to potentially convey all possible interpoint distances and angular relations.

5.1 Distance

Egocentric. The *egocentric distance* of a point is a primitive in the egocentric locational representation. The egocentric distance must be derived in the allocentric locational representation, since the navigator is without special status.

Point from Origin. The *distance of a point from an arbitrary origin* is a primitive in the allocentric representation. The distance of a point from an arbitrary origin must be derived in the egocentric representation.

Interpoint. In both representations, *distances between arbitrary pairs of points* (other than an origin or the ego) must be derived by computation.

5.2 Bearing

Egocentric. The *egocentric bearing* of a point is a primitive in the egocentric representation, but it requires computation within the allocentric representation. This computation can only be done if the ego is represented as an oriented object in the allocentric representation.

Point from Origin. The *bearing of a point from an origin* is a primitive in the allocentric locational representation. See the next section for the bearing of a point from the origin within an egocentric representation.

Interpoint. *Bearings between arbitrary points* can be computed from either an egocentric or allocentric representation. However, there is no external reference direction in the egocentric representation. A bearing between two arbitrary points could still be defined with respect to the navigator's axis of orientation; this would be the ego-oriented bearing. It could be computed from the distance and angle coordinates that are primitives of the egocentric representation, but since the reference direction (i.e., the navigator's axis of orientation) changes with the navigator's rotations, the ego-oriented bearing within an egocentric representation is intrinsically unstable.

5.3 Heading

Object Heading. The *heading of an arbitrary object* in space is the angle between its axis of orientation and some reference direction. Heading is the sole primitive in the allocentric heading representation. As was noted above, an object's heading can be computed as a derived parameter from primitives in the allocentric locational representation, if the object is represented as two (or more) points with an axis of orientation. The object's heading is then the bearing along the axis of orientation, with respect to the reference axis.

In the egocentric representation, it might be possible to use the navigator's axis of orientation as a reference axis, in order to define the heading of other objects. In this case, the object's heading would be equivalent to the ego-oriented bearing. However, like other ego-oriented bearings, this is intrinsically unstable, because the object's heading, so defined, would change as the navigator turned.

Navigator Heading. The *heading of the navigator* cannot be represented within egocentric space, since heading is defined in allocentric terms, and the reference axis within the egocentric representation always remains aligned with the navigator. That is, the navigator's heading is always zero in egocentric space. If, in the allocentric space, the navigator is represented as an object with an axis of orientation, then the navigator's allocentric heading can be computed.

6 Allocentric and Egocentric Representation in Rodents: A Systems Organization

Gallistel (1990) suggested that allocentric (in his terms, geocentric) maps of a spatial environment are constructed from two lower-level processes. One is the construction of an egocentric representation, which is assumed to result from early perceptual processes. The second is path integration, the process by which velocity or acceleration signals are integrated to keep track of a navigator's position in allocentric coordinates. Knowing their allocentric position in the space, and having the egocentric coordinates to other objects, navigators can build a map that allows the object-to-object relations to be represented allocentrically.

A functional system that connects egocentric and allocentric representations in rodents has been developed in more detail, with assumptions about neural localization, by Touretzky and Redish (Redish, 1997; Redish and Touretzky, 1997; Touretzky & Redish, 1996). As described by Touretzky and Redish (1996), the system has five interconnected components that convey information about spatial layout. Inputs to these components come from sensory (visual and vestibular) signals and motor efference copy. The components are as follows.

i) A *visual-perception component* provides egocentric coordinates of objects in the environment; this component is also assumed to determine the object type through pattern-recognition processing. The neural localization of this component is not considered in depth by the authors, but the relevant perceptual outputs presumably result from processing by both "what" and "where" streams in early vision. The spatial (cf. object type) information conveyed by this component corresponds to the primitives of an egocentric locational representation, as described here (i.e., egocentric bearing and distance).

ii) A *head-direction component* conveys the animal's heading in allocentric coordinates. This component is neurally instantiated by head-direction cells, which fire when the rat adopts a particular heading, regardless of its location. Head-direction cells that could serve this function have been found in several areas of the rat brain. The reference direction for heading is established by remote landmarks and/or vestibular signals.

(iii) A *path-integration component* provides an allocentric representation of the animal's position. In the rat, efference copy, vestibular signals, and optic flow could all contribute to path integration. The process has been studied in a wide variety of lower animals (reviewed in Gallistel, 1990; Maurer & Séguinot, 1995; Etienne, Maurer & Séguinot, 1996) and to some extent in humans (reviewed in Loomis, Klatzky and Golledge, in press). Redish (1997) suggested that the path integration component in the rodent involved a loop among several cortical areas.

(iv) A *local-view component* receives inputs from the visual-perception and head-direction components. It represents objects in space using the same coordinates as the egocentric representation -- that is, the distance and bearing of objects from the navigator -- but now the bearing is relative to an allocentric reference direction rather than to an axis oriented with the navigator. In order to determine the allocentric bearing, the navigator's heading, obtained from the head direction component, has been taken into account.

(v) The local view and path integrator components feed into a *place-code component*, which serves to associate them. The neural instantiation of the place code is a set of place cells, which fire when the animal is in a specific location (the cell's place field) without regard to its orientation. Place cells in the hippocampus of rodents have been widely studied since their discovery more than two decades ago (e.g., O'Keefe & Dostrovsky, 1971; O'Keefe, 1976; O'Keefe & Nadel, 1978). The response to place is not driven entirely by visual cues, since vestibular contributions have been demonstrated by the finding that place cells fire even in the dark (McNaughton, Leonard, & Chen, 1989).

For present purposes, the first three of these components are most relevant, since they clearly differentiate between egocentric and allocentric forms of locational representation in the rodent, and they specify an additional component, concerned with allocentric heading direction, that is necessary to compute one representation from another. The allocentric representation is instantiated in the dead-reckoning component within the system, and the egocentric representation is instantiated by the product of perception (visual, in this model, although contributions from other modalities are clearly possible).

The analysis of the rodent system supports the general proposal that an egocentric locational representation, allocentric locational representation, and allocentric heading component constitute distinct, interconnected functional modules that interact to produce higher-level representations and support functioning in space. The local view is a higher-level representation of a hybrid nature, in that its distance coordinate is egocentric (referred to the rat's viewing position), but its bearing coordinate is allocentric.

While the rodent's spatial system offers general support for distinctions among representations, one must be careful in extending the parallels to humans. One point worth noting is that the egocentric and allocentric representations proposed for the rodent reside at fairly low levels in the perceptual-cognitive stream. Humans, however, are capable of forming spatial representations through top-down, cognitive processes such as generative imagery and memory retrieval. The extent to which they do this appears to be constrained, however, as will be described below.

7 How Well do Navigators Form Representations of Spatial Parameters? Empirical Studies

There is a large body of research literature evaluating the ability of humans to represent various spatial parameters. Studies of particular interest are those in which subjects must maintain and even update a representation after it has been formed perceptually, rather than the more typical psychophysical task in which a response is made in the presence of an ongoing stimulus. On the whole, the literature is suggestive of a more accurate representation of egocentric parameters than allocentric parameters.

Consider first egocentric parameters. There has been considerable work on representing visually perceived *egocentric distance* over the course of open-loop travel without vision. This research indicates highly accurate perception and maintenance of the perceived representation for distances out to more than 20 m (see Loomis, Da Silva, Fujita, & Fukisima, 1992, for review). Recently, this work has been extended to auditorially perceived targets (Loomis, Klatzky, Philbeck and Golledge, in press). With respect to *egocentric bearing*, studies addressing this issue have used various perceptual modalities, including vision, audition, and touch. On the whole, they suggest that there is excellent ability to represent egocentric bearing after perceptual exposure and even to update it over the course of travel, without further perceptual input (e.g., Amorim, Glasauer, Corpinot, and Berthoz, 1997; Fukisima, Loomis, & DaSilva, 1997; Loomis et al., 1992; Rieser, 1989).

Allocentric parameters have not always been studied without ambiguity. Consider, for example, *allocentric heading*. Ideally, a study of allocentric heading perception should have some means of establishing a reference direction independent of the subject's orientation, relative to which the subject's heading in the space could be indicated. The reference direction could be defined by the geometry of a room, by a direction of travel, or by alignment of salient landmarks, and heading could be determined while the subject was stationary or moving . Typically, however, studies of heading first establish a reference direction aligning a target object with the subject's axis of orientation (or in some cases, direction of gaze). The subject then changes his or her position in the space and the ability to update heading is assessed from his or her ability to keep track of the target's azimuth (Berthoz, 1991; Bloomberg, Jones, Segal, McFarlane, & Soul, 1988). But heading is confounded with egocentric bearing in this situation.

An important study of people's ability to represent nonegocentric *interpoint distance* was conducted by Loomis and associates (1992). It asked subjects to match an interpoint interval in depth (saggital plane) so that it appeared to match an interpoint interval in the frontoparallel plane. Considerable distortion was evidenced by inequalities in the adjustments: Depending on the distance of the configuration from the subject, the sagittal interval was made greater than the frontal interval by up to 90%. The literature on cognitive mapping also indicates considerable error in distance perception, being subject, for example, to a filled-space illusion (Thorndyke, 1981) and to distortions resulting from higher-order units in a spatial hierarchy such as state boundaries (Allen, 1981; Kosslyn, Pick, & Fariello, 1974; Maki, 1981).

In order to assess representation of allocentric *interpoint bearings,* a study is needed in which people indicate the angle formed by a line from one object to another, relative to a reference axis. Lederman et al. (1985) assessed interpoint bearings within the haptic domain by having subjects reproduce the angle between a raised line that was traced on the table top, without vision, and a reference axis aligned with the table edge. The responses erred by being drawn toward the perpendicular by about 20%. The cognitive mapping literature also assesses people's ability to determine interpoint bearings, in this case from perceived and remembered visual displays. There is substantial tendency for error (e.g., Stevens & Coupe, 1978; Tversky, 1981).

8 Processes for Computing Egocentric and Allocentric Parameters

How are egocentric and allocentric parameters computed? Up to this point, I have focused on the nature of the parameters, rather than the underlying representations and processes that produce them.

Suppose, for example, a person is asked to estimate the distance between two objects in front of her. This is, by definition, a nonegocentric distance. But by what process is the response computed? There are at least four general types of processing that one can envision, depending on the type of representation that is accessed (allocentric or egocentric) and the type of process that is applied (abstract and symbolic, or imaginally, by a perceptual analogue). Of these processes, it seems more likely that people have access to computational machinery that will allow them to create and retrieve information from images than that they have an internal equivalent of the law of cosines. The imaginal process is itself certainly nontrivial. It subsumes two components: forming the appropriate image, and retrieving the information from it. If an allocentric image is to be used in computing the distance between two objects, the subject must somehow externalize the two objects; if an egocentric image is required, the subject must somehow have a representation of himself in the image at the same location as one of the objects.

If the content of an image is congruent with a person's current field of view (e.g., the perspective to be taken in the image matches the current visual perspective), perceptual processes will support formation of the requisite representation. But if there is a mismatch between the demands of the image and the subject's current perceptual field, some imaginal process must be performed that transforms the relative positions of person and/or objects. This has been called imaginal updating.

The demands of imaginal updating differ for images that constitute allocentric and egocentric representations. To update ego's position in an allocentric representation means to create new coordinates for ego only. To update ego's position in an egocentric representation means to compute primitive locational parameters for all the objects in the space, since these are defined relative to ego. If ego rotates, bearing parameters must be changed. If ego translates, both bearing and distance coordinates change.

9 Imagined Rotation Vs. Translation: Implications for Relation Between Representations

Seminal studies demonstrating the difference between imagined translations and rotations were performed by Rieser (1989). The subject first learned about a circular array of equally spaced objects while standing at the center of the circle. In the translation condition, the subject was then asked to imagine being at one of the objects, facing in the same direction as his or her current heading, and to point to each of the others from that new position. In essence, the subject was asked to indicate the bearing of each object from a new position, but relative to his or her current heading -- this is what we have called the ego-oriented bearing. Subjects were able to do this as accurately as if they were pointing from their current position (averaging 16° of error in both the translation and no-change conditions). Thus they showed ability to compute the ego-oriented bearing without physical movement. Moreover, their response times did not differ, depending on whether they were to remain stationary or mentally translate.

In the rotation condition, the subject was to imagine rotating from the current orientation, without translating. He or she was then to point to each object as if from the rotated orientation. This corresponds to computing a new egocentric bearing. Subjects found this difficult, producing relatively high error (18° to 37°) and manifesting response latencies that increased with the angular difference between their physical heading and the imagined reference direction.

Presson and Montello (1994) verified Rieser's results with simpler displays that equated rotations and translations with respect to the mean and variance of directional change. Easton and Sholl (1995) demonstrated that to some extent, the results of Reiser reflected the use of regular object arrays. They found that when subjects stood within regular arrays (e.g., circle, square), even if not at the center, then the errors and response latency for pointing responses after imagined translation (without rotation) did not depend on the translation distance. If the arrays were irregular, however, there was a tendency for errors and latency to increase with the distance from the actual location to the imagined response point. Even with the irregular arrays, however, imagined rotations produced substantially longer response latency and higher error than imagined translations (see also May, 1996).

Why do people have so much difficulty in reporting a new egocentric bearing after imagined rotation, when they can do quite well in reporting an ego-oriented bearing after imagined translation? It is important first to understand that there is a fundamental difference between the effects of translation and rotation:

(i) Translation (without rotation) changes the egocentric distances and egocentric bearings of objects, but does *not* change distances between objects or allocentric bearings, including ego-oriented bearings.

(ii) Rotation (without translation) changes the egocentric bearings of objects and ego-oriented bearings between objects, but does not change egocentric distances of objects, allocentric distances between objects, or allocentric bearings that are not defined by an ego-oriented axis.

A critical difference, then, is that under translation, ego-oriented bearings remain constant, whereas under rotation, they change. Egocentric bearings, on the other hand, must be updated whether rotation or translation occurs.

There are two general ways to account for the difficulty of rotation, relative to translation, depending on the type of representation that is assumed to be used, allocentric or egocentric. One might propose that an allocentric representation is used, and that the difficulty of updating under rotation reflects difficulties in using the imagined sensory cues to change ego-oriented, allocentric bearings. However, an egocentric representation could also be used. In this case the data indicate it is easier to update egocentric bearings under translation than rotation. Regardless of which source of difficulty is assumed, imagined updating with physical rotation is clearly problematic.

Klatzky, Loomis, Beall, Chance, and Golledge (in press) conducted a study that directly indicates a failure to update egocentric bearings in the absence of physical cues to rotation. The task was as follows: Subjects were asked to imagine walking two legs of a triangle and then to physically make the turn that would be required to face the origin. For a hypothetical example, illustrated in Figure 3, suppose subjects were asked to imagine moving forward 1 m, make a turn of 90° (call this the stimulus turn), and then move forward another 1m. At that point, they were to immediately make the turn that a person who physically walked the path would make, in order to face the origin. We will call their final heading, which would result from that turn, the response heading. (We dealt with response heading rather than turn, because what is important is the direction the subject finally pointed in, rather than how he or she got there.) The subject was to make the response while still standing at the origin, since he or she did not actually walk along the described path.

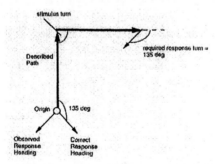

Figure 3. Task used by Klatzky et al. (in press). The subject learns about the first two legs of a triangular pathway while at the origin and then attempts to make the turn required to complete the triangle

Subjects made a highly consistent error: They overturned by an amount equal to the stimulus turn. In terms of our example, where the stimulus turn was 90°, the correct response turn would be 135° -- but the subjects would assume a heading corresponding to a turn of 225°, a +90° error. When a set of paths was used varying in stimulus-turn angle, the function relating the signed error in response heading to the value of the stimulus turn had a slope close to 1.0.

By consistently over-responding by the amount of the stimulus turn, subjects appeared to ignore the change of heading that occurred at the stimulus turn. That is, where a physically walking subject would make a response turn that took into account having already turned 90° between the first and second leg, the imaginally walking subjects did not take the stimulus turn into account. The same outcome was found in a version of the task where subjects were disoriented (by being randomly rotated) before beginning the experiment. This suggests that their failure to take the described turn into account was not due to their representing themselves as at a fixed heading in terms of the external room, since knowledge of their heading within the room should have been eliminated by the initial disorienting movement. It is likely that they represented their heading relative to features of the path itself, for example, using the first leg of the triangle, along which they faced, to define a reference direction.

We proposed that subjects did not register the change of heading that occurred at the stimulus turn, because they did not receive proprioceptive cues. Similar results were obtained in other conditions where proprioceptive signals were absent -- when subjects watched someone else walk, and when they viewed optic flow corresponding to the walked path, by means of a VR display. In contrast, subjects turned correctly when they made a physical turn corresponding to the stimulus turn. Two such cases were tested. In one condition, the subjects physically walked, which would produce efference, kinesthetic, and vestibular cues. In the other condition with a physical turn, they saw the VR display while seated on a rotating stool, and were turned by the experimenter at the point of the stimulus turn. In this latter condition, kinesthetic cues and efference would be absent, but vestibular signals (along with optic flow) would still be present. The contrasting results from the two VR conditions -- with and without a physical turn -- indicate that optic flow alone was not sufficient to change the representation of heading, and that vestibular signals were critical.

At the same time that the subjects did not respond correctly, the regular pattern in their responses indicated that they could represent the spatial layout of the pathway. And, the pattern of responses indicates that they must have represented something more -- their axis of orientation aligned with the first leg. Although they did not make the turn that a physically walking subject would make, their response was equivalent to turning the value of the ego-oriented bearing. The ego-oriented bearing would not be known if the subject's axis of orientation relative to the pathway was unknown.

The operative representation remains ambiguous. In using an egocentric representation, for example, the subject might (a) imagine standing at the end of the second leg, then (b) respond by turning the value of the egocentric bearing to the origin. In using an allocentric representation of pathway layout, the subject could (a) determine the ego-oriented bearing from the end of the second leg to the origin, and then (b) respond by turning the value of that bearing. While the operative representation is ambiguous, it clearly failed to incorporate the change of heading commensurate with the description of an imagined turn.

These results are consistent with the idea that proprioceptive signals accompanying physical rotation are input to an allocentric heading representation, which in turn provides information to locational representations that allow updating of bearings. This process fails, however, when the rotation is purely imaginary. In contrast, the translational movement of traveling the pathway -- either by imagination or by physical movement -- can be used to update a representation that permits the computation of the bearing from the end of the traveled pathway back to the origin.

10 Summary

In this chapter, I have attempted to clarify fundamental concepts related to allocentric and egocentric representations -- and also to indicate where ambiguity is inevitable. By starting with straightforward assumptions about three types of representation and their primitives, I have indicated what derived parameters can be computed, and with what degree of stability. A brief review of the literature cites studies indicating that egocentric parameters can be encoded accurately, along with studies indicating that allocentric parameters may be encoded with substantial error. The well-documented contrast between updating after imagined translation, as compared to rotation, appears to place constraints on the connectivity between different representational systems. Updating under translation requires reporting ego-oriented bearings from a new station point; updating under rotation requires reporting new egocentric bearings from the same station point. Imagination allows us to do the former but not the latter, because, it is argued, proprioceptive signals are essential to incorporating changes of heading into the operative representation, be it egocentric or allocentric.

References

1. Allen, G. L. (1981). A developmental perspective on the effects of "subdividing" macrospatial experience. Journal of Experimental Psychology: Human Learning and Memory, 7, 120-132.
2. Amorim, M., Glasauer, S., Corpinot, K., & Berthoz, A. (1997). Updating an object's orientation and location during nonvisual navigation: A comparison between two processing modes. Perception & Psychophysics, 59(3), 404-418.
3. Berthoz, A. (1991). Reference frames for the perception and control of movement. In J. Paillard (Ed.), Brain and space (pp. 81-111). New York: Oxford University Press.
4. Brewer, B., & Pears, J. (1993). Frames of reference. In R. Eilan, R. McCarthy, & B. Brewer (Eds.), Spatial representation: Problems in philosophy and psychology (pp. 25-30). Oxford: Blackwell.
5. Easton, R. D., & Sholl, M. J. (1995). Object-array structure, frames of reference, and retrieval of spatial knowledge. Journal of Experimental Psychology: Learning, Memory, and Cognition, 21, 483-500.
6. Etienne, A. S., Maurer, R., & Séguinot, V. (1996). Path integration in mammals and its interaction with visual landmarks. Journal of Experimental Biology, 199, 201-209.

7. Gallistel, C. R. (1990). The organization of learning. Cambridge, MA: MIT Press.
8. Klatzky, R. L., Loomis, J. M., Beall, A. C., Chance, S. S., & Golledge, R. G. Updating an egocentric spatial representation during real, imagined, and virtual locomotion. Psychological Science, in press.
9. Klatzky, R. L., Loomis, J. M., & Golledge, R. G. (1997). Encoding spatial representations through nonvisually guided locomotion: Tests of human path integration. In D. Medin (Ed.), The psychology of learning and motivation (Vol. 37, pp. 41-84). San Diego: Academic Press.
10. Klatzky, R. L., Loomis, J. M., Golledge, R. G., Cicinelli, J. G., Doherty, S., & Pellegrino, J. W. (1990). Acquisition of route and survey knowledge in the absence of vision. Journal of Motor Behavior, 22, 19-43.
11. Kosslyn, S. M., Pick, H. L., & Fariello, G. R. (1974). Cognitive maps in children and men. Child Development, 45, 707-716.
12. Lederman, S. J., Klatzky, R. L., & Barber, P. (1985). Spatial- and movement-based heuristics for encoding pattern information through touch. Journal of Experimental Psychology: General, 114, 33-49.
13. Lederman, S. J., & Taylor, M. M. (1969). Perception of interpolated position and orientation by vision and active touch. Perception & Psychophysics, 6, 153-159.
14. Levinson, S. C. (1996). Frames of reference and Molyneux's question: Crosslinguistic evidence. In P. Bloom, M. A. Peterson, L. Nadel, & M. F. Garrett (Eds.), Language and space: Language, speech, and communication (pp. 109-169). Cambridge, MA: MIT Press.
15. Loomis, J. M., Da Silva, J. A., Fujita, N., & Fukusima, S. S. (1992). Visual space perception and visually directed action. Journal of Experimental Psychology: Human Perception and Performance, 18, 906-922.
16. Loomis, J. M., Klatzky, R. L., & Golledge, R. G. (in press). Human navigation by path integration. In R. Golledge (Ed.), Wayfinding: Cognitive mapping and spatial behavior. Baltimore, MD: Johns Hopkins University.
17. Loomis, J. M., Klatzky, R. L., Golledge, R. G., Cicinelli, J. G., Pellegrino, J. W., & Fry, P. (1993). Nonvisual navigation by blind and sighted: Assessment of path integration ability. Journal of Experimental Psychology: General, 122, 73-91.
18. Loomis, J. M., Klatzky, R. L., Philbeck, J. W., & Golledge, R. G. (in press). Assessing auditory distance perception using perceptually directed action. Perception & Psychophysics.
19. Maki, R. H. (1981). Categorization and distance effects with spatial linear orders. Journal of Experimental Psychology: Human Learning and Memory, 7, 15-32.
20. Maurer, R., & Séguinot, V. (1995). What is modeling for? A critical review of the models of path integration. Journal of Theoretical Biology, 175, 457-475.
21. May, M. (1996). Cognitive and embodied modes of spatial imagery. Psychologische Beitraege, 38, 418-434.
22. McNaughton, B. L., Leonard, B., & Chen, L. (1989). Cortical-hippocampal interactions and cognitive mapping: A hypothesis based on reintegration of the parietal and inferotemporal pathways for visual processing. Psychobiology, 17(3), 230-235.
23. O'Keefe, J. (1976). Place units in the hippocampus of the freely moving rat. Experimental Neurology, 51, 78-109.
24. O'Keefe, J., & Dostrovsky, J. (1971). The hippocampus as a spatial map: Preliminary evidence from unit activity in the freely moving rat. Experimental Brain Research, 34, (171-175).

25.O'Keefe, J., & Nadel, L. (1978). The hippocampus as a cognitive map. Oxford: Clarendon Press.

26.Paillard, J. (1971). The motor determinants of spatial organization. Cahiers de Psychologie, 14, 261-316.

27.Philbeck, J. W., Loomis, J. M., & Beall, A. C. (1997). Visually perceived location is an invariant in the control of action. Perception & Psychophysics, 59(4), 601-612.

28.Pick, H. L. (1988). Perceptual aspects of spatial cognitive development. In J. Stiles-Davis, M. Kritchevsky, & U. Bellugi (Eds.), Spatial cognition: Brain bases and development (pp. 145-156). Hillsdale, NJ: Lawrence Erlbaum Associates.

29.Presson, C.C., & Montello, D.R. (1994). Updating after rotational and translation body movements: Coordinate structure of perspective space. Perception, 23, 1447-1455.

30.Redish, A. D. (1997). Beyond the cognitive map: A computational neuroscience theory of navigation in the rodent. Unpublished doctoral dissertation, Carnegie Mellon University, Pittsburgh, PA.

31.Redish, A. D., & Touretzky, D. S. (1997). Cognitive maps beyond the hippocampus. Hippocampus, 7, 15-35.

32.Rieser, J. J. (1989). Access to knowledge of spatial structure at novel points of observation. Journal of Experimental Psychology: Learning, Memory, and Cognition, 15(6), 1157-1165.

33.Soechting, J. F., Tong, D. C., & Flanders, M. (1996). Frames of reference in sensorimotor integration: Position sense of the arm and hand. In A. M. Wing, P. Haggard, & J. R. Flanagan (Eds.), Hand and brain: The neurophysiology and psychology of hand movements (pp. 151-167). San Diego, CA: Academic Press, Inc.

34.Stevens, A., & Coupe, P. (1978). Distortions in judged spatial relations. Cognitive Psychology, 10, 422-437.

35.Thorndyke, P. W. (1981). Distance estimation from cognitive maps. Cognitive Psychology, 13, 526-550.

36.Touretzky, D. S., & Redish, A. D. (1996). Theory of rodent navigation based on interacting representations of space. Hippocampus, 6, 247-270.

37.Tversky, B. (1981). Distortions in memory for maps. Cognitive Psychology, 13, 407-433.

The Route Direction Effect and Its Constraints

Karin Schweizer, Theo Herrmann, Gabriele Janzen & Steffi Katz

Universität Mannheim, L13, 15 (Zi 319), D-68138 Mannheim, Germany
{schw, fprg, janzen, katz}@rumms.uni-mannheim.de

Abstract. The route direction effect can be characterized as follows: When a person has learned a route and imagines that he or she is at a specific point along this route, then it is easier for him or her to imagine him or her-self at another point on the route which lies in the direction in which the route was learned than the other way around. This means that route knowledge includes information on the route if acquisition, i.e. the direction of acquisition is co-represented in cognitive maps. Within a network theory approach, the route direction effect can be conceived of as a specific asymmetry in the spread of activation. Priming-experiments were carried out to determine a time window for the experimental realization of asymmetrical activation spread. We were actually able to show that the route direction effect disappears when the Stimulus Onset Asynchrony is extended. This was to be expected because a longer SOA causes the target to be pre-activated, regardless of whether it represents an object which was positioned in the direction of acquisition, or not. In further experiments we showed that if route knowledge is gained via a gradient sequence, then normal ecological conditions of perception must have existed: The direction of perception must correspond to the direction of movement. Objects, as well as their surroundings, must be perceived within the normal optical flow. Otherwise the information about the direction of acquisition cannot be adequately incorporated, and the route direction effect does not occur.

1 Introduction: Route Knowledge and the Route Direction Effect

In this article we will deal with the following phenomenon: Assume you have walked through a town several times, each time following the same route. If you then imagine that you are standing at a certain point A on the route, it is much easier to imagine yourself being at a point B, which had followed A on your original route (= forward direction), than to imagine yourself being at a point B', which had preceded A on the original route (= backward direction). We call this phenomenon the *route direction effect*. Our assumption is that the direction in which a route is learned (= direction of acquisition) is internally represented. It is a part, or a characteristic of the internal structure of information depicting external spatial constellations.

The route direction effect belongs to the field of spatial cognition. We have the ability to acquire knowledge about the spatial structure of our environment. This spatial knowledge can be conceived of as a *cognitive map*. Pioneer work on this topic has been carried out, for example, by Appleyard (1970), Lynch (1960), Tolman (1948), Trowbridge (1913) and Shemyakin (1962). (For an overview see also Aginsky, Harris, Rensink and Beusmann, 1996; Buhl, 1996; Cohen, 1989; Downs and Stea, 1977; Engelkamp, 1990; Evans, 1980; Freksa and Habel, 1990; Herrmann, Schweizer, Janzen and Katz, 1997; May, 1992; Moore and Golledge, 1976; Schweizer, 1997; Siegel and White, 1975; Wagener-Wender, 1993; Wippich, 1985.)

Cognitive maps can involve varying degrees of elaboration depending on the acquisition situation: for example, one person (i) might have noticed only a few houses, walls and trees as well as some other orientation points, without any particular sequential order, on their first walk through the town. The things remembered were just "somewhere along the way". - (ii) or this person may have followed the route several times, but only this particular route. He or she then has route knowledge which enables him or her to follow the same route again and also to return to the starting point. This kind of route knowledge will be the topic of discussion later in this paper. - (iii) It is also possible to imagine that the person not only took this route, but also several other routes through the town. Now he or she has survey knowledge. He or she is able to switch from one route to another, to find new routes, short-cuts and diversions, etc.

The precise form of these spatial relationships is subject to controversy. There are generally two approaches being discussed at present: Proponents of the *computational* viewpoint assume that mental representations are encoded in an abstract, inaccessible form as a result of symbol manipulation. The *analogue* approach, on the other hand, assumes that information exists explicitly - is directly accessible - and concurs, in certain characteristics, with the (external) original image in the real world (e.g. Kosslyn, 1994; Lüer, Werner and Lass, 1995; Palmer, 1978; Pylyshyn, 1981; 1984; Rehkämper, 1990; 1993; Tack, 1995; Zimmer, 1993). These differences of opinion (imagery-debate) continue today, (see Sachs and Hornbach (Eds.), 1994). The assumption of analogue processes is often criticized because cognitive processes - for example visual imagination - can be influenced by wishes and assumptions of the individual (see Pylyshyn, 1981; 1984). In order to carry out experiments which can defend themselves against the criticism of this kind of influence, it is advisable to employ the *priming-paradigm*.

According to Glaser (1992) priming means two things (see also Schweizer, 1997): first, the *method*, which can be traced back to experiments by Beller (1971), Meyer and Schvaneveldt (1971) as well as to Posner and Mitchell (1967), and, secondly, the *effect* resulting from implementation of the method. The method involves presenting a respondent with two stimuli, usually with a very short delay. The stimulus which is presented (or processed) first, is called the *prime*, the stimulus which follows is called the *target*. The time between presentation of the prime and exposure to the target is called SOA (Stimulus Onset Asynchrony). It is assumed that presentation of a suitable prime leads to pre-activation of the target, which in turn leads to shortened reaction times to the target (priming effect). The strength of the pre-activation of the target, and therefore also the reduction of reaction times, depends to a large extent on the strength of the connection between prime and target. (We will not discuss further reasons for pre-activation here; see Neely, 1991.) The strength of the connection between prime and target itself depends, above all, on the similarity of prime and target: reductions in reaction time to the target increase proportionally to the similarity between prime and target. In this context we can distinguish, for example, figural and phonetic similarity from semantic similarity (cf. Glaser, 1992). When dealing with spatial knowledge networks, however, the strength of this connection may also depend on how big the distance between the points (locations) in the surroundings of the cognitive system is within which points are internally represented in the form of network nodes: If prime and target represent two points in the surroundings which are relatively close together, pre-activation of the target by the prime is stronger than when these points are far apart from one another. (see McNamara, 1992; 1994; Neely, 1991; Neely and Keefe, 1989; Ratcliff and McKoon, 1994.)

Therefore, the distance between objects represented in a cognitive network influences pre-activation of the target and by doing so the reaction times in priming experiments (see also Wagener-Wender and Wender, 1990; Wagener-Wender, 1993). These results can easily be integrated into the well-known body of results showing that the time required to mentally bridge the gap between two points in a spatial representation co-varies in direct proportion to the *Euclidean* (geometric, topographical) distance between these points (see Kosslyn, 1973; Kosslyn, Ball and Reiser, 1978; Pinker, 1980).

McNamara, Ratcliff and McKoon (1984) analyzed whether the subjective distance between two towns on a map is determined by the Euclidean distance, or also by the *route distance*. With route distance we mean the distance between two towns via a route drawn on a map. In priming experiments it has become clear that route distance (as well as the Euclidean distance) effects priming results. Wender and Wagener (1990) were also able to demonstrate specific effects of route distance on the activation spread from prime to target (see also Wagener-Wender, 1993).

In addition to the effects of Euclidean and route distance discussed so far, the above-mentioned *route direction effect* can also be demonstrated in priming experiments: If components of a spatial constellation on which route knowledge has been acquired are Euclidean (equally far apart) on the route of acquisition; then the facilitating effect of the prime is larger if prime and target follow one another in the direction of acquisition, than if they are presented in the opposite direction. The priming effect, with regard to the route of acquisition, is therefore larger in the forward than in the backward direction. - This effect is independent of Euclidean and route distances. We have been able to identify this effect repeatedly (Herrmann, Buhl and Schweizer, 1995; Herrmann, 1996).

In one of these experiments (see also Schweizer, 1997; Schweizer and Janzen, 1996), respondents were given route knowledge via films. (The films showed the spatial configuration illustrated in figure 1.) Respondents saw the films several times, and then took part in a priming phase during which the prime-target combinations, which had previously been shown as figures on flags along the route, were presented (see figure 1).

Amongst other things, the items were selected to fit the direction in which the route was to be followed: If object 3 was seen before object 4, then the prime-target combination was object 3-object 4 (forwards), and object 4-object 3 (backwards). During the priming phase respondents had to decide whether a presented target was an object they had seen before or not (= recognition task). We expected respondents to recognize forward items more quickly than backward items. Our expectations were fulfilled: respondents could react significantly faster when the target had appeared after the prime during acquisition than when the target had preceded the prime. We view this result, along with several others (Herrmann, Buhl and Schweizer, 1995; Katz, 1995), as evidence that the direction of acquisition is also represented in spatial knowledge networks.

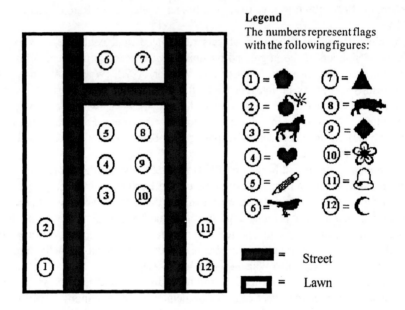

Legend
The numbers represent flags with the following figures:

Fig. 1. Ground-plan of the layout presented in the experiments of Schweizer (1997).

As well as in priming experiments this effect has been repeatedly demonstrated via cued-recall experiments (Bruns, 1997; Schweizer, Katz, Janzen and Herrmann, 1997) and also by presenting a moving virtual environment as well as the successive presentation of descriptions of objects (see below and also Herrmann, Schweizer, Buhl and Janzen, 1995). The route direction effect cannot be put down merely to the superiority of "forward associations" over "backward association" in the sense of classical serial learning of lists (cf. Ebbinghaus, 1985; Kluwe, 1990), it is a *genuinely "spatial"* effect. We tested this by presenting the pictures used in the experiment described above (figure 1) in sequence in the same position in the center of a computer screen. Respondents took part in the same priming phase described above. In this case the route direction effect disappeared completely under otherwise identical conditions (see also Herrmann, Schweizer, Buhl and Janzen, 1995; Schweizer, 1997; Schweizer, Janzen and Herrmann, 1995).

Obviously it is essentially important for the activation of the route direction effect that route knowledge has been acquired. The route is the internal representation of a specific sequence of various single objects at *various different* locations in space, which may also exist under a variety of perspectives: thus it is possible to internally represent a route from above, from a bird's-eye view. This is the case, for example, when somebody imagines viewing a street from a nearby hill, or out of a helicopter. We call this the observer perspective (O-perspective). It is, however, more common to remember having been in the middle of a situation, "on-the-spot". In this case you imagine a street as if you were walking down it yourself. We call this the field perspective (F-perspective) (this differentiation was introduced by Nigro and Neisser, 1983; see also Cohen, 1989, Frank and Gilovich, 1989).

The acquisition of knowledge in the field perspective is, however, accompanied by specific perceptual processes: when a person moves through a spatial configuration, a complex sequence of stimulus images is created on their retina. This is essential to their visual perception of motion. The stimulus background changes in a different way from the perceived objects; these also change in a variety of ways depending on their objective distance from the observer and sometimes depending on their own movements. An optical flow with its characteristic depth planes is the result. (On optical flow and the perception of movement see Gibson, 1950; 1966; 1979; Horn and Schunck, 1981; Kim, Growney and Turvey, 1996; Verri and Poggio, 1989.) The cognitive system calculates the relative distance of objects from one another from the relative size and direction of the movement of the images on the retina (= movement parallax; see Bruce and Green, 1985). The flow in the field of vision - whether we ourselves move and/or whether the objects around us move - allows us to recognize *objects*, we gain information about *what* we are seeing. We also have information about *where*, i.e. about the *location* of objects (see Ungerleider and Mishkin, 1982). Perception of movement is the result of the combination of stability and movement of objects. We assume that the internal representation of routes when one moves through a spatial configuration is created with considerable help of the optical flow. Routes are not, however, always determined by the optical flow.

In addition to the perspectives mentioned up to now, internally represented routes can also be characterized as follows: (i) They have a *sequential structure*. We can define routes as sequences of (primarily) visually represented objects at different locations, resp. of orientation points (landmarks). Routes are then internal representations which can be compared to weak orders of (discrete) "inner images", resp. image sequences. The image sequence stands in an almost veridical pictorial relationship to an attributable sequence of locations in the objective system environment. (ii) Routes can be conceived under certain circumstances as *continuous* sequences of visual representations (as gradient sequences). The route then takes the form of a continuous sequence of perceptual gradients with depth planes, i.e. of an optical flow, whilst the internal representation of discrete landmarks (or other discrete visual representations) can be found alongside this continuous informational structure. Gradients are generally understood as being essential to the visual perception of depth (i.e. Gibson, 1982; Nomura, Miike and Koga, 1991). The development of represented gradient sequences does not require the system itself to move. (The development of internally represented gradient sequences can be evoked in experiments via the presentation of film material; see also Herrmann, Buhl and Schweizer, 1995; Kato, 1982.) (iii) Routes can also be represented as *point of view sequences* (see also Herrmann, 1996; Schweizer, 1997; Schweizer and Janzen, 1996). (For a definition of the point of view as the origo of a retino-centered system see Herrmann, 1996; on points of view see also Tversky, 1991; Franklin, Tversky and Coon, 1992.) The mental representation of point of view sequences is the result of complex perceptual processes in which the visual perception of motionless and moving objects, other visual factors and their background - i.e. the perception of the complete optic flow - interacts with feedback, resp. re-afferences from eye-movements (and also head and neck movements), and other perceptual information such as gravity, rotation and pressure, (see also Bruce and Green, 1985; Paillard (Ed.), 1991, Prinz, 1990). The representation of routes as point of view sequences is generally accompanied by internally represented sequences of self-movements, which go beyond the point of view sequence. Co-representation of such movement sequences is based on feedback resulting from this activity, in particular from the muscular activity of the extremities. The difference between point of view and movement sequences can be demonstrated as follows: if you follow a route in the dark using only the sense of touch so that visual information is missing you can still create an internal representation which is based on your own movements and their proprioceptive feedback.

Regarding gradient sequences (= (ii)) we refer in the following discussion to the fact that, in everyday life, people mainly have to deal with the processing of *forward gradients*: Normally we perceive buildings or other objects as being at first small, and then growing larger as we approach them (Gibson, 1979). However, in spatial cognition *backward gradients* can also occur. This is, for example, the case if you are sitting in a train with your back to the direction of travel. Objects suddenly appear in your field of vision from behind and they are immediately large. Gradually they become smaller and less clear, until they disappear completely in the distance. An important characteristic of backward gradients is that the direction of perception is separated from the direction of movement. Given forward gradients, these directions are identical. The question is whether it is still possible to demonstrate the route direction effect when a route has been learned under conditions which include backward gradients. Or does the effect depend on normal route acquisition in the forward direction? We will further examine this question below (section 4). The question affects the extent to which the route direction effect is demonstrable under the condition that a route is learned as a sequence of gradients.

2 The Route Direction Effect and Asymmetrical Spread of Activation

We will now attempt a more precise theoretical formulation of the assumption that the direction in which a route is (first) learned is coded in cognitive maps. Cognitive maps can - like all other knowledge bases - be theoretically conceived of as an (internal, cognitive) *network*. This kind of network conception can take on a variety of different forms (see also Anderson, 1976; 1983; Helm, 1991; Herrmann, Grabowski, Schweizer and Graf, 1996; Hinton and Anderson (Eds.), 1989; Opwis and Lüer, 1992; Schade, 1992). All forms distinguish between *nodes* and *paths*. In a calculated effort to simplify matters we assume that the nodes in the networks of knowledge we are interested in represent hypothetical cognitive units. That means that the nodes in a cognitive network depicting an external spatial configuration are disjunctive spatial components in specific material locations. The nodes are connected loosely or closely to many other nodes in the network to different extents. Activation can spread via the connections between nodes (see among others McClelland and Rumelhart, 1986).

Other specific pieces of information on the sequence of spatial components during acquisition which are interesting in the context of the route direction effect (direction of acquisition) can be viewed in the sense of a *spread of activation* within networks. Important here is that the strength of the spread of activation from node A to node B is related to a specific connecting weight which need not necessarily correspond to the strength of the spread of activation from B to A. The direction of acquisition manifests itself in an *asymmetrical spread of activation*. This means that, in the sense of associations with variable strengths, it is easier to get from A to B than from A to B' (see above).

Theoretical attempts to systematize the issue, like the one above, can only lead us nearer to our goal if they (via the addition of adequate assumptions) lead to the development of empirically testable considerations. What consequences can we glean - without already going into concrete experimentation (see below) at this point - from the hypothetical asymmetry of the spread of activation? We distinguish between the following three:

1. When a node A is activated a specific level of activation is more quickly reached given a spread of activation from A to B if the spatial element *B,* corresponding to node B, occurred after the spatial element *A,* corresponding to node A, during acquisition compared to the opposite case where the spatial element *B* was learned before *A.*

2. The asymmetry of the spread of activation does not occur if the level of activation of node B is measured very quickly after the activation of A has been established; in this case the spread of activation from A to B has not yet reached node B: independent of the fact of whether the spatial element *B* corresponding to B was placed before or after the spatial element *A* corresponding to A. - In the same vein, it is not possible to demonstrate asymmetry of the spread of activation if the level of activation of node B is measured a long time after the activation of node A has been established; in this case the spread of activation from A to B has already reached the node. In order to demonstrate asymmetry of the spread of activation a definable time window exists (Schweizer, in preparation).

3. Factors determining the spread of activation can not only be analyzed via retrieval of spatial knowledge, but also via acquisition of the represented knowledge components. We assume that perception processes (as described in section 1) play an important role in the representation of routes and, therefore, also in the coding of the route of acquisition. Characteristics of the acquisition situation play a decisive role in the generation of a specific asymmetry in the connections of the theoreti-

cally assumed network. In perceptual situations from which an optical flow can be extracted (see above), this can be regarded as a constitutive factor in the coding of the direction of acquisition. We can check this by destroying or deforming the optical flow, all other things being equal. The asymmetry of the spread of activation regarding route direction effect is only very weak, resp. unidentifiable, when we succeed in deforming the optical flow.

The first of these conclusions could be confirmed in a multitude of experiments on the route direction effect (see above, section 1). We therefore arrive at two main approaches to further analysis of the route direction effect: variation of the acquisition situation (= conclusion 3) and the retrieval situation (= conclusion 2). We will now describe experiments on the retrieval situation before turning our attention to the acquisition situation.

3. The Retrieval Situation

We carried out three experiments (experiments 1, 2 and 3) to test conclusion 2, although the test of the first part of the conclusion (very short time space) still remains to be done. We varied the time interval between disappearance of the prime and presentation of the target, i.e. the inter-stimulus interval (ISI) changed systematically from 250 over 500 to 750 ms. The corresponding SOAs were 350, 600 and 850 ms. (These figures were in part the result of a pre-test which was used to set the length of presentation of the prime. The results showed that respondents were able to recognize images which had been shown in a virtual setting after 50 to 80 ms. In order to take account of lack of attention during the priming phase, we chose a somewhat higher presentation time of 100 ms and an ISI of 250 ms, which had been shown to be suitable in other experiments (Schweizer, 1997).)

The resulting SOA in experiment 1 was then 350 ms. Experiment 1 was conceived as a control experiment in order to identify a suitable time window in which to evoke the route direction effect. In line with our assumptions the effect was then to be systematically destroyed in experiments 2 and 3. This resulted in a variation in SOA from 350 ms over 600 ms to 850 ms. Table 1 gives an overview of the experiments used to test conclusion 2.

Table 1. Experiments on the retrieval situation.

Experiment 1	Experiment 2	Experiment 3
SOA 350 ms	SOA 600 ms	SOA 850 ms
n = 20	n = 20	n = 20

Experimental procedure. A virtual environment including a film sequence (frame rate: 14 frames per second) was generated with the help of the computer graphic software "Superscape VRT". We will refer repeatedly in the following text to this set-up. Figure 2 shows a plan of the spatial configuration the virtual environment was based on. We used a u-shaped space with 12 objects. The u-shape appeared to have a length of 49 meters in relation to the simulated eye height of the observer (1,70 meters). The objects were articles found in an office, and were standing on small pedestals. The room was introduced to respondents as a museum for the office equipment of famous people. The film sequence started at "pot-plant" and led "clockwise" past the individual objects. The objects introduced in the film could be combined to make prime-target pairs which could be classified according to the direction of acquisition.

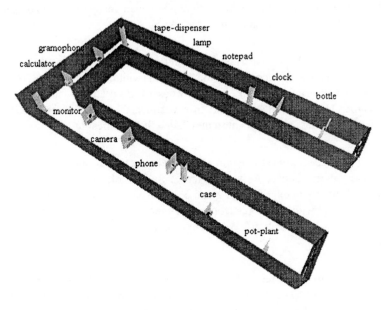

Fig. 2. Layout of experiments 1-3 and 5-7.

The experiments were divided into a *cognition phase*, a *priming phase* and a *treatment test*. During the *cognition phase* respondents were shown the film as often as was necessary for them to learn the order in which the objects appeared. They were instructed to imagine that they were walking through the illustrated scene. The *priming phase* followed. The prime-target pairs described above were shown in succession on a computer screen. The presentation time for the prime was 100 ms. According to the experiment being carried out a different inter-stimulus interval followed (see above) before the target appeared. The target disappeared after the respondent had reacted. There was an interval of 1000 ms between respondent reaction and the next prime. The images used as targets were either those which had been seen during the cognition phase, or unknown objects (distractor stimuli). As primes we only used images of objects which had already been seen. The respondents' task was to decide whether the image presented had been in the original scene or not (recognition task). The respondent had to press one of two keys for yes or no. We measured respondents' reaction times as well as the reactions themselves (yes/no).

After the priming phase a treatment test was carried out. This involved the respondents laying out model material according to the spatial configuration they had become acquainted with from the film (placing technique). This enabled us to choose, *ex post facto,* those respondents whose layouts corresponded with the spatial, resp. temporal order of the stimuli. (After the priming phase respondents were instructed to re-build the layout from memory.) The treatment test was introduced to ensure that the reaction time differences measured during the priming phase were really based on adequate route knowledge (Herrmann, Buhl and Schweizer, 1995).

Variables examined: In all experiments the independent variable was the prime-target pair (items) which varied according to the direction (within factor)[1]. The SOA was also varied through experiments 1 to 3 - as already described - (between factor).

Altogether 71 respondents took part in experiments 1 to 3. 11 respondents had to be excluded from the analysis because they failed the treatment test.

Data analysis: The recorded reaction times were corrected and averaged across the distances as reported above. As we had forecast a lack of, resp. a weakening of a directional hypothesis in experiments 2 and 3, the data were analyzed in an analysis of variance, together with the scores from experiment 1 with the between factor "SOA" and the within factor "direction". We had expected a significant interaction between the experiments.

Results and discussion: The analysis of variance delivered a marginal main effect of the factor SOA ($F(2,57) = 2.65$; $p = .079$), a strong main effect of direction ($F(1,57) = 11.04$; $p < .005$) and an obvious interaction between both factors ($F(2,57) = 3.07$; $p = .054$). The individual reaction time averages for the forward and backward items of the experiments 1 to 3 are shown in figure 3.

---

[1] The items also variied according to the route distance. As we did not find any indication to systematic changes in reaction times according to distances we neglected the variable distance in all experiments reported here. The analyses of all experiments refer to a mean computed of all variations of route distance.

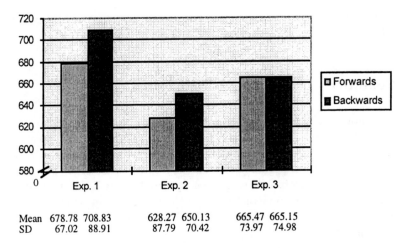

	Forwards					
Mean	678.78	708.83	628.27	650.13	665.47	665.15
SD	67.02	88.91	87.79	70.42	73.97	74.98

Fig. 3. Mean reaction times and SD of Experiments 1,2 and 3.

In order to test the impact of the route direction effect on the individual levels of the SOA variation, post-hoc t-tests were calculated for each experiment. Table 2 gives an overview of the results.

Table 2. Summary of the post-hoc tests.

	t-test ($df = 19$) (one-tailed)
Experiment 1	$t = 3.10; p < .005$
Experiment 2	$t = 2.08; p = .026$
Experiment 3	$t = -0.05; p = .479$

As expected, respondents in experiment 1, i.e. with an *SOA of 350 ms*, were able to react to items in the route of acquisition faster than to items in the opposite direction. At an *SOA of 600 ms* it was also possible to evoke a route direction effect. When the time window was increased by 250 ms in the next experiment (*SOA 850 ms*) the asymmetric spread of activation was not be found.

Our second conclusion regarding an asymmetrical spread of activation could be confirmed. This asymmetry can no longer be observed when the activation level of the target (node B) is measured a relatively long time (850 ms) after induction of activation of the prime (node A); in this case we assume that the spread of activation from A to B, both in the forward and the backward direction, has already reached B. The route direction effect is strongest given an SOA of 350 ms but is also evoked after an average SOA of 600 ms. The graphic illustration of the results shows further that a longer SOA reduces the reaction times. This marginal result may also be a reference to the assumption that the spread of activation depends on the time lapse between presentation of prime and target. Here we have to make the additional assumption that the level of activation adds up. In this case the shorter reaction times for longer SOAs could also be seen as an indication that both components (are already activated, and that a facilitation effect of the forward direction can no longer make itself visible.

4. Conditions of the Acquisition Situation

In this section we will describe experiments carried out to test conclusion 3. We have already explained that a deformation of the optic flow ought to have a negative impact on the generation of specific asymmetries. If the optic flow can somehow be deformed the route direction effect ought to be very weak if not invisible under otherwise identical conditions. As described above gradients generally exist in the forward direction. This leads to the assumption that coding of the direction of acquisition should be harder given backward gradients (Janzen, 1996). We compared forward and backward gradients in experiment 4.

Experimental procedure: To produce the described forward and backward gradients two films were shot which lead through the layout described in section 1 from both sides (see figure 1.) One film began with object 1 and led to object 12, whilst the other film began with object 12 and led to object 1. We also produced backward versions of both films, thus enabling us to produce backward gradients. Therefore we had 4 different films: (i) film A with forward gradients, (ii) film B with forward gradients, (iii) film A with backward gradients and (iv) film B with backward gradients. The experiment was carried out with 98 students from the University of Mannheim. Respondents were allocated randomly to one of the four films.

In the priming phase which followed the respondents were shown the items in the route of acquisition and in the opposite direction as described above. (For the analysis we separated the reaction times of those respondents who had seen the films with forward gradients from those who had seen the films with backward gradients respectively to receive two experimental condition (condition 1: film A and B with forward gradients, condition 2: film A and B with backward gradients).) We only analyzed the reaction times of respondents who were able to complete the following placing technique to a large extent, i.e. to rebuild the scene with the correct spatial configuration and order of stimuli (n = 80).

Results and discussion: The analysis of variance resulted in an obvious interaction between the type of gradient (forward vs. backward) and the direction (items in, vs. opposite, the route of acquisition) $(F(1,78) = 4.24; p < .05)$, which is due to the fact that the route direction effect can be evoked by forward gradients $(F(1,39) = 5.74; p < .05)$, not, however, by backward gradients $(F(1,39) = 0.36; p = .55)$. This means that, given the former, the route of acquisition is included in the coding of the mental representation of the layout. We can assume that, in the latter case, coding of the route of acquisition is made difficult by the deformation of the situation during perception. The reaction times are shown in figure 4.

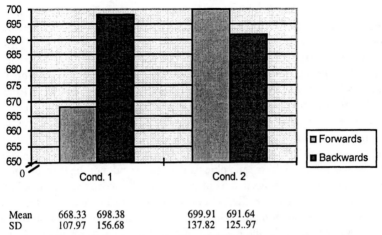

Mean	668.33	698.38	699.91	691.64
SD	107.97	156.68	137.82	125..97

Fig. 4. Mean reaction times and SD of experiment 4.

In order to further support our assumption that a deformation of the optic flow influences the resulting spatial representation we distinguished between object and environmental gradients. When route knowledge is being established out of the field perspective (see section 1) then (i) objects you approach grow continually larger *and* (ii) the surroundings demonstrate the dynamics of motion characteristic of the optical flow. (i) Deformation of *object gradients* involves presenting them at a constant size with an unchanged optical flow. (ii) Deformation of *environmental gradients* was achieved by allowing the objects to grow larger but by removing all motion from the environment. Regarding the route direction effect as a genuinely space-related effect we assume that environmental gradients are more important than object gradients. This means that the route direction effect should be weaker, or more likely to even disappear when environmental gradients (= (ii)) are deformed than when object gradients (=(i)) are deformed. We studied this in experiments 5 to 7.

For these experiments we returned to the virtual museum layout used in experiment 1. Experiment 1 can be regarded as a control experiment as the film sequences employed both types of gradient corresponding to a real-life situation. Table 3 gives an overview of the systematic variations of object and environmental gradients.

Table 3. Experiments on the acquisition situation.

	Presence of object gradients	Absence of object gradients
Presence of environmental gradients	- Experiment 1 - n = 20	- Experiment 5 - (holograph museum) n = 20
Absence of environmental gradient	- Experiment 6 - (night-watch with torch) n = 20	- Experiment 7 - (night-watch with emergency lighting) n = 20

32

Experimental procedure in experiments 5 to 7: In experiment 5 the environmental gradients corresponded to those of the first experiment. However, the object to be learned was only visible when the respondent had almost reached the corresponding object in the museum. This meant that the size of the objects was held constant. We instructed respondents to imagine that they were in a holograph museum where you could only see the objects on display once you were standing directly in front of them.

In *Experiment 6* respondents saw the objects along with gradients, i.e. objects in the distance were small but visible and grew larger as the respondents approached them. In this case the background was black (see also Schweizer, 1997). Respondents were to imagine that they were on the night-watch and that they only had a torch to show the way, they could see the objects themselves but not the surroundings.

In *Experiment 7* we showed stills of the objects. Between the individual pictures respondents saw a black screen. Respondents were told to imagine that they were walking through a museum at night. Whenever they stood in front of an object on display the emergency lighting in the ceiling came on for a certain time period. In this case there were neither object nor environmental gradients.

74 respondents took part in experiments 5 to 7. After the treatment test 60 respondents were included in the analysis.

Results and discussion of experiments 5 to 7: In order to test the hypothesis regarding direction we calculated a three-dimensional analysis of variance on the data measured by Experiments 1, 5, 6 and 7. The between factor was the factor object and environmental gradients, within factor was the direction. The results showed a marginal statistic tendency towards the factor direction ($F(1,76) = 2.87$; $p = .094$), and towards the three-way interaction between object gradients, environmental gradients and direction ($F(1,76) = 2.49$; $p = .118$). Figure 5 gives an overview of the reaction time averages.

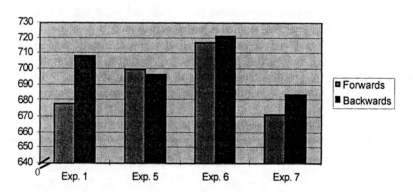

	Exp. 1		Exp. 5		Exp. 6		Exp. 7	
Mean	678.78	708.83	699.58	696.55	717.32	721.82	672.12	683.68
SD	67.02	88.91	113.27	101.86	94.06	104.34	92.37	92.50

Fig. 5. Mean reaction times and SD of experiments 1, 5, 6, and 7.

In order to test the impact of the route direction effect on the individual levels we once again carried out post hoc t-tests. The table below shows the results.

Table 4. Post-hoc t-tests.

	t-test (df = 19) (one-tailed)
Experiment 1	t = 3.10; p < .005
Experiment 5	t = -0,23; p = .410
Experiment 6	t = 0,31; p = .381
Experiment 7	t = 0.90; p = .189

The results partly confirmed our expectations regarding the route direction effect. A route direction effect can only be evoked when environmental gradients are available. In contrast with our expectations, however, object gradients do appear to be necessary to produce asymmetrical spread of activation. The route direction effect only occurs if both described types of gradient are given. If the perceptual situation is deformed in some way, and therefore does not correspond to a real-life situation, then no representation of the direction of acquisition takes place. This is illustrated by the marginal three-way interaction but most of all by a comparison of the t-tests. When routes are learned as a sequence of gradients the route direction effect only occurs when the optical flow was undisturbed during acquisition.

5. Conclusions

The route direction effect can be characterized as follows: When a person has learned a route and imagines that he or she is at a specific point along this route, then it is easier for him or her to imagine him or her-self at another point on the route which lies in the direction in which the route was learned than the other way around. This means that route knowledge includes information on the route of acquisition, i.e. the direction of acquisition is co-represented in cognitive maps.

The subject of this article is to show some of the conditions under which the route direction effect can be demonstrated and under which we can assume co-representation of the direction of acquisition in cognitive maps. Within a network theory approach the route direction effect can be conceived of as a specific asymmetry in the spread of activation (see section 2). It is possible to infer which conditions have to be fulfilled in order to evoke this asymmetry, resp. the route direction effect.

Up to now we have tested three conclusions which resulted from the assumption of an asymmetrical activation spread: (i) one conclusion on the demonstrability of the route direction effect via priming experiments, (ii) a conclusion on time windows and (iii) a conclusion regarding the necessary perceptual conditions during route learning.

(i) Reaction time latencies of prime-target pairs in priming experiments are shorter when both prime and target represent objects placed in the forward direction of acquisition than in the backward direction. This effect cannot merely be explained as a result of serial learning. The route direction effect has much more to do with the fact that prime and target represent objects which were present in different locations during learning. - We have not followed up on this result in detail here as it is explained elsewhere (see section 1).

(ii) One of the topics of this article (see section 3) was based on first results of an attempt to determine a time window for the experimental realization of asymmetrical activation spread. We were actually able to show that the route direction effect disappears when the Stimulus Onset Asynchrony is extended. This was to be expected because a longer SOA causes the target to be pre-activated, regardless of whether it represents and object which was positioned in the direction of acquisition, or not. - We are currently testing the corresponding prediction regarding very short SOAs.

(iii) The route direction effect can be demonstrated when the acquired route knowledge is in the form of a simple sequence of images as gradients or as a point of view sequence (see section 1). If route knowledge is gained via a gradient sequence then normal ecological conditions of perception must have existed: The direction of perception must correspond to the direction of movement. Objects as well as their surroundings must be perceived within the normal optical flow. Otherwise the information about the direction of acquisition cannot be adequately incorporated, and the route direction effect does not occur (see section 4).

It is very common to find serial effects which benefit the forward direction. It is easier to count, spell and read forwards than backwards (Fraisse, 1985, Block, 1989) The route direction effect discussed here has some peculiarities not shared by this ubiquitous fact. It is a specifically space-related effect: objects whose mental representations serve as primes and targets must not only follow one another on the route but also be positioned at different locations which are strongly related to them. Representations of objects on a route must be generated on the basis of normal spatial perception; backward gradients, even given a normal order of objects, do not lead to a normal asymmetrical activation spread and, therefore, do not lead to the route direction effect. A comparable dependence on ecological conditions of perception or even on conditions of spatial cognition cannot be inferred for normal serial learning.

6. References

Aginsky, V., Harris, C., Rensink, R., & Beusmans, J. (1996). *Two strategies for learning in a driving simulator, Technical Report CBR TR 96-6* . Cambridge: Cambridge Basic Research.

Anderson, J. R. (1983). *The architecture of cognition.* Cambridge, M.A.: Harvard University Press.

Anderson, J. R. (1976). *Language, memory and thought.* Hillsdale, NJ.: Erlbaum.

Appleyard, D. (1970). Styles and methods of structuring a city. *Environment and Behavior,* 2, 100-116.

Beller, H. K. (1971). Priming: effects of advance information on matching. *Journal of Experimental Psychology, 87,* 176-182.

Block, R. A. (1989). Models of psychological time. In R.A. Block (Ed.) *Cognitive models of psychological time* (pp. 1-35). Hillsdale, NJ: Erlbaum.

Bruce, V., & Green, R. R. (1985). *Visual perception, physiology, psychology and ecology*. London: Erlbaum.

Bruns, A. (1996). *Zum Einfluß des Blickpunkts beim kognitiven Kartieren*. Unveröffentl. Diplomarbeit, Mannheim: Universität Mannheim.

Buhl, H. M. (1996). *Wissenserwerb und Raumreferenz. Ein sprachpsychologischer Zugang zur mentalen Repräsentation*. Tübingen: Niemeyer.

Cohen, G. (1989). *Memory in the real world*. Hove: Erlbaum.

Downs, R. M., & Stea, D. (1977). *Maps in minds*. New York: Harper & Row.

Ebbinghaus, H. (1985). *Über das Gedächtnis. Untersuchungen zur experimentellen Psychologie (1. Aufl., Leipzig 1885)*. Darmstadt: Wissenschaftliche Buchgesellschaft.

Engelkamp, J. (1990). *Das menschliche Gedächtnis*. Göttingen: Verlag für Psychologie.

Evans, G. W. (1980). Environmental cognition. *Psychological Bulletin, 88*, 259-287.

Fraisse, P. (1985). *Psychologie der Zeit*. München: Ernst Reinhardt.

Frank, M. G., & Gilovich, T. (1989). Effect of memory perspective on retrospective causal attributions. *Journal of Personality and Social Psychology, 57*, 399-403.

Franklin, N., Tversky, B., & Coon, V. (1992). Switching point of views in spatial mental models. *Memory & Cognition, 20*, 507-518.

Freksa, C., & Habel, C. (Eds.). (1990). *Repräsentation und Verarbeitung räumlichen Wissens*. Berlin: Springer.

Gibson, J. J. (1950). *The perception of the visual world*. Boston: Houghton Mifflin.

Gibson, J. J. (1966). *The senses considered as a perceptual problem*. Boston: Houghton Mifflin.

Gibson, J. J. (1979). *The ecological approach to visual perception*. Boston: Houghton Mifflin.

Glaser, W. R. (1992). Picture naming. *Cognition, 42*, 61-105.

Helm, G. (1991). *Symbolische und konnektionistische Modelle der menschlichen Informationsverarbeitung. Eine kritische Gegenüberstellung*. Berlin: Springer.

Herrmann, Th. (1996). Blickpunkte und Blickpunktsequenzen. *Sprache & Kognition, 15*, 159-177.

Herrmann, Th., Buhl, H. M., & Schweizer, K. (1995). Zur blickpunktbezogenen Wissensrepräsentation: Der Richtungseffekt. *Zeitschrift für Psychologie, 203*, 1-23.

Herrmann, Th., Grabowski, J., Schweizer, K. & Graf, R. (1996). Die mentale Repräsentation von Konzepten, Wörtern und Figuren (pp. 120-152). In J. Grabowski, G. Harras & Th. Herrmann (Eds.), *Bedeutung - Konzepte - Bedeutungskonzepte*. Opladen: Westdeutscher Verlag.

Herrmann, Th., Schweizer, K., Buhl, H. M., & Janzen, G. (1995). *Ankereffekt und Richtungseffekt - Forschungsbericht* (Arbeiten der Forschungsgruppe „Sprache und Kognition", Bericht Nr. 54). Mannheim: Universität Mannheim.

Herrmann, Th., Schweizer, K., Janzen, G., & Katz, S. (1997). *Routen- und Überblickswissen -Konzeptuelle Überlegungen -* (Arbeiten des Mannheimer Teilprojekts „Determinaten des Richtungseffekts" im SPP Raumkognition, Bericht Nr.1). Mannheim: Universität Mannheim.

Hinton, G. G., & Anderson, J. A. (Eds.). (1989). *Parallel models of associative memory*. Hillsdale, N.J.: Erlbaum.

Horn, B. K. P., & Schunck, B. G. (1981). Determing optical flow. *Artificial Intelligence, 17*, 185-203.

Janzen, G. (1996). *Wahrnehmungsrichtung und Bewegungsrichtung: Haben Gibsonsche Gradienten einen Einfluß auf den Richtungseffekt?* Unveröffentl. Diplomarbeit. Mannheim: Universität Mannheim.

Kato, K. (1982). A study on route learning: the simulation technique by videotape presentation. *Tohoku Psychologica Folia, 41*, 24-34.

Katz, S. (1995). *Der Ankereffekt und der Richtungseffekt*. Unveröffentl. Diplomarbeit, Mannheim, Universität Mannheim.

Kim, N.-G., Growney, R., & Turvey, M. T. (1996). Optical flow not retinal flow is the basis of wayfinding by foot. *Journal of Experimental Psychology: Human Perception and Performance, 22*, 1279-1288.

Kluwe, R. H. (1990). Gedächtnis und Wissen. In H. Spada (Ed.), *Lehrbuch Allgemeine Psychologie* (pp. 115-187). Bern: Huber.

Kosslyn, S. M. (1973). Scanning visual images: some structural implications. *Perception & Psychophysics, 14*, 90-94.

Kosslyn, S. M. (1994). *Image and brain: the resolution of the imagery debate*. Cambridge, MA: MIT-Press.

Kosslyn, S. M., Ball, T. M., & Reiser, B. J. (1978). Visual images preserve metric spatial information: evidence from studies of image scanning. *Journal of Experimental Psychology: Human Perception and Performance, 4*, 47-60.

Lüer, G., Werner, S., & Lass, U. (1995). Repräsentationen analogen Wissens im Gedächtnis. In D. Dörner & E. van der Meer (Eds.), *Das Gedächtnis. Probleme - Trends - Perspektiven* (pp. 75-125). Göttingen: Hogrefe.

Lynch, K. (1960). *The image of the city*. Cambridge: The Technology Press & Harvard University Press.

May, M. (1992). *Mentale Modelle von Städten. Wissenspsychologische Untersuchungen am Beispiel der Stadt Münster*. Münster: Waxmann.

McClelland, J. L., Rumelhart, D. E. (1986). *Parallel distributed processing. Explorations in the microstructures of cognition*. (Vol. 2: Psychological and biological models). Cambridge: MIT-Press.

McNamara, T. P. (1992). Priming and constraints it places on theories of memory and retrieval. *Psychological Review, 99*, 650-662.

McNamara, T. P. (1994). Theories of priming: II. Types of primes. *Journal of Experimental Psychology: Learning, Memory, and Cognition, 20*, 507-520.

McNamara, T. P., Ratcliff, R., & McKoon, G. (1984). The mental representation of knowledge acquired from maps. *Journal of Experimental Psychology: Learning, Memory, and Cognition, 10*, 723-732.

Meyer, D. E., & Schvaneveldt, R. W. (1971). Facilitation in recognizing pairs of words: evidence of a dependence between retrieval operations. *Journal of Experimental Psychology, 90*, 227-234.

Moore, G. T., & Golledge, R. G. (Eds.). (1976). *Environmental knowing: theories, research, and methods*. Stroudsburg, P. A.: Dowden, Hutchinson & Ross.

Neely, J. H. (1991). Semantic priming effects in visual word recognition. A selective review of current findings and theories. In D. Besner & G. W. Humphreys (Eds.), *Basic processes in reading. Visual word recognition* (pp. 264-337).

Neely, J. H., & Keefe, D. E. (1989). Semantic context effects on visual word processing: A hybride prospective-retrospective processing theory. In G. H. Bower (Ed.), *The psychology of learning and motivation* (Vol. 24, pp. 202-248). New York: Academic Press.

Nigro, G., & Neisser, U. (1983). Point of view in personal memories. *Cognitive Psychology, 15*, 467-482.

Nomura, A., Miike, H., & Koga, K. (1991). Field theory approach for determining optical flow. *Pattern Recognition Letters, 12*, 183-190.

Opwis, K., & Lüer, G. (1992). Modelle der Repräsentation von Wissen. In D. Albert & K. H. Stapf (Eds.), *Gedächtnispsychologie: Erwerb, Nutzung und Speicherung von Informationen (Enzyklopädie der Psychologie, Bereich C, Serie II, Bd.4)* (pp. 337-431). Göttingen: Hogrefe.

Paillard, J. (1991) (Ed.). Brain and space. Oxford. Oxford University Press.

Palmer, S. E. (1978). Fundamental aspects of cognitive representation. In E. Rosch & B. B. Lloyd (Eds.), *Cognition and categorization* (pp. 259-303). Hillsdale, N.J.: Erlbaum.

Pinker, S. (1980). Mental imagery and the third dimension. *Journal of Experimental Psychology: General, 109,* 354-371.

Posner, M. I., & Mitchell, R. F. (1967). Chronometric analysis of classification. *Psychological Review, 74,* 392-409.

Prinz, W. (1990). Wahrnehmung. In H. Spada (Ed.), *Lehrbuch Allgemeine Psychologie* (pp. 25-114). Bern: Huber.

Pylyshyn, Z. W. (1981). The imagery debate: analogue media versus tacit knowledge. *Psychological Review, 88,* 16-45.

Pylyshyn, Z. W. (1984). *Computation and cognition: Toward a foundation for cognitive science.* Cambridge: MIT-Press.

Ratcliff, R., & McKoon, G. (1994). Retrieving information from memory: Spreading-activation theories versus compound-cue theories. *Psychological Review, 101,* 177-184.

Rehkämper, K. (1990). Mentale Bilder - analoge Repräsentation. In C. Habel & C. Freksa (Eds.), *Repräsentation und Verarbeitung räumlichen Wissens* (pp. 47-67). Berlin: Springer.

Rehkämper, K. (1993). Picture yourself in a boat on a river - Über die Bildhaftigkeit mentaler Repräsentationen. *Kognitionswissenschaft, 3,* 117-126.

Schade, U. (1992). *Konnektionismus: Zur Modellierung der Sprachproduktion.* Opladen: Westdeutscher Verlag.

Schweizer, K. (1997). *Räumliche oder zeitliche Wissensorganisation? Zur mentalen Repräsentation der Blickpunktsequenz bei räumlichen Anordnungen.* Lengerich: Pabst Science Publishers.

Schweizer, K. (in prep.) Primingeffekte bei Routenwissen - eine Theorie zur Aktivationsausbreitung bei mentalen Raumrepräsentationen.

Schweizer, K., & Janzen, G. (1996). Zum Einfluß der Erwerbssituation auf die Raumkognition: Mentale Repräsentation der Blickpunktsequenz bei räumlichen Anordnungen. *Sprache & Kognition, 15,* 217-233.

Schweizer, K., Janzen, G., & Herrmann, Th. (1995). *Richtungseffekt oder serielles Lernen - ein Nachtrag* (Arbeiten der Forschungsgruppe „Sprache und Kognition", Bericht Nr. 55). Mannheim: Universität Mannheim.

Schweizer, K., Katz, S., Janzen, G., & Herrmann, Th. (1997). *Route knowledge and the effect of route direction: A comparison of priming and cued-recall* (Arbeiten des Mannheimer Teilprojekts „Determinanten des Richtungseffekts" im SPP Raumkognition, Bericht Nr. 2). Mannheim: Universität Mannheim.

Shemyakin, F. N. (1962). Orientation in space. In B. G. Ananyev et al. (Ed.), *Psychological science in the U.S.S.R.* (Vol 1, pp. 186-255). Washington, DC: Department of Commerce, Office of Technical Services.

Siegel, A. W., & White, S. H. (1975). The development of spatial representations of large-scale environments. In H. R. Reese (Ed.), *Advances in child development and behavior* (pp. 10-55). New York: Academic Press.

Tack, W. H. (1995). Repräsentationen menschlichen Wissens. In D. Dörner & E. van der Meer (Eds.), *Das Gedächtnis. Probleme - Trends - Perspektiven* (pp. 53-74). Göttingen: Hogrefe.

Tolman, E. C. (1948). Cognitive maps in rats and men. *Psychological Review, 55,* 189-208.

Trowbridge, C. C. (1913). Fundamental methods of orientation and imaginary maps. *Science, 38,* 888-897.

Tversky, B. (1991). Spatial mental models. *The Psychology of Learning and Motivation, 27,* 109-145.

Ungerleider, L. G., & Mishkin, M. (1982). Two cortical visual systems. In D. J. Ingle, M. A. Goodale, & R. J. W. Mansfield (Eds.), *Analysis of visual behavior* (pp. 549-586). Cambridge: MIT-Press.

Verri, A., & Poggio, T. (1989). Motion field and optical flow: Qualitative properties. *IEEE Transactions on Pattern Analysis and Machine Intelligence., 11,* 490-498.

Wagener-Wender, M. (1993). *Mentale Repräsentation räumlicher Informationen.* Bonn: Holos.

Wagener-Wender, M., & Wender, K. F. (1990). Expectations, mental representation, and spatial inferences. In A. C. Graesser & G. H. Bower (Eds.), *The psychology of learning and motivation* (Vol. 25, pp. 137-157). New York: Academic Press.

Wender, K. F. & Wagener, M. (1990). Zur Verarbeitung räumlicher Informationen: Modelle und Experimente. *Kognitionswissenschaft, 1,* 4-14.

Wippich, W. (1985). *Lehrbuch der angewandten Gedächtnispsychologie.* (Vol. 2). Stuttgart: Kohlhammer.

Zimmer, H. D. (1993). Modalitätsspezifische Systeme der Repräsentation und Verarbeitung von Informationen: Überflüssige Gebilde, nützliche Fiktionen, notwendiges Übel oder zwangsläufige Folge optimierter Reizverarbeitung. *Zeitschrift für Psychologie, 201,* 203-235.

Spatial Information and Actions

Silvia Mecklenbräuker, Werner Wippich, Monika Wagener, and Jörg E. Saathoff

Department of Psychology
University of Trier, D - 54286 Trier, Germany
{mecklen, wippich, monika, evert}@cogpsy.uni-trier.de

Abstract. The present study investigates connections between spatial information and actions. Only a few studies on this topic have been published to date. Moreover, changes in spatial representations have not been examined following the connection with actions. Our results show that previously acquired spatial information (i.e., information about locations along a learned route) can be associated with imagined or with symbolically performed actions. However, we have not yet found any evidence that spatial representations are altered by the formation of these associations. We conclude that more sensitive and, perhaps, implicit measures of spatial representations are needed in order to detect the presumed action-dependent changes. Furthermore, the salience of the actions and their connection to the spatial environment should be considered as important variables.

1   Introduction

One of the requirements for human beings to move around without many problems in their environments is the ability to store information about spatial arrangements. These spatial layouts are acquired by different means. The most natural procedure is navigating through the environment, looking at objects and distances, and other parts of the real world. Because humans are social beings, communication about the environment is another possibility for acquiring spatial information. Most commonly, in large-scale environments, maps are used to communicate locations and routes. In small-scale environments, like a room with furniture or a shelve with objects, verbal descriptions are quite frequent. When persons learn the locations of objects in an environment, they typically also acquire nonspatial information, for instance, information about activities which have been performed at particular locations. The present research investigates interactions of this action information and spatial information. The questions we would like to answer are whether this nonspatial action information is embedded in the spatial mental representation, whether memory for actions is influenced by spatial information and vice versa, that is, whether memory for spatial settings is influenced by actions.

The present research can be seen as an attempt to combine spatial cognition research and research on memory for actions. Thus, the next sections of the introduction will deal with the following topics: spatial mental representations and the integration of nonspatial information, memory for actions, and connections between spatial information and actions. Because we did not use only traditional explicit measures of memory but also so-called implicit measures, one section will discuss implicit and explicit memory measures. Presentation of our empirical work follows, and we finish our paper with some conclusions and suggestions for future research.

1.1 Spatial Mental Representations and the Integration of Nonspatial Information

Nearly all human activities require that people are able to store information about spatial arrangements. Therefore, in cognitive psychology, mental representations (cf. e.g., Herrmann, 1993; Tack, 1995) of spatial information are assumed. Spatial representations might be distinguished in many ways (for an overview, see McNamara, 1986). For example, when looking at the acquisition of spatial knowledge, that is, the construction of a spatial representation, the distinction between route and survey knowledge (or configurational knowledge as called by Hirtle & Hudson, 1991) recurs frequently. With origins in developmental models (Siegel & White, 1975), there are now numerous studies trying to find the properties that distinguish both types and how one type relates to the other (e.g., Carleton & Moar, 1982; Evans & Pezdek, 1980; Hirtle & Hudson, 1991; Rothkegel, Wender, & Schumacher, this volume; Thorndyke, 1981; Thorndyke & Hayes-Roth, 1982). "Route knowledge is characterized by the knowledge of sequential locations without the knowledge of general interrelationships. In contrast, configurational knowledge is characterized by the ability to generalize beyond learned routes and locate objects within a general frame of reference" (Hirtle & Hudson, 1991, p. 336). Survey knowledge is easy to obtain when looking at a map which provides an overview of a space that is normally too large to be seen at once. Route knowledge is acquired by navigating through the environment. However, frequent travel on a particular route can lead to survey knowledge. It is important to note that the cognitive map acquired through some kind of communication (map or verbal instructions) might be different from the route or survey knowledge acquired by frequent travel on a particular route or browsing around in a city (Easton & Sholl, 1995; Sholl, 1987). In the study reported here we tried to induce route knowledge. The participants had to navigate through the environment, which was a large-scale one. In contrast to a small-scale environment like configurations of objects in a room or a map, a large-scale environment is defined by the impossibility to view all points at one time when standing in the environment (e.g., Acredolo, 1981). This means that barriers (as in our study) or boundaries are present to restrict the visual field (e.g., Cohen, Cohen, & Cohen, 1988), or that special devices, like a headpiece with adjustable blinders, are used (e.g., Sadalla & Staplin, 1980). In this case, people have to acquire information sequentially. Therefore, the acquisition of route knowledge is favored. In order to decide what kind of knowledge was acquired, the participants in our study were required to give estimations of both route distances and of Euclidean distances. It is known that estimations of route distance are favored by route knowledge whereas estimations of Euclidean distances are favored by survey knowledge (Thorndyke & Hayes-Roth, 1982; see also Hirtle & Hudson, 1991; Sholl, 1987).

Space is filled with objects and objects have many properties in addition to their spatial interrelations with other objects. When people learn the locations of objects in an environment, they typically also acquire nonspatial (e.g., semantic) information about the objects. For instance, students do not only learn the locations of buildings on a campus, but also a lot of nonspatial information such as which departments are housed in which buildings. An important - though still neglected - topic in current spatial cognition research concerns the interdependence in memory of spatial and nonspatial information.

There is no doubt that spatial memories can be influenced by nonspatial information. Several studies could demonstrate effects of nonspatial information on intentional learning of spatial layouts and on spatial problem solving (e.g., Hirtle & Mascolo, 1986; McNamara, Halpin, & Hardy, 1992; McNamara & LeSueur, 1989; Sadalla, Staplin, & Burroughs, 1979). Sadalla et al. (1979), for example, have shown that routes containing high-frequency names are estimated as longer than routes of equivalent objective distance containing low-frequency names. Hirtle and Mascolo (1986) could demonstrate effects of semantic clustering on the memory of spatial locations. Participants had to learn artificial maps in which the labels affixed to the locations fell into two semantic clusters: names of city buildings and names of recreational facilities. Places in the same semantic cluster were close to each other on the map. However, one city building was closer to the cluster of recreational facilities than to the other city buildings, and vice versa. Distance estimations showed that the odd names were displaced in memory toward their fellow category members. These results provide evidence that the semantic labels attached to clusters can produce mental clusters and that these clusters can alter a person's memory of spatial locations.

McNamara et al. (1992) pointed out that these findings necessarily imply a connection between visual-spatial and linguistic-nonspatial information but that they do not imply that verbally coded information has been integrated in a spatial representation. For example, the semantic relations shared by a set of objects could cause people to misrepresent inter-object spatial relations (e.g., Hirtle & Mascolo, 1986) without being encoded as part of the spatial representation. McNamara et al. (1992) conducted a series of experiments to investigate whether people could integrate nonspatial information about an object with their knowledge of the object's location in space. In some experiments, participants learned the locations of cities on a fictitious road map; in other experiments they were already familiar with the locations of buildings on a campus. A critical subset of the locations on the map and on the campus could be divided into pairs that were close together and pairs that were far apart. In the second phase of the experiments, the participants learned fictitious facts about the cities or the buildings. The third phase consisted of a priming task. Priming has proven to be an effective tool for investigating properties of spatial memory (e.g., Clayton & Chattin, 1989; McNamara, 1986; McNamara, Ratcliff, & McKoon, 1984; Wagener & Wender, 1985). It is particularly useful for investigating knowledge integration because different findings (e.g., McNamara, Hardy, & Hirtle, 1989; Ratcliff & McKoon, 1981) indicate that it is primarily an automatic process. Consequently, priming should be informative about the structure and the content of memory rather than about retrieval strategies (cf. e.g., McNamara et al., 1992). In the McNamara et al. (1992) study, location judgments were required in the priming phase. The facts and either city or building names appeared sequentially on a computer terminal screen; the participants' task was to decide whether each city or building was in one region of the space (e.g., in County A in the case of a city) or another (e.g., in County B), and whether each fact was associated with a city or building in region A or B. It could be shown that the previously learned facts strongly influenced the location judgments. Responses were faster and/or more accurate when a city or a building name was primed by a fact about a neighboring city or building than when a city or building name was primed by a fact about a distant city or building. These results indicate that the spatial and nonspatial information were encoded in a common memory representation.

The current research was conducted to examine - as far as we know, for the first time - interactions of spatial information and actions. It has been shown (e.g., Quinn, 1994; Rieser, Guth, & Hill, 1986) that motor components, that is, movements of the whole body connected with orientation changes without vision, have an impact on the mental representation of spatial configurations. In our research we will not focus on this aspect of actions. With *actions* we mean simple activities as studied in action memory research. The next section briefly summarizes this research.

1.2 Memory for Simple Actions

In a typical study of memory for simple actions such as *to smoke a pipe*, participants are required verbally to perform actions in order to get the actions under the control of the experimenter. Hence, the problem emerges that performance of actions is experimentally studied in the context of verbal utterances. Consequently, only the additional effect of performing an action in comparison with solely listening to the description of the action can be studied. Action phrases such as *smoke a pipe* are retained much better when participants encode the phrases by self-performing the denoted actions than when they learn the phrases under standard verbal learning instructions, that is, by verbal tasks. This enactment effect or effect of self-performed tasks (SPTs) has been extensively replicated for both recall and recognition (see Cohen, 1989; Engelkamp, 1991; Engelkamp & Zimmer, 1994, for reviews). The SPT effect is independent of whether imaginary or real objects are used in performing the tasks (Engelkamp & Zimmer, 1994). Moreover, enactment of the denoted action seems to be decisive because the SPT effect can be observed even when the participants close their eyes while performing the actions. Thus, seeing the task performed by oneself is not necessary for the production of the SPT effect (Engelkamp, Zimmer, & Biegelmann, 1993).

Despite much research, the reason for the SPT effect remains unclear. According to Cohen (e.g., Cohen, 1983), the effect is the result of automatic encoding processes. However, it is unclear how a presupposed automatic processing might explain the excellent encoding of performed actions. In elaborating this view, Bäckman, Nilsson, and Chalom (1986) assume that the performed actions are multimodal events which provide the participants with information about form, color, etc. of the objects and about movement. According to this explanation, after enacting, the participants may benefit from the automatic, multimodal, rich item-specific encoding of the actions. Finally, in a multimodal and more detailed approach, Engelkamp and Zimmer (1994) have argued that the motor information provided by enactment is a critical component of the SPT effect. This hypothesis is supported by selective interference experiments in which memory after SPTs was selectively impaired by secondary SPTs (e.g., Saltz & Donnenwerth-Nolan, 1981). In addition, Engelkamp, Zimmer, Mohr, and Sellen (1994) demonstrated that the performance of an action during learning, and again, before making a recognition decision, enhanced the SPT effect.

If the enactment effect is - at least also - due to motor information, this may have important implications with respect to connections between actions and spatial information. There is evidence that a motor encoding enhances an *item-specific* encoding, that is, the performed action is represented as a single event more extensively and more distinctly. In contrast, a *relational* encoding, which provides connections with other actions or with other information (e.g., spatial information), is not promoted by performing actions. In other words, motorically encoded information is abounding in item-specific information but lacks relational properties (for the distinction between item-specific and relational information, see Hunt & Einstein, 1981; Zimmer & Engelkamp, 1989). Consistent with such assumptions, it could be shown that cued recall for actions does not exceed - or may even be lower than - free recall. For example, such a cue-failure effect could be demonstrated by Engelkamp (1986). The finding that a relational encoding is not promoted by performing actions is not compatible with the intuitive idea that spatial information may influence our memories for actions. For instance, seeing, imagining, or even speaking about a certain place may evoke autobiographical memories of actions carried out there. Thus, the present research examined whether connections between actions and spatial information can be observed under certain conditions.

1.3 Connections between Spatial Information and Actions

In our opinion, investigations of the interdependence between spatial information and actions in memory will offer new insights not only for spatial cognition research. As we see it, spatial cognition research has focused on the conceptualization and investigation of spatial representations and has neglected the purpose and the benefits of spatial representations. Spatial representations do not only play an important role for navigational decisions. When going beyond the traditional framework of the standard dependent variables such as distance estimations or direction judgments, it can be generally stated that spatial representations improve or enlarge our potential to perform adequate and more or less purposeful actions in complex environments.

Whereas the study of interactions of spatial information and actions is a neglected topic, some experimental research has focused on the influence of activities on the *development* of spatial representations (for a review, see e.g., Cohen, 1985). The results of developmental studies conducted by Cohen and colleagues (Cohen & Cohen, 1982; Cohen, Cohen, & Cohen, 1988) show the utility of performing activities for the construction of spatial representations. For example, Cohen and Cohen (1982) and Cohen et al. (1988) assessed the effects of simply walking among locations versus performing isolated activities at locations versus performing activities that were functionally linked across locations (through a letter writing and mailing theme). First and sixth graders had to walk through a novel environment constructed in a large, otherwise empty classroom. In the Cohen and Cohen (1982) study, the entire spatial setting could be viewed from single vantage points (small-scale environment). In contrast, Cohen et al. (1988) used barriers and, consequently, the environment qualified as a large-scale one. The results (i.e., distance estimates) suggest that when the environment can be viewed in its entirety, the theme serves to facilitate the acquisition of spatial knowledge for the entire space. However, when the environment is segmented by barriers, the theme serves to aid primarily the knowledge of the spatial information linked by the theme. Cohen et al. (1988) emphasize that any activity that focuses the child's attention on the relatedness of locations in large-scale space increases the accuracy of the child's representations of those related locations compared to those locations not related by an activity.

So far activities seem to affect the construction of spatial representations. But there may also be a reverse effect: Spatial information may influence our memories for actions. Most of us would share the assumption that if we have to remember a series of actions, knowing the names of the places where the actions took place should enhance memory of the actions (in the sense of the famous loci mnemonic). This intuitive idea, however, is not compatible with the finding that a relational encoding, which provides connections with other actions or with other information (e.g., spatial information), is not promoted by performing actions.

A study by Cornoldi, Corti, and Helstrup (1994) is highly relevant for the present research. The authors expected that imagined places seem to be good cues for remembering actions, both because actions are typically performed in relation to specific places and because imagery is a powerful encoding system for integrating information (Paivio, 1986). In a first experiment, they contrasted memory for enacted versus imagined actions in association with imagined places. Contrary to expectations, a cue-failure effect under the enactment condition - cued recall for actions did not exceed free recall - could also be observed when spatial locations (imagined places) were used as cues for recalling the actions. In contrast, cued recall exceeded free recall in the imagery group. In another experiment, an attempt was made to strengthen the connection between places and actions (in all experiments, obvious associations between places and actions were avoided) by requiring the participants to find reasons why the actions were performed at the assigned places. This encoding instruction eliminated the cue-failure effect. Thus, it is possible to find conditions where imagined places can work as successful cues for recalling performed actions. According to Cornoldi et al., the give-a-reason instruction probably has facilitated both item-specific and relational encoding. However, when no attempts are made to strengthen the connection between places and actions, a cue-failure effect is observed. This effect indicates "... that motor programs do not easily link with each other or with other types of information codes" (Cornoldi et al., 1994, p. 326).

One aim of the present research was to examine whether a cue-failure effect would be obtained under conditions where participants were required to encode actions in relation to previously acquired spatial representations and not only to arbitrary and isolated places as in the study of Cornoldi et al. Whether new associations between locations and actions were acquired under such conditions was tested not only with traditional explicit measures of memory but also with implicit measures.

1.4   Implicit versus Explicit Measures

Studies on spatial cognition have not only used different procedures of acquiring spatial knowledge, but also a diversity of methods for assessing spatial knowledge. For example, in the study by Taylor and Tversky (1992), participants studied maps of different environments ranging from a single building to an enclosed amusement park with several buildings to a small town. In the test phase, some participants had to draw the map from memory whereas the other participants had to give a verbal description of the map. Many other tasks have been used to study spatial representations (e.g., Evans, 1980). Some of the more common ones have been distance estimation (cf. Rothkegel, Wender, & Schumacher, this volume), location and orientation judgments (e.g., Wender, Wagener, & Rothkegel, 1997), and navigation. However, some authors have questioned whether these tasks can be considered the most informative about the mental representation of spatial knowledge (e.g., Liben, 1981; Siegel, 1981). Siegel (1981) has suggested that tasks should be used that minimize performance demands so that the contents of spatial representations can be assessed more accurately. Although the aforementioned tasks differ in many aspects, they all share one common property: Following the experimental or pre-experimental spatial experiences the participants had made, these spatial experiences were explicitly referred to before the participants had to express their spatial knowledge. Thus, from the perspective of cognitive psychology, spatial knowledge has typically been assessed with so-called explicit memory tasks. In recent years, however, implicit tests have become very popular in memory research.

It is important to note that we use the terms "implicit" and "explicit" memory in a purely *task oriented* manner. The defining criterion of the distinction between explicit and implicit tests is the instruction participants are given at the time of test: In an explicit test they are told to recollect events from an earlier time; in an implicit test they are told simply to perform a task as well as possible, with no mention made of recollecting earlier experiences in performing the task. For example, in an implicit test participants are asked to complete word stems or word fragments with the first meaningful word that comes to mind. The finding is that prior presentation of a word increases the likelihood of that word being used to complete a stem or a fragment. This facilitation of performance in implicit memory tests is termed *(repetition) priming* (for reviews on implicit memory research, see Graf & Masson, 1993; Roediger & McDermott, 1993).

Many findings suggest that repetition priming *primarily* reflects automatic, that is, *incidental* or *unintentional* (not consciously controlled) uses of memory. It is important to note that we do not equate "automatic" with "unconscious". We agree with Richardson-Klavehn and Gardiner (e.g., Richardson-Klavehn & Gardiner, 1995) who emphasize the distinction between memorial state of awareness (conscious vs. unconscious) and retrieval intention (intentional vs. unintentional). Postexperimental queries indicate that in an implicit memory task "normal" student subjects notice, at least for some items, the study-test relationship. Nevertheless, most participants follow the implicit test instructions and do not adopt intentional retrieval strategies (see Roediger & McDermott, 1993). Thus, it is justified to speak of an unintentional (incidental) and less conscious use of memory in implicit memory tasks.

In our opinion, information that is assigned to spatial representations is normally used automatically (i.e., unintentionally) rather than consciously controlled. Everyday experiences emphasize the importance of familiarity, that is, of automatic influences of prior experiences. For example, when finding our way along once-familiar routes, we usually remain confident as long as the surroundings seem familiar. Without such a feeling of familiarity, we conclude that we have taken a wrong turn (cf. Anooshian & Seibert, 1996; Cornell, Heth, & Alberts, 1994). Only if there are problems, for example, if we had lost our way, consciously controlled uses of memory will be likely. Such considerations suggest the use of implicit measures in spatial cognition research.

Implicit tests were also seldom used in action memory research. Whereas an SPT effect could be observed consistently in all kinds of explicit tests, SPTs may have little or no influence on implicit memory test performance (Engelkamp, Zimmer, & Kurbjuweit, 1995; Nyberg & Nilsson, 1995). Thus, the SPT effect may presuppose an intentional mode of retrieval from memory. Considered in this way and presupposing that spatial knowledge is used most often in an incidental or less conscious mode of retrieval, it is not at all self-evident that simple actions have an impact on spatial representations.

One important question of the present research was whether new associations between locations and actions could be acquired. This question was examined by using explicit as well as implicit testing procedures (see Graf & Schacter, 1985, for details concerning the paradigm of priming of new associations). In our experiment, in the study phase unrelated location-action pairs (e. g., *museum - to squeeze the grapefruit; post office - to fold the handkerchief*) were presented. In the implicit test phase, the verbs of the action phrases were given in the context of a location. There were two different conditions. In the identical (context) condition, the action (i.e., the verb) was paired with the same location as in the study phase (e.g., *museum - to squeeze*). In the recombined condition, it was paired with another location from the study list (e.g., *post office - to squeeze*). In addition, new location-verb combinations (from an alternate study list) were presented in order to provide a baseline measure. Participants were required to spontaneously associate a suitable object to the presented location-verb pairings, for example, they were asked what they could "squeeze at the museum". Priming of new associations is demonstrated when an object is associated more often to its corresponding verb in the identical than in the recombined condition.

Using this paradigm, we could demonstrate a connection between enacted actions and locations in a *pilot study*. The main aim of this study was to examine whether newly acquired associations between symbolically performed actions and locations could be observed when the locations were not arbitrary and isolated places as in the study by Cornoldi et al. (1994), but places on a certain route presented in the form of a simple map. Because it was a pilot study, participants were not required to learn the map which showed 24 locations connected via a black line indicating a route. Participants in an enactment condition were asked to symbolically perform (i.e., without concrete objects) certain actions at certain locations. Participants in an imagery condition were required to imagine themselves performing the actions at the locations. The order of the locations followed the route on the map which the participants were able to view. The map was removed for the tests of memory for new associations. Following the implicit test (for a description, see above), all participants were given an explicit cued-recall test with the location names and verbs serving as cues for remembering the studied objects.

Table 1. Mean Proportion of Objects Associated (Implicit Test) or Recalled (Explicit Test) as a Function of Location Type in the Two Encoding Conditions (Pilot Study)

Test	Implicit Test		Explicit Test	
Location type	Identical	Recombined	Identical	Recombined
Imagery	0.30	0.11	0.47	0.24
Enactment	0.30	0.19	0.47	0.36

The results showed that memory for action phrases was influenced by spatial information. When we presented the verb as a cue together with the same location name as in the encoding phase, the corresponding objects were associated (implicit test) or recalled (explicit test) more often than when locations and verbs were recombined (for the means, see Table 1). Thus, we could demonstrate newly acquired associations between spatial information (i.e., locations) and actions. As expected, cued-recall performance generally exceeded implicit performance, but the patterns of results were very similar. Moreover, in contrast to the findings of Cornoldi et al. (1994), whether the actions had been imagined or symbolically performed in relation to the locations had only a minor effect. This is a surprising result because imagery is seen as a powerful encoding system for integrating information (e.g., Paivio, 1986), whereas the formation of new relations is not enhanced by motor encoding as can be seen in the cue-failure effect (Cornoldi et al., 1994). One possible cause for these contradictory findings might be the kind of locations used. We used locations on a certain route while Cornoldi et al.'s places were arbitrary. Furthermore, there were differences in the paradigms used. The first one concerns the type of test. Cornoldi et al. employed only explicit tests (free and cued recall). In a further experiment of our pilot study, we were able to show that the type of test could not be held responsible for the findings. When we employed only explicit tests, we once again could not observe any differences between the encoding conditions. The clear effect of the type of location (identical vs. recombined) was replicated. The second difference concerns the cues used in the cued-recall test. In the Cornoldi et al. study, only the location names served as cues and participants had to remember the whole action phrases. In our experiment, the location name and the verb were provided as cues and participants only had to remember the object.

Because we were mainly interested in whether connections between actions and *previously acquired* spatial information could be demonstrated, participants in our main study had to acquire spatial knowledge.

2 Study: Spatial Information and Actions

The study consisted of several phases. In the *first phase*, participants had to acquire spatial knowledge by learning a route with different locations. In the *second phase*, memory for the newly acquired spatial information was assessed. Different spatial measures were employed because they may be differentially sensitive for the detection of possible action-dependent changes in the spatial representation. In the *third phase*, the so-called *action phase*, simple actions were related to the previously learned spatial information (i.e., the locations on a route). Actions had to be imagined or they had to be symbolically performed in association with locations. In *the fourth phase*, memory for newly acquired associations between actions and locations was assessed with implicit and explicit memory tests. In the *fifth* and final *phase*, the assessment of spatial knowledge was repeated in order to detect possible action-dependent changes.

We were interested in the following questions:

• Under which *action conditions* can associations between actions and spatial information be observed? Are imagined actions connected with spatial information more easily than enacted actions? (cf. Cornoldi et al., 1994)

• Is it possible to demonstrate a connection between actions and spatial information with *implicit measures* of memory? That is, can significant priming of new associations be obtained?

• If new associations can be shown, then there is the *important question* of whether there are *changes in spatial representations* indicating an integration of spatial and non-spatial information.

In the experiment, a large-scale environment with barriers present was used. Thus, the acquisition of route knowledge was favored. The participants had to walk through a kind of maze in order to get all the information for the following tests. Self-controlled exploration was limited to ensure an equal exposure time for all participants (see Procedure). In a way, the setting resembled a simulation of a real environment because we reduced the setting to the necessary information needed for the impending tasks.

The experiment also included a control group whose participants were not required to perform or to imagine any actions in the environment. Furthermore, and more importantly, spatial knowledge was assessed both before and after the action phase in a different room without any map or other perceptual aids. Thus, we were able to gain some evidence about spatial representations and a possible change in the representations following the performance or imagination of actions in the setting.

2.1 Method

Participants and Materials. In this experiment, 50 students from the University of Trier participated in 90-minute sessions and were paid for their services. The data from two participants were incomplete because they were not able to learn the spatial setting in 30 minutes. Sixteen participants were randomly allocated to one of three different groups (enactment, imagery, and control).

Fig. 1. Perspective view of the corridor network

As a *spatial setting* we constructed a maze with transportable walls. A perspective view of the maze is shown in Figure 1. The complete path, which participants had to walk, was 21 m long. To exclude orientation clues besides the intended ones, we covered the walls with slightly opaque plastic foil that diffused the incoming light. This resulted in a network of corridors or ducts with six decision points. We posted 18 signs with names of common locations (e.g., *museum*, *cinema*) on both the left and right walls of the corridors. Signs for the critical pairs in distance estimation were always posted on the left walls. Critical pairs of names were assigned with 0, 1, or 2 intervening signs (i.e., locations). An example for a pair with no intervening location on a straight path is *café - cinema,* and an example for a pair with no intervening location on a cornered path is *gas station - museum* (see Figure 2). Hence, there were two independent variables regarding the spatial setting and knowledge test: 3 (number of intervening locations) x 2 (type of path). The physical distance between critical pairs was measured from a point in front of the signs in the middle of the corridor resulting in approximately 2 m for straight paths and approximately 2.7 m for cornered paths. To ensure that participants viewed the signs from exactly the point that was used to measure the physical distance, viewing positions were marked on the floor with duct tape crosses. Also, six decision points were marked with black tape.

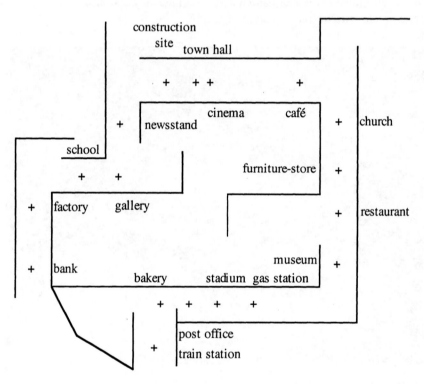

Fig. 2. Simplified map of the corridor network. A cross symbolizes the participants' viewpoint at a particular location

Spatial knowledge was assessed by requiring participants' estimations of route distances and of Euclidean distances. The same pairs of items with different orders of presentation on sheets of papers were used for both estimations: six critical and four filler pairs. The estimations were obtained using a ratio estimation technique. This was accomplished by displaying a reference line on each answer sheet. At the beginning and at the end of the reference line, the names of the first (*train station*) and the last (*bank*) location in the network of corridors were mentioned for the estimation of route distances. For the estimation of Euclidean distances, the two locations that were the farthest apart regardless of the route were mentioned. Participants knew that the distance between the given locations should be taken as a reference point when estimating the distances of the other pairs. To accomplish these tasks they were asked to indicate, with a slash on a connecting line between the pairs of names, the distance that corresponded to the remembered distance. The materials used in the two spatial testings (before and after the action phase) were identical.

Thirty-six action phrases of the kind used in subject-performed tasks (such as *to squeeze the grapefruit*) were selected for the *action phase*. All of the verbs and object nouns that were used in describing the actions were unique. Pretesting had shown that the pre-experimental associations between the verbs and the nouns were low (less than 7% of the target nouns were elicited by the verb cues). Each of the 18 location names was randomly combined with two of the actions. Bizarre combinations were avoided. These two sets of location-action pairs served as the two study lists. Thus, location names were identical in both lists whereas the actions were unique. Across participants, each list was used equally often as the study list.

For the *implicit memory test*, four test lists were constructed, two versions for each study list. Each test list consisted of 36 items, 18 new and 18 old location-verb pairs. The new items provided a baseline measure and consisted of pairs of new locations and the 18 nonstudied verbs (from the alternate study list). The 18 old items, pairs of studied locations and verbs, were divided into two categories, *identical* and *recombined*. If the item was identical, the combination of location and verb was identical to the combination in the study phase. The recombined items were constructed by recombining the remaining nine studied locations and verbs. The set of verbs combined with either an identical or a recombined location varied across the two versions of the test lists. The verbs that were identical in one version were recombined in the other version, and vice versa. The presentation order was random with the restriction that no more than four old or new items followed each other. The four lists for the *explicit test* (the second memory test) consisted of the 18 old items already used in the implicit test arranged in a new random order.

Procedure and Design. The experiment was divided into five phases: Learning of the corridor network, first spatial testing, action phase (associating actions with locations), memory tests for new associations (consisting of an implicit test followed by an explicit test), and second spatial testing. The different phases involved changes of experimental rooms: the corridor network (or maze) and the test room.

The *learning phase* took place in the corridor network which had to be traveled three times. The main focus was on learning the route through the corridors but no instruction was given to learn the signs with the location names. During the first walk through, the experimenter directed the participants. After that, the participants had to find their way through the corridors on their own although the experimenter always stayed close behind them. After three completed walks the participants received an explicit learning instruction concerning the location signs. They were told that during their walk through the corridors they would have to give the names for the subsequent locations. The learning criterion was to name all the locations correctly two times in a row.

After successful completion of this phase, the *first spatial testing* took place in another, adjacent room. The route distances were estimated always before the Euclidean distances. The participants were instructed to imagine standing in front of the location sign that was mentioned first on the left side of the answer sheet. Then they were asked to imagine the distance to the second mentioned location on the right side of the answer sheet. For route distances they were explicitly instructed to imagine the distance they traveled, for Euclidean distances the instruction was to imagine the direct distance from a bird's eye perspective. As mentioned above, a ratio estimation technique was used.

The first spatial testing was immediately followed by the *action phase*. Participants were told that we wanted to know how easily they could connect locations with actions. Subsequent memory testings were not mentioned (incidental learning). Participants received their instruction in the test room but left the room to walk through the maze again. At every location sign, the participants had to stop and turn toward the sign. Then the experimenter read aloud one action phrase. Participants in an enactment condition were asked to symbolically perform (i.e., without concrete objects) certain actions at certain locations. Participants in an imagery condition were required to imagine themselves performing the actions at the locations as if they could observe their own performance from the outside. After performing or imagining each action, the participants had to rate how easily they could connect location and action. In addition, a control condition was assessed. Here, the participants were simply instructed to estimate the probability of finding the locations in a typical town with about 20,000 inhabitants. When the walk was completed, the experimenter guided the participants back to the test room where the memory tests took place.

First, the *implicit test* was given. A list containing location names and verbs describing simple actions was presented (by a tape-recorder). Participants were required to spontaneously associate appropriate objects which could be used for performing these actions at these locations. After presentation of each location-verb pair (e.g., *museum - to squeeze*), they had to write down the first two objects that came to mind. Two objects were required to increase the likelihood that participants would associate objects other than the most common ones. The item rate was 10 seconds.

Following the implicit memory test, all participants were given a written cued-recall test (i.e., an *explicit test*. Location names and verbs were presented as cues for remembering the studied objects. The participants were informed that only some of the locations could help them to remember the objects and that therefore they should concentrate on the verbs. A maximum of five minutes was allowed for this task.

The experiment comprised a 3 x 2 mixed factorial design for both the implicit and the explicit memory task. The between-subjects factor was *encoding conditions* (enactment, imagery, control) and the within-subjects factor was *location type* (identical vs. recombined).

Finally, the participants completed the *second spatial testing* in the test room. Concerning these testings the experiment comprised a 3 x 2 x 2 x 3 mixed factorial design. The between-subjects factor was the same as before. The within-subjects factors were constituted by the *time of testing* and by the variation of the test materials: *type of path* (straight, cornered), *number of intervening locations* (0, 1, 2).

2.2 Results

First, we will report the results of the spatial testings. Then the results of the memory tests will be described. The significance level was set at α = .05 for all statistical tests.

Spatial Measures. We measured the length of the lines from the first name of a pair to the slash on the line for each pair on the answer sheets. As a next step, these measures were converted into meters and compared with the actual physical distance. These difference scores were the dependent variables in the two analyses of variance (ANOVAs). First, the results for the estimations of route distances will be described. Most importantly, the encoding conditions failed to show any reliable effect on the estimation data. Hence, there is no evidence for a change in the mental representation following the action phase, $F < 1$ for the interaction between the encoding conditions and the time of testing (before or after the action phase). In Table 2 the mean difference scores are displayed as a function of time of testing (first vs. second test), type of path, and number of intervening locations. As expected, the number of intervening locations had a reliable effect on the distance estimates: The more locations were displayed along the route, the more the distance was overestimated, $F(2,90)$ = 26.3. The same is true for the straight vs. cornered paths: Participants overestimated the route distances to a greater extent when the path between two locations contained a corner, $F(1,45)$ = 9.19. Furthermore, a reliable interaction between the time of testing and the number of intervening locations was observed. The effect of the number of intervening locations was less pronounced in the second test, $F(2,90)$ = 4.28. No other main effects or interactions proved to be reliable.

Table 2. Mean Differences between Estimates and Physical Route Distances as a Function of Test Phase, Type of Path, and Number of Intervening Locations

Test	Phase I			Phase II		
# of locations	0	1	2	0	1	2
straight path	0.04	1.44	2.63	0.68	1.62	2.20
cornered path	1.04	2.38	2.74	1.62	2.27	3.01
average mean	0.54	1.91	2.69	1.15	1.95	2.61

The computation of the estimates of Euclidean distances followed the same rationale (for the mean difference scores, see Table 3). Once again, the encoding conditions did not affect the estimation data. Regarding the number of intervening locations, we observed an interesting reliable effect that, at a first glance, seemed to be opposed to the influence of the intervening locations on the estimates of the route distances: The estimates grew increasingly more precise with more intervening locations, $F(2,86) = 16.76$. This means that paths without intervening locations were estimated as shorter, and this is in compliance with the results of the route distances. In addition, we found a reliable main effect of the time of testing: Participants underestimated distances more in the second testing than in the first one, $F(1,43) = 7.69$. The influence of the type of path was only marginally significant. There is a tendency that straight paths were underestimated more than cornered paths. In general, it should be noted that the estimates of Euclidean distances were surprisingly good. This result suggests that the participants also acquired survey knowledge. Furthermore, the Euclidean distances were underestimated in all conditions. This is important to mention because for pairs on straight paths, physical route distances and Euclidean distances were the same.

For estimations of both route and Euclidean distances, almost identical patterns of results were obtained when the ratio between the estimated distance and the physical distance was used as the dependent measure.

Table 3. Mean Differences between Estimates and Physical Euclidean Distances as a Function of Test Phase, Type of Path, and Number of Intervening Locations

Test	Phase I			Phase II		
# of locations	0	1	2	0	1	2
straight path	-1.05	-0.69	-0.33	-1.05	-0.60	-0.59
cornered path	-0.62	-0.34	-0.21	-0.90	-0.60	-0.35
avererage mean	-0.84	-0.52	-0.27	-0.98	-0.60	-0.47

Implicit Memory Measures. The baseline for the association of the predesigned objects to new verbs was very low in both encoding conditions (less than 4% of the targets were generated). The mean scores for the old items are given in Table 4. As expected, the participants in the control condition associated few of the old objects (mean proportion: .02), and significantly more old objects were associated after enactment or after imagery, $t(30) = 7.77$ and $t(30) = 4.96$ compared to the control group. An ANOVA with encoding conditions (enactment vs. imagery) and location type (identical vs. recombined) as independent variables showed a clear effect of location type. Significantly more studied objects were generated for the verbs with identical location than for the verbs with recombined location, $F(1,30) = 22.59$, $MSE = 0.02$. All other effects can be disregarded, $F < 2.00$. Thus, the results revealed reliable priming of new associations for both encoding conditions.

Explicit Memory Measures. The mean proportions of items recalled are also summarized in Table 4. An ANOVA showed reliable effects of encoding conditions and location type. More objects were recalled after enactment than after imagery, $F(1,30) = 9.19$, $MSE = 0.08$, and cued recall performance was higher when the verb cue was paired with an identical location than when verb and location were re-combined, $F(1,46) = 28.96$, $MSE = 0.02$. The interaction approached significance, $F(1,30) = 3.06$, p < 0.10. The effect of location type was more pronounced in the imagery condition as compared to the enactment condition.

Table 4. Mean Proportion of Objects Associated (Implicit Test) or Recalled (Explicit Test) as a Function of Location Type in the Two Encoding Conditions

Test	Implicit Test		Explicit Test	
Location type	Identical	Recombined	Identical	Recombined
Imagery	0.33	0.15	0.60	0.37
Enactment	0.40	0.24	0.74	0.65

2.3 Discussion

The results show that memory for action phrases is influenced by spatial information. In the implicit as well as in the explicit test of memory, we have demonstrated newly acquired associations between spatial information (i.e., locations) and actions. More-over, whether the actions had been imagined or symbolically performed only had a minor effect on this pattern of findings. Explicit memory measures revealed a general SPT effect. The findings underscore the possibility of using a SPT procedure to "in-fuse" memory for action phrases with spatial information. This is an important mes-sage because our results show that memorial information about simple activities can be enriched with spatial information. This enrichment is revealed even with implicit measures. The latter result may be due to the fact that a combined and integrated memory representation has been formed at encoding. However, on the basis of our findings we cannot definitely say which processes are responsible for the observed new associations. We assume that a connection between spatial knowledge and nonspatial information was established at encoding. However, we cannot rule out that only verbal information, that is, the names of the locations and the action phrases, was connected.

Concerning our main hypothesis that the spatial representation should be influenced and changed by performing or imagining actions in the environment, we have to admit that we could not observe even the slightest effect. Nevertheless, the other independent variables proved to affect the distance estimates. These data support the assumption that a mental representation of the distance along a route was formed that is related to the *quantity of information* stored when traveling this route (cf. Sadalla & Staplin, 1980; Sadalla et al., 1979). The number of intervening locations reliably affected the estimates (both route and Euclidean distances): Paths with more locations that had to be remembered were estimated as longer than paths of equivalent objective distance with fewer locations. Furthermore, when walking along cornered paths, participants had to make a turn on their way. This is an increase in the information that has to be processed, too. Hence, cornered paths were estimated as longer than straight paths. However, we observed a reduced effect of these independent variables on the second test. At second time of testing, participants had traveled the complete route once more getting additional information about the locations. This additional information led to a weaker influence of the number of intervening locations (route distance) and to a general minimizing of estimates (Euclidean distance). A possible explanation would be that the amount of route information was higher in the second test but retrieval of that information was influenced by the general familiarizing of the location names during the additional walk.

3   Further Studies

It is important to note that a replication of our experiment with a virtual environment of the corridor network led to nearly identical findings. Moreover, our main findings - associations between locations and actions, but no impact of actions on spatial representations - were replicated in a third experiment that only included an enactment condition and implicit measures of memory. This experiment differed in several aspects from the previous ones. One difference concerned the spatial setting. We used a map of a fictitious town. The locations were presented sequentially on a computer monitor and a window restricted the view. Furthermore, when performing actions at locations, participants *imagined* following the previously learned route. In the real and virtual environment, they walked through the maze again. Moreover, in the third experiment, we varied the *salience* of the actions. "Salience" was operationalized by ratings of the difficulty (with respect to the movements) to perform an action. On the route to be learned there were two crucial points consisting of three successive locations where participants had to perform salient actions. We expected that more salient actions may have a greater impact on spatial representations than less salient ones. Because filled distances are estimated as longer than empty ones (Sadalla et al., 1979), it was assumed that distances that included locations with salient actions were overestimated in the second spatial testing as compared to the first one. This expectation was not confirmed. Different spatial measures (estimations of route distances, time estimations of route distances, drawing a map) were not influenced by the salience of the actions. However, the results of a new spatial measure suggest that it may be useful to employ topological measures in future studies. The new spatial task required participants to associate locations from the previously learned route to a list of standard spatial items (such as *zoo* or *pedestrian crossing*). In the second spatial testing, more participants associated locations which had been connected with salient actions.

4 Conclusions

The present research strived to extend our knowledge of memory for actions by having participants perform or imagine simple actions at spatial locations. Furthermore, memory for actions was assessed with traditional explicit measures as well as with implicit measures of memory. We have gained firm evidence that spatial information can be combined and associated with information about activities. Furthermore, our findings reliably demonstrate the priming of new connections between spatial and nonspatial information with implicit measures of memory. It remains to be seen, however, which processes and which type(s) of information may underlie the observed effects. Selective interference experiments as well as systematic variations of the retrieval conditions should be useful in this regard. SPTs are multimodal events and the explicit inclusion of spatial information enriches the multimodality of such events. Consequently, there are many ways to investigate these questions.

Our results do not only demonstrate priming of new associations but also the formation of a spatial representation of the environment. The latter conclusion is based on the finding that systematic relations between certain independent variables (such as the number of intervening locations on the route) and the spatial measures (i.e., distance estimations) were found. It is somewhat disappointing, then, that we could not detect any change in the presumed spatial representation depending on performing or imagining actions in the environment. This finding might suggest that there is no integration of spatial information and actions in memory. Because such an integration was shown for other nonspatial information (e.g., McNamara et al., 1992), we think that before rejecting our main hypothesis it is useful to identify and to probe further conditions which may favor an integration.

One point is that there are many ways to measure hypothesized spatial representations. The results of our third experiment suggest that it may be useful to employ topological measures rather than metric ones. More importantly, it may be necessary to develop new (and perhaps more implicit) measures of spatial knowledge in order to tackle this problem more adequately. Perhaps the connection between spatial knowledge and memorial information about actions is more loose (and divisible) than expected by us so that an explicit measurement of spatial knowledge may facilitate or instigate a separation of otherwise more or less integrated spatial and nonspatial information. Another explanation would be that the actions selected for our experiments were simple ones and, more importantly, that their relation to the locations was chosen to be arbitrary. It is more than a speculation to assume that a systematic variation of the actions sampled (e.g., of their complexity and salience) and, even more importantly, a closer relation between actions and locations may lead to a different pattern of findings. For instance, in future studies, participants will be required to give reasons for performing the actions at the assigned places (cf. Cornoldi et al., 1994), and - in other experiments - actions related by a theme will be used (cf. Cohen et al., 1988).

Furthermore, we will vary the cues used in the implicit and explicit memory tests. In the present experiments, a location name and the verb of the action phrase served as cues for associating (implicit test) or remembering (explicit test) objects. In future experiments, we will present location-object pairs as cues for producing verbs or action phrases for associating location names. Associations are not necessarily symmetric. The discrepancy in our findings (newly acquired associations between locations and actions, but no action-dependent changes in spatial representations) may also be due to the fact that locations were associated with actions, but not actions with locations.

Finally, it should be mentioned that the performed or imagined actions have been placed in the context of a simplified spatial setting. Furthermore, the acquisition of route knowledge was favored. A more natural spatial context (such as a room with many objects and clear actional functions) may show a greater dependency on information about actions in what concerns the construction of a spatial representation.

Acknowledgments

This research was supported by a grant from the Deutsche Forschungsgemeinschaft (German Research Foundation) to Silvia Mecklenbräuker and Werner Wippich (Me 1484/2-1). The main experiment was conducted together with the research group of Karl F. Wender (We 498/27-1). We would like to thank Claus C. Carbon, Matthias Conradt, Shawn Hempel, Susanne Koritensky, and Kai Lotze for their assistance with data collection, and Bernd Leplow and an anonymous reviewer for helpful comments on an earlier version of this paper.

References

Acredolo, L.P. (1981). Small- and large-scale spatial concepts in infancy and childhood. In L.S. Liben, A.H. Patterson, & N. Newcombe (Eds.), *Spatial representation and behavior across the life span* (pp. 63-81). New York: Academic Press.

Anooshian, L.J., & Seibert, P.S. (1996). Diversity with spatial cognition: Memory processes underlying place recognition. *Applied Cognitive Psychology, 10*, 281-299.

Bäckman, L., Nilsson, L.G., & Chalom, D. (1986). New evidence on the nature of the encoding of action events. *Memory & Cognition, 14*, 339-346.

Carleton, L.R., & Moar, I. (1982). Memory for routes. *Quarterly Journal of Experimental Psychology: Human Experimental Psychology, 34A*, 381-394.

Clayton, K.N., & Chattin, D. (1989). Spatial and semantic priming effects in tests of spatial knowledge. *Journal of Experimental Psychology: Learning, Memory, and Cognition, 15*, 495-506.

Cohen, R. (Ed.). (1985). *The development of spatial cognition*. Hillsdale, NJ: Erlbaum.

Cohen, R., Cohen, S.L., & Cohen, B. (1988). The role of functional activity for children's spatial representations of large-scale environments with barriers. *Merrill-Palmer Quarterly, 34*, 115-129.

Cohen, R.L. (1983). The effect of encoding variables on the free recall of words and action events. *Memory & Cognition, 11*, 575-582.

Cohen, R.L. (1989). Memory for action events: The power of enactment. *Educational Psychological Review, 1*, 57-80.

Cohen, S.L., & Cohen, R. (1982). Distance estimates of children as a function of type of activity in the environment. *Child Development, 53,* 834-837.

Cornell, E.H., Heth, C.D., & Alberts, D.M. (1994). Place recognition and way finding by children and adults. *Memory & Cognition, 22,* 633-643.

Cornoldi, C., Corti, M.A., & Helstrup, T. (1994). Do you remember what you imagined you would do in that place? The motor encoding cue-failure effect in sighted and blind people. *Quarterly Journal of Experimental Psychology, 47A,* 311-329.

Easton, R.D., & Sholl, M.J. (1995). Object-array structure: Frames of reference, and retrieval of spatial knowledge. *Journal of Experimental Psychology: Learning, Memory, and Cognition, 21,* 483-500.

Engelkamp, J. (1986). Nouns and verbs in paired-associate learning: Instructional effects. *Psychological Research, 48,* 153-159.

Engelkamp, J. (1991). *Das menschliche Gedächtnis.* Göttingen: Hogrefe, 2. Aufl.

Engelkamp, J., & Zimmer, H.D. (1994). *The human memory: A multi-modal approach.* Seattle, WA: Hogrefe and Huber.

Engelkamp, J., Zimmer, H.D., & Biegelmann, U. (1993). Bizarreness effects in verbal tasks and subjects performed tasks. *European Journal of Cognitive Psychology, 5,* 393-415.

Engelkamp, J., Zimmer, H.D., & Kurbjuweit, A. (1995). Verb frequency and enactment in implicit and explicit memory. *Psychological Research, 57,* 242-249.

Engelkamp, J., Zimmer, H.D., Mohr, G., & Sellen, O. (1994). Memory of self-performed tasks: Self-performing during recognition. *Memory & Cognition, 22,* 34-39.

Evans, G.W. (1980). Environmental Cognition. *Psychological Bulletin, 88,* 259-287.

Evans, G.W., & Pezdek, K. (1980). Cognitive mapping: Knowledge of real-world distance and location information. *Journal of Experimental Psychology: Human Learning and Memory, 6,* 13-24.

Graf, P., & Masson, M.E.J. (Eds.). (1993). *Implicit memory: New directions in cognition, development, and neuropsychology.* Hillsdale, NJ: Erlbaum.

Graf, P., & Schacter, D.L. (1985). Implicit and explicit memory for new associations in normal and amnesic subjects. *Journal of Experimental Psychology: Learning, Memory, and Cognition, 11,* 501-518.

Herrmann, T. (1993). Mentale Repräsentation - ein erläuterungsbedürftiger Begriff. In J. Engelkamp & T. Pechmann (Hrsg.), *Mentale Repräsentation* (S. 17-30). Bern: Huber.

Hirtle, S.C., & Hudson, J. (1991). Acquisition of spatial knowledge for routes. *Journal of Environmental Psychology, 11,* 335-345.

Hirtle, S.C., & Mascolo, M.F. (1986). Effect of semantic clustering on the memory of spatial locations. *Journal of Experimental Psychology: Learning, Memory, and Cognition, 12,* 182-189.

Hunt, R.R., & Einstein, G.O. (1981). Relational and item-specific information in memory. *Journal of Verbal Learning and Verbal Behavior, 20,* 497-514.

Liben, L.S. (1981). Spatial representation and behavior: Multiple perspectives. In L.S. Liben, A.H. Patterson, & N. Newcombe (Eds.), *Spatial representation and behavior across the life span* (pp. 3-36). New York: Academic Press.

McNamara, T.P. (1986). Mental representations of spatial relations. *Cognitive Psychology, 18,* 87-121.

McNamara, T.P., & LeSueur, L.L. (1989). Mental representations of spatial and nonspatial relations. *Quarterly Journal of Experimental Psychology, 41A,* 215-233.

McNamara, T.P., Halpin, J.A., & Hardy, J.K. (1992). The representation and integration in memory of spatial and nonspatial information. *Memory & Cognition, 20,* 519-532.

McNamara, T.P., Hardy, J.K., & Hirtle, S.C. (1989). Subjective hierarchies in spatial priming. *Journal of Experimental Psychology: Learning, Memory, and Cognition, 15,* 211-227.

McNamara, T.P., Ratcliff, R., & McKoon, G. (1984). The mental representation of knowledge acquired from maps. *Journal of Experimental Psychology: Learning, Memory, and Cognition, 10,* 723-732.

Nyberg, L., & Nilsson, L.-G. (1995). The role of encactment in implicit and explicit memory. *Psychological Research, 57,* 215-219.

Paivio, A. (1986). *Mental representations. A dual coding approach.* New York: Oxford University Press.

Quinn, J.G. (1994). Towards the clarification of spatial processing. *Quarterly Journal of Experimental Psychology: Human Experimental Psychology, 47A,* 465-480.

Ratcliff, R., & McKoon, G. (1981). Automatic and strategic priming in recognition. *Journal of Verbal Learning and Verbal Behavior, 20,* 204-215.

Richardson-Klavehn, A., & Gardiner, J.M. (1995). Retrieval volition and memorial awareness in stem completion: An empirical analysis. *Psychological Research, 57,* 166-178.

Rieser, J.J., Guth, D.A., & Hill, E.W. (1986). Sensitivity to perspective structure while walking without vision. *Perception, 15,* 173-188.

Roediger, H.L., & McDermott, K.B. (1993). Implicit memory in normal human subjects. In H. Spinnler & F. Boller (Eds.), *Handbook of neuropsychology* (Vol. 8, pp. 63-131). Amsterdam: Elsevier.

Sadalla, E.K., & Staplin, L.J. (1980). An information storage model for distance cognition. *Environment and Behavior, 12,* 183-193.

Sadalla, E.K., Staplin, L.J., & Burroughs, W.J. (1979). Retrieval processes in distance cognition. *Memory & Cognition, 7,* 291-296.

Saltz, E., & Donnenwerth-Nolan, S. (1981). Does motoric imagery facilitate memory for sentences? A selective interference test. *Journal of Verbal Learning and Verbal Behavior, 20,* 322-332.

Sholl, M.J. (1987). Cognitive maps as orienting schemata. *Journal of Experimental Psychology: Learning, Memory, and Cognition, 13,* 615-628.

Siegel, A.W. (1981). The externalization of cognitive maps by children and adults: In search of ways to ask better questions. In L.S. Liben, A.H. Patterson, & N. Newcombe (Eds.), *Spatial representation and behavior across the life span* (pp. 167-194). New York: Academic Press.

Siegel, A.W., & White, S.H. (1975). The development of spatial representations of large-scale environments. In H.W. Reese (Ed.), *Advances in child development and behavior* (Vol. 10, pp. 9-55). New York: Academic Press.

Tack, W.H. (1995). Repräsentation menschlichen Wissens. In D. Dörner & E. van der Meer (Hrsg.), *Das Gedächtnis. Probleme - Trends - Perspektiven* (S. 53-74). Göttingen: Hogrefe.

Taylor, H., & Tversky, B. (1992). Spatial mental models derived from survey and route descriptions. *Journal of Memory and Language, 31,* 261-282.

Thorndyke, P.W. (1981). Spatial cognition and reasoning. In J.H. Harvey (Ed.), *Cognition, social behavior, and the environment.* Hillsdale, NJ: Erlbaum.

Thorndyke, P.W., & Hayes-Roth, B. (1982). Differences in spatial knowledge acquired from maps and navigation. *Cognitive Psychology, 14,* 560-589.

Wagener, M., & Wender, K.F. (1985). Spatial representations and inference processes in memory for text. In G. Rickheit & H. Strohner (Eds.), *Inferences in text processing* (pp. 115-136). Amsterdam: North-Holland.

Wender, K.F., Wagener, M., & Rothkegel, R. (1997). Measures of spatial memory and routes of learning. *Psychological Research, 59*, 269-278.

Zimmer, H.D., & Engelkamp, J. (1989). Does motor encoding enhance relational information? *Psychological Research, 51*, 158-167.

The Impact of Exogenous Factors on Spatial Coding in Perception and Memory

Jörg Gehrke and Bernhard Hommel

Max-Planck-Institute for Psychological Research, Leopoldstr. 24, D-80802 Munich,
Germany
e-mail: {gehrke, hommel}@mpipf-muenchen.mpg.de

Abstract. In the course of acquiring knowledge about layouts and maps spatial
information can undergo considerable changes and distortions, which
systematically affect knowledge-based judgments of human observers. In the
literature, these biases have been attributed to memory processes, such as
memory encoding or retrieval. However, we present both theoretical reasons
for, and first empirical evidence that at least some biases originate already in
perception, that is, much earlier in the processing stream than commonly
believed. Human subjects were presented with visual map-like layouts, in
which objects were arranged to form two different spatial groups. When asked
to estimate distances between object pairs and to verify statements about spatial
relations, verification times, but not distance estimations, were affected by
group membership: Relations between members of the same group were
verified quicker than those between members of different groups, even if the
Euclidian distance was the same. These results did not depend on whether judgments were based on perceptual or memory information, which suggests that
perceptual, not memory processes were responsible.

1   Introduction

Spatial cognition is of central importance for a wide range of human everyday
activities, such as reaching and grasping an object, typing on a keyboard, or finding
one's way home. To achieve good performance in such tasks, our cognitive system
does not only need to register and integrate relevant portions of the available spatial
information, but also to retrieve and use already acquired and stored information from
short-term and long-term memory. Interestingly, there is strong evidence that spatial
information undergoes considerable changes on its way from the sensory surface to
memory, often distorting the original information in systematic ways (for overviews
see McNamara, 1991; Tversky, 1981). In the literature, such distortions have been
often attributed to memory processes, such as the encoding of spatial information
(e.g., McNamara & LeSuer, 1989), its retrieval (e.g., Sadalla, Staplin, & Burroughs,

1979), or both (Tversky, 1991). However, in the present paper we entertain the hypothesis that at least some distortions might originate already from perception, not memory, hence much earlier in the processing stream than hitherto assumed. To motivate our hypothesis, we will briefly review some evidence for that complex visual structures are coded in a hierarchically fashion in both perception and memory. Memory distortions are often ascribed to hierarchical representation, so that such a commonality suggests that memory distortions may merely reflect the perceptual organization of stimulus information. We than report, as an example of our research, an experiment that investigated whether and how perceptual similarities between perceived and to-be-memorized elements of a map-like display affect perception- and memory-based judgments of spatial relations. To anticipate, our data will in fact provide preliminary evidence that the structure of memory representations is already formed in perception, a finding that calls for a reinterpretation of a considerable part of previous observations.

1.1   Hierarchical Coding in Memory

There is a big deal of evidence supporting the idea that spatial relations are coded hierarchically in memory. For instance, Maki (1981) had participants to verify sentences describing the spatial relation between pairs of american cities ("City A is west of City B" or "City A is east of City B"), and observed that, as one might expect, verification time was a decreasing function of Euclidian inter-pair distance. However, this was only true for cities that belonged to the same state (e.g., Alamo and Burlington, North Dakota), but not for cities located in different states (e.g., Jamestown, North Dakota, and Albertville, Minnesota). Such findings might indicate that information about cities and states is hierarchically organized, so that cities are stored as elements of superordinate state categories. If so, comparing elements from the same category should be in fact easier the more discriminable (i.e., distant) the elements are; however, judgments about elements from different categories might be often based on category membership, hence influenced by the spatial relationship between categories (i.e., states), so that within-category discriminability does not (or not that much) come into play.

Further evidence for hierarchical structures in memory comes from experiments made by Stevens and Coupe (1978). These authors presented their subjects with to-be-memorized artificial maps each containing two cities (e.g., city x and city y) that fell in different superordinate regions (e.g., Alpha county Beta county). In a congruent condition, the spatial relation between the cities matched the relation between the counties, e.g., city x (located in Alpha county) was to the west of city y (located in Beta county) and Alpha county was to the west of Beta county. In an incongruent condition, the relationship between cities was the opposite of that between counties, e.g., city x was to the west of city y and Alpha county was to the east of Beta county. When subjects made directional judgments about the two cities, systematic errors were observed with incongruent conditions producing more errors than congruent conditions. According to Stevens and Coupe, this is because participants used their knowledge about superordinate relations in judging the subordinated cities, so that the judged relations were distorted to conform with the relation of the superordinate geographical units.

A similar type of bias can also be demonstrated for real-world locations, as was shown by Hirtle and Jonides' (1985) study on the cognitive representation of landmarks in the city Ann Arbor, Michigan, (e.g., city hall, central cafe). Protocols of the free recall of landmarks were used to (re-) construct individual clusters, separately for each subject, and the validity of these clusters was then tested by means of a spatial-judgement task (i.e., distance estimation). As expected, distances within a cluster were judged smaller than distances across clusters.

In experiments reported by Hirtle and Mascolo (1986), participants memorized maps in which place names fell into two 'semantic' cluster: names of recreational facilities (e.g., Golf Course or Dock) and names of city buildings (e.g., Post Office or Bank). Locations were arranged in such a way that, although places belonging to the same semantic cluster were spatially grouped on the map, the Euclidian distance of one recreational facility was shorter to the cluster of the city buildings than to any other recreational facility, and vice versa. However, when subjects were asked to estimate inter-object distances on the basis of memory information, they showed a clear tendency to (mis)locate these critical places closer to their fellow category members then to members of the other cluster.

Taken altogether, these results provide strong evidence that global nonspatial relations between objects induce the formation of hierarchical object clusters in memory, thereby distorting certain inter-object spatial relations, or at least the judgments made about these relations.

1.2 Hierarchical Coding in Perception

The available results from memory studies provide strong evidence for the assumption that information about spatial configurations is not cognitively represented in a one-to-one correspondence, but seems to be at least partly organized in a hierarchical fashion. However, it is far from being settled which processes are responsible for such an organization. An obvious candidate are memory processes, which may work to reduce the perceptual information to minimize storage costs, optimize later retrieval, and so forth. But hierarchical coding may also be a result of perceptual processes, which may not only register sensory evidence but actively integrate it into a structured whole. If so, hierarchical coding in memory would tell us not so much about memory principles but about perceptual organization.

In fact, several authors have argued that complex visual structures are perceptually coded in a hierarchical fashion. For instance, Navon (1977) tested the idea that global structuring of a visual scene precedes analysis of local features. Participants were presented with large letters (the global level) made of small letters (the local level), and they were to recognize either the global or the local letter level. There were two important outcomes: First, it took more time to identify the global than the local letter, showing that global identification is easier than local identification. Second, the congruence between global and local letter produced asymmetric effects, that is, global identification was more or less independent of the identity of the local letters, while local identification was much easier if global and local letters were identical than if they were incongruent. This latter finding supports the notion that local analysis is always preceded by global processing, while global information can be extracted without local analysis. Obviously, visual structures are perceptually represented in a hierarchical fashion and this hierarchy affects informational access.

More evidence for the hierarchical clustering of visual information has been found by Baylis and Driver (1993), who had their subjects to judge the relative height of object features that were part of the same or of different visual objects. Although the distance between the features was held constant, the judgements were made faster when both features were part of the same rather than different objects. The authors argued that codes of features of the same object, including their spatial relations, make up a single representational cluster, with different clusters (i.e., object representations) being hierarchically organized. If so, judging features of different objects requires switching between cluster levels while judging features of the same object does not, so that between-level judgments are slower than within-level judgments. Obviously, these argument follow exactly the same lines as those of Maki (1981), although Baylis and Driver refer to perception, while Maki refers to memory. This strenghtens our suspicion that the way complex configurations are represented in perception and memory may be similar, or even identical.

1.3   Present Study

Taken altogether, the evidence suggests that perceptual coding processes do not only affect perceptually-based judgments, but may also determine the way perceptual information is stored, thus indirectly affecting memory-based judgments. This implies that the distortions and clustering effects observed so far in memory tasks may not so much reflect organizational principles of memory processes, but rather be a more or less direct consequence of distortions and clustering tendencies in perception.

The present study investigated this hypothesis by comparing perceptually-based and memory-based judgments of the same stimulus layout, a visual map-like configuration of eight houses and two ponds. As shown in Figure 1, these objects were visually grouped in such a way to induce a subdivision of the configuration into two perceptual (and/or memory) clusters. We asked participants to perform two "spatial" tasks, the unspeeded estimation of Euclidian distances—a task very common in memory experiments—and the speeded verification of sentences describing spatial relations (e.g., "is house A left of house B")—a task often used in perceptual experiments. We had our participants to perform these tasks under three conditions in three consecutive sessions: In the *perceptual* condition, the configuration was constantly visible; in the *memory* condition, participants first memorized the configuration and then performed without seeing it; and in the *perceptual/memory* condition, the configuration was again visible, so that both perceptual and memory information was available.

We expected both tasks to reveal the same pattern of results, hence the cognitive clustering of the configuration should affect distance estimations as well as verification times. In particular, our prediction was as follows: Objectively identical Euclidian distances between two given objects should be estimated smaller when objects were elements of the same than of different visual group. If so, this would suggest that the objects were in fact clustered and that this clustering led to the distortion of the objective spatial information. In the same vein, we expected the verification of spatial relations to proceed more quickly if the to-be-judged object pair belonged to the same as compared to different visual groups. If so, this would support the idea that (inter-) object information is hierarchically represented, so that within-cluster information can be accessed more quickly than between-cluster information.

2 Method

2.1 Participants

Twenty adults, 11 females and nine males, were paid to participate in the experiment. They reported having normal vision or corrected-to-normal vision, and were unaware of the purpose of the study.

2.2 Apparatus and Stimuli

Stimuli were presented via a video beamer (BARCODATA 700) on a 142 × 109 cm projection surface and participants were seated in front of the surface with a viewing distance of about 170 cm. The data acquisition was controlled by a personal computer. Participants made their responses by pressing a left or right sensor key with the corresponding index finger.

Stimuli were map-like configurations of eight houses, displayed at the same locations for each participant. The houses were 15 × 15 cm in size and were arranged into two groups, each centered around a pond (see Figure 1). Each house was named by a consonant-vowel-consonant nonsense syllable, such that there were no obvious phonological, semantic, or functional relations between the names associated with the to-be-judged location pairs. The name-to-house mapping varied randomly between subjects.

Eight horizontal location pairs were chosen for distance estimations and location judgements. Three of these pairs were separated by 300 mm (D_{300}: C-D, E-F, D-E), four by 600 mm (D_{600}: C-E, D-F, A-B, G-H), and one by 900 mm (D_{900}: C-F). Two of the pairs (C-D, E-F) had a pond in between. A small set of diagonal pairs was used as fillers; judgments for these pairs were no further analyzed.

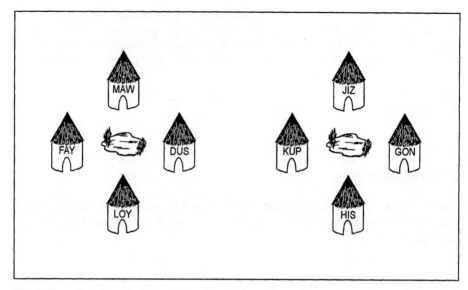

Fig. 1. Example of a stimulus configuration. Each configuration consisted of eight houses that were named by a nonsense syllable. For clarity we will in this paper use the letters A-H to indicate particular locations (location 'A': MAW; location 'B': JIZ; location 'C': FAY; location 'D': DUS; location 'E': KUP; location 'F': GON; location 'G': LOY; location 'H': HIS). Note that the house in a particular location had a different name for each participant.

2.3 Design

The experiment consisted of three experimental sessions (*perceptual, memory,* and *perceptual/memory* condition). Each session was divided into one experimental block for location judgements and another block for distance estimations, with task order being balanced across participants. A set of 256 judgements was composed of eight repetitions of each of the possible combinations of eight experimental pairs, two relations (*left of, right of*), and two orders of location within the pair (*A-B, B-A*). Forty-four judgments on distractor pairs were added to the set. Half of the participants responded yes and no by pressing the left and right response key, respectively, while the other half received the opposite response-key mapping. A set of 48 distance estimations was composed of three repetitions of each of the possible combinations of eight experimental pairs and two orderings of location within the pair. Twelve further pairs served as fillers.

2.4 Procedure

Each participant participated in three experimental sessions on three consecutive days. The stimulus configuration for a given participant was the same in each session. In the first session (*perceptual* condition), the configuration was visible throughout the whole experiment. The second session (*memory* condition) started with an acquisition phase, in which the participants first memorized the positions and syllables of the displayed houses and were then tested on their memory. This memory test was performed in front of a blank projection surface. The third session (*perceptual/memory* condition) was identical to the first session with respect to the display, hence the configuration was visible all the time.

Distance estimations. Sixty pairs of house names (48 critical distance pairs and 12 filler pairs) were displayed one pair at a time in the upper center of the projection surface. The names were displayed in adjacent positions, separated by a short horizontal line. Another horizontal line of 70 cm in length was shown below the names and participants were explained that this line would represent the width of the whole projection surface. It was crossed by a vertical pointer of 5 cm in length, which could be moved to the left or right by pressing the left and right response key, respectively. For each indicated pair, participants were required to estimate the distance between the corresponding objects (center to center) by adjusting the location of the pointer accordingly, and then to verify their estimation by pressing the two response keys at the same time. They were instructed to take as much time as needed for each estimation. The time for each estimation of the distances was measured.

Location judgements. A series of 300 (256 critical and 44 filler) to-be-verified locational statements was presented to each participant, one statement at a time. In each trial, a fixation cross appeared for 500 msec in the top center of the display. Then the statement appeared, consisting of the names of two objects and a relation between them, such as "FAY left of DUS" or "DUS right of FAY". Participants were instructed to verify (or falsify) the sentence by pressing the 'yes' or 'no' key accordingly; the assignment of answer type and response key was counterbalanced across participants. The sentence stayed on the projection surface until the response was made, but instructions emphasized that participants should respond as quickly and as accurately as possible. After an intertrial interval of 1,000 ms the next trial appeared. In case of an incorrect keypress, the error was counted without feedback and the trial was indexed. All indexed trials were repeated immediately after the 300th trial until no indexed trial exists or until the same error on the same trial was made four times.

Acquisition. The second session always started with the acquisition of the stimulus configuration. The configuration was presented to the participants, who had unlimited time to memorize the locations and names of the displayed objects. Then the configuration disappeared and the participants were sequentially tested for each object. A rectangle of an object's size appeared in the lower right corner of the display, together with an object name in the lower left corner. Using a joystick, participants moved the rectangle to the exact position of the named object. After pressing the left and right key simultaneously, the computer recorded the position of the rectangle, the projection surface was cleared, and the next test trial started. There were eight such trials, one for each object, in a random order. If an object was mislocated for more than 2.5 cm, the whole procedure was repeated from the start.

3   Results

From the data of the *distance-estimation task,* mean estimates in cm were computed for each participant and condition. On average, estimates increased with real distance: 395 mm for D_{300} pairs, 782 mm for D_{600} pairs, and 1002 mm for D_{900} pairs. Figure 2 shows the estimated distances across sessions. Estimates took about 25 sec on average and there was no indication of any dependence of estimation latency on session or real distance.

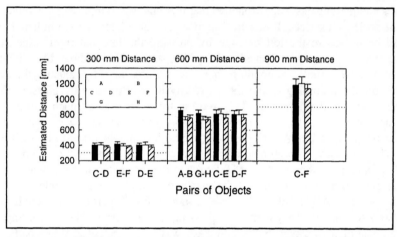

Fig. 2. Mean estimated Euclidian inter-object distance as a function of real distance across sessions/conditions (solid: perceptual condition; white: memory condition; hatched: perceptual/memory condition). The dotted line indicates real distances.

The relevant comparison was among the D_{300} pairs—the within-cluster pairs C-D and E-F and the between-cluster pair D-E—because these had identical Euclidian distances but different types of visual "cluster membership". However, an ANOVA with the factors session (condition) and pair did not reveal any significant main effect or interaction, hence, no systematic distortions were observed for object pairs spanning two vs. one clusters (see Figure 3).

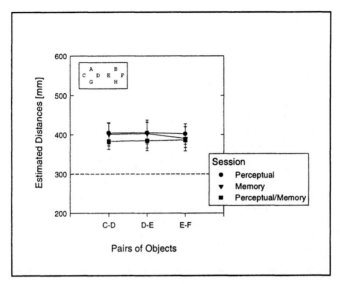

Fig. 3. Mean estimated Euclidian distances across session and equally-distant pairs of objects (C-D, D-E and E-F).

In the *location-judgement task*, error rates were low (< 4%) and the respective trials were excluded from analyses. Reaction times (RTs) from correct trials were analyzed by means of an ANOVA with the within-subjects factors session/condition and distance (D_{300}, D_{600}, and D_{900}). All three sources of variance were highly significant: the main effects of session, $F(2,18) = 133.25$; $p < .001$, and distance, $F(2,18) = 44.71$; $p < .001$, and the interaction, $F(4,16) = 6.68$; $p < .001$. As shown in Figure 4, verification times decreased over sessions and with increasing real distance. There was no difference between the inter-object distances D_{300} and D_{600} in the memory and perception/memory condition (critical difference of the Scheffé-test: 233 ms, p < .05).

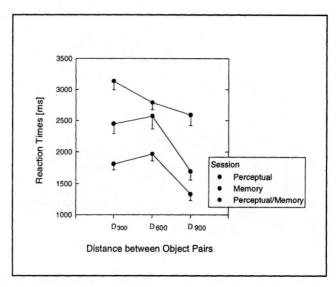

Fig. 4. Mean reaction times of judged spatial propositions across session/condition (perceptual condition, memory condition and perceptual/memory condition) and Euclidian distance between the judged objects (D_{300}: 300 mm; D_{600}: 600 mm; D_{900}: 900 mm).

As with distance estimations, an ANOVA was conducted on RTs for the equally-distant D_{300} pairs (C-D, E-F, and D-E) with session/condition and pair as within-subject factors. This time, two sources of variance were highly significant: the main effects of session, $F(2,18) = 102.27$; $p < .01$, and of pair, $F(2,18) = 73.37$, $p < .01$. The interaction failed to reach the significance level. Post-hoc analyses of the session effect (critical difference of Scheffé test: 238 ms) showed that RTs decreased from session to session (perceptual: 3132 ms; memory: 2446 ms; perceptual/memory: 1806 ms; see Figure 5). More interesting, however, was the analysis of differences between pairs. The Scheffé test yielded a critical difference of 218 ms, indicating significantly longer RTs for the between-clusters pair D-E (3041 ms) as compared to the within-cluster pairs C-D (2115 ms) and E-F (2227 ms).

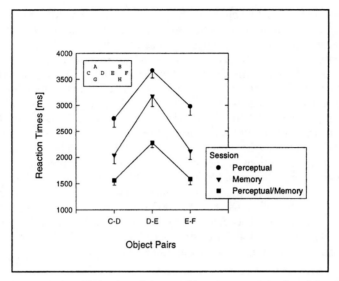

Fig. 5. Mean reaction times of judged spatial propositions across session/condition (perceptual condition, memory condition and perceptual/memory condition) and equally-distant object pairs.

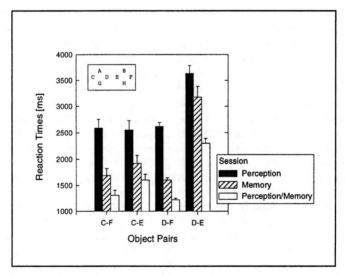

Fig. 6. Mean reaction times of judged spatial propositions across session/condition (perceptual condition, memory condition and perceptual/memory condition) and equally-distant objects pairs that consisted elements of two induced perceptual cluster.

One problem with this analysis is that the time to verify spatial propositions related to pairs C-D and E-F might be affected by the extreme left and right positions of the objects C and F. To control for this influence, we composed pairs consisting of elements of both perceptual clusters. The pairs C-E, D-F and C-F included at least one extreme spatial position, whereas the pair D-E contained adjacent object positions. If the spatial information is hierarchically organized then no differences in response times should occur ($RT_{C-E} = RT_{C-F} = RT_{D-E} = RT_{D-F}$) because judgements of each pair are based on the same path lenght within the hierarchy. On the other hand, if the organization of the spatial information would follow a linear formation, then RTs depended on the Euclidian distance between the elements of each pair ($RT_{C-F} < RT_{C-E}$, $RT_{D-F} < RT_{D-E}$). The ANOVA on the RTs for pairs C-F, C-E, D-F, and D-E yielded two significant main effects of session, $F(2,18) = 216.18$; $p < .001$, and pair, $F(3,17) = 60.83$; $p < .001$. Post-hoc tests revealed that RT constantly decreased across the three sessions/conditions (perception: 2849 ms; memory: 2098 ms; perception/memory: 1608 ms). Moreover, and this speaks to the spatial representation of the objects, the pair with adjacent elements (D-E: 3036 ms) was associated with higher RTs than the other pairs (C-F: 1863 ms; C-E: 2025 ms; D-F: 1816 ms). In addition, the interaction between session and pairs reached significance ($F(6,14) = 4.89$, $p < .01$) which is solely based on longer RTs for pair D-E under the perception, memory and perception/memory conditions (see Figure 6).

4 Discussion

The aim of this study was to examine how spatial information is coded in perception and memory and, in particular, to test the hypothesis that the spatial information is hierarchically organized in perception as well as in memory. We used two tasks which are very common in perceptual and memory studies and expected converging results in both tasks. The results we obtained are somewhat mixed.

First, the distance estimations showed that participants slightly overestimated the physical distances presented on the projection surface, which was true for all distances tested. This stands in opposition to previous observations that people tend to underestimate short as compared to long distances (McNamara, Ratcliff, & McKoon, 1984). In contrast to distance estimations, there was no reliable difference between estimation latencies in the D_{300}, D_{600} and D_{900} condition. This finding is inconsistent with some models of distance estimations from maps (e.g., Thorndyke, 1981), which propose that estimation time is a direct function of how long it takes to scan from one map element to the other. One reason for this inconsistency could have to do with the estimation procedure used in our experiment. Note that participants did not directly respond by pressing digits on a keyboard but they moved a vertical line back and forth on the projection surface without any time limit. Possibly, the time needed for this adjustment outlastet and, in a sense, overwrote the effect of scanning time.

Second, no difference was observed between estimations under perceptual and memory conditions. On the one hand, both the perceived and memorized distance between two objects increased linearly with their physical distance and the observed deviations were very similar under all conditions. This close correspondence between perceptually- and memory-based judgments suggests that the underlying processes and representation on which they operate are very similar if not identical. On the other hand, the visual grouping manipulation did not produce any systematic distortions of distance estimations. Although there are many possible explanations for this finding, three immediately come to mind. One is that the configuration simply induced no cognitive clustering of the objects. Given that clustering effects were obtained in the judgement latencies discussed below, this is an unlikely explanation. Another account might be based on the assumption that distance estimations and verifications of spatial relations are tapping into different processes. Although such an account would have important implications for research on spatial memory—where both measures are usually treated as equivalent—we are unable to judge its viability on the basis of the present results. Finally, one might assume that the strong symmetry of our stimulus display was responsible for the absence of systematic effects on distance estimations. In this context, the outcome of a recent experiment of ours (Heidemeier, Gehrke, & Hommel, in preparation) might turn out to be of considerable interest. There we used the same distance-estimation procedure as in the present experiment and had people judge vertical, horizontal, and diagonal inter-object distances. However, we added random spatial jitter to each object position, which resulted in a more asymmetrical configuration as compared to the present stimulus material. This time, we did observe cluster-related distortions in estimated Euclidian distance, suggesting that people use different estimation strategies for judging symmetrical and asymmetrical configurations.

The major aim of this study was to answer the question of whether spatial information is coded in the same—presumably hierarchical—way in perception and memory. The analysis of the time to verify spatial propositions provided some evidence for hierarchical organization induced by our visual-grouping manipulation: RTs were shorter for object pairs within (e.g., C-D and E-F) than between visual groups (e.g., D-E). Interestingly enough, this data pattern was found in all sessions, which can be taken to support the view that perceptually- and memory-based judgements were based on the same cognitive representation. A possible objection against such a conclusion could be based on the assumption that the shorter RTs for within-cluster judgments are mainly due to the inclusion of objects at extreme left and right locations (C and F), which because of their outstanding positions might facilitate the perception and/or retrieval of the corresponding object information. And, indeed, the verification times were shortest for distance D900 (see Fig. 4), which was related to the left- and rightmost object in the configuration. However, we have pointed out that this objection can be rejected on the basis that no RT differences were observed between the pairs C-E, C-F, and D-F under all conditions (see Fig. 6). Therefore, we think that there is some reason to maintain and pursue the idea that the coding of spatial configurations is—or at least can be—hierarchical, and that this is so in perception as well as in memory. If so, this suggests that phenomena of cognitive clustering as observed in studies on spatial memory may not so much reveal the logic of memory processes, but rather reflect the principles of coding and organization of spatial information in the perceptual system. If this argument is correct, then we can expect exogenous, perceptually relevant (Gestalt) factors, such as grouping by color or shape, to affect and possibly determine the way spatial information is coded and stored — a line of research we are pursuing in our lab. Figure 7 illustrates the theoretical framework that is suggested by the outcome of the present study. Perceptual processes come first and organize a given stimulus configuration into clusters that depend on feature-related similarities between objects. The emerging representation is then the basis for memory coding and other processes. That is, in contrast to traditional views, clustering is not a memory (-specific) process.

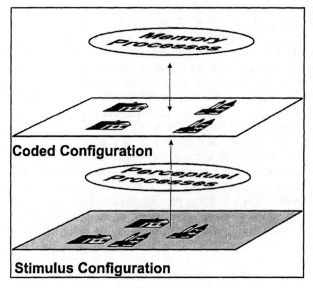

Fig. 7. Illustration of proposed relationship between perceptual and memory processes for the coding of spatial information. Perceptual processes come first and organize the stimulus configuration (indicated by the clustering of the objects within the coded configuration). The so far coded configuration is the basis for memory processes.

To conclude, our findings suggest a high degree of coherence between the processing of spatial information in perception and memory and therefore stress the importance of perceptual mechanism in the area of spatial cognition. However, more research is clearly needed to investigate the interdependence between perceptual and memory processes and representations.

References

Baylis, G. C., Driver, J.: Visual attention and objects: Evidence for hierarchical coding of location. Journal of Experimental Psychology: Human Perception and Performance 19 (1993) 451-470

Berendt, B., Barkowsky, T., Freksa, C., Kelter, S.: Spatial representation with aspect maps (this volume)

Heidemeier, H., Gehrke, J., Hommel, B.: Distance estimations in perception and memory under speed conditions. Ms. in preparation

Hirtle, S. C., Hudson, J.: Acquisition of spatial knowledge for routes. Journal of Enviromental Psychology 11 (1991) 335-345

Hirtle, S. C., Jonides, J.: Evidence for hierarchies in cognitive maps. Memory and Cognition 13 (1985) 208-217

Hirtle, S. C., Mascolo, M. F.: Effect of semantic clustering on the memory of spatial locations. Journal of Experimental Psychology: Learning, Memory, and Cognition 12 (1986) 182-189

Maki, H.: Categorization and distance effects with spatial linear orders. Journal of Experimental Psychology: Human Learning and Memory 7 (1981) 15-32

McNamara, T. P.: Memory's view of space. The Psychology of Learning and Motivation 27 (1991) 147-186

McNamara, T. P., Halpin, J. A., Hardy, J. K.: Spatial and temporal contributions to the structure of spatial memory. Journal of Experimental Psychology: Learning, Memory and Cognition 18 (1992) 555-564

McNamara, T. P., Hardy, J. K., Hirtle, S. C.: Subjective hierarchies in spatial memory. Journal of Experimental Psychology: Learning, Memory, and Cognition, 15, (1989) 211-227

McNamara, T. P., LeSuer, L. L. Mental representations of spatial and nonspatial relations. Quarterly Journal of Experimental Psychology 41 (1989) 215-233

McNamara, T. P., Ratcliff, R., McKoon, G. The mental representation of knowledge acquired from maps. Journal of Experimental Psychology: Learning, Memory and Cognition 10 (1984) 723-732

Navon, D.: Forest before trees: The precedence of global features in visual perception. Cognitive Psychology 9 (1977) 353-383

Thorndyke, P. W.: Distance estimations from cognitive maps. Cognitive Psychology 13 (1981) 526-550

Stevens, A., Coupe, P.. Distortions in judged spatial relations. Cognitive Psychology 10 (1978) 422-427

Tversky, B.: Spatial mental models. The Psychology of Learning and Motivation 27 (1991) 109-145

Judging Spatial Relations from Memory

Rainer Rothkegel, Karl F. Wender, and Sabine Schumacher

Department of Psychology
University of Trier, 54286 Trier, Germany
{rainer, wender, schumacher}@cogpsy.uni-trier.de

Abstract. Representations and processes involved in judgments of spatial relations after route learning are investigated. The main objective is to decide which relations are explicitly represented and which are implicitly stored. Participants learned maps of fictitious cities by moving along streets on a computer screen. After learning, they estimated distances and bearings from memory. Response times were measured. Experiments 1 and 2 address the question of how distances along a route are represented in spatial memory. Reaction times increased with increasing number of objects along the paths, but not with increasing length of the paths. This supports the hypothesis that only distances between neighboring objects are explicitly encoded. Experiment 3 tested whether survey knowledge can emerge after route learning. Participants judged Euclidean distances and bearings. Reaction times for distance estimates support the hypotheses that survey knowledge has been developed in route learning. However, reaction times for bearing estimates did not conform with any of the predictions.

1   Introduction

Psychological research on spatial memory has used a number of different methods. One particular method is the judgment of distances between locations. The present chapter reports results from three experiments. These experiments tested specific hypotheses about how subjects perform distance estimates from memory in the context of route learning. Before turning to the experiments we briefly discuss the logic behind the use of distance estimation in research on spatial memory.

In a general sense, space consists of objects and distances in between. In mathematics a *metric space* is defined as a set of objects (i.e., points) and a function assigning a positive real number to each pair of points. This function has to satisfy three axioms (symmetry, positive definiteness, and the triangle inequality). Therefore, a space can be fully described if the objects and the inter-object distances are given. Hence it is not unreasonable that psychological research on space perception and on spatial memory has focused a great deal on judgment of distances.

The general situation can be characterized as follows: On the one side we have the *physical space*. Or, to be more precise, we have those aspects of physical space that a theory under consideration selects as being relevant for spatial behavior. Second, we have a set of processes that extract information from physical space. Third, we have storage processes that build a *memory representation* from visual perception. And finally, in a psychological experiment, there is a set of *decoding processes* used by the subject to fulfill the experimenter's requirements. Hopefully, these processes also reveal something about the form of the mental representation cognitive psychologists are interested in.

Distance judgments have received so much attention because researchers hope that it might be possible to "measure" subjective space or to "measure" the mental representation using distance judgments as a measuring device. If we knew the structure of the mental representation, then we could see, for instance, how spatial memory is used in navigation. We also may be able to present spatial information more efficiently, design better instruments, and develop help systems like geographic information systems.

There are, however, serious problems. These problems arise because, at the outset, there are too many unknowns. It will not be possible, without additional assumptions, to identify both the structure of the representation and the processes working thereupon. This is possibly the reason why psychologists have been looking for analog representations. A great deal is known about the processes which transform physical space into perceptual representations. If we can assume that the processes working on the mental representation resemble the perceptual processes to some extent, then interpretation of results becomes much easier.

Research on distances in mental representations, henceforth cognitive distances, has taken two approaches. One is judgments of distances. Several methods have been used to obtain distance judgments. The experiments reported in this chapter are examples of this approach and the discussion presented here will focus on distance judgments.

The second approach uses what has been called the *spatial priming paradigm*. By measuring response times in recognition experiments researchers tried to identify cognitive distances (McNamara, Ratcliff, & McKoon, 1984; Wagener & Wender, 1985). More recently, however, this approach has undergone some severe criticism (Clayton & Habibi,1991; Wagener, Wender, & Rothkegel, 1997).

1.1   Types of Distances

From a psychological point of view we have to distinguish between several kinds of distances because it appears that quite different psychological processes are involved when these distances are assessed. With respect to *distance perception* we restrict ourselves to the visual modality here. It is obvious, of course, that spatial information can be obtained through other senses as well (like hearing or touching). The visual modality has been used in most investigations. It must be mentioned, however, that kinesthetic sensations, as experienced while walking, also play a role in perception of distances. Furthermore, there is apparently an interaction between kinesthetics and vision (Wagener, 1997), and activities carried out during learning may also have an influence (Mecklenbräuker, Wippich, Wagener, & Saathoff, this volume).

Distance perception: People do not perceive distances per se, rather distances are seen between objects. This gives rise to the first distinction between *egocentric* and *exocentric* distances. Egocentric distances are distances originating from the observer. That is, the observer perceives the distance between himself or herself and one point in the environment.

Exocentric distances refer to distances between two objects other than the observer. However, it must be specified what constitutes an object. Distance perception and perception of length are closely related. If we speak, for example, of the length of a rectangle such as a sheet of paper, we mean the distance between the two corners of one of its sides. This is where perception of exocentric distances and perception of *size* become almost indistinguishable.

There is evidence that egocentric and exocentric distances are perceived differently. For example, Loomis, Da Silva, Philbeck, & Fukusima (1996) found that egocentric distances in the range from 1.5 m to 12 m were judged quite accurately. This was measured by having participants walk, with their eyes closed, to an earlier viewed location. Yet in the same experiment, observers produced substantial errors in the comparison of exocentric distances.

Research on egocentric distances goes back to the 19th century. For egocentric distances, several variables have been identified that contribute to the perception of depth, that is, to the perception of egocentric distances. Most important are the vergence of the eyes, the visual angle, binocular disparity, the texture gradient, and changes in retinal position produced by movements, the so-called optical flow (cf. Baird, 1970; Foley, 1980; Gogel, 1993; Gilinsky, 1951; Cutting, 1996). For the present context, two main results are of interest: (1) visual space is not a linear transformation of physical space (Luneburg, 1947); (2) short physical distances are overestimated, whereas longer distances are underestimated. If one fits a power function between physical distances and distance estimates, the exponent of the function is frequently less than one (Kerst, Howard, & Gugerty, 1987; Wender, Wagener-Wender, & Rothkegel, 1997).

If we accept that spatial memory contains analog representations, an interesting question is: Which of these depth cues play also a role when judging distances from memory? As of yet, not much research has been done along these lines.

Exocentric distances are apparently not just differences of egocentric distances. This is not even the case when the endpoints of the exocentric distance to be judged lie on a straight line originating from the observer (Loomis et al., 1996).

With respect to exocentric distances further cases have to be distinguished. Exocentric distances corresponding to a line in a horizontal plane or to a line in a vertical, frontal parallel plane are judged differently from distances corresponding to lines in oblique planes. The latter, so-called 3D-lines, are obviously more difficult to judge and judgments are reportedly more in error.

With regard to perception, another important distinction has to be made. First, there are distances that can be seen at once by the observer. That is, the observer can perceive the whole distance without changing his or her position, even without moving the eyes. These distances have been called *perceptual distances* (Baird,

1970). They have to be to distinguished from distances which require the observer to change position, turn around, or travel around some visual barriers.

Perceptual distances have to be further subdivided. There is evidence that distances close to the observer, approximately within reach, are judged differently than longer distances. For longer distances, there may be another relevant distinction between points not too far away and distances that are really far such as buildings close to the horizon and astronomical objects (Higashiyama & Shimono, 1994). It appears that untrained observers are almost unable to judge very long perceptual distances with some degree of accuracy. There may exist an upper bound corresponding to the largest distance that can be reliably judged by an observer (Gilinsky, 1951).

In contrast to perceptual distances there are distances that cannot be perceived without moving around. Examples would be distances within a neighborhood, or on a campus, in the center of a small town, or even in a large supermarket. To evaluate such distances, obviously memory comes into play. Such distances have been called *environmental distances* by Montello (1988).

Again, for environmental distances, it matters how long they are. Small distances that are within walking range will be judged differently from distances that are experienced during a longer drive or flight. Apparently, not only the length of the distance matters but also the way it is experienced, i.e., the mode of transportation over the distance.

1.2 Modes of Learning

According to our hypothesis, distance estimation from memory is made using the mental representation of spatial information. Insofar as the spatial representation depends on the mode of learning, distance estimates will be affected. There are several ways in which spatial information can be learned. Perhaps the most natural way is by navigating through the environment (by walking, driving, etc.). It is conceivable that different modes of navigation result in different mental representations.

Knowledge of environmental distances, which is acquired by direct experience in environments, may be derived from multiple, partially redundant information sources. These sources are (1) number of environmental features, (2) travel time, and (3) travel effort or expended energy (cf. Montello, 1997).

Number of features has been the most frequently discussed source of environmental distance information, where features would be any kind of object in the environment that is perceptible - visually or in any other modality - during locomotion. Substantial empirical support exists for number of environmental features as an important source of distance information (cf. Sadalla, Staplin, & Burroughs, 1979).

Travel time also seems to be an important piece of information for environmental distance, for example, separation between places is often expressed in temporal terms. Surprisingly, nearly all of the empirical evidence on the relationship of travel time to subjective distance is negative (e.g., Sadalla & Staplin, 1980). It should be noted that nearly all of the studies concerned with the influence of travel time have been carried out with small experimental configurations and short temporal duration. Also, it should be noted that in a model of travel time and subjective distance, subjective speed should be considered.

Travel effort, that is, the amount of effort or energy a person expends while traveling through an environment, is a third potential source of information. Journeys that require more effort might be judged longer in distance. Although this idea is appealing, there is little clear evidence for the role of effort as a source of distance information. Possibly, to demonstrate the influence of travel effort, longer trips under a larger spatiotemporal scale may be necessary.

The role of environmental features in distance knowledge is most strongly emphasized by the existing empirical evidence. Further empirical support is needed to investigate the role of travel time and travel effort. In this regard the investigation of environments with larger spatiotemporal scales seems necessary.

A second mode of learning spatial information is from texts. A text may be explicitly written as a route description, yet also narratives describing some events convey spatial information because events take place in space (and time). There has been a substantial amount of research on how people acquire spatial information from texts (Bower & Morrow, 1990; Taylor & Tversky 1992; van Dijk & Kintsch, 1983; Wagener & Wender, 1985). The general conclusion has been that the same mental representations are built regardless of whether the information is perceived by viewing or by reading.

A third way to learn spatial information is by reading maps. This can be a very efficient way when entering a new area. There are, of course, individual differences. Reading a map immediately leads to survey knowledge. A lot of psychological research has been done by using maps as stimulus materials. The question is whether map reading and traveling lead to the same representation. Many researcher implicitly have assumed that they do. Nonetheless, there are also authors with a different opinion (c.f., Chown, Kaplan, & Kortenkamp, 1995).

Finally, psychological experiments have used different techniques to present spatial stimuli like photographs, slide shows, video tapes, and more recently, virtual reality. There is some research comparing the different modes of presentation although many questions are still unresolved.

1.3   Methods of Judgment

The results of distance estimation from perception or from memory do furthermore depend on the experimental method that is used to obtain the estimates. We can distinguish between verbal and nonverbal methods. In verbal methods, the subject has to respond by providing an estimate either directly on a scale like meters or in comparison to a second stimulus as in ratio or magnitude estimation. In contrast there are nonverbal methods where observers have to choose between several comparison stimuli or have to produce an analog estimate (using a caliper for example) or have to walk a distance. There is evidence that verbal and nonverbal techniques do not give the same results and that nonverbal techniques are more accurate (Leibowitz, Guzy, Peterson, & Blake, 1993).

1.4 Survey Knowledge from Route Learning

Spatial knowledge can be obtained by navigating through an environment or a configuration of objects (Antes, McBride, & Collins, 1988), it can be acquired from maps (Denis, 1996) or by reading or hearing verbal descriptions of spatial settings (Franklin & Tversky, 1990; Morrow, Bower, & Greenspan, 1989; Wagener & Wender, 1985). These different sources of information about spatial layouts may lead to differences in the resulting spatial knowledge.

According to a widely adopted model (Siegel & White, 1975) spatial learning in an environment usually takes three consecutive steps. At first, landmarks are learned and then their spatial and temporal connection. The connection of landmarks leads to route knowledge which is contrasted to configurative or survey knowledge (Evans, 1980; Hirtle & Hudson, 1991; Levine, Jankovic, & Palij, 1982; Moar & Carleton, 1982; Siegel & White, 1975; Stern & Leiser, 1988). Additional effort is necessary for the development of survey knowledge from route knowledge.

A new environment (e.g., after having moved to a new city) is usually learned by navigating through this environment. This implies that the environment cannot be seen as a whole, but different objects or landmarks will be learned in a certain order.

Learning a new environment from a map, on the other hand, permits the direct retrieval of spatial relations between objects or landmarks without reference to the routes connecting them (Thorndyke & Hayes-Roth, 1982). As the acquisition of information and the information itself are different for route learning and map learning, it may be hypothesized that these differences will show in the spatial representation. Using a priming technique, some researchers found effects for different sequences of learning objects in a spatial configuration (Herrman, Buhl, & Schweizer, 1995; Wagener-Wender, Wender, & Rothkegel, 1997).

Increased experience through traveling affects the content of the memory representation. As one travels and becomes more familiar with a variety of routes through an environment, points of intersection for multiple routes may be identified. Along with knowledge about route distances and knowledge of compass bearings along the routes, a reorganization of the spatial representation into a survey representation may be supported. Thus, direct retrieval of spatial relations between points (landmarks) without reference to the routes connecting them seems possible (e.g., Thorndyke & Hayes-Roth, 1982). Empirical evidence supporting this notion has been presented by Appleyard (1970) and Golledge and Zannaras (1973) in natural environments. For instance, survey knowledge improved with longer residence in a community and also in experimental settings (Allen, Siegel, & Rosinski, 1978; Foley & Cohen, 1984). Thorndyke and Hayes-Roth (1982) compared judgments of distances, orientations and locations of objects in an office building for secretaries and research assistants who worked in the office building and students to whom the office building was not familiar. Whereas the secretaries and research assistants had acquired their knowledge of the building solely from navigation, the students had acquired knowledge of locations solely from studying a map. Employees who had worked at the office building for only a short time could judge Euclidean distances between different rooms in the building only by estimating the lengths of the component legs on the routes connecting the rooms and the angles between different legs on the route. Students who had acquired their knowledge about the building through a map had no problems in judging Euclidean distances. The Euclidean distance judgements of employees who had worked in the building for a longer time resembled the results of the students who had learned the map.

So far, the results support the notion that survey knowledge may develop out of route knowledge. But this view is also criticized in recent publications. Montello (in press) argues against a strict sequence of modes of representations. In his view, survey knowledge can develop in parallel to route knowledge (see also Chown, Kaplan, & Kortenkamp, 1995). Bennet (1996) goes one step further in questioning the evidence for survey representations. According to his view, there is no study that demonstrates conclusively that mental maps exist at all, in either humans or animals.

1.5   Implicit Versus Explicit Representations

The question of which aspects of space are preserved in mental representations is an important topic in the research on spatial representations. The possible answers range from topological spaces, where only neighborhood relations are encoded, to richer metric spaces, where spatial relations are represented at an interval scale level (or even higher).

Although in this chapter some results concerning these questions are reported, the main focus is slightly different. The primary question addressed here is not at what scale level spatial relations are represented in mental representations, but rather how they are stored and retrieved. Therefore, we introduce a distinction between explicit and implicit representations of spatial relations.

If a spatial relation is explicitly encoded, there is a chunk in memory from which the spatial relation of interest can be read out immediately. There is no need to integrate different information, and the information is already separate from other, irrelevant information. An example of an explicitly encoded spatial relation would be the proposition "the distance between Bonn and Trier is 150 km".

In contrast, if a spatial relation is implicitly encoded, there is no chunk in memory that stores this and only this relation. Rather, the information has to be computed (in a loose sense of the word) by integrating relevant bits of information while ignoring irrelevant information. One example of an implicitly stored spatial relation is distance information in a mental image of a spatial configuration. In this case, distances could be computed by mental scanning from one point to the other, and taking the time needed as a measure for the distance in question (e.g., Kosslyn, Ball, & Reiser, 1978). It is important to note, however, that the implicit-explicit dichotomy does not map to the distinction between analog and propositional representations. It is easy to construct examples of propositional representations, where some of the spatial relations are stored implicitly. For instance, if route knowledge is represented by a set of propositions encoding distances between neighboring objects, distances between objects that are not direct successors on the route are implicitly encoded. They have to be computed by integrating interobject distances between neighboring objects along the path. If Euclidean distances have to be computed, angles along the paths must be taken into account (Thorndyke & Hayes-Roth, 1982).

In the experiments reported below, the time needed to judge spatial relations is used to test hypotheses concerning representations and retrieval processes of spatial relations. Depending on the representation-process pair, different variables should affect reaction time for these judgments.

2   Experiment 1

Experiment 1 addresses two issues, a methodological one and a theoretical one. The methodological question we tried to answer with this experiment was whether reaction times for distance estimations are useful for testing hypotheses about representations and processes involved in this task. The theoretical question was, given that reaction time is a suitable measure, what types of representations and processes are actually involved in distance judgments.

In research on spatial memory, little use has been made of reaction times for spatial judgments (e.g. McNamara et al., 1984). Most studies in this field dealing with reaction times are spatial priming studies where participants had to decide as fast as they could whether an item was present in a previously learned configuration or not (McNamara, 1986; McNamara, Hardy, & Hirtle, 1989; McNamara, Ratcliff, & McKoon, 1984; Wagener & Wender, 1985; Wagener-Wender, 1993; Wender & Wagener, 1986; Wender & Wagener, 1990). While these studies were conducted to test hypotheses about spatial representations, the judgments do not necessarily reflect spatial judgments since it was sufficient to remember only whether the presented item was in the learned set or not (Clayton & Habibi, 1991; Hermann, Buhl, & Schweizer, 1995; Sherman & Lim, 1991; Wagener, Wender, & Rothkegel, 1997).

In Experiment 1, participants were asked to judge distances between objects along the shortest possible path. Path length and number of objects on the path were used as independent variables. Depending on the representation and processes involved in this task, different outcomes would be expected. If distances are explicitly encoded, reaction time should be independent of both path length and number of objects. If distances have to be computed by combining explicitly stored distance information for neighboring objects (henceforth *summation model*), reaction time should increase with increasing numbers of objects. If distances are estimated using a simple mental scanning process, reaction time should be positively related to path length. Thorndyke´s (1981) *analog timing model* would predict increasing reaction times with both increasing number of objects and increasing path length.

The design of Experiment 1 also allows us to examine the scale level of the metric object positions along a path are represented in. If positions are represented on an ordinal scale, distance estimates should only be a function of the number of objects along a path. If, in contrast, positions are represented on an interval scale, distance estimates should be affected by the length of the path.

2.1 Method

Participants. A total of 35 Persons (20 female, 15 male) participated in the experiment. Most of them were psychology students at the University of Trier. They were given course credit for participation.

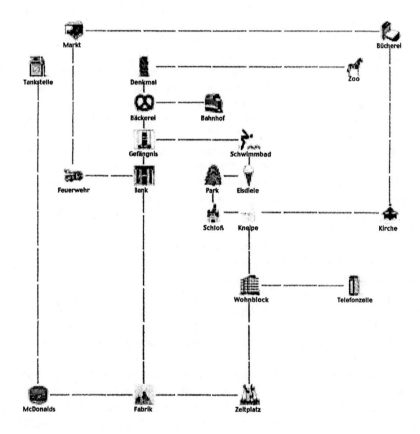

Fig. 1. Spatial layout used in Experiment 1

Material. The spatial layout that participants were required to learn was a map of 21 objects. The objects were small pictures with names listed below them. All objects were items that can occur in a town. The objects were connected by dashed lines symbolizing streets. There was at least one route connecting every pair of objects.

The experiment was carried out on a Macintosh PowerPC 7100 with a 14-inch Apple color monitor connected to it.

In the navigation phase, a small part of the map was displayed in a 6.5 by 6.5 cm large window. Maximally two objects could be visible at once in this window. A small black dot was present in the center of the window, symbolizing a taxi. By pressing the arrow keys on a Macintosh extended keyboard, participants could "move" the taxi up, down, left, or right along the dashed lines on the map. Movement was simulated by scrolling the visible part of the map, while the location of the window and the location of the dot relative to the window was kept constant.

Design. The main dependent variable in this experiment was reaction time for distance judgments. In addition, distance judgments themselves were analyzed. The major independent variables were path length and number of objects. Path length refers to the shortest distance between two objects along the route. Number of objects refers to the number of objects between two objects along the shortest possible route. Both variables were varied independently in three steps, resulting in a 3 x 3 factorial design. Path length was either 3, 6, or 9 units. Number of objects was either 0, 1, or 2. Both factors were within-subject factors. The map was constructed in a way to ensure that two critical location pairs existed for every combination of path length and number of objects.

Each participant received the same map of locations and paths, but objects were assigned randomly to the locations for each participant. In the distance estimation task, all participants had to estimate distances for the same set of location pairs, but the order of items was randomized for each participant.

Procedure. Experiment 1 consisted of four phases: navigation phase, learning check, distance estimation, and map drawing.

Participants were told to imagine that they had moved into a new town and that they wanted to work as a taxi driver in this town. Therefore they had to learn the shortest routes from any object to any other object.

Navigation Phase. Participants were allowed to move freely along the streets with the goal of learning the shortest routes between the locations in the town. They were also told they had to estimate route lengths at a later time. There were no time restrictions for the navigation phase, but participants were told that we expected learning to last about 30 minutes.

Learning Check. Immediately after the navigation phase, participants´ knowledge of the map was tested. On each trial a pair of object names was presented. Participants´ task was to write down the names of all objects along the shortest path connecting the presented objects on a sheet of paper. They were told to write down the names in the order of appearance along the path. After writing down the names they had to hit the return key to see the correct solution. They had to check their answer against the solution and press the "R" key if the answer was correct. In this case the next items were presented. If the answer was wrong, they had to press the "F" key. Whenever they hit the "F" key, participants were automatically put back into the navigation phase, where they had to visit all objects along the correct path before returning to the learning check. They were told they could press the return key as soon as they had visited all objects along that path or that they could also stay longer in the navigation phase if they wanted to improve their knowledge about the map. The learning check consisted of 9 probes. In the first set of 3 probes, the path between the presented objects contained 1 object. The solution for the second set of 3 probes encompassed 2 objects, and for the third set, three objects were included.

Distance Estimation. In the distance estimation task, participants were presented two objects, one at a time. First, the anchor object appeared in a dialog window. Participants were instructed to imagine the position of the anchor in the map and to press the return button as soon as they were fully concentrated. After the return button was pressed, the dialog window disappeared, and two seconds later the target object appeared in a new dialog window. Now participants had to estimate the distance between the anchor and the target as quickly as possible. They were told to use the length of one dash found in the dashed lines that symbolized the streets as the unit of measurement, that is, they were told to estimate how many dashes were on the shortest possible route between the anchor and the target object. To avoid counting strategies during the navigation phase, participants were told the unit of measurement only at the beginning of the distance estimation phase. They were instructed to respond verbally and to press the return button simultaneously. Reaction time was measured from the onset of the target stimulus presentation until the return button was pressed. After the return button was pressed, the dialog containing the target object disappeared and a third dialog was presented, where participants had to enter their estimates on the keyboard. Each participant had to answer 21 probes. The first 3 probes served for practice purposes and were not included in the analysis.

Map Drawing. At the end of the experiment participants were asked to draw a map of the routes on a sheet of paper.

2.2 Results

Informal inspection of the route maps showed that most maps were in close correspondence to the stimulus maps. Only the maps of four participants showed a strong deviation from the stimulus map, therefore, their data were removed from further analysis. After inspection of the response time distributions, an outlier criterion of 12,000 ms was set.

Figure 2 shows mean reaction time as a function of path length and number of objects. For the smallest route length, number of objects does not have any effect on reaction times. For route lengths 6 and 9 reaction time increases with increasing number of objects. An ANOVA with the factors "path length" and "number of objects" yielded a significant effect for the number of objects factor, $F(2,23) = 24,97$, $p < .001$. There was no significant effect for the path length factor, $F(2,23) = 0.50$, $p = .61$, but the interaction reached significance, $F(4,21) = 6.58$, $p = .001$.

The same analysis was conducted for distance estimates as the dependent variable. Figure 3 shows mean distance estimates as a function of path length and number of objects.

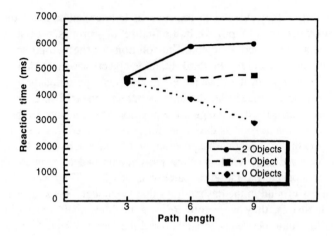

Fig. 2. Mean reaction times as a function of path length and number of objects on the path in Experiment 1.

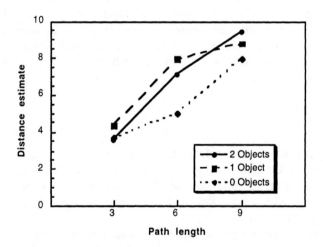

Fig. 3. Mean path length estimates as a function of path length and number of objects on the path in Experiment 1

Path length estimates were analyzed in the same way as reaction times. Figure 3 shows that distance estimates are sensitive to the actual path length, $F(2,29) = 163,43$, $p < .001$, and the number of objects, $F(2,29) = 23.19$, $p < .001$. The interaction was also significant, $F(4,27) = 12.27$, $p = .001$.

Figure 3 also shows that short distances were overestimated (by 0.86 on average) while long distances were underestimated (by 0.31 on average).

2.3   Discussion

The results of Experiment 1 show that reaction times for distance judgments are sensitive to variations in the spatial relations to be estimated (e.g. Baum & Jonides, 1979; Denis & Zimmer, 1992; McNamara et al., 1984). Although the reaction times and their variances are much higher than in simple binary choice tasks as in spatial priming studies, they contain enough systematic variation to show reliable effects.

The effect of number of objects is in line with Thorndyke´s (1981) analog timing model as well as with the summation model. But in addition to the effect of number of objects, the analog timing model predicts an increasing reaction time with increasing path length which is not supported by the data. Thus, concerning the main effects, the summation model is the only one supported by the data. Yet the summation model does not predict an interaction between number of objects and path length. In fact, none of the models mentioned above predicts this interaction. We could not think of any model predicting an interaction between path length and number of objects that predicts no main effect of path length. Since one cannot rule out the possibility that the interaction is due to the specific map used in this experiment, we decided to determine whether the effect can be replicated with a different map. This was done in Experiment 2.

Although the effect of the number of objects on reaction time is in line with the hypothesis that distances along a route are determined by summing distances between neighboring objects along that route, a possible alternative explanation cannot be ruled out by these results. It might still be the case that all spatial relations for all possible object pairs are explicitly encoded, but longer time is needed to retrieve this information if more objects are on the path to be estimated. This hypothesis would predict the same pattern of results as the summation model. Therefore, the models cannot be tested against each other with the data from Experiment 1. This issue is also addressed in Experiment 2.

The analysis of path length estimates supports the notion that object positions along the route were represented at a scale level higher than an ordinal scale. Path length estimates were sensitive to variations of actual path lengths while controlling for effects of the number of objects on the path. The analysis also shows that path length estimates tend to increase with increasing number of objects on the path (although two data points are not in line with this; see Figure 3). This result is in accordance with the view that the number of environmental features or "clutter" affects distance estimates (Kosslyn, Pick, & Fariello, 1974; Sadalla, Staplin, & Burroughs, 1979; Thorndyke, 1981). The effect that short distances were overestimated while long distances were underestimated is also in line with evidence from other studies (e.g., Björkman, Lundberg, & Tärnblom, 1960; McNamara & LeSueur, 1989; Wender, Wagener-Wender, & Rothkegel, 1997).

Both effects can be explained with a modified form of the *uncertainty hypothesis* (Radvansky, Carlson-Radvansky, & Irwin, 1995; see Berendt & Jansen-Osmann, 1997, for an alternative model). According to the uncertainty hypothesis, exponents below one in the power function relating estimated distances to physical distances are due to the fact that information about some distances may not be available and has to be provided by guessing. These guesses show a tendency to avoid extreme responses and favor more moderate responses. The result is a regression toward the mean.

To account for the number of objects effect, the uncertainty hypothesis has to be slightly modified. If distances are estimated by summing up distances between neighboring objects along the route, forgetting applies only to these elementary distances. If some of the elementary distances are forgotten and have to be guessed, the sum of elementary distances along a route should also regrade toward the mean. In addition, this modification also allows an explanation of the number of objects effect. If a certain proportion of the elementary distances are forgotten and have to be guessed, they are independent of the actual distances. Therefore, estimated path lengths should increase with increasing numbers of objects along the paths.

3 Experiment 2

Experiment 2 was designed with two goals in mind. The first goal was to replicate the findings of Experiment 1 with a different spatial layout by using the same factors as in Experiment 1. The second goal was to test one further prediction of the hypothesis that distances are estimated by combining elementary distances.

If the calculation of a distance in one trial involves elementary distances that were already retrieved in the previous trial, distance estimation should be faster compared to a condition where in the previous trial an unrelated set of distances was retrieved. This *repetition effect* may be due to easier retrieval of the elementary distances already used before, but it might also be the case that the effort of combining distances is reduced because some of the calculations have already been performed in the previous trial and the result is still remembered.

If a repetition effect could be demonstrated, this could also rule out a possible alternative explanation of the results of Experiment 1. If an increase in reaction time caused by an increasing number of objects on the path to be estimated is due to slower accessibility of explicitly stored distance information, there should be no repetition effects for overlapping paths in subsequent trials.

3.1 Method

Participants. A total of 47 persons (23 female, 24 male) served as participants. They were paid for their participation.

Material. The map used in Experiment 2 was similar to the map used in Experiment 1.

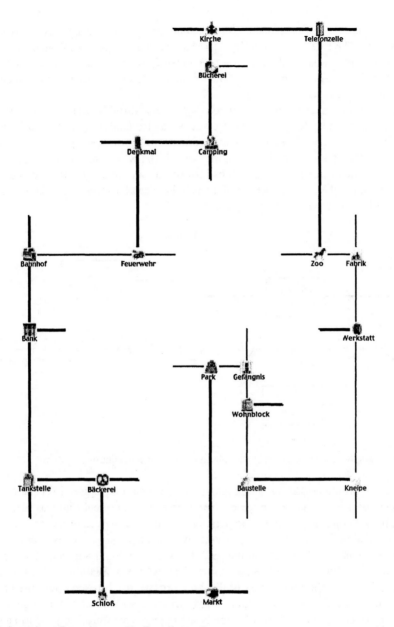

Fig. 4. Spatial configuration used in Experiment 2

There was one closed route connecting all objects. The objects were a subset of the objects used in Experiment 1. They were displayed in the same manner. Each object was placed on an intersection of the street system. One of the streets at an intersection was a dead end, the other street was part of the closed route. All dead ends were of the same length. Streets were symbolized using a thick black line with a dashed white line in the middle. As in Experiment 1, only a small part of the configuration was visible at one time. The sides of the visible square had the same length as a dead end.

Experiment 2 was conducted on a Macintosh PowerPC 7200 computer with a 17 inch Apple 1710 AV color monitor. The technique used for simulating movement was the same as in Experiment 1.

Design. As in Experiment 1, the main dependent variable was reaction time for distance judgments. Judgments themselves were also analyzed. The most important independent variable used in this experiment was path overlap. In the *subroute condition,* the path to be estimated in the critical trial was a part of the path estimated in the previous (preparation) trial. For instance, in an alphabetically ordered list of locations along the path, the length of the path between locations B and D would have to be estimated in the critical trial (see Figure 5).

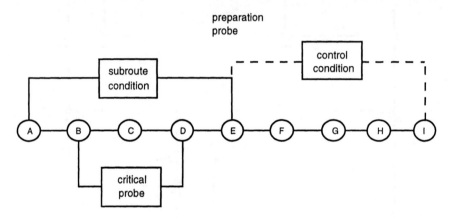

Fig. 5. Example of the paths used in the subroute condition and control condition

In the subroute condition, the preparation trial would have asked for the distance between locations A and E. In the *control condition*, the critical trial was identical to the subroute condition, only the preparation trial differed. To rule out the hypothesis that all distances are represented explicitly and only access times increase with increasing number of objects on the path, the subroute condition should differ from the control condition only in one aspect, namely, the availability of partial route information from the preparation trial. Therefore, we tried to keep possible priming effects constant. Since reaction times were measured with onset of the target object, priming effects are only critical for the target object. Therefore, we tried to keep the nearest distance between the objects presented in the preparation trial and the target object in the critical trial constant. For the example mentioned above, this means that in the preparation trial of the control condition, the path length between locations E and I had to be estimated.

For the critical items, the major independent variables used in Experiment 1 were also used in Experiment 2. Path length was varied in two steps (the length of 3 vs. 6 dead ends, i. e., 14 vs. 28 cm). Number of objects was varied in two steps as well (0 vs. 1 object on the path). All variables were varied independently resulting in a 2 x 2 x 2 factorial design, using within-subjects variation on all factors. There were two critical item pairs for each combination of path length and number of objects. Each critical item pair was presented twice, once in the subroute condition and once in the control condition. The second time a critical item pair was used in a probe, anchor and target were reversed. To reduce repetition effects, the order of probes was randomized for each participant separately for the first and second occurrence, and all item pairs were estimated once before the second occurrence of an item pair. There were a total of 35 probes: 16 critical probes, 16 probes for the preparation trials, and 3 additional training items.

Procedure. *Navigation Phase.* The navigation phase was identical to Experiment 1.

Learning Check. After participants completed the navigation phase, their knowledge of the distances in the spatial layout was tested. Three objects from the city were displayed in a dialog box on the computer screen, one at the top and two at the bottom. They had to judge which of the two objects at the bottom was closer (on the path) to the object at the top. After participants selected an object, they were given feedback. If the answer was correct, the next question appeared. If the answer was incorrect, they were automatically returned to the navigation phase. They had to renavigate to all three objects displayed in the question, but they were also told that they could stay longer in the city to explore it further if so desired. After exploring the map again, they could proceed with the learning check by hitting the return key. The learning check consisted of 10 questions.

Distance Estimation. The distance estimation procedure was identical to Experiment 1 with one exception. Participants were told to use the length of the dead end streets as a unit of measurement, specifically, they were told to estimate how many dead ends would fit into the path connecting a given object pair.

Map Drawing. As in Experiment 1, participants were asked to draw a map of the objects and streets.

3.2 Results

For reaction times on critical items an ANOVA with factors "path overlap", "path length", and "route distance" was computed. Reaction times in the subroute condition (M=4631 ms) were shorter than reaction times in the control condition (M=5044 ms), $F(1,46)=6.04$, $p<.02$.

Reaction times were also shorter for paths with no objects (M=4445 ms) than for paths running through one object (M=5229 ms), $F(1,46)=13.63$, $p=.001$. There was no main effect for the route length factor, $F(1,46)=.46$. None of the interactions were significant.

The correspondence between estimated path lengths and actual path lengths was quite low. The correlation coefficient computed over all estimates had a mean of .58 and a standard deviation of .30. To test whether the pattern of results changes when only participants with good knowledge of the maps are taken into account, the ANOVA was repeated with the subset of participants with correlation coefficients higher than .50, thus leaving 32 subjects for the analysis. The pattern of results for this subset showed no substantial deviation from the results for the whole group of subjects.

As in Experiment 1, estimates of path lengths were analyzed as well. While there was no significant difference between the subroute condition and the control condition, $F(1,46)=.45$, distance estimates increased with increasing path length, $F(1,46)=36.86$, $p<.001$, and with increasing number of objects on the path, $F(1,46)=7.54$, $p=.009$. None of the interactions reached significance (all $F<.84$). As in Experiment 1, small distances were overestimated (by 0.40 on average) while large distances were underestimated (by 1.16 on average).

3.3   Discussion

Experiment 2 was designed to replicate the findings of Experiment 1 and to test one further prediction. As in Experiment 1, there was a reliable main effect of number of objects on the path to be estimated. The finding that path length has no effect on reaction times is also in line with the results of Experiment 1. Only in one aspect do the results of Experiment 2 differ from Experiment 1. The interaction between path length and number of objects could not be replicated. The interaction found in Experiment 1, therefore, might be due to the specific map used.

With regard to the effect of path overlap, the results corroborate the hypothesis that distances are estimated by summing up distances between neighboring objects. Since priming effects were held constant between the subroute condition and the control condition, this result is at variance with the notion that all interobject distances are stored explicitly and that only access times differ.

With regard to the distance estimates, Experiment 2 replicated the basic results of Experiment 1. As already noted in the discussion of Experiment 1, these results can also be quite simply explained in connection with the summation model.

Taken together with the results of Experiment 1, we conclude that in path learning, distances between neighboring objects are represented explicitly, while distances between objects further apart have to be computed by summing up elementary distances.

4   Experiment 3

Experiment 3 was designed to test whether survey knowledge can emerge from a route learning task by using reaction times for spatial judgments. Participants were asked to judge Euclidean distances after learning a configuration in which Euclidean distance and path length were varied independently. Thorndyke and Hayes-Roth (1982) found evidence that route learners had no survey knowledge in an initial stage of learning. Error patterns of distance judgments, bearing judgments, and positional judgments revealed that participants had to combine the legs of routes to come to an estimate. After extended practice however, their error patterns came close to the ones of map learners.

Experiment 3 uses reaction times to distinguish between route representations and survey representations. If Euclidean distances are estimated by mentally combining the legs of the connecting path, reaction time should be an increasing function of the number of objects on the path. If, on the other hand, a survey representation has been developed where Euclidean distances can be estimated by mental scanning, reaction time should increase with increasing Euclidean distance.

In addition to the verbal distance judgments used in Experiment 1, bearing estimates produced by mouse movements were introduced. This technique was used because we hypothesized that it is a nontrivial task to translate a mentally represented distance into a verbally reported number. We felt that judging bearings on a 360 degree scale using the computer mouse might tap the participants´ knowledge more directly. If reaction times for bearing estimates showed the same results as reaction times for distance estimates, this would corroborate the results from the distance estimates (Montello & Pick, 1993; Sholl, 1987).

Participants learned a map and were subsequently asked to judge spatial relations, that is, distances and bearings. For some critical test items, route distance and Euclidean distance were varied independently.

4.1 Method

Participants. A total of 46 Persons (26 female, 20 male) participated in the experiment. Most participants were psychology students at the University of Trier. They were given course credit for their participation.

Material. A map was constructed as the learning configuration that consisted of 14 objects. The objects were a subset of the items used in Experiment 1. They were displayed in the same way. Each object was connected with two neighboring objects by a dashed line. The lines should symbolize paths connecting the objects. The path formed a closed route. An object was placed on every turn of the route. Experiment 3 was conducted on a Macintosh PowerPC 7100 computer with a 14-inch Apple color monitor.

Design. The main dependent variable used in Experiment 3 was reaction time for spatial judgments (distance estimates and bearing estimates). The estimates themselves were analyzed as well. Type of judgment, path length, and Euclidean distance were the major independent variables used in this experiment. Euclidean distance was varied in two steps. The short Euclidean distance was 6.4 cm, the long one 12.8 cm. As in Experiment 1, path length refers to the shortest distance between two objects along the path. Because it is not entirely clear which aspects of the route affect reaction times if Euclidean distance is estimated by combining distance information along the route, the number of objects along the route and the number of turns were varied together with path length. The short route was 27 cm long, went past two objects, and had three turns. For the long route, these three variables were doubled in value. Type of judgment, Euclidean distance, and path length were varied independently, resulting in a 2 x 2 x 2 factorial design. All factors were varied within subjects.

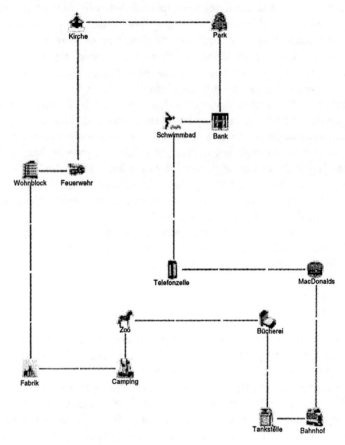

Fig. 6. Spatial configuration used in Experiment 3

Because we wanted to keep the map as simple as possible, only one critical location pair existed in the map for each combination of path length and Euclidean distance. This implies that only four distance estimates and four bearing estimates could be used for the factorial analysis. Because this could mean that the data are not stable enough to register possible effects, additional location pairs were used for distance estimates and bearing estimates. In these additional probes, Euclidean distance and path length were not varied independently. Therefore, they could not be submitted to an ANOVA; instead it was planned to analyze them using partial correlation coefficients. In total, there were 27 probes, three training probes, four probes for the factorial analysis, and 20 additional probes.

To make sure that the bearings to be estimated are independent from the distances in the critical item pairs, the map was rotated in 90 degree steps between subjects meaning that the whole route system for participant 2 was the same as the one for participant 1, except it was rotated by 90 degrees. As in Experiment 1, objects were placed randomly on the locations for each participant. The order of probes was also randomized for each participant and both types of judgments.

Procedure. The experiment consisted of 5 phases: navigation phase, learning check, distance estimation, bearing estimation, and map drawing.

Navigation Phase. At the beginning of the experiment, participants were told that they had to familiarize themselves with a city so that they would be able to estimate crow flight distances and bearings from memory at a later stage of the experiment. They were instructed to use the arrow keys on a Macintosh extended keyboard to navigate through the city. They were also told they could navigate in any direction and change direction as often as they wished. They could explore the city as long as they wished, but were told we expected the learning phase to take about 30 minutes.

Learning Check. The learning check was identical to Experiment 2, with the exception that participants had to base their judgments on Euclidean distance rather than on path length.

Distance Estimation. The distance estimation task was also identical to Experiment 1, again with the exception that participants had to judge Euclidean distances instead of path lengths.

Bearing Estimation. In the bearing estimation task, participants had to judge the direction of a target object compared to an anchor object. On each trial, the anchor object was presented until participants hit the return key. Participants were told to press the return key only after they could imagine the position of the anchor object in the map and were fully concentrated. After the return key was pressed, the anchor object disappeared and a second dialog box appeared with the target object at the top and a bearing gauge at the bottom. The bearing gauge consisted of a circle and a line originating at the center of the circle. The end of the line followed mouse movements. Participants were instructed to move the end of the line out of the circle in the direction to where the target object was situated compared to the anchor object and to press the mouse button as soon as the direction of the line corresponded to the remembered bearing. A similar technique has been used by Shelton and McNamara (1997) to assess bearing estimates after varying amounts of imagined rotation of the observers from the original viewpoints.

Map Drawing. As in Experiment 1, participants were asked to draw a map of the objects and the connecting route on a sheet of paper.

4.2   Results

To eliminate subjects with poor configurational knowledge from the analysis, distance estimates were correlated with actual distances in the configuration for each subject. Subjects with correlation coefficients lower than .50 were excluded, leaving 28 subjects for further analysis. Informal inspection of the map drawings revealed that the excluded subjects were also the ones with the lowest correspondence of the drawn maps with the stimulus maps.

For the critical items an ANOVA with factors "type of judgment", "path length", and "Euclidean distance" was computed. Only the Euclidean distance factor reached (marginal) significance, $F(1,27)=4.08$, $p=.053$. Reaction times for distance estimates and bearing estimates were higher for location pairs with large Euclidean distances than for location pairs with small Euclidean distances.

In addition to the ANOVA, correlation analyses were conducted for the entire set of items (excluding training items). Partial correlations of reaction times for distance judgments and bearing judgments with path length as the predictor were computed in which the effect of Euclidean distance was partialled out. Likewise, Euclidean distance was used as a predictor while partialling out effects of path length. Reaction times for distance judgments increased with increasing Euclidean distance, $r=.41$, $p<.01$. Path length did not reveal any influence, $r=-.05$, $p=.74$. For bearing estimates, the partial correlation yielded a marginally significant decrease of reaction times with increasing Euclidean distance, $r=-.29$, $p=.08$. Again, route distance did not show any linear effect, $r=.16$, $p=.54$. To examine whether the decrease of reaction times with increasing Euclidean distance in the bearing judgments goes along with decreasing errors, absolute deviations of the bearing estimates from the actual bearings were correlated with Euclidean distances. This analysis revealed a decrease of estimation errors with increasing Euclidean distance, $r=-.40$, $p<.01$.

4.3 Discussion

The distribution of correlations between participants´ distance estimates and actual distances shows that participants had difficulties judging Euclidean distances after route learning. This is not surprising, since combining distances and angles along a route to compute Euclidean distances is a nontrivial task. Since participants with correlation coefficients lower than .50 were excluded from further analysis, the results only apply to participants who were able to achieve this goal more or less satisfactorily. However, the main question pursued in Experiment 3 was not whether persons are able to judge Euclidean distances accurately after a route learning task, but rather whether persons use a route representation to form Euclidean distance estimates, or are able to form a survey representation before estimating Euclidean distances. In both cases, distances and angles along a path have to be combined mentally. The difference only concerns whether the results of these computations are integrated into the spatial representation or not. If the Euclidean distance estimates have to be computed from distances and angles along a route for every judgment, the time needed for the judgment should increase with increasing number of information to be integrated. In contrast, if participants are able to form a survey representation in form of a mental image, they are able to judge spatial relations between pairs of objects by mental scanning. In this case, the time needed to come to an estimate should be a function of Euclidean distance. Indeed, the results of the distance estimation task support the notion that participants were able to form a survey representation. Both the analysis of variance and the regression analysis show increasing reaction times with increasing Euclidean distance and no effects of route distance.

With regard to the bearing estimation task, the picture is not so clear. While in the analysis of variance there is no evidence that the pattern of results in the bearing estimation task is different from the distance estimation task, the correlation analysis shows a decrease of reaction time with increasing Euclidean distance. The decrease of errors with increasing Euclidean distance shows that this is not an artifact in form of a speed-accuracy tradeoff. Thus, for bearing estimates, the results of the analysis of variance clearly contradict the results of the correlation analysis. This leads to the question of which analysis can be trusted more. On the one hand, the data used in the ANOVA are better controlled for possible artifacts. For instance, by the between-subjects rotation of the entire map the factors are not confounded with the bearings to be estimated. On the other hand, much more data are used in the correlation analysis. The results of the correlation analysis support neither the predictions for survey representations nor for route representations.

One possible explanation might be that in spatial representations, the positions of objects are represented in areas of uncertainty. This is claimed by Giraudo and Pailhous (1994). This implies that bearings and distances between objects also have intervals of uncertainty. For distances, the intervals of uncertainty are independent of the distances themselves (as long as the uncertainty regions do not overlap). In contrast for bearings, the intervals of uncertainty decrease with increasing distances. If participants tried to keep a certain level of accuracy independently of the distance, this could mean that reaction times for bearing estimates increase with decreasing distance. Although this explanation is highly speculative, it provides a possible account for the dissociation of reaction times for distance estimates and bearing estimates.

The results of bearing estimates parallel the results of the distance estimates with regard to the effect of path length. None of the analyses revealed any effect of path length. Thus, while there is no support for route representations, there is at least some support for survey representations. However, this should not be taken as evidence that survey knowledge arises spontaneously whenever people learn routes. It is quite conceivable that people navigating more complex routes that are not closed without the goal of being able to judge Euclidian distances and bearings never develop a survey representation.

5   Conclusions

The experiments reported above use reaction times for spatial judgments to test hypotheses about the representation of spatial relations. All experiments show that reaction times for these judgments are sensitive to variations in the spatial properties of the relations to be judged.

Experiment 1 and 2 deal with distance estimates along a route. The results support the notion that in route learning, only distances between neighboring objects are represented explicitly, while distances between objects that are not direct successors on the route have to be mentally computed.

Experiment 3 uses reaction times for bearing estimates and distance estimates to test whether survey knowledge can emerge in a route learning task. While reaction times for distance estimates supported this notion, reaction times for bearing estimates neither conformed to the predictions made for survey representations nor to the predictions made for route representations. Thus, at least for the dissociation in reaction times for distance estimates and bearing estimates, one cannot help but state that further research is needed to resolve this issue.

Acknowledgments

We wish to thank Bettina Berendt, Theo Herrmann, and Berndt Krieg-Brückner for reviews of the manuscript. For their help in preparing, conducting, and evaluating the experiments we would like to thank Claus Carbon, Frauke Faßbinder, Arndt Görres, Petra Hoppe, Ulla Kern, Simone Knop, Claire Koch, Oliver Lindeman, Wolfgang Steinhoff, and Maria Wilmer.

References

Allen, G. L., Siegel, A. W., & Rosinski, R. R. (1978). The role of perceptual context in structuring spatial knowledge. *Journal of Experimental Psychology: Human Learning and Memory, 4*, 617-630.

Antes, J. R., McBride, R. B., & Collins, J. D. (1988). The effect of a new city traffic route on the cognitive maps of its residents. *Environment and Behavior, 20*(1), 75-91.

Appleyard, D. (1970). Styles and methods of structuring a city. *Environment and Behavior, 2*, 100-118.

Baird, J. C. (1970). *Psychophysical analysis of visual space.* Oxford: Pergamon Press.

Baum, D. R., & Jonides, J. (1979). Cognitive maps: Analysis of comparative judgments of distance. *Memory and Cognition, 7*(6), 462-468.

Bennet, A. T. (1996). Do animals have cognitive maps? *Journal of Experimental Biology, 199*, 219-224.

Berendt, B., & Jansen-Osmann, P. (1997). Feature accumulation and route structuring in distance estimations - an interdisciplinary approach. In S. C. Hirtle & A. U. Frank (Eds.). *Spatial Information Theory* (pp. 279-296). Berlin: Springer.

Björkman, M., Lundberg, I., & Tärnblom, S. (1960). On the relationship between memory and percept: A psychophysical approach. *Scandinavian Journal of Psychology, 1*, 136-144.

Bower, G. H., & Morrow, D. (1990). Mental models in narrative comprehension. *Science, 247*, 44-48.

Chown, E., Kaplan, S., & Kortenkamp, D. (1995). Prototypes, location, and associative networks (PLAN): Towards a unified theory of cognitive mapping. *Cognitive Science, 19*, 1-51.

Clayton, K., & Habibi, A. (1991). Contribution of temporal contiguity to the spatial priming effect. *Journal of Experimental Psychology: Learning, Memory, and Cognition, 17*(2), 263-271.

Cutting, J. E. (1996). Wayfinding from multiple sources of local information in retinal flow. *Journal of Experimental Psychology: Human Perception and Performance, 22*, 1299-1313.

Denis, M. (1996). Imagery and the description of spatial configurations. In M. de Vega, M. J. Intons-Peterson, P. N. Johnson-Laird, M. Denis, & M. Marschark (Eds.), *Models of visuospatial cognition* (pp. 128-197). New York: Oxford University Press.

Denis, M., & Zimmer, H. D. (1992). Analog properties of cognitive maps constructed from verbal descriptions. *Psychological Research, 54*, 286-298.

Downs, R. M., & Stea, D. (1977). *Maps in minds*. New York:

Evans, G. W. (1980). Environmental Cognition. *Psychological Bulletin, 88*(2), 259-287.

Foley, J. E., & Cohen, A. J. (1984). Working mental representations of the environment. *Environment And Behavior, 16*(6), 713-729.

Foley, J. M. (1980). Binocular Distance Perception. *Psychological Review, 87*(5), 411-434.

Franklin, N., & Tversky, B. (1990). Searching imagined environments. *Journal of Experimental Psychology: General, 119*(1), 63-76.

Gilinsky, A. S. (1951). Perceived size and distance in visual space. *Psychological Review, 58*, 460-482.

Giraudo, M. D., & Pailhous, J. (1994). Distortions and fluctuations in topographic memory. *Memory and Cognition, 22*(1), 14-26.

Gogel, W. C. (1993). The analysis of perceived space. In S. C. Masin (Eds.), *Foundations of Perceptual Theory* (pp. 113-182). Amsterdam: Elsevier Science Publishers B.V.

Golledge, R., & Zannaras, G. (1973). Cognitive approaches to the analysis of human spatial behavior. In W. Ittleson (Eds.), *Environment and cognition* New York: Academic Press.

Herrmann, T., Buhl, H. M., & Schweizer, K. (1995). Zur blickpunktbezogenen Wissensrepräsentation: Der Richtungseffekt. *Zeitschrift für Psychologie, 203*, 1-23.

Higashiyama, A., & Shimono, K. (1994). How accurate is size and distance perception for very far terrestrial objects? *Perception and Psychophysics, 55*(4), 429-442.

Hirtle, S. C., & Hudson, J. (1991). Acquisition of spatial knowledge for routes. *Journal of Environmental Psychology, 11*(4), 335-345.

Kerst, S. M., Howard, J. H., & Gugerty, L. J. (1987). Judgment accuracy in pair-distance estimation and map sketching. *Bulletin of the Psychonomic Society, 25*(3), 185-188.

Kosslyn, S. M., Ball, T. M., & Reiser, B. J. (1978). Visual images preserve metric spatial information: Evidence from studies of image scanning. *Journal of Experimental Psychology Human Perception and Performance, 4*(1), 47-60.

Kosslyn, S. M., Pick, H. L., & Fariello, G. R. (1974). Cognitive maps in children and men. *Child Development, 45*, 707-716.

Leibowitz, H. W., Guzy, L. T., Peterson, E., & Blake, P. T. (1993). Quantitative perceptual estimates: Verbal versus nonverbal retrieval techniques. *Perception, 22*(9), 1051-1060.

Levine, M., Jankovic, I. N., & Palij, M. (1982). Principles of spatial problem solving. *Journal of Experimental Psychology General, 111*(2), 157-175.

Loomis, J. M., Da-Silva, J. A., Philbeck, J. W., & Fukusima, S. S. (1996). Visual perception of location and distance. *Current Directions in Psychological Science, 5*(3), 72-77.

Luneburg, R. K. (1947). *Mathematical Analysis of Binocular Vision*. Princeton: Princeton University.

McNamara, T. P. (1986). Mental representations of spatial relations. *Cognitive Psychology, 18*(1), 87-121.

McNamara, T. P., Hardy, J. K., & Hirtle, S. C. (1989). Subjective hierarchies in spatial memory. *Journal of Experimental Psychology Learning, Memory, and Cognition*, *15*(2), 211-227.

McNamara, T. P., & LeSueur, L. L. (1989). Mental representations of spatial and nonspatial relations. *Quarterly Journal of Experimental Psychology Human Experimental Psychology*, *41*, 215-233.

McNamara, T. P., Ratcliff, R., & McKoon, G. (1984). The mental representation of knowledge acquired from maps. *Journal of Experimental Psychology Learning, Memory, and Cognition*, *10*(4), 723-732.

Moar, I., & Carleton, L. R. (1982). Memory for routes. *Quarterly Journal of Experimental Psychology Human Experimental Psychology*, *34A*(3), 381-394.

Montello, D. R. (1988) *Route information and travel time as bases for the perception and cognition of environmental distance*. Unpublished doctoral dissertation, Arizona State University, Tempe, AZ.

Montello, D. R. (1997). The perception and cognition of environmental distance: direct sources of information. In S. C. Hirtle & A. U. Frank (Eds.), *Spatial Information Theory: a theoretical basis for GIS*. Berlin: Springer.

Montello, D. R. (in press). A New Framework for Understanding the Acquisition of Spatial Knowledge in Large-Scale Environments. In: M.Egenhofer & R.G.Golledge (Eds.), *Spatial and Temporal Reasoning in Geographic Information Systems*. Oxford University Press.

Montello, D. R., & Pick, H. L. (1993). Integrating knowledge of vertically aligned large-scale spaces. *Environment and Behavior*, *25*(4), 457-484.

Morrow, D. G., Bower, G. H., & Greenspan, S. L. (1989). Updating situation models during narrative comprehension. *Journal of Memory and Language*, *28*(3), 292-312.

Prinz, W. (1992). Wahrnehmung. In H. Spada (Eds.), *Allgemeine Psychologie* (pp. 25-114). Bern: Verlag Hans Huber.

Radvansky, G. A., Carlson Radvansky, L. A., & Irwin, D. E. (1995). Uncertainty in estimating distances from memory. *Memory and Cognition*, *23*(5), 596-606.

Sadalla, E. K., & Staplin, L. J. (1980). The perception of traversed distance: Intersections. *Environment and Behavior*, *12*(2), 167-182.

Sadalla, E. K., Staplin, L. J., & Burroughs, W. J. (1979). Retrieval processes in distance cognition. *Memory and Cognition*, *7*, 291-296.

Shelton, A. L., & McNamara, T. P. (1997). Multiple views of spatial memory. *Psychonomic Bulletin & Review*, *4*(1), 102-106.

Sherman, R. C., & Lim, K. M. (1991). Determinants of spatial priming in environmental memory. *Memory & Cognition*, *19*(3), 283-292.

Sholl, M. J. (1987). Cognitive maps as orienting schemata. *Journal of Experimental Psychology: Learning, Memory, and Cognition*, *13*(4), 615-628.

Siegel, A. W., & White, S. H. (1975). The development of spatial representations of large-scale environments. In H. W. Reese (Eds.), *Advances in Child Development and Behavior* New York: Academic.

Stern, E., & Leiser, D. (1988). Levels of spatial knowledge and urban travel modeling. *Geographical Analysis*, *20*, 140-155.

Taylor, H. A., & Tversky, B. (1992). Spatial mental models derived from survey and route descriptions. *Journal of Memory and Language*, *31*(2), 261-292.

Thorndyke, P. W. (1981). Distance estimation from cognitive maps. *Cognitive Psychology*, *13*(4), 526-550.

Thorndyke, P. W., & Hayes Roth, B. (1982). Differences in spatial knowledge acquired from maps and navigation. *Cognitive Psychology*, *14*(4), 560-589.

van Dijk, T. A., & Kintsch, W. (1983). *Strategies of discourse comprehension*. New York: Academic Press.

Wagener, M., & Wender, K. F. (1985). Spatial representations and inference processes in memory for text. In G. Rickheit & H. Strohner (Eds.), *Inferences in text processing* (pp. 115-136). Amsterdam: North-Holland.

Wagener-Wender, M. (1993). *Mentale Repräsentationen räumlicher Informationen*. Bonn: Holos.

Wagener-Wender, M., Wender, K. F., & Rothkegel, R. (1997). Priming als Maß für das räumliche Gedächtnis. In C. Umbach, M. Grabski, & R. Hörnig (Eds.), *Perspektive in Sprache und Raum* (pp. 11-34). Wiesbaden: Deutscher UniversitätsVerlag.

Wagener, M. (1997). Memory for text and memory for space: Two concurrent memory systems? In P. Olivier (Ed.). *Language and space*. Proceedings of a workshop at the Fourteenth National Conference on Artificial Intelligence (AAAI 97) (pp. 121-129). Providence, Rhode Island.

Wender, K. F., & Habel, C. (1995). Verarbeitung und Repräsentation räumlicher Informationen. In K. Pawlik (Eds.), *Bericht über den 39. Kongreß der Deutschen Gesellschaft für Psychologie in Hamburg 1994* (pp. 867-869). Göttingen: Hogrefe.

Wender, K. F., & Wagener, M. (1986). Mental representation and inference. In F. Klix & H. Hagendorf (Eds.), *Human memory and cognitive capabilities* (pp. 353-359). Amsterdam: Elsevier.

Wender, K. F., & Wagener, M. (1990). Zur Verarbeitung räumlicher Informationen: Modelle und Experimente. *Kognitionswissenschaft, 1*, 4-14.

Wender, K. F., Wagener-Wender, M., & Rothkegel, R. (1997). Measures of spatial memory and routes of learning. *Psychological Research, 59*, 269-278.

Relations Between the Mental Representation of Extrapersonal Space and Spatial Behavior[1]

Steffen Werner, Christina Saade & Gerd Lüer

University of Göttingen
Georg-Elias-Müller Institute of Psychology, Gosslerstr. 14, D-37073 Göttingen
<swerner,csaade,gluer@uni-goettingen.de>

Abstract. We propose that mental representations of extrapersonal space are largely determined by the information that is required for actions within that space. Three different aspects of mental representations of space are briefly discussed. First, the physical size of a space restricts the set of behaviors that can be performed within the space. Mental representations of space therefore should differ in important respects depending on the perceived physical size of the referent space. Second, different reference systems can be used to code spatial information. Third, the demands on accuracy of representing spatial information differ between tasks. We posit that specialized mental representations are best suited to accomodate the different demands posed by actions within a given space. The benefit of different formats of mental representations, especially of analog representation, is addressed from a theoretical perspective. We argue that although different representational systems can be constructed that possess similar behavioral characteristics, analog representations in some cases lend themselves to more directly testable predicitions than other representations.

1. Introduction

Humans, like all other mobile organisms, live and act in a spatial environment. Both the perception of our own body, and the perception of extrapersonal entities include spatial relations as one of their key elements. We usually perceive ourselves as part of some environment, often at a specific location, position, and orientation. Our actions are equally linked to spatial characteristics of the environment. Whether it be simple motoric behaviors such as grasping a cup of coffee on the table, more complex ones like avoiding collisions while walking down the street, or planning elaborate routes to

---

[1] We wish to thank Jörn Diedrichsen, Gerhard Strube and Nicole Wellman for their helpful comments on earlier versions of this article. We also thank Jörn Diedrichsen, Jörg Gehrke, Friederike Lux, Meret Neumann and Stefanie Wolf for their help in conducting the studies reported herein. This work was funded by a grant by the German Research Foundation (We 1973/3-1).

find the shortest way home, most actions must take spatial constraints of the environment into account.

The kind of spatial information required to perform specific actions differs by the type of action at hand. For example, the localization of a painful sensation on one's foot can be achieved solely using information about the relative position of the sensation with regard to the foot, however touching the same spot with one's hand to pull out a thorn requires the computation of the relative position of ones hand in relation to the foot in three-dimensional coordinates. Grasping a band-aid, on the other hand, requires knowledge of the three-dimensional location of oneself and one's hand in relation to the three-dimensional location of the band-aid in the external world. Planning the way to the pharmacy, finally, entails knowledge of a sequence of locations and the proper spatial actions at each point to safely arrive at the destination. Insomuch as all these behaviors rely on very different kinds of spatial information, it will be argued that different forms of mental representation are most economical depending on the task that an organism has to perform. In particular, the accessibility of information and thus the efficiency of mental processing should be different depending on the task demands. For example, given that in a certain situation one needs to catch a falling object, fast and veridical access to the probable trajectory of the object is essential. In other tasks, such as the route finding behavior described above, information about the complete route does not need to be easily accessible. It suffices to make the right spatial decision at the relevant locations (see for example Kuipers, 1982; Krieg-Brückner et al., this volume).

Several different kinds of representations or representational systems have been proposed for the physical and mental representation of space. In the external world, globes, maps according to different projections, graphical and verbal route descriptions, and many more systems have been developed to represent physical space and spatial analogs to non-spatial dimensions. The number of proposed mental representations of space is similarly high and reflects at least partially the influence of physical (external) representations on the theorizing about mental processing (for this general point see also Gigerenzer, 1988). The "mental map" metaphor, for example, includes a wide range of assumptions comparing mental representations to physical maps (for a discussion see Kuipers, 1982). A related theoretical concept is the "mental image" in the sense of a two-dimensional visuo-spatial buffer that can be inspected and scanned comparable to a physical picture or map (Kosslyn, 1980). Route descriptions and sequential way-finding behavior can be modeled by ordered lists or more complex graph structures (Kuipers, 1982; Schölkopf & Mallot, 1995). To accommodate spatial biases and distortions, hierarchically organized representations of spatial knowledge have been proposed (Hirtle & Jonides, 1985; Stevens & Coupe, 1978; Tversky, 1981). Other approaches focus in more detail on the processes that embody spatial constraints. In their seminal work, Shepard and colleagues, for example, have provided compelling evidence that mental transformations, such as mental rotation or mental folding, follow similar time constraints as their physical counterparts (Shepard & Cooper, 1982; Shepard & Metzler, 1971). Neurophysiological evidence corroborates their view that mental rotation resembles a continuous change of a mental representation, similar to a physical rotation (Georgopoulos et al., 1989; Georgopoulos & Pellizzer, 1995).

In the following chapter we will propose that mental representations of extrapersonal space are largely determined by the information that is required for actions within that space. Three different aspects of spatial representations are discussed. First, the physical size and scale of a space constrains the set of behaviors which are usually performed within it. Spaces of different physical size therefore might require mental representations that differ in important respects. Second, different reference systems can be used to code spatial information. We will argue that the representation of spatial information in a certain reference system should be dependent on the intended action based on that information. Third, the demands on accuracy of representing spatial information differ between tasks. The question of whether the representation of space is veridical or biased will be addressed in light of different spatial tasks.

In a last section of this chapter we will address the question of whether the assumption of different formats of mental representations, especially of analog representation, is beneficial from a theoretical perspective. We will argue that although different representational systems can be constructed that display similar behavioral characteristics, analog representations might provide clearer and more testable predicitions than other representations.

2. Spatial representation and action

The idea that actions determine at least partially what and how information about the physical world is mentally represented is hardly surprising. After all, the purpose of mental representation is to enable an organism to appropriately act and react to external events.

One of the most common conceptual distinctions in spatial cognition concerns the separation of landmark, route, and survey knowledge (Siegel & White, 1975; Werner et al., 1997). The motivation for distinguishing between these three different kinds of spatial knowledge is quite obvious. Landmark knowledge concerns the information that is available about certain objects or places. This kind of information is particularly useful when one needs to identify a location as a place that one has visited before. It therefore serves as the basic building block of environmental spatial knowledge. Route knowledge, on the other hand, consists of the knowledge of how different locations are connected. Neither the connections between all pairs of locations need to be known, nor the relative directions or distances between them. Route knowledge, thus, can be quite coarse. In a simple model, Kuipers (1982; see also Schölkopf & Mallot, 1995) exemplified how route knowledge can be based on simple association of views or places with certain actions. In his model, a place can be defined as a collection of views that result from rotation in the same position. A path consists of the association of two specific places and a certain action. Following a route just requires activating the correct place representation at the right time and recalling the adequate action to get to the next place. These kind of models are very economical representations of paths in environments with a limited number of places, sparse connectivity between those places, and easily specifiable actions (such as "turn right", "go up the hill", etc.). At the level of survey knowledge, on the other hand, spatial information is integrated into a coherent representation and information about the relative distance and/or direction between two points is available (for a detailed model of how survey knowledge might be acquired see Poucet, 1993). Only this third kind of spatial knowledge enables an organism to plan different routes on paths not yet travelled. Evidently, the types of spatial information required for navigational tasks differ significantly depending on the tasks that the organism has to perform (e.g. planning or route execution). Other spatial activities, such as moving in a well-known room or playing basketball or soccer, might involve even more detailed spatial representations of the action space (see below). The distinction between landmark, route, and survey knowledge is designed to fit different behaviors and different stages in the acquisition of spatial information (Siegel & White, 1975).

In addition, recent approaches to the mental representation of spatial information have provided evidence that visuo-spatial information relating to the same distal stimulus can potentially be represented in distinct processing subsystems. In a number of studies, Milner and Goodale (1995, Goodale et al., 1991) provided evidence that the representation of spatial information differs depending on the task to be performed. When asked to perceptually match the tilt of a hand-held card and a randomly oriented slot, for example, their patient D.F. was not able to reproduce the tilt of the slot that was being presented to her. Although both the slot and the card were visible at the same time, her responses were randomly oriented. However, if she was asked to put the card through the tilted slot, the rotation of her hand closely corresponded to the tilt of the opening and she was able to insert the card into the slot with normal accuracy. Milner and Goodale interpret these results as evidence for two distinct visuo-spatial processing systems: an action based motor system and a visual system. Along the same lines, Aglioti, Goodale, and DeSouza (1995) were able to demonstrate that non-clinical subjects produce similar dissociations of motor behavior and visual processes in a different task. When asked to judge the size of two circles of equal size that are surrounded by a ring of either smaller or larger circles, people usually report that one of them looks larger than the other (Ebbinghaus/Titchener illusion). When the same subjects are faced with a different task, in which they have to grasp a disk of a certain size that is surrounded by other smaller or larger discs accordingly, the aperture of the grasp is unaffected by this visual illusion. While in vision the inducing circles produce an erroneous size perception, the grasp is not fooled. Both behaviors therefore seem to rely on different representations of the distal stimuli. A similar point is made by Loomis and colleagues (1992). In their experiments, subjects had to either visually match distances in depth with distances in frontal view, or had to walk blindfolded towards previously seen targets in a distance. Their results showed a strong bias towards an overestimation when matching distances in depth with distances in frontal view (the distances in depth had to be considerably larger), whereas subjects were well able to walk towards the targets without visual feedback. The authors conclude that distortions in the mapping to visual space do not occur in motoric tasks.

In the following sections we will focus on three different aspects of spatial knowledge and how they might relate to spatial actions: the physical size of the space that is to be represented, the role of reference systems in the mental representation of space, and the veridicality of spatial representations.

2.1 The size of the space

The actions we are able to perform depend partly on the size of the space that we act in. Small spaces, such as a sheet of paper or a desk, allow easy drawing, reaching, grasping, or other direct spatial manipulations. Common behaviors in medium-sized spaces, such as a room or a small portion of an open field, include walking to objects or throwing things. Spatial behaviors in large spaces, such as buildings or a city, are usually more complex, require planning, and are sequentially executed. Grüsser (1983), for example, distinguishes four basic units of extrapersonal space based on the actions performed within them. The grasping space defines the part of the environment surrounding a person in which direct manipulations of objects are possible. Its size differs depending on the size and the arm-span of a particular person. Through skilled use of tools, this space can be extended to an instrumental grasping space. This extension assumes that tools, such as screwdrivers, ice-hockey sticks, or brooms allow a skilled user to experience space not with respect to the limits of their body but with respect to the extension of their body by means of their tool. A second unit of space is the near-distant action space—the space a person can easily move in. Grüsser operationally defines its size as the immediate surroundings that one can walk to when blindfolded without feeling insecure about ones position (also compare Thomson, 1983). At larger distances, the near-action space gradually changes to a far-distant action space in which distance is largely perceived through visual and auditory cues. Finally, the visual background is perceived as the boundary of extrapersonal space which is undifferentiated with respect to depth cues. Interestingly, not only do the possible actions differ with increasing size of extrapersonal space, but the perceptual quality of the space also changes with size. As Grüsser observes, „the number of different modalities and types of receptors involved decreases from the grasping space to the visual background" (p. 331). Thus, depending on the size of a space, different modalities might play an important role in representing the spatial layout of the situation.

The question of the scale of a space has received some attention in the spatial cognition literature. On the one hand, the physical size of a space, as in Grüsser (1983), can be used to determine whether it is small or large compared to the organism under scrutiny. This simple and intuitive definition stresses the importance of actual, physical size of the referent space with respect to the actions that an organism could potentially perform in or on that space. On the other hand, the common distinction between small-scale and large-scale spaces is not necessarily related to the physical size of a space (e.g. Montello, 1993). Rather, it is based on the way that information about the space is acquired or potentially could be acquired. Whereas small-scale spaces are supposed to be apprehensible as a whole ("viewed at a glance"), large-scale spaces are defined as those which can only be experienced by integrating separate segments of the space, usually by combining local observations from different vantage points (Montello, 1993; Siegel, 1981). Clearly, this definition intends to differentiate slow and complex, sequential learning of spatial information, such as learning the location of buildings in a city, from quick and instant learning of spatial configurations, like viewing a small room from a corner or studying a map or a model (see McDonald & Pellegrino, 1993, for an overview).

However, problems arise when the space to be learned is itself only a representation of the actual referent space (e.g. Presson, DeLange, & Hazelrigg, 1989) or when the perception of the space is not compatible with its size (e.g. looking at a city from an airplane). A common example of this is the acquisition of spatial information through the use of maps and models. Although their actual size is usually fairly small, the space that they represent can be of any size. In these cases it is unclear whether the size of the space experienced or the size of the referent space is essential for later behavior.

For our purposes, the more intuitive notion of small and large spaces in the sense of Grüsser (1982) is of greater importance because it is more closely linked to actions and behavioral significance. Two unrelated research findings illustrate the connection between the size of a space and the actions performed therein:

In a set of experiments, Presson, DeLange and Hazelrigg (1989; Presson & Hazelrigg, 1984; Sholl & Nolin, in press) had subjects study different multi-segment paths. In some conditions, the paths to be learned were presented as lines of 1m-3m length, whereas in other conditions subjects studied a small map of the same layout. Subjects viewed the paths from one viewpoint. Later, subjects' directional judgements for different points on the path were tested from different positions on the studied path. As their main result, Presson et al. found that reaction times and accuracy of the directional judgements were not influenced by the direction the subject was facing during the testphase if the viewed paths were large (orientation-free encoding). If, however, the subjects had studied a small path or a small map of the path, the direction a subject was facing during the test phase strongly influenced time and accuracy of their judgements. When the orientation of the subject was in alignment with the familiar orientation during the study phase, better performance resulted than when they were not aligned (orientation-specific encoding). Presson et al. interpret their results as suggesting that larger spaces are perceived as navigable and thus encoded differently from smaller, non-navigable spaces. Large spaces, in this view, afford movement, which might lead to an "experience in terms of action and exploration" (Presson et al., 1989, p. 896; for the concept of affordances see Gibson, 1979). Sholl and Nolin (in press) further specify these results by demonstrating that a specific (horizontal) perspective and on-site testing are necessary for orientation-free performance.

As a second example at a different spatial scale, the neurophysiological results of Graziano and Gross (1995) also point into the direction of a close link between the characteristics of a space, the actions being performed therein, and the mental representation of that space. In single-cell recording studies with macaque monkeys they were able to show that the activity of certain (bimodal) neurons in the ventral premotor cortex was dependent on specific combinations of visual input and the position of the monkey's limbs. Only when a visual stimulus was presented in the appropriate location relative to the limb did the cells respond. For example, if a neuron's receptive field was determined holding the arm of the animal in one position and the arm of the animal was moved later, the receptive field of the neuron moved with it. Detailed analyses of the receptive fields of these neurons imply that they only react if the stimuli are within a certain distance of the specific bodypart. It thus seems tempting to conclude that visual space is coded in terms of its relevance to potential actions of the animal and with respect to the specific bodypart involved because "it would be useful to have a visual coordinate frame fixed to every part of the body surface, for the purpose of hitting, grasping, or avoiding visual stimuli in extrapersonal space." (Graziano & Gross, 1995, p. 1032; see also Milner & Goodale, 1995).

The above examples provide evidence suggesting that different mental representations of extrapersonal space exist according to the size of the space and the actions that may be performed in that space. Of course these results do not challenge the existence of more general mental representations of space which could be used across many different situations. In the following two subsections we will discuss two aspects of mental representations, the reference system used and the veridicality of spatial memories, as potential grounds to distinguish different mental representations. In each section, empirical results of our recent experiments will be discussed.

2.2 Reference systems

In most situations, spatial behavior must take into consideration the spatial relationships between different objects (including the self). To unambiguously specify the location or direction of objects, a reference system is required that relates the location of an object to some other known location or direction. Locations and directions can be specified in ego- or allocentric reference systems that are either global or local with respect to the relevant space. Egocentric reference systems specify spatial information according to the location and perspective of the observer. In situations such as grasping, kicking, or eating, more specific egocentric reference systems can be used (e.g. the hand, the foot, or the head of the observer). Allocentric reference systems, on the other hand, use an external frame of reference that is independent of the observer's position or orientation. For example, if one were to describe one's own position to a second person, an allocentric reference system that related one's position to other reference objects would be most useful ("in the basement", "in front of my desk"). In addition to the distinction of ego- and allocentric reference systems, the scope of a reference system must be considered. A reference system is global if the positions and orientations of all objects or locations within the relevant space can be expressed in terms of the same referent. For objects on the surface of the earth, the geographical reference system of longitude and latitude is a global reference system. However, it is not a global reference system when the positions of other planets or stars need to be included. In that sense, the earth-bound reference system is local. A more typical local reference system can be found when talking about the position of someone's furniture (the local reference system used is commonly the room the furniture is in). Using this kind of reference system (which, incidentally, is global for all the objects it contains) it would be easy to determine whether two apartments have a similar setup of the living room. However, it would be almost impossible to determine whether the television sets in two apartments faced the same direction in geographical terms (for a more detailed introduction to reference systems see Berthoz, 1991; Klatzky, this volume).

Reference systems differ in their usefulness for specific actions. For example, the movement of one's hand to an object can be specified in a global, allocentric reference system. If both the position and orientation of the hand and the position and orientation of the object are known, the movement of hand to object can be computed. However, if the relative location and orientation of the object were known in hand- or arm-centered coordinates, the required movement could be specified directly and with less computational effort (for similar examples see Graziano & Gross, 1995). In grasping, pointing, walking towards a single object, etc., an egocentric representation of the situation is adaptive since the movement eventually has to be stated in egocentric terms. Learning the configuration of objects within a room, in contrast, is less dependent on the position of the observer. Here, the relation of the objects to each other and with respect to the room are more important. In planning routes, different reference frames at different locations can be used (for example Poucet, 1993; Werner et al., 1997). However, if one wants to find a shortcut or needs to know the position of certain locations on the way, a global reference system for all locations is most useful.

The influence of different reference systems on the mental representation of space has been found in several studies. In her work, Sholl (1987) demonstrated that subjects' reaction times to indicate the direction of either a familiar building on campus or a city in the northeastern United States were differently influenced by the orientation of the subjects. While judgements of directions for buildings were equally fast, no matter which way a subject was oriented, the reponses for cities took longer if the subject was oriented away from north. This implies that the geographical relations under scrutiny in her study were represented differently from the navigable and well known environment of the campus. Sholl's results fit nicely with the above mentioned findings of Presson and Hazelrigg (1984; Presson, DeLange, & Hazelrigg, 1989; Sholl & Nolin, in press) that different reference systems (orientation-free and orientation-specific) are used for small and large spaces. This assumes that the judgement of geographical relations were actually based on map-knowledge that the subjects had acquired prior to the experiment.

Egocentric coding of space in mental imagery tasks was demonstrated by Hintzman, O'Dell and Arndt (1981, see also Easton & Sholl, 1995; Rieser, 1989). In a series of experiments, participants learned a layout consisting of multiple objects. During a test phase, their memory of the locations of the objects was assessed. Results indicated a strong association between the direction that an object was located with respect to the participant (e.g., in front, to the left) and reaction times to correctly name the object. Whereas objects which were imagined in front were quickly identified, objects on the sides were increasingly slower towards the back. Objects imagined directly behind participants were identified almost as fast as in the front. These findings clearly indicate that participants in these experiments represented the layout of the objects during the test phase at least partially in terms of an egocentric reference system (for similar results using layouts in geographical space see Boer, 1991). More recently, Franklin and Tversky (1990; Bryant & Tversky, 1992; also see Claus et al., this volume) found additional evidence that participants have differential access to objects whose positions they learned from texts. According to their (egocentric) spatial framework model, access to locations differs with respect to the three body axes: above-below, front-back, and left-right. Consistent with their predictions, reaction times in these experiments have been found to be fastest for above-below, followed by reactions to front-back judgments, whereas left-right judgments have been slowest (see also Sholl, 1995, for a detailed model).

Although the above findings stress the influence of an egocentric reference frame, the potential role of the containing space on the acessibility of remembered spatial information has largely been ignored. To address this gap in the literature, we conducted a series of experiments using a paradigm similar to that of Hintzman et al. (1981).

2.2.1 Empirical evidence for the combined influence of different reference systems

Method: In one of our experiments, 11 participants learned the configuration of eight unrelated objects (e.g. cactus, basket, dumbells) within a nearly square room (approx. 2m x 2m). Each participant was seated in the center of the room on a swiveling chair and was surrounded by eight objects, arranged circularly around the participant. Four objects were placed near the center of the walls, and the other four were positioned on the diagonals near the corners of the room (see figure 1 A). Participants studied the layout for five minutes, during which they were permitted only to turn themselves while remaining seated. After the learning phase, participants were escorted to a second room and placed in front of a computer. Next, participants were instructed to imagine themselves in the center of the room they had just exited, facing one of eight different orientations (either a certain wall or the corner between two walls). After the participant had imaged the specified view, a direction was specified acoustically by means of one of eight loudspeakers surrounding the participant, indicating the appropriate direction by their location. Participants were instructed to name the object as quickly as possible that would have been in that direction in the studied room (relative to the imagined position and view).

Results and Discussion: The main results of these experiments can be summarized as follows (see also figure 1 B). An egocentric coding of object locations, similar to that reported in Hintzman et al. (1981), was observed. Whereas objects in front were always identified the fastest, objects on either the left or right were identified considerably slower ($F(7,10)=15.87$, $p<.001$). Objects located directly behind participants were identified more quickly than those to the right or left, but more slowly than those in the front. Error rates also differed by the direction probed ($F(7,10)=3.71$, $p<.002$).

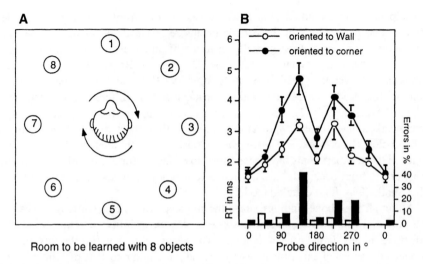

Fig. 1. A. Layout of the room to be learned. The subject is seated in the center of the room and is restricted to rotational movements on the chair. The numbered circles indicate the location of eight objects that had to be studied. **B.** Reaction times and error rates for the naming of objects in a probed direction while imagining being oriented towards a wall or a corner of the room.

In addition to the effects of egocentric coding of location, however, our findings suggest that a room-based reference system was also in effect. Specifically, a strong main effect of imagined direction at test was found (F(1,10)=45.75, p<.001). When participants were asked to imagine themselves facing a wall, reaction times were significantly faster than when they were asked to imagine facing a corner of the room (2321 ms vs. 3135 ms respectively). A similar pattern is observable in the error rates. Participants had much less trouble identifying objects correctly when oriented towards a wall than towards a corner of the room (5.1 % vs. 13.1 %, F(1,10)=32.30, p<.001). This effect cannot be due solely to egocentric coding but must at least in part be dependent on the reference frame imposed by the room. Both kinds of reference systems, egocentric and room-based, therefore, appear to play a role in remembering the spatial configuration presented in our studies (for a similar point see v. Wolff, 1997). As recent findings from Shelton and McNamara (1997) demonstrate, differences between the two orientations (towards wall or corner) may be due to "view-dependent" spatial memories. This explanation assumes that the structure of a room influences the way in which people view the room during study by making orientations towards walls more salient. However, besides the influence of the external reference frame during the learning phase, different reference systems might also interfere during recall or in imagery tasks (see Werner & Wolf, in preparation).

Depending on the type of space and behavior to be executed, different types of reference systems (or combinations of them) might be appropriate and thus determine the kind of mental representation used. In a series of ongoing studies we are investigating the generality of these finding for different sizes of spatial layouts. Of special interest to our research are layouts that are either confined to the grasping space of the organism, or are part of a participants' everyday knowledge of the spatial layout of the city they live in. Within grasping space, the role of room-based reference systems or other allocentric reference systems may be less pronounced than in a larger room because of the functional significance of egocentric coding in the former space. With respect to larger spaces a different prediction can be made. Since configurations of objects in large spaces are typically learned by travelling along specific routes within them, place-dependent reference frame effects similar to the ones within the room might be expected.

2.3 Biases, distortions, and hierarchical organization of spatial information

To be of any use for action, mental representations of space need to have a certain degree of accuracy. When reaching for an object, for example, only the general direction and distance of the object have to be specified at the outset of the movement, whereas later adjustments and fine tuning of the movement can be based on visual control if the object is visible or on haptic feedback if it is not visible. Since, in these cases, the position of the object and hand can be updated by external information, the physical world can act as an external spatial memory for action (e.g. O'Regan, 1992). This is even more pronounced in navigation tasks, where the direction of a goal location might only be vaguely specified. Only after a number of places have been visited, will the direction of the goal become clearer. In contrast, touching one's own nose with the index finger while blindfolded relies on a mental representation of those two points relative to each other in three-dimensional coordinates. That people are able to perform these tasks indicates that mental representations can be quite accurate.

Accuracy, in this sense, consists of two independent components. The first concerns the deviation of the average remembered (or reported) location of an object from its true location. If the mean response falls on the true location, the responses are considered unbiased. If, however, responses have a systematic tendency to err in one direction, the responses are said to be biased towards that direction. The second component takes the variability of responses into account. The more variable the responses, the less accurate they are.

Especially the bias of spatial memories has attracted a great amount of research. Stevens and Coupe (1978; also Hirtle & Jonides, 1985; Tversky, 1981) demonstrated that geographical relations between cities were strongly biased with respect to the superordinate geographical entities that contained them. Seattle, for example, is incorrectly thought of as being southwest of Montreal because the United States border Canada to the north. Results such as this have led researchers to propose that spatial representations are hierarchically organized and that superordinate relations strongly influence the responses on subordinate (contained) objects' locations. That the hierarchical organization of spatial information is not limited to geographical knowledge was demonstrated by McNamara, Hardy, and Hirtle (1989). They had subjects learn the layout of a large number of objects lacking inherent structure or physical boundaries. Later, subjects were asked to recall the objects several times, and their recall protocols were analyzed to find reappearing groupings within their protocols. Their results showed that objects were grouped in spatially organized regions and that objects within the same region were judged to be closer to each other and primed each other more than objects which belonged to different regions (even if the objective distance between the objects was the same). Another source of bias was demonstrated by Sadalla, Burroughs, and Staplin (1980). They showed that the distance between reference points, e.g. points of interest that are visited often and other, non-reference points, are not judged symmetrically. Non-reference points were judged as being closer to reference points than vice versa.

Although much of the discussion on spatial representations has mainly focused on the fact that spatial knowledge is biased and does not veridically represent physical space, Huttenlocher, Hedges, and Duncan (1991) have proposed a general model that accomodates veridical and biased spatial knowledge within a common framework. In their model, spatial information is represented at two different levels. At the level of fine-grain coding, the location of an object is represented in absolute terms without bias but with a certain inexactness. The inexactness depends on different factors, for example the precision of encoding and loss of information from memory. Spatial information, at the same time, is represented in terms of categories that are associated with certain prototypical spatial locations. Estimation of a location depends on the recollection of both the fine-grained value and the category, which are then weighted with respect to the inexactness that is associated with them. Although spatial recollections that are solely based on the fine-grained value will be unbiased, they may nonetheless be very inaccurate due to great variability. Categorical information, on the other hand, even if systematically biased towards a prototypical location, may more closely represent the true location if the variability is less than that for fine-grained information.

Depending on the task demands, fine-grained and categorical coding could be weighted differently. In tasks that require memory for approximate configurations of objects, for example, categorical coding is an efficient and robust way of coding spatial locations. If, however, only one or few object locations have to be remembered as accurately as possible, high precision in encoding the corresponding fine-grained values would be most valuable. In their empirical work, Huttenlocher et al. (1991) provide evidence that, even in a simple task such as reporting the location of a single point within a circle, significant categorical biases occur. Subjects seemingly divide the circle into quadrants along the horizontal and vertical axes and report the location of the dot biased towards the prototypic location (the centroid of the quadrant). Because subjects in these experiments viewed the original circle on a screen or on cards and had to report the location by drawing it onto a separate sheet of paper, it is unclear whether the same finding would hold true if subjects were asked to directly respond to the stimulus location. In a number of experiments, however, Huttenlocher and colleagues have found evidence that similar biases can also be observed in other spatial behaviors aimed directly at a spatial location (e.g. Huttenlocher, Newcombe, & Sandberg, 1994).

To test whether different spatial behaviors performed in spaces of appropriate size are subject to comparable distortion effects we conducted three experiments. Similarly to the differential use of reference systems for certain spatial behaviors, characteristic distortions might be expected depending on the task demands. We thus tried to investigate distortions in different spatial tasks. Subjects either had to reproduce a visually presented dot location on a computer screen, point (with the eyes closed) to locations where a dot had just appeared within their grasping space, or walk (blindfolded) to objects previously seen in short walkable distances.

2.3.1 Experiments compairing distortions across different spatial actions

Method: To compare systematic biases over different spatially directed behaviors and at different spatial scales we conducted three experiments. Our goal was to make the spatial task as similar as possible across the experiments.

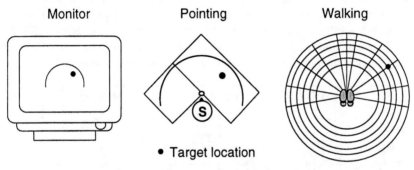

Fig. 2. Experimental settings for the three experiments. Reproduction of a dot location on a monitor (left), pointing with closed eyes to a previously seen dot location on a table top (middle), and walking blindfolded to an object location up to 10 m away (right).

In the first experiment, 11 subjects were seated in front of a computer monitor and viewed a 200° circle-segment on the screen subtending a visual angle of approximately 10° (see figure 2, left). A dot was displayed at one of 76 possible locations inside the circle-segment (4 distances x 19 angles) for 1 second. After the stimulus had disappeared for 2 seconds, the subjects had to reproduce the dot location within the circle-segment as accurately as possible by moving an identical dot across the monitor with the mouse. Each dot location was tested twice. Errors in direction and distance were measured.

In a second experiment, 13 new subjects were seated in front of two tables forming an L-configuration (see figure 2, middle). The same stimulus configurations as in the previous experiment were projected onto the table-top. The distance of the circle-segment to the subject was approx. 50 cm with the subject's index finger resting in the center of the circle. All 76 dot locations were within reaching distance of the subjects inside the circle-segment. Again, subjects viewed the circle-segment and the dot for 1 second and were then instructed to close their eyes. After a variable delay of 2s or 6s a sound signaled subjects to point to the dot-location as accurately as possible by putting the tip of their indexfinger at the location where the dot had been presented.

120

To avoid haptic feedback subjects were not allowed to touch the table before they reached their final target position. Again, each location was tested 2 times. The position of the index-finger and the resulting directional error and distance error were determined using an OPTOTRAK-system (Northern Digital, 250 Hz) and a sensor on the tip of the subject's index finger[2].

In the third experiment, a new group of 10 subjects was tested in a modified, two-dimensional "walking without vision" paradigm (Loomis et al., 1992; Rieser et al., 1990; Thomson, 1983). Subjects had to walk to a single object while blindfolded. The targets ranged in distance between 5 m to 10 m from the subjects' original location. The targets were located in different directions with respect to the orientation of the subject ranging from 100° to -100° to the sides (see figure 2, right). Altogether, 6 distances in combination with 11 directions were tested. To reduce the number of trials per subject, only every other distance was tested for each subject except for the 0° condition, resulting in 36 trials. Subjects viewed the targets for a few seconds with their feet remaining stationary. When objects were located slightly behind them or towards the sides, subjects were allowed to turn their head and upper body while their feet had to remain stationary. The endpoint of the movements were marked and later the distance and angular deviation of the final point walked was determined.

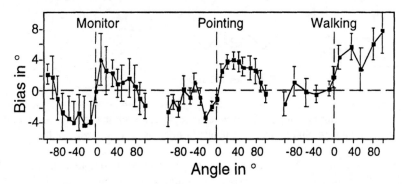

Fig. 3. Directional biases in the three experiments (positive values indicate angles to the right) in respect to the original direction.

Results and Discussion: The results concerning the angular deviations in all three experiments are shown in figure 3. Since errors and biases in distance are less comparable across the different spatial actions reported here we will focus only on angular errors. In all experiments a significant (p<.05) distortion-pattern in respect to the angle probed was found. When reproducing a dot location within a circle-segment on a computer monitor a clear pattern emerged: when dots lay either on the vertical radius (0°) or near ±90° to the left or right, subjects' responses were unbiased. However, locations close to the vertical were strongly biased away from the vertical, overestimating the deviation from 0°. A similar pattern is evident for the pointing task. When pointing to a location within a circle-segment, subjects erred by

---

[2] We thank Jörg Gehrke, Bernhard Hommel, and the Max-Planck-Institute for Psychological Research, Munich, for their help and assistance in carrying out this experiment.

overestimating the angle between straight-ahead and the true dot location. Only around 100° were subjects' pointing judgments unbiased. On the left hand side the pattern clearly deviates from the one found in the Monitor-experiment. However, as figure 2 (middle) shows, the crease between the two tables probably served as a strong reference point that enabled subject's to improve their pointing accuracy. Since the subjects did not have any visual or haptic feedback during the pointing movement, simple correction strategies were not available to them. The results of the third experiment also show a tendency of a systematic bias depending on the angle that had to be walked by the subjects. While subjects in general had the tendency to veer to the right (positive angles) they were fairly accurate when the object was directly in front of them. Again, small deviations from 0° were overestimated in comparison to straight ahead. To both sides the error in the direction walked increased with increasing angle with an exception at -80° and 60°. This general increase in angular deviation might be due to the accumulation of errors while performing full body rotations. Since in all other epxeriments subjects did not change their position this might also explain the continuing increase of angular error for large angles (>80°). Taken together these results demonstrate systematic biases for all three spatial behaviors considered. While the pattern of biases in walking-without-vision experiments still remains unclear, the reproduction of point locations and the pointing to actual object locations in grasping space showed very consistent similarities. This finding is particularly intriguing for pointing movements within grasping space. The coding of a dot location for pointing movements was originally expected to exhibit only little bias. The results, in contrast, indicate that a similar bias occured no matter whether an actual movement or a visual adjustment procedure were performed. This implies that both tasks either tap into a common mental representation or that similar mechanisms are at play in both situations. One reason for the similarity of both results might lie in the 2s - 6s delay before subjects were allowed to start the movement. As Elliott and Madalena (1987) pointed out, the accuracy of motor behaviors such as manual aiming without vision strongly depends on the delay of the movement. In their study reaching accuracy deteriorated within the first 2 seconds and was stable for delays up to 10 seconds. The lack of any difference between the 2 s and 6 s delay conditions in our experiment also points into this direction.

In further experiments we are planning to use a more controlled spatial situation to compare different spatial behaviors and their susceptibility to distortions. In addition, the time frame of spatial behaviors will be taken into account.

So far, our discussion has mainly focused on the relation between different characteristics of mental representations and the actions that depend on them. In this last section, we will discuss ways in which different formats of mental representation may provide valuable information for increasing our understanding of spatial behavior.

3. The format of mental representations of extrapersonal space

In the field of visuo-spatial cognition, the debate about different formats of mental representation has played an important role. In the imagery debate (Block, 1981; Kosslyn & Pomerantz, 1977; Pylyshyn, 1981) the format of mental representations of visuo-spatial information has been predominant. Whereas some theorists suggest that a spatial medium is used to represent visuo-spatial information during imagery, others stress the fact that a universal propositional code is equally suited to model human behavior. Kosslyn's construct of a visual buffer, for example, emerged as a leading metaphor in research on mental imagery (Kosslyn, 1980) and has recently been refined to incorporate neurophysiological findings (Kosslyn, 1994). In its best-known version (Kosslyn, 1980), the visual buffer consists of a two-dimensional matrix of simple cells that can either be turned on or off (similar to pixels with a decay function). Mental images are thought to consist of patterns of activation within the visual buffer and can either be instantiated by processes of retrieving information stored in long-term memory, or through (visual) perceptual processes. Transformations of the image, such as rotating, rearranging, or scaling, are easily performed within the visual buffer. To interpret the mental images, processes like scanning and inspecting have been proposed. Most importantly, the two-dimensional structure of the visual buffer allows for a straightforward implementation of these processes. In articifical intelligence, Funt's (1980) WHISPER system was based on similar ideas, and more recent approaches have used comparable models to simulate spatial events in simple physical contexts (Gardin & Meltzer, 1989) or imagery (Glasgow & Papadias, 1992).

However, assumptions about representational structures or formats have been criticized. Anderson (1978), for example, emphazises the indeterminacy of mental representations. Because proposed formats of mental representation do not unambiguously specify how information is accessed and processed, only the combination of a representation with its corresponding processes can be sensibly analyzed (referred to as representational systems). By pairing a representation with a suitable process, on the other hand, any other representational system can be simulated that relies on the same kind of information. Hence, if different representational formats can lead to near-identical behavior, the question arises as to why one should differentiate between different types of mental representations.

We suggest that postulating different formats of mental representation can help us analyze and understand human spatial cognition. In this view, different formats of hypothesized representations are associated with behaviorally observable differences in mental processing of information and are thus of empirical value. Especially the concept of „analog" representations, as it will be introduced below, may lead to testable predictions in the field of spatial cognition.

Anderson's view, as described above, rightly claims that on logical grounds different pairs of representations and processes can lead to identical behavior given that they both rely on the same information. However, as Pylyshyn (1979) has already pointed out, this argument does not address an important aspect of mental representations. Different kinds of mental representations are usually postulated within theories because they naturally imply certain algorithms or heuristics that differ from alternative accounts. These *intended interpretations* of a representation do not have to be formalized but are a non-arbitrary consequence of the representational structure. It is only through these intended interpretations that a representation gains explanatory and empirical power. Sloman (1985, 1995) makes a similar point. In his view, representations differ in their usefulness for specific tasks. Although different representations might represent the same information, the format of a representation determines efficient algorithms or inference methods. This is particularly evident in analog representations. According to Sloman (1985), analog representations represent properties and relations of objects by using properties and relations of the structure of the representation, instead of referring to explicit names, relations, or symbols. The structure of the representation therefore makes explicit some aspect of the information to be represented (Palmer, 1978). Unlike many authors have claimed, analog representations, in this view, do not have to be continuous, they can include symbolic reference to objects, and they don't have to be fully isomorphic to the things they represent (Sloman, 1995). They do, however, represent at least part of the information based on intrinsic or „built in" characteristics of the representing structure (also compare Lüer, Werner, & Lass, 1995).

Figure 4 illustrates one way to summarize the results of experiments dealing with the access to objects in different locations as it was described in an earlier section. When remembering objects at different locations surrounding an observer, access to locations in front is faster than to the back, while locations on either side of the observer are least accessible.

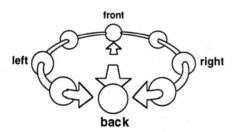

Fig. 4. Egocentric access to object locations. Subjects have fast access to both objects in front and behind. Access increases in difficulty as objects to either left of right lie further to the back.

Different models can be used to describe these results. One model is already implied by figure 3. Objects might be represented in the form of a circular list, linking neighboring object locations together. To access information about the identity of an object at a certain location, the list would have to be searched from a starting point (front) and times would increase linearly with the number of intervening objects. The faster access to objects' identities directly behind an observer requires the additional assumption of special access to that list location or separate coding. This representational format can be considered analogical because the structure of a list readily implies increasing search times—it is one of the intended interpretation of a list structure.

One drawback of this model, however, involves the fact that the number of free parameters (list structure, priviliged access for front and back) is close to the number of datapoints that need to be explained. It is also unclear how more complex configurations, for example objects at different distances but the same direction, might be represented in the model. Nonetheless, interesting empirical questions can be derived even from even this simple, exemplary model. For example, it is unclear whether the list structure links together objects or specific locations because the two are confounded in both the experiments reported above and the model conceived here. Experiment 10 of Hintzman, O'Dell, and Arndt (1981) addressed this question by showing that access times were not reduced when some locations were left empty. These findings suggest that the structure of the list is based on physical location and not solely on the order of objects.

An alternative account of these data might try to explain the increase in access time by referring to mental transformation processes, such as mental rotation (for a discussion see Hintzman et al.). In this case, the access to object locations should increase monotonically with increasing angular difference to the front. Evidently, the predictions of the linked list model and the mental rotation model are very similar. However, they both stress different points. Whereas the mental rotation model implies a continuous representation (or discontinuous in small steps), the linked list implies a discontinuous representation. Of course these (intended) implications don't follow logically from the specifications of the representational structure. However, introducing these kind of models is reasonable only if the intended implications are functionally important.

Of course, specifying the format of a mental representation only makes sense if it has strong implications for the ways in which information is processed. Analog representations, if introduced in this way, naturally lead to empirically testable predictions by means of their intended implications. This is in marked contrast to non-analog forms of representations. Especially in the case of propositional or „logical" representations (Sloman, 1985), any empirically testable hypothesis has to be introduced explicitly by specifying arbitrary characteristics of the processes that operate on the represented information.

The popularity of analog representations in psychological theorizing (and also in computer programming) might lie in the ease by which human beings are able to use them in complex problem solving tasks (of which theorizing is one). Much like well-designed interfaces, an analog representation already implies what kind of operations and transformations are easily performable and how they can be carried out, which allows researchers to easily derive testable hypotheses. On the other hand, a potential problem of analog representations might lie in the fact that they overly restrict theories of mental representation. Because general representational systems (i.e. propositional representations) are more versatile, many models phrased in propositional terms are not expressable in terms of analog representations. If actions require different kinds of spatial knowledge representation, as was argued in the preceeding sections, a variety of representational formats might be most fruitful for the actions considered. Only further research will show whether the pragmatic use of analog representations will outweigh their potential restrictions.

References

Aglioti, S., Goodale, M.A., & DeSouza, J.F.X. (1995). Size-contrast illusions deceive the eye but not the hand. *Curr. Biology, 5*, 679-85.

Anderson, J.R. (1978). Arguments concerning representations for mental imagery. *Psychological Review, 85*, 249-277.

Berthoz, A. (1991). Reference frames for the perception and control of movement. In J. Paillard (Ed.), *Brain and Space* (pp. 81-111). Oxford: Oxford University Press.

Boer, L.C. (1991). Mental rotation in perspective problems. *Acta Psychologica, 76*, 1-9.

Block, N. (1981). *Imagery*. Cambridge: MIT-Press.

Bryant, D.J. & Tversky, B. (1992). Assessing spatial frameworks with object and direction probes. *Bulletin of the Psychonomic Society, 30*, 29-32.

Claus, B., Eyferth, K., Gips, C., Hörnig, R., Schmid, U., Wiebrock, S., & Wysotzki, F. (this volume). Reference frames for spatial inference in text understanding.

Easton, R.D. & Sholl, M.J. (1995). Object-array structure, frames of reference, and retrieval of spatial knowledge. *Journal of Experimental Psychology: Learning, Memory, and Cognition, 21*, 483-500.

Elliott, D. & Madalena, J. (1987). The influence of premovement visual information on manual aiming. *The Quarterly Journal of Experimental Psychology, 39 A*, 541-559.

Franklin, N. & Tversky, B. (1990). Searching imagined environments. *Journal of Experimental Psychology: General, 119*, 63-76.

Funt. B.V. (1980). Problem-solving with diagrammatic representations. *Artificial Intelligence, 13*, 201-230.

Gardin, F. & Meltzer, B. (1989). Analogical representation of naive physics. *Artificial Intelligence, 38*, 139-170.

Georgopoulos, A.P. & Pellizzer, G. (1995). The mental and the neural: Psychological and neural studies of mental rotation and memory scanning. *Neuropsychologia, 33*, 1531-1547.

Georgopoulos, A.P., Lurito, J.T., Petrides, M., Schwartz, A.B., & Massey, J.T. (1989). Mental rotation of the neuronal population vector. *Science, 243*, 234-236.

Gibson, J.J. (1979). *The ecological approach to visual perception*. Boston: Houghton-Mifflin.

Gigerenzer, G. (1988). Woher kommen Theorien über kognitive Prozesse? *Psychologische Rundschau, 39*, 91-100.

Glasgow, J. & Papadias, D. (1992). Computational imagery. *Cognitive Science, 16*, 355-394.

Goodale, M.A., Milner, A.D., Jakobson, L.S., & Carey, D.P. (1991). A neurological dissociation between perceiving objects and grasping them. *Nature, 349*, 154-156.

Graziano, M.S.A. & Gross, C.G. (1995). The representation of extrapersonal space: A possible role for bimodal, visual-tactive neurons. In M.S. Gazzaniga, (Ed.), *The cognitive neurosciences* (pp. 1021-1034). Cambridge: MIT Press.

Grüsser, O.-J. (1983). Multimodal structure of the extrapersonal space. In A. Hein & M. Jeannerod (Eds.), *Spatially oriented behavior* (pp. 327-352). New York: Springer.

Hintzman, D.L., O'Dell, C.S., & Arndt, D.R. (1981). Orientation in cognitive maps. *Cognitive Psychology, 13*, 149-206.

Hirtle, S.C. & Jonides, J. (1985). Evidence of hierarchies in cognitive maps. *Memory and Cognition, 13*, 208-217.

Huttenlocher, J., Hedges, L.V., & Duncan, S. (1991). Categories and particulars: Prototype effects in estimating spatial location. *Psychological Review, 98*, 352-376.

Kosslyn, S.M. & Pomerantz, J.R. (1977). Imagery, propositions, and the form of internal representations. *Cognitive Psychology, 9*, 52-76.

Kosslyn, S.M. (1980). *Image and mind.* Cambridge: Harvard University Press.

Kosslyn, S.M. (1994). *Image and brain.* Cambridge. MIT-Press.

Krieg-Brückner, G., Röfer, T., Carmesin, H.-O., & Müller, R. (this volume). A taxonomy of spatial knowledge for navigation and its application to the Bremen autonomous wheelchair.

Kuipers, B. (1982). The "map in the head" metaphor. *Environment and Behavior, 14*, 202-220.

Loomis, J.M., Da Silva, J.A., Fujita, N., & Fukusima, S.S. (1992). Visual space perception and visually directed action. *Journal of Experimental Psychology: Human Perception and Performance, 18*, 906-921.

Lüer, G., Werner, S. & Lass, U. (1995). Repräsentation analogen Wissens im Gedächtnis. In E. van der Meer and D. Dörner (Hrsg.), *Das Gedächtnis.* Göttingen: Hogrefe.

McDonald, T.P. & Pellegrino, J.W. (1993). Psychological perspectives on spatial cognition. In T. Gärling & R.G. Golledge (Eds.), *Behavior and environment: Psychological and geographical approaches.* Amsterdam: Elsevier.

McNamara, T.P., Hardy, J.K., & Hirtle, S.C. (1989). Subjective hierarchies in spatial memory. *Journal of Experimental Psychology: Learning, Memory, and Cognition, 15*, 211-227.

Milner, A.D. & Goodale, M.A. (1995). *The visual brain in action.* Oxford: Oxford University Press.

Montello, D.R. (1993). Scale and multipe psychologies of space. In A.U. Frank & I. Campari (Eds.), *Spatial information theory: A theoretical basis for GIS* (pp. 312-321). Berlin: Springer.

O'Regan, J.K. (1992). Solving the real mysteries of visual perception: The world as an outside memory. *Canadian Journal of Psychology, 46*, 461-488.

Palmer, S.E. (1978). Fundamental Aspects of cognitive representation. E. Rosch & B.B. Lloyd (Eds.), *Cognition and categorization* (pp. 259-303). Hillsdale: Erlbaum.

Poucet, B. (1993). Spatial cognitive maps in animals: New hypotheses on their structure and neural mechanisms. *Psychological Review, 100*, 163-182.

Presson, C.C. & Hazelrigg, M.D. (1984). Building spatial representations through primary and secondary learning. *Journal of Experimental Psychology: Learning, Memory, and Cognition, 10*, 723-732.

Presson, C.C., DeLange, N., & Hazelrigg, M.D. (1989). Orientation specificity in spatial memory: What makes a path different from a map of the path? *Journal of Experimental Psychology: Learning, Memory, and Cognition, 15*, 887-897.

Pylyshyn, Z.W. (1979). Validating computational models: A critique of Anderson's indeterminacy of representation claim. *Psychological Review, 86*, 383-394.

Rieser, J.J. (1989). Access to knowledge of spatial structure at novel points of observation. *Journal of Experimental Psychology: Learning, Memory, and Cognition, 15*, 1157-1165.

Rieser, J.J., Ashmead, D.H., Talor, C.R., & Youngquist, G.A. (1990). Visual pereption and the guidance of locomotion without vision to previously seen targets. *Perception, 19*, 675-689.

Sadalla, E.K., Burroughs, W.J., & Staplin, L.J. (1980). Reference points in spatial cognition. *Journal of Experimental Psychology: Human Learning and Memory, 6*, 516-528.

Schölkopf, B. & Mallot, H.A. (1995). View-based cognitive mapping and path planning. *Adaptive Behavior, 3*, 311-348.

Shelton, A.L. & McNamara, T.P. (1997). Multiple views of spatial memory. *Psychonomic Bulletin & Review, 4*, 102-104.

Shepard, R.N. & Cooper, L.A. (1982). *Mental images and their transformations.* Cambridge: MIT-Press.

Shepard, R.N. & Metzler, J. (1971). Mental rotation of three dimensional objects. *Science, 171*, 701-703.

Sholl, M.J. & Nolin, T.L. (in press). Orientation specificity in representations of place. *Journal of Experimental Psychology: Learning, Memory, and Cognition.*

Sholl, M.J. (1987). Cognitive maps as orienting schemata. *Journal of Experimental Psychology: Learning, Memory, and Cognition, 13*, 615-628.

Sholl, M.J. (1995). The representation and retrieval of map and environment knowledge. *Geographical Systems, 2*, 177-195.

Siegel, A.W. & White, S.H. (1975). The development of spatial representations of large-scale environments. In H.W. Reese (Ed.), *Advances in child development and behavior* (Vol. 10). New York: Plenum Press.

Siegel, A.W. (1981). The externalization of cognitive maps by children and adults: In search of ways to ask better questions. In L.S. Liben, A. Patterson, & N. Newcombe (Eds.), *Spatial representations and behavior across the life span: Theory and application* (pp. 167-194). New York: Academic.

Sloman, A. (1985). Why we need many knowledge representation formalisms. In M. Bramer (Ed.), *Research and development in expert systems* (pp. 163-183). New York: Cambridge University Press.

Sloman, A. (1995). Musings on the roles of logical and nonlogical representations in intelligence. In J. Glasgow, N.H. Narayanan, & B. Chandrasekaran (Eds.), *Diagrammatic reasoning: Cognitive and computational perspectives* (pp. 7-32). Cambridge: MIT-Press.

Stevens, A. & Coupe, P. (1978). Distortions in judged spatial relations. *Cognitive Psychology, 10*, 422-437.

Thomson, J.A. (1983). Is continuous visual monitoring necessary in visually guided locomotion? *Journal of Experimental Psychology: Human Perception and Performance, 9*, 427-443.

Tversky, B. (1981). Distortions in memory for maps. *Cognitive Psychology, 13*, 407-433.

v. Wolff, A. (1997). Der Einfluß von Raumachsen auf das egozentrische Umraummodell. In W. Krause, U. Kotkamp, & R. Goertz (Eds.), *KogWis97: Proceedings der 3. Fachtagung der Gesellschaft für Kognitionswissenschaft* (241-243). Friedrich-Schiller-Universität Jena.

Werner, S., Krieg-Brückner, B., Mallot, H.A., Schweizer, K. & Freksa, C. (1997). Spatial cognition: The role of landmark, route, and survey knowledge in human and robot navigation. In M. Jarke, K. Pasedach, & K. Pohl (Eds.), *Informatik '97* (41-50). Berlin: Springer.

Werner, S. & Wolf, S. (in preparation). The effects of egocentric and allocentric frames of reference on the mental representation of extrapersonal space.

Representational Levels for the Perception of the Courses of Motion

A. Eisenkolb[1], A. Musto[2], K. Schill[1], D. Hernández[2], W. Brauer[2]

[1] Ludwig-Maximilians-Universität München,
Goethestrasse 31, 80336 München, Germany
amadeus@imp.med.uni-muenchen.de
[2] Technische Universität München,
Arcisstrasse 21, 80333 München, Germany

Abstract. The problem of representation and processing of motion information is addressed from an integrated perspective covering the range from early visual processing to higher-level cognitive aspects.

A spatio-temporal memory is presented as indispensible representational prerequisite for the recognition of spatiotemporal gestalt. We assume that this structure is replicated on different processing-levels in the visual system mirroring its hierarchical structure. Thus, each level requires a different representation for spatio-temporal information.

As a first step, we present a two-layered architecture for the qualitative representation of motion trajectories: The vectorial layer is quite accurate and allows switches between deictic and intrinsic frame of reference. The propositional layer is more abstract and reveals similarities and regularities of motion paths which will be useful for motion prediction.

First psychophysical experiments indicate that information about direction and position are not stored independently but merely in form of a spatio-temporal compound.

1 Introduction

The understanding of motion perception is essential for the comprehension of the human visual system. At the same time this understanding is important for the development of artificial systems like user interfaces in multimedia databases or mobile robots equipped with computer vision devices and developed to cooperate and interact with humans. The analysis and modeling of processes and representational levels involved in motion processing is in the focus of our research (Schill et al. 1998). This is approached in an interdisciplinary way:

Psychophysical experiments on the limits and capabilities of the visual system to deal with dynamic information shall provide evidence for the modeling of early visual processes. On the other hand, we try to model and analyze higher cognitive processes in motion processing. A major point of our research interest concerns the possible interdependencies between these modeling stages, like relationships between the representational requirements on a short term memory range and qualitatively expressed motion information.

Our paper starts with an overview of the research on the processing of motion information in the visual system. This overview reflects the variety of frameworks with which the different traditions have conceptualized motion information — a fact which made it difficult to analyze which results are already available for modeling. However, a thorough analysis of these approaches reveals that the aspect of spatio-temporal processing we are interested in, namely how the temporal stream of spatial input information is mapped into some representational structure to provide a basis for the analysis and processing of complex spatio-temporal patterns, has reached only little attention.

Since motion representation and reasoning with motion has astonishingly not come to the same attention and tradition in the qualitative reasoning community as in psychophysics, few approaches can be found and they are quite heterogeneous. Therefore we review this literature only shortly in chapter 3.2.

In chapter 3, first results on the qualitative representation and on the analysis and modeling of the perceptional stage are summarized. The guideline for the experiments is a spatio-temporal memory model (Schill and Zetzsche 1995a) which has been shown to provide necessary prerequisites for the representation and processing of spatio-temporal information. Furthermore, this model allows us to explain a variety of results obtained in experiments on early vision memory stages. A brief description of the model is provided at the beginning of section 3.1 followed by an experiment to investigate the encoding of spatio-temporal patterns over time.

On the higher-level representational stage, we want to provide a simple but effective means for the representation of the course of motion which allows description at several levels of granularity, abstraction and accuracy and is suitable for classification and compact storage of motion events as well as for prediction of the future position of a moving object. We have several contraints for our representation:

1. We want our representation to be qualitative. This has to do with representational economy: a qualitative representation provides mechanisms for representing only those features that are unique or essential, whereas a quantitative representation allows representation of all those values that can be expressed with respect to a predefined unit. A qualitative representation often can reflect and compensate vague data better than a quantitative one, which often computes at an unnecessary level of accuracy. Third, qualitative representations are easily parametrizable for different scales and can also easily provide different levels of granularity, abstraction and accuracy.

2. We want our representation to use the shape of the course of motion as important (but not unique) representational primitive, since this is one of its most outstanding qualitative features. We claim that this feature (together with certain dynamic characteristics) allows easy detection of patterns and periodicity, which is important for the task of prediction. Another reason is that the shape of the course of motion is one of the key features in structural

similarity detection, which will be important for the tasks of classification and prediction.

3. We want our representation to abstract from special properties of the moving object (like extension, shape, purpose) and of the environment (like what buildings, streets, etc. there are) and from events that are associated with the motion event. Our formalism shall represent only spatio-temporal variation of point-like objects in (at least for now) 3d-space-time and shall allow for representing motion paths to simplify point (2). This abstraction shall be useful for representation of a great variety of motion events and furthermore reflects that motion is one of the central issues in visual cognition.

In other words, we want to provide a mechanism for the representation of motion trajectories in 2D space, like known from differential geometry, but only using qualitative means. We describe this qualitative representation in chapter 3.2.

For experimental study of the processing of courses of motion also of a more complex nature, a systematic variation of the experimental conditions requires a measure which defines such a complexity. Chapter 3.3 comprises some considerations on similarity and complexity of courses of motion, where results and needs of both the computational and the psychophysical research merge.

We will conclude with a summary and an outlook on our future research.

2 State of the Art in Psychological Research

In this chapter we will briefly outline the motion literature with respect to the underlying concepts of spatio-temporal information processing.

Computational models The most explicit models of motion processing are provided by computational models. For overviews see, e.g., (Adelson and Bergen 1985; Nakayama 1985; Grzywacz et al. 1994). The earliest model was the Reichardt detector, a local motion detector. The construction principles of local visual motion detectors are basically or directly based on Reichardt's first correlator or formally equivalent to it, as in the case of filter models, in that they differ from it only in how the delay-and-compare principle is realized. In the most simple case a local motion detector has two inputs at two adjacent spatial locations. By delaying one input for some short time and a subsequent comparison of the delayed and the undelayed input a measure for motion in form of a local-momentary direction signal is achieved. For details see Reichardt (1957), Adelson and Bergen (1985), van Santen and Sperling (1985), Watson and Ahumada (1985). While local motion detection algorithms allow for a sampling of the visual signal in a purely continuous mode (filter property), so called matching algorithms are "designed to make predictions about stimuli presented as frames", i.e., they work on a discrete description of spatio-temporal stimuli (Adelson and Bergen 1985). Matching models can be conceived of as global correspondence-detectors, where certain "salient" features can serve as matching points. The correspondence problem is a direct consequence of that conceptualization of spatio-temporal information. Another advantage of local motion detection models is that their front-end can be associated with anatomico-physiological properties like receptive fields tuned to a certain spatial frequency or, originally, with two adjacent facets of an insects compound eye. A principal problem, of course, is as to how that distributed local-momentary direction information is linked together on subsequent processing stages. It also turned out that bilocal motion detectors can only respond to luminance-defined spatio-temporal change but are "blind" for "higher-order" characteristics of the visual spatio-temporal signal like those occuring with texture-defined motion stimuli, see (Lu and Sperling 1995). Because the local motion detection principle plays a central role in many models to account for data of psychophysical experiments, e.g., velocity integration see (Van Doorn and Koenderink 1982a, 1982b; Werkhoven et al. 1992) and because they are the most explicit among all motion models they might become interesting for our research when a neural or hardware realization of the memory model has to be developed.

Psychophysics A closer look at the literature on motion research reveals that concepts concerning the processing of dynamic information in a time span exceeding the temporal range given by the delay properties of the diverse correlator models or filter models were not considered in an explicit manner.

In cases where time spans greater than 80-100 ms are involved, psychophysical results are usually interpreted in terms of processing- or likewise, integration time needed by the first visual processing stages to encode a certain local-momentary variable of the incoming motion information, e.g., velocity. There is a surprising lack of explicit theories about what happens with the spatio-temporal information once its local-momentary properties have been calculated.

This is true for the computational-oriented approaches whose aim was principally the modeling of elementary properties like motion-detection, but also for a big part of psychophysics. Psychophysical research, so far, was concerned primarily with finding just notifiable differences for the detection of changes of velocity (velocity discrimination), the investigation of direction selectivity and sensitivity (McKee and Watamaniuk 1994), or even more basic minimum motion thresholds, e.g. Again, the implication was that information about velocity or direction is basically available in form of a local-momentary scalar as a result of the spatio-temporal filtering characteristics of the respective neural circuitry.

The basic assumption that underlies virtually all psychophysical work done in the last 30 years can be summarized as follows: Motion information is processed by a unique, specialized mechanism. Motion information is not inferred (at least at a very early visual stage) from positional changes over time but is a "quality per se". For an overview of physiological evidence on this position and some historical notes see (Nakayama 1985). Within this research tradition the general questions to be answered are: Which variables of visual motion, e.g., velocity, acceleration, direction of motion, are processed and how (number, tuning, and independency of channels)? (see Grzywacz et al. (1994)). Subsequently, besides the exploration of the visual system's response properties to motion variables, much effort has been spent on clarifying the question as to whether motion variables are processed by the same units (channels) as the static pendants, e.g., orientation (static) processing vs. direction (dynamic) processing. Detection and discrimation thresholds were measured in a variety of experimental paradigms. McKee and Watamaniuk (1994) mentioned the minimum threshold (smallest detectable displacement), direction discrimination, and speed discrimination as basic psychophysical data on early motion processing.

A considerable amount of research has been conducted in order to clarify the fundamental spatio-temporal parameters at the input stage of the motion system (Van Doorn and Koenderink 1982a, 1982b). These results indicate that many detectors exist in any retinal region for different velocities but to our knowledege no psychophysical empirical data are available as to which extent and whether information about direction, position, velocity and acceleration is represented over time to enable recognition of spatio-temporal patterns that are defined by a dynamic change over time.

To exemplify our point of critique of the conception of the sensory basis of motion information, as it predominates psychophysical research, imagine the following: Consider a highly visible dot moving in the frontoparallel plane stimulating detectors depending of velocity and direction of movement. Banks of correlators provide the necessary information about the retinal location; the spatial and temporal spacing of the input stages of the correlators provide a velocity scalar. It is one question as to whether retinal image motion is the necessary and sufficient prerequisite (see Rock (1975) for a critique on this point), another question is more far-reaching, namely, as to how higher dimensional spatio-temporal properties like acceleration are coded and processed. It seems obvious that the acceleration profile of dynamical events has to be accounted for *over time* in some way enabling human beings and animals to perform recognition of a complex spatio-temporal Gestalt in a remarkably short time and with high precision. Freyd (1987) refers to this fact "as the perceptual system's natural competence with information carried dynamically". Examples of this competence are given by Johansson (1975), see below, and Poizner et al. (1981). While there is general agreement on the motion variables velocity and direction (measured as discrimination threshold or Weber fractions) literature on acceleration provides contradictory conclusions, both in theory and empirical data. From a theoretical standpoint Grzywacz et al. (1994) conclude that acceleration as a second derivative (the derivative of velocity over time) "is a variable that the visual system should seriously consider ignoring", because any method that measures derivatives is practically unstable ("suscetible to noise"), which holds to some extent also for speed. Rosenbaum (1975) comes to another conclusion. Based on his experimental results to motion extrapolation he argues for a "direct" and accurate encoding of velocity and acceleration. This in turn, is contradicted by the conclusion Werkhoven et al. (1992) draw from a signal-theoretical analysis of the delay-properties of the bilocal motion detector. The empirical data provided in that study give evidence that there do not exist "specific acceleration detectors", in the sense that a temporal derivative of velocity over time is calculated for detecting speed modulations. There is an ongoing debate about the coding of velocity and acceleration in the literature of cognitive psychology. Two standpoints can be distinguished, one claiming that acceleration and velocity are coded exclusively "directly" (Rosenbaum 1975), the other one proliferating evidence for the existence of "inferred" extrapolation of motion, the latter one being more "cognitive" rather than based on sensory characteristics (Peterken, Brown, and Bowman 1991).

Psychophysical data can in principle receive theoretical foundation by relating the experimental data to elementary motion processing units as the Reichardt detector, e.g., thus giving the notion "direct coding" a concrete meaning whereas in cognitive psychology the need of representing higher derivatives of motion information is claimed on a merely phenomenological basis. To our knowledge, however, no experimental data is provided as to which degree of precision higher order dynamics have to be made available to the system, and, furthermore, no satisfying hypothesis exist as to how the problem of keeping different points of time, necessary for the recognition of patterns that are defined by a quite complicated change in the time domain, is solved.

Apparent motion Another approach used to characterize the time character-istics of human motion processing in more detail was investigation of apparent motion (AM). Korte found out that time and distance of discretely presented dots that give rise to "good" AM are in a lawful relationship, as expressed by the 3 laws of AM by Korte, see, e.g., (Anstis 1986).

Braddicks work (Braddick 1974) indicated that two motion processes exist, a short (temporal and spatial) range process and a long (temporal and spatial) range process. The first, short range process was identified with the mechanism responsible for processing "real motion", the long range process was thought to be performing some inference of motion information on the basis of the compar-ison of positional information.

Strong evidence for a more complicated spatio-temporal relationship between what is available in stimulus and what is perceived is given by Shepard's ex-periments to path guided apparent motion (Shepard 1983). Shepard displayed blurred grey-paths with systematically varied curvature during the time inter-vall that separated the presentation of two dots. As a result he could present a revised version of Korte's law of AM where it is not the Euclidean distance which determines the optimal timing for perceiving AM, but the path length between the dots. The path could even have the shape of a closed circle thereby leading to the perception of AM around a circular path.

To our knowledge the path-guided AM-Paradigm was not used to investigate the coding of spatial properties over time further. Moreover, no structural require-ments about how spatial information of different points in time is represented has been formulated that allow for a detailed quantitative exploration.

Ecological Psychology One branch of psychology concerned with the processing of motion is often labelled ecological psychology or the ecological approach
to perception. As a prominent example, Johansson's experiments (Johansson
1976) showed the human observers capacity to distinguish accurately in a quite
short time different human motion patterns that were formed by a sequence of
frames each consisting of a swarm of light dots on a dark background. To understand the point-like stimulus, imagine a person with small lamps attached at its
joints (knee, angle, shoulder etc.) performing some movement in a completely
dark room. The stimulus consisted of a number of frames registered from such a
scene. Johansson showed that observers identified correctly running or walking
persons within 400 ms. On the basis of these results Johansson formulated the
principle of invariant vectors based on projective geometry.

Another "holistic" (Hochberg 1986) approach was Restle's Minimum Coding
principle, which principally states that humans prefer among several possibilities to perceive that pattern that has the minimal code – an alternative formulation of the principle of simplicity, for a discussion see (Hochberg 1986; Cutting
and Proffitt 1982). The problem with higher-level principles is the fact that a
quantification of the effects has never been systematically undertaken, probably
because of the difficulties in setting up a theoretical framework that allows for
both the integration and formal treatment of higher-level principles and an experimental investigation that yields quantitative data.

An approach where the visual short term memory is referred to as the underlying representational structure, where visual processes may act on dynamical
representations, is the work related to representational momentum.

Initially, the finding was that after viewing a photographic action scene the
memory of human observers is distorted in the direction of motion implied in
the picture (e.g. a man jumping from a wall, (Freyd 1987)). This distortion of
the represented event was identified by a *memory shift* that was indicated by
the observer's preference for identifying the previously displayed picture with a
testpicture where the implied motion was continued.

In numerous articles the effect of representational momentum was investigated in
more detail, for instance the time course of representational momentum (Freyd
and Johnson 1987; Finke, Freyd, and Shyi 1986). Instead of photographic snapshots the employed stimulus was mostly a rectangle, shown across time (with
large interstimulus intervals[1] to prevent the rise of a motion percept) in several
orientations around its center such that a rotation was implied. The rectangle
was formed by dots to prevent orientation information caused by aliasing of
oblique lines displayed on a raster device.

Finke and Shyi (1988) clarified the relation between representational momentum and perceptual anticipation in that an expected object's course does not
follow from a strict analogy to physical momentum. Verfaillie and d'Ydewalle
(1991) could show that the measured memory shift which is an indicator of the
representation of the previously displayed dynamic event (implied rotation of
an rectangle), is determined by it's long term history rather than by the pure

---

[1] The time interval between two subsequent presented stimuli.

physical momentum. Since then the "momentum analogy holds only in a highly abstract sense".

The work related to representational momentum is appealing to us because on a time scale from 100 ms up to 10 s cognitive phenomena of the representation of dynamic information and the extrapolation of motion information can be investigated with a powerful paradigm.

3 Results

3.1 Paradigms of Psychophysical Experimentation and Modeling

Spatio-temporal Memory Model In the review of current approaches to motion information processing (see previous section) we have shown that one implicit assumption is common to all approaches to motion processing: The assumption that the course of the internal motion infomation corresponds directly to the external motion signal. In other words, the internal signals, i.e. information, change in accordance to the variation of external signals, and at any chosen instant of external time a collinearly related instant of time of the input sequence is available internally. In contrast to these collinear concepts Schill and Zetzsche (1995a) have suggested a spatio-temporal memory model based on what they call an orthogonal representation of temporal information. Their concept assumes that at any given instant of external time, features belonging to a past sequence are internally available, i.e., there exists an internal representation of time which is orthogonal to the physical time, see Fig. 1 for illustration.

The orthogonal concept is not only a metaphor of representation, but allows experimental predictions like on "iconic memory" (immediate memory span, partial report superiority, e.g.). Furthermore it resolves inconsistencies arising within the collinear view with respect to the interpretation of experimental results on backward masking (Schill and Zetzsche 1995b). With this new interpretation conclusions about the performance (sequential vs. parallel processing) of the visual system have to be revisited. The orthogonal representation can be achieved by the mapping of time into simultaneously available, spatially distributed properties (such as the instantaneous spike rates of an array of neurons). A consequence of the mapping of time into space is the resulting discreteness of the internal representation of time, i.e. internal time is spatially "quantized". There are two basic approaches of how the spatio-temporal model can be implemented in neural or technical hardware: as a shiftregister or as an ordered or cascaded delay line structure. Both types lead to a quite similar type of representation of spatio-temporal information, but have to be differentiated with respect to their modes of processing. The shift register solution can be thought of as a stack of frames with each frame representing the spatial input pattern at a certain moment in time. In this case a kind of pacemaker has to be assumed which initiates the periodic shifts resulting in a discrete mode of processing. In contrast to this, the delay line solutions imply a continuous mode of information processing (cf. Uttley, 1954)

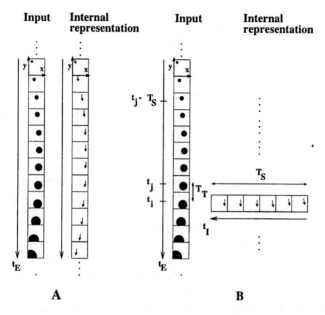

Fig. 1. A: The internal representation of a moving object according to the collinear concept. The input is shown on the left side with time running from top to bottom. Its representation is shown (in abstract form) on the right side. **B:** How the orthogonal representation of the spatial input during a recent period is achieved at a moment of external time $t_E = t_i$. (Physical time is denoted as external in order to avoid confusion. All variables and states in our model are functions of t_E). The orthogonal representation enables the events that occur during a recent time interval (with duration T_s), to be simultaneous accessible within the system. Because of an unavoidable transduction time T_T in the system, the most recent frame currently available corresponds to the spatial input at the external time $t_E = t_j = t_i - T_T$. So the oldest frame corresponds to the spatial input at external time $t_E = t_j - T_S$. The direction of internal time in the memory is from right to left

Psychophysical experiments From the foregoing investigations evidence has been provided for the existence of a spatio-temporal memory. However, this investigations give us no information about the specific properties being stored in this memory. Is it the luminance itself, or some more complicated type of spatio-temporal feature? In fact, we would assume that there is not a single spatio-temporal memory at some early stage of the visual system, but rather that this basic structure is replicated on the different levels of processing. In analogy to the parallel increase in the complexity of the spatial features being encoded and in the size of the receptive fields of neurons in the higher cortical areas, we would also assume that the spatio-temporal features become more complex, and that the temporal granularity is increased when we proceed towards the higher processing stages. Traditionally, spatial properties are beeing seen as more relevant than temporal ones. For this reason empirical information on these issues is extremely limited. While it is known that neurons on the lower levels are encoding elementary spatio-temporal properties like the local instantaneous velocity, investigations of the encoding of more complex spatio-temporal features are rare. This prompted us to start a series of psychophysical experiments on the perception of elementary spatio-temporal patterns by human observers. As a first step, experiments have been started in which thresholds for discrimination of the direction of straight motion paths are measured.

Direction discrimination of motion paths To further identify the features used in spatio-temporal processing we devised an experiment with a simple spatio-temporal configuration. We investigated to which extent an interstimulus interval (ISI) and a spatial displacement (d) influence the discrimination of the direction of the movement of two sequentially presented motion paths. The rationale of this experiment was: for a comparison both paths have to be neurally encoded. The question was: How do spatial and temporal separation influence the thresholds for direction discrimination and what conclusions can be drawn with respect to the representation of spatio-temporal information?

Method Per trial two paths were presented which were formed by well visible black dots moving on a white background of a computer screen. The paths were temporally separated by a variable ISI which was chosen among 6 possible ISI's ranging from -150 to 1000 ms (independent variable 1). The speed was fast (15.6 deg/s) and was held constant for all trials. Each path had a duration of 300 ms and subtended 4.7 deg visual angle. A displacement of the origin of the paths was tested against no displacement (independent variable 2). Two direction-conditions were employed: In one session thresholds for direction discrimination

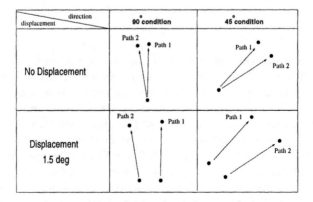

Fig. 2. Conditions of displacement and direction of the experiment

were measured for one displacement condition (diplacement vs. no displacement) and all six ISI's (-150 ms - 1000 ms) with both direction-conditions. Ss had to decide whether the first or second path appeared to be more advanced in clockwise direction. For 3 subjects thresholds were measured in 16 sessions (8 sessions for the no-displacement-condition and 8 sessions for the displacement-condition).

Results Except at an ISI of -150 ms where the both paths overlapped temporally the ISI does not appear to have an influence on the performance of the direction discrimination. However, a displacement of the origin of the paths causes a consistent threshold elevation as shown in Figure 3. Thus the manip-

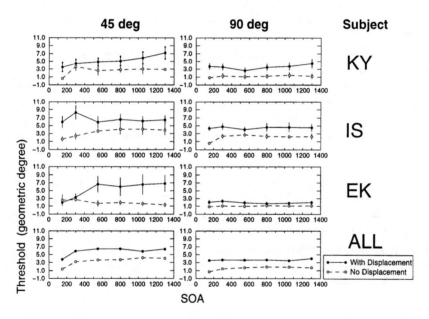

Fig. 3. Results of the experiment: For 3 subjects thresholds were measured in 16 sessions each. Additionally, we measured thresholds of further 4 Ss in 4 sessions each. In the last row the data of 7 Ss are averaged

ulation of the loci of the paths by a spatial displacement of their origins influences the threshold for direction discrimination. The dependency of direction discrimination from information about the location suggests a unified representation in the sense that motion information is internally available in form of a spatio-temporal compound. Based on these results, further experiments will be conducted to investigate the human observer's ability to use more complicated dynamic information of motion paths for discrimination. As an example we are currently investigating the influence of the dynamical profile of a motion path on orientation discrimination by accelarating and decelarating the moving dot. If the motion paths do not differ in global parameter (duration, average speed) but only on some local properties (accelaration at time ti) and if they can be discriminated to some extent we conclude that its the local path information giving rise to the discriminability.

3.2 Architecture of a Qualitative Model

Proposal: a two-layer qualitative representation of the course of motion As mentioned before, we want to provide a simple but effective means for the representation of the course of motion which allows description at several levels of granularity, abstraction and accuracy and is suitable for compact storage of motion events as well as for prediction of the future position of a moving object.

To this end, we propose a two-layered qualitative representation of motion. A vectorial representation at the lower level is qualitative, but still quite accurate and has a fine granularity. A propositional representation at the higher level is closer to natural language, more abstract, less accurate and has coarser granularity, but reflects something of the semantics of the course of motion, in that it identifies meaningful subsequences in the vectorial representation and sums them up to a higher structure. "Semantics" and "meaningful" have, of course, to be understood in a certain context: In our case, the semantic property of a course of motion is its shape and a subsequence is meaningful, if it constitutes a shape that is element of the predefined shape vocabulary. In a traffic scenario, the semantic of a course of motion would be the driving maneuvers it consists of; a meaningful subsequence would be a single maneuver.

Please note that the two representations coexist and that we don't throw away any information we have got at the lower level. The transformation to the higher-level representation provides additional (semantic) information.

Vectorial representation With regard to later technical applications, we assume that we have an external observer and at first adopt a deictic point of view with a fixed scan rate. From a neuroanatomical point of view, the scan rate is assumed to be event-triggered and realized by a relaxating oscillator (Pöppel and Schill 1995; Pöppel 1997), which might be taken into consideration for further development of our qualitative model.

Furthermore, the fixed scan rate corresponds to a discrete processing of spatial information, which is in accordance with one possible mode of processing in our spatio-temporal memory model. Similar to the spatio-temporal memory model, the discrete processing lets us map time into space.

Thus, we need only the two components distance and direction of the movement, because at a fixed scan rate distance and velocity are equivalent. To model standstill, we need a distance 0 and an direction 0.

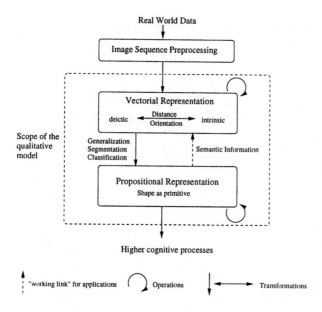

Fig. 4. A two-layer qualitative representation of motion

As quantity space for the deictic frame of reference, we can use e.g. $D_D = \{$very-close, close, medium-dist, far, very-far, $0\}$, $O_D{}^2 = \{$north, south, east, west, $0\}$. These directions in 2D-space correspond to the directions on a map.

If two subsequent qualitative motion vectors (QMV's) are equal at our level of granularity, we represent this by incrementing a counter. This leads to shorter vector sequences. A description of a course of motion is a QMV sequence:

$$\langle D_1, O_1 \rangle^{i_1} \langle D_2, O_2 \rangle^{i_2} \ldots \langle D_n, O_n \rangle^{i_n}$$

e.g.

```
<0 0>1 <close east>5 <close north>2 <close west>3 <close south>1
<medium-dist south>1 <medium-dist east>1 <far east>1 <0 0>1
<far east>2 <close east>1
```

This example sequence was generated with our graphical input tool and its trajectory looked like depicted in figure 5 when drawn with the computer mouse.

Now we can abstract this from scale and orientation. To this end, we transform the absolute, deictic frame of reference into an relative, intrinsic one, i.e., in each starting point of a vector we only look at the change in direction and distance relative to the last vector.

---

[2] O like orientation; orientation of a movement in space = direction

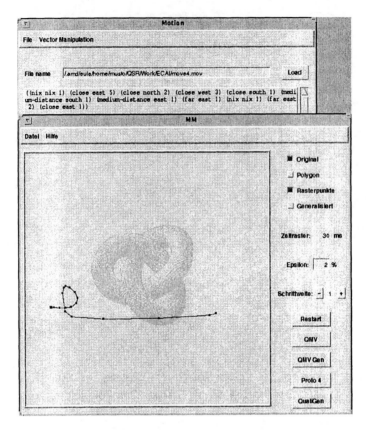

Fig. 5. Example sequence generated with the graphical input tool

An intrinsic frame of reference is not only useful for comparing QMV sequences independent from scale and orientation, but also for certain applications where locomotion has to be taken into account like robot motion planning. A transformation between the frames of reference is necessary where, e.g., explorative (intrinsic) data from locomotion has to be transformed into a (deictic) survey map. Our approach supplies full flexibility, since deictic and intrinsic frame of reference can be combined freely, i.e. direction can be represented intrinsic, and distance deictic, and so on. Furthermore, all transformations are reversible.

As quantity space for the intrinsic frame of reference, we can, e.g., use $D_I = \{\ll, <, \prec, =, \succ, >, \gg\}$, $O_I = \{\texttt{forward}, \texttt{backward}, \texttt{left}, \texttt{right}\}$. To achieve orientation independence, we define the direction of the first movement always as forward. The direction of the next movement will be in relation to the last one, as it is in route descriptions ("Go up to the shop, then turn left, at the third crossing turn right ...").

To transform the distances into relative distances, we first have to calculate really "absolute" distances from the distance information in the QMV and the counter information: if an object has moved **very-far** in one time step, it could have covered the same distance as an object that has moved **close** in ten subsequent time steps. The difference is only in the velocity: The first object has covered the distance much faster. This has to be resolved. If we don't want to lose the velocity information at this point, we must add a third component to our QMV, namely qualitative velocities. We can calculate the mapping of counter/distance-information into distance/velocity-information via a simple table look-up. <0 0>^{i}-vectors are ignored in all these transformations.

So the transformation of our example sequence looks like this:

<0 0>^{1} <close east>^{5} <close north>^{2} <close west>^{3} <close south>^{1}
<medium-dist south>^{1} <medium-dist east>^{1} <far east>^{1} <0 0>^{1}
<far east>^{2} <close east>^{1}

$\rightarrow$ transforming directions[3]:

<0 0>^{1} <close forward>^{5} <close left>^{2} <close left>^{3} <close left>^{1}
<medium-dist forward>^{1} <medium-dist left>^{1} <far forward>^{1} <0 0>^{1}
<far forward>^{2} <close forward>^{1}

$\rightarrow$ resolving distances and velocities:

<0 0>^{1} <medium-dist forward slow> <close left slow>
<close left slow> <close left slow> <medium-dist forward medium-vel>
<medium-dist left medium-vel> <far forward fast> <0 0>^{1}
<far forward fast> <close forward slow>

$\rightarrow$ transforming distances:

<0 0>^{1} <dist forward slow> <$\prec$ left slow> <$\approx$ left slow>
<$\approx$ left slow> <$\succ$ forward medium-vel> <$\approx$ left medium-vel>
<$\succ$ forward fast> <0 0>^{1} <$\approx$ forward fast> <$\ll$ forward slow>

Since the distance information in the first vector is irrelevant for a scale-independent representation, we substitute it with a meaningless value like "dist". Nevertheless, if we want these transformations to be reversible, we have to store the distance and direction information of the first vector. Pragmatically, we can do this in the first vecor itself and ignore the values when comparing intrinsic QMV sequences.

---

[3] Please note that the counters have a slightly different meaning in the orientation-independent representation: They don't indicate that the same vector is repeated i times, but that there was no change in distance and direction for i scan cycles.

Please note that the actual values in the QMV's are by no means fixed; what granularity will be chosen and what will be transformed to what may vary greatly depending on the application. The findings of our psychophysical experiments will indicate how the quantity spaces of time, space, velocity, etc. correspond to the real-world quantitative values for the purpose of modeling human cognitive capacity. Practical applications like robot motion planning may require different values, depending on the kind and size of the space the robot is moving in.

Propositional representation Using all the information we got from the different transformations of the QMV sequence, we can construct from our vectorial representation a propositional one which focuses mainly on the shape of the course of motion. This will hopefully turn out to be useful for detecting periodicity in the course of motion, which will be needed for prediction of further movements.

For the propositional representation, we need to identify a fixed vocabulary of movement shapes, like $S = \{straight\text{-}line, \ left\text{-}turn, \ right\text{-}turn, \ u\text{-}turn, \ loop, \ \ldots\}$. At this layer, a motion representation is a sequence of movement shapes and relations between the shapes:

$$\text{Shape}_1 \langle \text{rel}_{11}, \ldots \text{rel}_{1n} \rangle \text{Shape}_2 \langle \text{rel}_{21}, \ldots \text{rel}_{2n} \rangle \ldots \text{Shape}_m,$$

where $\text{Shape}_i \in S$.

The relations could describe differences in magnitude, orientation and velocity between the motion segments.

If $\{left\text{-}turn, \ right\text{-}turn, \ u\text{-}turn\text{-}left, \ u\text{-}turn\text{-}right, \ \ldots\}$ are contained in the movement vocabulary, we don't need any relational description of directional change, because any directional change could be modeled as a shape segment of its own. So, the only relation between the segments that is left will be change in magnitude and maybe change in velocity, although at the moment we are not sure whether we will need the velocity information at this level at all.

The QMV sequence in our example would look like this:

$$straight\text{-}line \ \langle \prec \rangle \ loop\text{-}left \ \langle \ll \rangle \ straight\text{-}line.$$

For the task of prediction, a propositional representation has to be built on the fly, and therefore a hierarchical construction of the propositional vocabulary is very useful. In our current approach, where we discriminate only four directions, we have four primitives at the propositional layer: *straight-line*, *left-turn*, *right-turn* and *direct-u-turn*, from which we construct other motion shapes like *loop-left* or *u-turn-right*.

How the layers are linked The crucial step in constructing a propositional representation from the vectorial one is the identification of meaningful subsequences in the QMV sequence, which correspond to the shapes in our propositional vocabulary. This involves the problems of generalization, segmentation and classification of a QMV sequence.

The propositional representation is constructed in three steps. First, the course of motion is generalized to smooth the curve and eliminate minimal irregularities. Then, the QMV sequence is segmented in pieces that stand for a straight line, a left turn and a right turn. The problem here is the determination of beginning and end of the turns. Last, the segments are combined and classified to more complex shapes.

To implement generalization, we first had to develop some means to calculate with QMV's, a QMV linear space which is described in (Stein 1998). The generalization algorithm works in this linear space and yields as a result a smoothed motion path that deviates only ϵ percent from the path of the original QMV sequence. ϵ is variable; a very big ϵ leads to a straight line. At the moment, we can only generalize the spatial component of the QMV sequence, so the dynamic aspect of the course of motion is ignored in the subsequent steps too. Integration of motion dynamics into generalization, segmentation and classification is a topic for future work.

To construct the shape based propositional representation from the vector based representation properly, we have to perform a segmentation of the QMV sequence. The basic step for this is to find out where one shape ends and the next begins. The input for the algorithm is a generalized QMV sequence, so we do not have to worry about minimal movements; all movements below a minimal size are filtered by generalization. Therefore we can take each turn in the QMV sequence to be a curve in the propositional representation. The problem is the size of this curve, we do not know where a curve starts and ends. This means that we would not know, for example, whether a QMV sequence has to be segmented into a *left-turn*, *straight-line* and a *left-turn*, or into an *u-turn-left*.

Our segmentation algorithm transforms a QMV sequence to the basic shapes straight line, 90° curves to right and left and direct u-turns in different sizes, where the size of a curve is its radius. More complex shapes are built by combination of these basic ones. See (Stein 1998) for further details.

The identification of the vocabulary for more complex shapes will be guided by experimental results.

Since we eventually want to use our representation also for prediction of movements, collision prediction, etc., the two layers have to be linked closely and on the fly. That means, the construction of the propositional representation must begin with the first motion we observe and must be revised with every further motion observed. Additionally, we have to keep track of what parts in each representation correspond to each other in order to be able to make prediction on a level of finer granularity than the propositional one. The advantage of the propositional representation for the task of prediction is that the more accurate representation is already pre-analyzed and parts of it are classified, so that it will be easier to detect patterns and periodicity.

Related Work Although motion processing is an important issue for human visual cognition, it has reached until now only little attention in the qualitative reasoning community. In the following, we give a short review of related work in the field, which is quite heterogeneous.

Logic based approaches like (Galton 1995; Del Bimbo, Vicario, and Zingoni 1992, 1993, 1995) see motion mostly as change of position over the course of time. A motion event is often represented only by the positions where it began and where it ended; the motion trajectory is not represented.

Constraint based approaches like (Kim 1992; Faltings 1987) describe mostly motion within a closed system: a machine, a system of linkages, etc. These approaches concentrate on configurations (of machine parts, of space, etc.) and allow for reasoning over the possibility of certain motion events (e.g. can this part of the machine move to the left?) or what consequences a certain motion will have (e.g. if I turn this wheel to the right, in what direction does another wheel turn?).

Forbus (1983), Forbus, Nielsen, and Faltings (1990) cover the topic of motion of point-like masses in space (FROB). Space is divided in "places", i.e. regions of equal character. Motion is change of position in this space, enhanced with a description of the kind of motion the object is performing (e.g. "fly", "collide"). Since the authors conjecture that motion cannot be represented solely by qualitative means, they combine the qualitative description with "metric diagrams", where metric information is represented. This representation of motion doesn't abstract from the environment, but depends on the properties of the "places" the object occupies: it makes a representational difference if the object passes over a slope or falls in a hole.

Hybrid approaches like (Mohnhaupt 1990; Fernyhough 1996; Fernyhough, Cohn, and Hogg 1997; Herzog and Rohr 1995) are more heterogeneous than the aforementioned categories. Fernyhough (1996), Fernyhough et al. (1997) use a similar qualitative approach as we do, but consider relations between moving objects as representational primitives.

Mohnhaupt (1990) constructs generic trajectories from dynamic scenes, which are represented in chain code. Then, descriptions are constructed which use perceptual primitives like position, velocity, acceleration as basic description units. There are some primitives which stay invariant for some specific motion events, and these are of interest for the generic model of this event.

From this model that uses analogue measures, a propositional representation is created. This representation is then used for event recognition, natural language communication, and long term storage. For generating the propositional representation, the generic model is segmented into "meaningful" subparts that can be described with predicates.

This two-layered approach was inspiring for our work and we stick with these two representational levels. Nevertheless, we do this in a different way than

Mohnhaupt (1990), because we use qualitative means for trajectory representation from the beginning and abstract from the environment. Furthermore, the "meaning" of meaningful subparts of the trajectory is not provided by the inherent meaning of the motion event (like "driving a car", where subparts can have the meaning of "bending off" or "overtaking"), but by the shape of the motion trajectory. This guarantees representational independence of the moving object.

Qualitative shape description Most approaches on qualitative shape description use representational primitives like regions or axes, that aren't suitable for our purposes because they don't allow for the representation of a motion trajectory, e.g., because they can only describe closed shapes or objects with at least 2-dimensional expansion.

The contour-based approaches in (Jungert 1993, 1996) could be more promising for our purposes, since we could use their representational primitives to describe a motion path, which can be seen as a contour (only that it needs not to be closed) and enhance them with primitives that can be used to describe motion dynamics. Unfortunately, these approaches depend rather closely on the availability of exact positional data.

Conclusion Some of these formalisms show much more quantitative aspects than the framework presented in this paper. Furthermore, most of these formalisms do not focus on the representation of the course of motion but on the representation of positional or configurational change.

The qualitative shape descriptions reviewed do not allow for the representation of motion trajectories, or if they do, use exact positional data.

None of the approaches deals with classification and similarity of courses of motion, so no measures of complexity or similarity can be found.

3.3 Similarity and Complexity

The respective results of psychophysical experimentation and computational modeling yielded some considerations on similarity and complexity measures for courses of motion. We will motivate this shortly considering some further experiments we plan to investigate the human observer's ability to use more complicated dynamic information of motion paths for discrimation.

As an example, a dot moving along a curved path will be presented, accelerating and decelerating in a previously fixed sequence. At each trial two such paths will be presented in succession at different locations on the screen while asking the observer whether the two presented paths are identical, irrespective of the different locations on the screen.

If the motion paths do not differ in global parameters (duration, average speed) but only in some local properties (acceleration at time t_i) and if they can be discriminated to some extent we conclude that it is the local path-information giving rise to the discriminability. By varying the complexity of the acceleration-deceleration sequences we will measure thresholds concerning the limits of using complex dyamic information.

The results of these experiments will indicate how a similarity measure for the qualitative representation of courses of motion that models human cognitive capacity must look like. We will illustrate this later with a similarity measure for QMV-sequences.

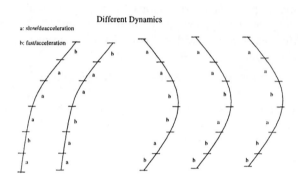

Fig. 6. Some examples of curved motion paths. All paths have the same average velocity but differ in how fast and slow pieces follow each other. The aim of this experiment is twofold: firstly, we want to determine thresholds for path discriminability depending on the time course of internal availability of velocity and velocity changes. Secondly, the thresholds we determine will enable us to ascertain the suprathreshold-level of changes of motion variables when we want to investigate their role for discrimination and prediction of motion with increased comlexity

To reach conclusions about the capabilities and especially the limits of the system a systematic variation of the complexity of the motion paths with respect to both its spatial as well as its temporal properties would be desirable. Based on the considerations in fig. 7, we can define a complexity measures for propositional represented motion trajectories.

Similarity Based on our generalization algorithm on QMV sequences we can define ϵ-**similarity**: Two QMV sequences are ϵ-similar if ϵ is the least factor that leads to the same QMV sequences applying generalization. That means that the two QMV sequences deviate at most ϵ percent. The courses of motion are the more similar, the smaller ϵ is. Since our first experiments indicate that discrimination of courses of motion is better if they originate from the same point, we take this into account with a translation component δ in our similarity measure:

Let a, b be two QMV sequences: $\text{SIM}(a, b) = \epsilon - \delta$, where ϵ is the similarity factor without regard of origin and δ is the translation component. This reflects

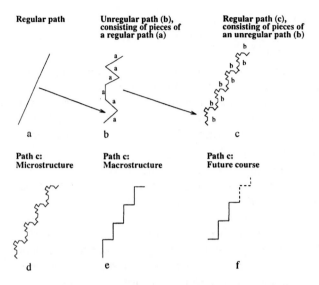

Fig. 7. Path "a" constitutes an elementary building block. Path "b", consisting of pieces of path "a" is unregular while path "c", constructed as a sequence of path-b-pieces, is regular. Note that the long-term history of a path can be arbitrarily complex. The construction principle of nesting motion information to gain insight in the quantitative temporal and spatial limits of motion processing is, on principle, not restricted to the "static" aspect of a motion path (shape). If higher order motion variables are a necessary prerequisite for recognizing spatio-temporal patterns which is an open question, it should be expected that acceleration information is also available in the spatio-temporal memory

that a and b are less discriminable, if δ is larger than zero. How δ has to be instantiated and the relation between ϵ and δ will be a topic of further research on discrimination thresholds with complex stimuli.

Complexity Besides the experimental questions of our study we are concerned with the exploration of meaningful measures of information or likewise complexity or redundancy of motion information under certain well defined conditions.

With well defined conditions we mean that one has to be careful in applying a complexity measure to a stimulus. Whereas in computer science, e.g., complexity is formally conceptualized and refers to computational complexity in space and time, in psychology and related fields the notion of complexity may comprise a variety of different meanings. For a behavioral biologist, e.g., task complexity might mean how challenging a procedure is for an animal to obtain a reward. The experimenter could determine the complexity by measuring the elapsed time or the number of erroneous trials before an animal finds a correct solution.

In psychopysics complexity refers mostly to stimulus complexity. Random dot patterns are often said to be complex because "meaningful" patterns can be perceived only when presented in a way that a matching process can succesfully extract something, by stereoscopic or temporal subsequent presentation (Braddick 1974; Julesz 1960).

A measure of complexity in our context has to be computed easily, must be useful and should express what we mean when we call a motion path more or less complicated. The notion of complexity plays a central role in the theory of dissipative systems but the measures of complexity used in this field are not easily to compute and furthermore don't exhibit direct relations to our problem.

The possibility of conceiving a path as a sequence of fast and slow subsections thereby forming a string where the characters of the string denote fast or slow pieces respectively has already been illustrated in Fig 6. This concept could possibly be extended by other properties like acceleration/deceleration or direction (see Fig. 7), by this forming a grammar of complexity with respect to temporal or dynamical properties of a course of motion. A string-based description of a motion path enables in principle to adopt the Kolmogorov complexity. The complexity of a path could as well be expressed in some way as result of stochastic process but its appliability has to be investigated further.

The basic problem of the formal qualitative approach to the description of motion paths is to abstract from the micro-structure of the motion path to extract a more general desription (forest vs tree-problem). While in the experimental part the shape of a path is well defined by its formal description, the qualitative description has to deal explicitly with the problem of detecting structural similarties within a path. This leads us to the conclusion that complexity measures are best applied to the propositional representation of the course of motion, because this representation abstracts from the microstructure, is already preanalyzed and classified according to certain similarity criteria. The complexity of a course of motion depends then on the number of different shapes it has and how regular the repetition of these shapes is. Additionally, single shapes can be weighted according to their intrinsic complexity, which depends on from how many primitive shapes they are constructed (cf. chapter 3.2).

4 Summary and Outlook

The spatio-temporal memory model developed so far relies on the assumption that a basic requirement for the representation of motion information is the mapping of the external course of time events into simultaneously available spatially distributed properties of these events. The model already enables to predict experimental results on early vision memory stages as the immediate memory span and partial report superiority. Furthermore it resolves inconsistencies with respect to backward masking arising with classical models like that to iconic memory (see also Schill and Zetzsche (1995a)).

In order to reach more detailed conclusions about the representation and processing of spatio-temporal information at early vision stages experiments have been provided. Here the question was the nature of the access to locus and direction information. As a first step we measured thresholds for the discrimination of the direction of moving dots separated in space and time. The results so far indicate that direction discrimination of different motion paths is not influenced by a temporal separation of more than 1000 ms. In contrast to that, a spatial separation of the two motion paths leads to a pronounced elevation of threshold. The higher threshold-level is also not influenced by the temporal separation. These results indicate a non-independency of the representation of direction and locus. A question that arises is, as to why the limited resources of the spatio-temporal memory do not influence the threshold from SOA's of 800 ms and higher. A possible explanation was given in terms of the coding efficiency of the subsequent memory stages. This might lead in the case of straight motion paths with constant speed and direction to a performance that is unaltered for a wide range of SOA's.

Our qualitative approach on motion representation and inference on motion has revealed that the combination of different representations is useful to realize different levels of abstraction, granularity and accuracy. Therefore we proposed a two-layered qualitative representation of motion, where one level provides rather detailed and accurate information and the other level reflects semantic properties of a course of motion, but is coarse and inaccurate. The model provides at the vectorial level the possibility to switch between deictic and intrinsic frame of reference, which may be important for future work with regard to application. Generation of a deictic representation out of an intrinsic one is important when, e.g., a survey map shall be computed from explorative data. For this purpose, some further representational means must be devolped that link the motion trajectory to its environment, e.g. via landmarks.

Combination of the findings of our psychophysical experiments and the qualitative modeling have yielded a similarity measure on QMV sequences that reflects that humans can discriminate motion trajectories less if they do not originate from the same point. Our work so far on the specification of the memory model has shown that a concept for describing the complexity of a course of motion is desirable. The availability of such a measure would allow for a variation of the experimental conditions as systematical as possible. We can define a complexity measure on our qualitative propositional representation that takes into account regularity and abstraction from the microstructure of the course of motion. However, we must determine whether this measure is also suitable for describing complexity with respect to the visual processing system. Experiments will reveal how the computational and the psychophysical notion of complexity correspond.

In our further research on the specification of the spatio-temporal model we will next determine the limits of detecting and discriminating spatial and temporal variations of a presented complex motion course. Results of these investigations, like e.g. the threshold to discriminate paths of moving objects will be incorporated in the qualitative model in that they will be used to constrain maximal granularity of the quantity spaces.

Variations of the duration with which the courses of motion are presented will be conducted to find out how the integration of spatio-temporal information is performed. The way how different time scales with possibly different spatio-temporal resolutions are linked is also of potential value for the definition of operators for concatenation and combination for the qualtitative modeling task. Another important next question will be to what extend we are able to deal with dynamic spatio-temporal information, where e.g. different parts of the shape of a course will have different velocities.

A topic for future research on the qualitative model will be the instantiation of the QMV representation with values like indicated from the psychophysical experiments. This will reveal whether this representation is suitable for the modeling of human cognitive capacity. Another question is whether an intrinsic representation of a QMV sequence should provide direct access to more antecedent vectors than only one to achieve a better memory span in accordance to our spatio-temporal memory model.

Applicational aspects like described above lead to a variety of further research topics, e.g. how other qualitative constraints like landmarks can be incorporated in the formalism.

Research was supported by a grant of the Deutsche Forschungsgemeinschaft (DFG, "Raumkognition") to W. Brauer, D. Hern'andez and K. Schill.

References

Adelson, E. and Bergen, J. (1985). Spatiotemporal energy models for the perception of motion. *Journal of the Optical Society of America*, A2, 284–299.

Anstis, S. M. (1986). Motion perception in the frontal plane. In Boff et al. (1986), chapter 16.

Del Bimbo, A., Vicario, E., and Zingoni, D. (1992). A spatio-temporal logic for image sequence coding and retrieval. In *Proceedings IEEE VL'92 Workshop on Visual Languages*, Seattle, WA.

Del Bimbo, A., Vicario, E., and Zingoni, D. (1993). Sequence retrieval by contents through spatio temporal indexing. In *Proceedings IEEE VL'93 Workshop on Visual Languages*, Bergen, Norway.

Del Bimbo, A., Vicario, E., and Zingoni, D. (1995). Symbolic description and visual querying of image sequences using spatio-temporal logic. *IEEE Transactions on Knowledge and Data Engineering*, 7(4), 609–622.

Boff, K., Kaufman, L., and Thomas, J., editors (1986). *Handbook of Perception and Human Performance*. John Wiley.

Braddick, O. (1974). A short-range process in apparent motion. *Vision Research*, 14, 519–527.

Cutting, J. E. and Proffitt, D. R. (1982). The minimum principle and the perception of absolute, common, and relative motion. *Cognitive Psychology*, 14, 211–146.

Faltings, B. (1987). Qualitative kinematics in mechanisms. In *Proceedings IJCA-87*, pages 436–442, Detroit.

Fernyhough, J., Cohn, A. G., and Hogg, D. C. (1997). Event recognition using qualitative reasoning on automatically generated spatio-temporal models from visual input. In *Proc. IJCAI-97 Workshop on Spatial and Temporal Reasoning*, Nagoya.

Fernyhough, J. H. (1996). Qualitative reasoning for automatic traffic surveillance. In *Proc. 10. International Workshop on Qualitative Reasoning*, pages 40–42. AAAI Press, Lake Tahoe, California.

Finke, R. A., Freyd, J. J., and Shyi, G. C.-W. (1986). Implied velocity and acceleration induce transformations of visual memory. *Journal of Experimental Psychology: General, 2*, 175–188.

Finke, R. A. and Shyi, G. C.-W. (1988). Mental extrapolation and representational momentum for complex implied motion. *Journal of Experimental Psychology: Learning, Memory, and Cognition, 14*(1), 112–120.

Forbus, K., Nielsen, P., and Faltings, B. (1990). Qualitative kinematics: A framework. In Weld, D. S. and de Kleer, J., editors, *Readings in Qualitative Reasoning about Physical Systems*, pages 562–567. Morgan Kaufmann Publishers Inc., San Mateo, California.

Forbus, K. D. (1983). Qualitative reasoning about space and motion. In Gentner, D. and Stevens, A. L., editors, *Mental Models*, pages 53–73. Lawrence Erlbaum, Hillsdale, NJ.

Freyd, J. J. (1987). Dynamic mental representations. *Psychological Review, 4*, 427–438.

Freyd, J. J. and Johnson, J. Q. (1987). Probing the time course of representational momentum. *Journal of Experimental Psychology, 13*(2), 259–268.

Galton, A. (1995). Space, time, and movement. Unpublished manuscript distributed at the Bolzano School on Spatial Reasoning.

Grzywacz, N. M., Harris, J. M., and Amthor, F. R. (1994). Computational and neural constraints for the measurement of local visual motion. In Smith and Snowden (1994), chapter 2, pages 19–50.

Herzog, G. and Rohr, K. (1995). Integrating vision and language: Towards automatic description of human movements. In *Proc. of the 19th Annual German Conference on Artificial Intelligence, KI-95*.

Hochberg, J. (1986). Representation of motion and space in video and cinematic displays. In Boff et al. (1986), chapter 22.

Johansson, G. (1975). Visual motion perception. *Scientific American, 232*(6), 76–88.

Johansson, G. (1976). Spatio-temporal differentiation and integration in visual motion perception. *Psychological Research, 38*, 379–393.

Julesz, B. (1960). Binocular depth perception of computer generated patterns. *Bell Syst. tech. J., 39*, 1125–1126.

Jungert, E. (1993). Symbolic spatial reasoning on object shapes for qualitative matching. In Frank, A. U. and Campari, I., editors, *Spatial Information Theory. A Theoretical Basis for GIS. European Conference, COSIT'93*, Marciana Marina, Italy, Volume 716 of *Lecture Notes in Computer Science*, pages 444–462. Springer, Berlin,Heidelberg,New York.

Jungert, E. (1996). A qualitative approach to recognition of man-made objects in laser-radar images. Unpublished manuscript.

Kim, H.-K. (1992). Qualitative kinematics of linkages. In Faltings, B. and Struss, P., editors, *Recent Advances in Qualitative Physics*, pages 137–151. The MIT Press, Cambridge, MA.

Lu, Z. and Sperling, G. (1995). The functional architecture of human visual motion perception. *Vision Research*, *35*(19), 2697–2722.

McKee, S. P. and Watamaniuk, S. N. J. (1994). The psychophysics of motion perception. In Smith and Snowden (1994), chapter 4, pages 85–114.

Mohnhaupt, M. (1990). Eine hybride Repräsentation von Objektbewegungen: Von analogen zu propositionalen Beschreibungen. In Freksa, C. and Habel, C., editors, *Repräsentation und Verarbeitung räumlichen Wissens*. Springer, Berlin.

Nakayama, K. (1985). Biological Image Motion Processing: A Review. *Vision Research*, *25*(5), 625–660.

Peterken, C., Brown, B., and Bowman, K. (1991). Predicting the future position of a moving target. *Perception*, *20*, 5–16.

Poizner, H., Bellugi, U., and Lutes-Driscoll, V. (1981). Perception of american sign language in dynamic point-light displays. *Journal of Experimental Psychology: Human Perception and Performance*, *7*, 430–440.

Pöppel, E. (1997). A hierarchical model of temporal perception. *Trends in Cognitive Sciences*, *1*(2), 312–320.

Pöppel, E. and Schill, K. (1995). Time perception: Problems of representation and processing. In Arbib, M. A., editor, *The Handbook of Brain Theory and Neural Networks*, pages 987–990. The MIT Press.

Reichardt, W. (1957). Autokorrelations-auswertung als Funktionsprinzip des Zentralnervensystems. *Zeitschrift für Naturforschung*, *12b*, 448–457.

Rock, I. (1975). *An Introduction to Perception*. Macmillan, New York.

Rosenbaum, D. A. (1975). Perception and extrapolation of velocity and acceleration. *Journal of Experimental Psychology: Human Perception and Performance*, *1*(4), 395–403.

Schill, K. and Zetzsche, C. (1995a). A model of visual spatio-temporal memory: The icon revisited. *Psychological Research*, *57*, 88–102.

Schill, K. and Zetzsche, C. (1995b). Why some masks do not mask: Critical reevaluation of a standard psychophysicical paradigm. In *Perception*, Volume 24, page 23.

Schill, K., Zetzsche, C., Brauer, W., Eisenkolb, A., and Musto, A. (1998). Visual representation of spatio-temporal structure. In Rogowitz, B. and Papas, T., editors, *Human Vision and Electronic Imaging*. Proceedings of SPIE, 3299. to appear.

Shepard, R. N. (1983). Path-guided apparent motion. *Science*, *220*, 632–634.

Smith, A. T. and Snowden, R. J., editors (1994). *Visual Detection of Motion*. Academic Press, San Diego.

Stein, K. (1998). Generalisierung und Segmentierung von qualitativen Bewegungsdaten. Master's thesis, TU München.

Van Doorn, A. J. and Koenderink, J. J. (1982a). Spatial properties of the visual detectability of moving spatial white noise. *Journal of Experimental Brain Research, 45*, 189–195.

Van Doorn, A. J. and Koenderink, J. J. (1982b). Temporal properties of the visual detectability of moving spatial white noise. *Experimental Brain Research, 45*, 179–182.

van Santen, J. P. H. and Sperling, G. (1985). Elaborated reichardt detectors. *J. Opt. Soc. Am. A, 2*(2), 300–320.

Verfaillie, K. and d'Ydewalle, G. (1991). Representational momentum and event course anticipation in the perception of implied periodical motions. *Journal of Experimental Psychology, 17*(2), 302–313.

Watson, A. B. and Ahumada, A. J. (1985). Model of human visual-motion sensing. *J. Opt. Soc. Am., 2*, 322–342.

Werkhoven, P., Snippe, H. P., and Toet, A. (1992). Visual processing of optic acceleration. *Vision Research, 32*(12), 2113–2329.

How Space Structures Language[1]

Barbara Tversky and Paul U. Lee

Stanford University Department of Psychology, Bldg. 420
Stanford, California 94305-2130

Abstract. As Talmy has observed, language schematizes space; language provides a systematic framework to describe space, by selecting certain aspects of a referent scene while neglecting the others. Here, we consider the ways that space and the things in it are schematized in perception and cognition, as well as in language. We propose the Schematization Similarity Conjecture: to the extent that space is schematized similarly in language and cognition, language will be successful in conveying space. We look at the evidence in both language and perception literature to support this view. Finally, we analyze schematizations of routes conveyed in sketch maps or directions, finding parallels in the kind of information omitted and retained in both.

1 Introduction

Language can be effective in conveying useful information about unknown things. If you are like many people, when you go to a new place, you may approach a stranger to ask directions. If your addressee in fact knows how to get to where you want to go, you are likely to receive coherent and accurate directions (cf. Denis, 1994; Taylor and Tversky, 1992a). Similarly, as any Hemingway reader knows, language can be effective at relating a simple scene of people, objects, and landmarks. In laboratory settings, narratives relating scenes like these are readily comprehended. In addition, the mental representations of such scenes are updated as new descriptive information is given (e. g., Glenberg, Meyer, and Lindem, 1987; Morrow, Bower and Greenspan, 1989). Finally, times to retrieve spatial information from mental representations induced by descriptions are in many cases indistinguishable from those established from actual experience (cf. Franklin and Tversky, 1990; Bryant, Tversky, and Lanca, 1998). Contrast these successful uses of language with another one. You've just returned from a large party of both acquaintances and strangers. You try to describe someone interesting whom you met to a friend because you believe the friend knows this person's name. Such descriptions are notoriously poor. In fact, in some situations, describing a face is the surest way to reduce memory for it (Schooler and Engstler-Schooler, 1991). Why is it that language is effective for conveying some sorts of spatial information but not others?

---

[1]We are grateful for the insightful comments of an anonymous reviewer.

The answer may lie in the way that language structures space. In 1983, Leonard Talmy published an article with that title which has rippled through cognitive psychology and linguistics like a stone skipped on water. In it, he proposed that language "schematizes" space, selecting "certain aspects of a referent scene...while disregarding the remaining aspects."(p. 225). For example, a term like "across" can apply to a set of spatial configurations that do not depend on exact metric properties such as shape, size, and distance. Use of "across" depends on the global properties and configuration of the thing doing the crossing and the thing crossed. Ideally, the thing doing the crossing is smaller than the thing being crossed, and it is crossing in a straight path perpendicular to the length of the thing being crossed. Thus schematization entails information reduction, encoding certain features of a scene while ignoring others. Talmy's analysis of schematization focused on the fine structure of language, in particular, closed-class terms, and less on the macroscopic level of sentences, paragraphs and discourse that uses a language's large set of open-class lexical items as elements. Closed-class grammatical forms include "grammatical elements and categories, closed-class particles and words, and the syntactic structures of phrases and clauses." (p. 227). Despite their syntactic status, they express meanings, but only limited ones, including space, time, and perspective, important to the current issues, and also attention, force, causation, knowledge state, and reality status. Because they appear across languages, they are assumed to reflect linguistic, hence cognitive, universals.

Not only language, but also perception and conception, which Talmy has collectively called 'ception, schematize space and the things in it (Talmy, 1996). In the following pages, we first examine how 'ception schematizes. Then, we go on to examine how the schematization of 'ception maps onto language. There is no disputing that language is a powerful clue to 'ception, that many of the distinctions important in 'ception are made in language, some in closed-class terms, others in lexical items. Yet, there are notable exceptions. As observed earlier, people are poor at describing faces, though excellent at recognizing them, a skill essential for social interaction. In contrast, routes and scenes are more readily conveyed by language despite the fact that, like faces, routes and scenes consist of elements and the spatial relations among them. Here, we propose a conjecture, the Schematization Similarity Conjecture: To the extent that language and 'ception schematize things similarly, language will be successful at communicating space.

To understand how 'ception schematizes space is to understand that perception is not just bottom-up, determined by the stimulus input alone, but is in addition top-down, conditioned by what is already in the mind, momentarily and longterm. Therefore, any generalizations based on schematizations of space necessarily lead to oversimplifications. One of these is ignoring context. It has long been clear, but is sometimes overlooked, that how people perceive of, conceive of, and describe a scene is deeply affected by a wealth of nonindependent factors, including what they are thinking, how they construe the scene, the goals at hand, past experience, and available knowledge structures.

Despite the fact that language and 'ception always occur in a context, there seem to be levels of schematization that hold over many contexts. People do not reinvent vocabulary and syntax at every encounter. If they did, communication would not be possible. Schematization in language and in 'ception is always a compromise; it must be stable enough for the general and the venerable, yet flexible enough for the specific and the new. In the following sections, we will review the existing research on how both 'ception and language schematize space and objects in it, abstracting certain features and ignoring others. This review of schematization will be schematic itself. It will be an attempt to give the "bottom line," the general aspects of objects and space most critical to our understanding of them. The evidence comes from many studies using different techniques and measures, that is, different contexts. Some of this evidence rests on language in one way or another. Ideally, evidence based purely on perception could be separated from evidence resting on language in order to separate the schematization of perception alone from that influenced by language. But this is probably not possible. For one thing, using non-linguistic measures is no guarantee that language is not implicitly invoked. With these caveats in mind, let us proceed to characterize how the things in the world and the spatial relations among them are schematized.

2   Figures, Objects, Faces

When we look at the world around us, we don't see it as a pattern of hues and bright-nesses. Rather, we perceive distinct figures and objects. For human perceivers, then, space is decomposed into figures and the spatial relations among them, viewed from a particular perspective. Similarly, figures can be decomposed into their parts and the spatial relations among them. Our experience of space, then, is not abstract, of empty space, but rather of the identity and the relative locations of the things in space.

2.1   Figures

There are two major questions in recognition of the things in space. First, how do we get from retinal stimulation to discernment of figures? This is the concern of the *Figures* section. Next, how do we get from a view-dependent representation to a view-independent representation? This is the concern of the *Objects* section. One of the earliest perceptual processes is discerning figures from background (e. g., Hochberg, 1978; Rock, 1983). Once figures are identified, they appear closer and brighter than their backgrounds. In contrast to grounds, figures tend to have closed contours and symmetry, so the Gestalt principles of figurality, including continuity, common fate, good form, and proximity, all serve as useful cues. Thus, the eye and the brain look for contours and cues to figurality in pursuit of isolating figures from grounds. An-other way to put this is that figures are schematized as contours that are likely to closed and likely to be symmetric.

Language for Figures. The distinctions that Talmy elucidates begin with figure and ground. Talmy borrows these terms from their use in perception and Gestalt psychology described above. Just as perception focuses on figures, so does language, according to Talmy. He argues that language selects one portion of a scene, the figure, as focal or primary, and describes it in relation to another portion, the ground, and sometimes in addition in relation to a third portion of the scene. We say, for example, "the horse is by the barn" or "the horse is near the trough in front of the barn." The figure is conceived of as geometrically simpler than the ground, often only as a point. It is also usually smaller, more salient, more movable, and more recent than the ground, which is more permanent and earlier. Although the ground is conceived of as geometrically more complex than the figure, the ground, too, is schematized, as indicated in English by prepositions, a closed-class form. For example, "at" schematizes the ground to a point, "on" and "across" to a two-dimensional surface, "into" and "through" to a three-dimensional volume.

A comparison between 'ception and language of figures shows a number of similarities and differences. Both divide the world into figures and ground, introducing asymmetries not present in the world per se. In 'ception, figures appear closer and brighter than grounds, becoming more salient. In language, figures are the primary objects currently salient in attention and discourse. Nevertheless, the object that is figural in perception may not be figural in language. An example comes from unpublished eye movement data collected by Griffin (Z. Griffin, 1998, personal communication). In scanning a picture of a truck about to hit a nurse, viewers fixate more on the truck, as the agent of the action. Yet, the nurse is the figure in viewers' descriptions of the scene. In addition, figures in 'ception are conceived of as shapes with closed contours and often symmetric, yet in language, they are often reduced to a point in space.

2.2 Objects

The human mind does not seem content with simply distinguishing figures from grounds; it also identifies figures as particular objects. But objects have many identities. What we typically sit on can be referred to as a desk chair, or a chair, or a piece of furniture. Despite the possibilities, people are biased to identify objects at what has been called the "basic" level (e.g., Brown, 1958; Murphy and Smith, 1982; Rosch, 1978). This is the level of chair, screwdriver, apple, and sock rather than the level of furniture, tool, fruit, and clothing, or the level of easy chair, Phillips-head screwdriver, delicious apple, and anklet. This is the level at which people seem to have the most information, indexed by attribute lists, relative to the number of alternative categories that must be kept in mind.

Many other cognitive operations also converge at the basic level. It is the level at which people are fastest to categorize instances (Rosch, 1975), the level fastest to identify (Murphy and Smith, 1982), the level people spontaneously choose to name, the highest level of abstraction for which an outline of overlapped shapes can be recognized, the highest level for which there is a common set of behaviors, and more (Rosch, 1978; Rosch, Mervis, Gray, Johnson, and Boyes-Braem, 1976). The basic level, then, has a special status in perception, in behavior, and in language (Tversky,

1985; Tversky and Hemenway, 1984). Rosch (1978) suggested that the natural breaks in labeling are based in the natural breaks in objects as we perceive them given our perceptual apparatus. Features of objects are not uniformly distributed across classes of objects. Instead, features of objects are correlated, that is, things that have feathers and beaks also lay eggs and fly.

The natural level for identifying objects, then, is the basic level. Arriving at view-independent representations of objects requires more than the visual input alone; it also requires some more general knowledge about the objects in question (e. g., Marr, 1982). As for figures, contour and symmetry characterize particular objects, but with greater specificity. Basic objects, such as couches and socks, can be recognized from a set of overlapping instances, standardized for size and viewpoint (Rosch, et al., 1976). Shapes of different kinds of socks are quite similar, but quite different from shapes of other objects even from the same category, such as shirts or ties. Furthermore, objects are most easily recognized when they are viewed from a canonical orientation, upright, and typically 3/4 view (Palmer, Rosch, and Chase, 1981). This view is one that presents the greatest number of features characteristic of the object. In many cases, those characteristic features are parts of the object (Biederman, 1987; Tversky and Hemenway, 1984); the greater the number of object parts detectable, the easier the identification of the object (Biederman, 1987). Parts have a dual status in cognition. On the one hand, they are perceptually salient as they are rooted in discontinuities of object shape (e. g., Biederman, 1987; Hoffman and Richards, 1984). On the other hand, different parts have different functions and serve different purposes to humans (Tversky and Hemenway, 1984). Parts are at once components of perception and components of function and facilitate inferences from appearance to behavior. Symmetry, too, is used to identify specific objects. Viewers interpret asymmetric nonsense figures as upright, off-center views of symmetric objects (McBeath, Schiano, and Tversky, 1997). 'Ception, then, schematizes specific figures, that is, objects, as shapes, composed of parts, and most likely upright and symmetric.

Language for Objects. Objects are typically named by open-class terms, thus not considered by Talmy. Perhaps individual objects are not an inherent part of the structure of language because there are so many of them and many of those are context specific. The place-holder for individual objects, nouns or subjects, is, of course, part of language structure as are various operations on them, such as pluralizing. Nevertheless, there are clues to way objects are conceived in the ways that names for objects are extended. Shape seems to be a primary basis for categorization as well as for extension of object terms, in both children's "errors" and adults' neologisms (Clark, 1973; Clark and Clark, 1979; Bowerman, 1978a, 1978b). There are old examples, like "stars" and "hearts" that are not really shaped like stars or hearts. And there are new examples, such as the body types loved by cardiologists--"pear-shaped"--and that disparaged by cardiologists--"apple-shaped,"--affectionately called simply "pears" and "apples."

2.3 Faces

Faces are a special kind of object in several ways. Recognition of faces is most typically at the level of the individual, not at the level of the class. For example, when we talk about identifying or recognizing a face, we mean recognizing that a specific face is the current president of the United States and not his brother. In contrast, when we talk about recognizing an object as a chair, we're usually not concerned with whose chair or even what type of chair. Of course, we need to identify some objects other than faces at the level of the individual. But identifying my house or car or jacket is facilitated by features such as locations or color or size, and such features may not facilitate identifying specific faces. Faces, in addition, are not integral objects in and of themselves, they are parts of other objects, human or otherwise. Recognizing faces is dependent on internal features, not just an outline shape. This is why we see faces not only in the moon, which has the proper outline, but also in cars, which do not. Furthermore, the features need to be in the proper configuration. Changing the overall configuration leads to something that is not a face, and even altering the relative distances among properly configured features diminishes resemblance substantially (cf. Bruce, 1988). For identifying individuals, in addition to configuration of features, the shapes of component features are also important, and those shapes are not regular. Similar to objects, 3/4 views are best recognized in faces (e. g., Hagen & Perkins, 1983; Shapiro & Penrod, 1986), perhaps because a 3/4 view gives better information about important component features, such as shape of nose, chin, and forehead. Even more than for objects, orientation is important in faces; upside down faces are considerably harder to recognize than right side up (e. g., Carey and Diamond, 1987; Yin, 1969). Turning objects upside down seems to be more disruptive to objects with irregular internal features such as faces than to objects with horizontal and vertical internal features like houses. Schematization of individual faces, then, is far more precise, entailing orientation as well as configuration and shapes of internal features.

Language for Faces. As noted earlier, faces are often perceived at the level of the individual. Similarly, they are referred to by open class terms, that is, names of individuals. In contrast to names for objects, when names of individuals are extended, it is typically personality traits or personal history that is extended, not shape as for objects, in fact, not appearance at all (cf. Clark and Clark, 1979). Identifying faces requires 'ception of subtle spatial relations among the parts (e. g., Bruce, 1988). Language, however, schematizes spatial relations in cruder categories, such as above, below, front, back, near, far, between, and among. Finer distinctions can be made but in the technical language of measurement. In addition, estimates of fine measurement are frequently unreliable (e. g., Leibowitz, Guzy, Peterson, and Blake, 1993). Thus, the puzzle that language is adequate for conveying routes but inadequate for describing faces is solved. The spatial relations usually needed for getting around are readily captured by language but the subtle spatial relations needed for identifying faces are not readily schematized by language.

2.4 Summary of Figures, Objects, and Faces

For detecting figures, contour (especially closed contour) and symmetry are among the diagnostic features. For objects which are figures identified at the basic level, specific contours or shapes that are decomposable into parts are characteristic, along with orientation and symmetry. For faces which are parts of objects identified at the level of an individual, the internal configuration and shapes of features is critical, in addition to orientation and symmetry. Returning to language, note that figures and objects are named by open-class terms, as are grounds. These refer to classes of things, and, interestingly, are sometimes extended to refer to shapes (as in "pear-shaped"). Faces, by contrast, are called by names that refer to individuals, not classes, much like street addresses, and that have no perceptual interpretation other than reference to the individual. Names for objects and faces, though less schematized than closed-class terms, are nevertheless schematized. A table is a table regardless of point of view, of color, of material, of location, to a large extent of size. This is not to say that people cannot or do not remember individual objects with their specific features and locations, but that people generally think about and refer to objects more abstractly.

3 Spatial Relations

Thus far, we have discussed the elements in space and their schematization in 'ception and language. Knowledge and schematization of space also entail the spatial relations among elements. In fact, we observed that entities can be decomposed into parts and the spatial relations among them, and that as entities are identified at more specific levels, the spatial relations among the parts become more critical. In this section we turn to the schematization of spatial relations. In perceiving a scene, figures are not just discerned and identified, they are also located. Figures are not located in an absolute way, but rather relative to other reference figures and/or a frame of reference. We note, for example, that we left the car by a particular street sign or that we buried the family heirlooms in the middles of a circle of trees. Locating figures relatively makes sense if only because perception of a scene is necessarily dependent on a particular viewpoint, yet a view-independent representation of a scene is desirable in order to recognize a scene or object from other viewpoints. Reference objects and reference frames serve to schematize the locations of figures. Memory for orientations and locations of dots, lines, or figures is biased toward reference objects or frames (e. g., Howard, 1982; Huttenlocher, Hedges, and Duncan, 1991; Nelson and Chaiklin, 1980; Taylor, 1961; Tversky, 1981).

How are reference objects and frames selected? Proximity, salience, and permanence are influential factors (Tversky, 1981; Tversky, Taylor, and Mainwaring, 1997). Domain, semantic, and pragmatic factors, such as current goals and recent experience, can also affect choice. Reference objects are other figures in the same scene as the target object whereas reference frames tend to surround the scene, the set of figures, in some way.

Natural borders and axes often serve as reference frames, such as the sides of a room, the sides of a piece of paper, the land and the sky. Horizontal and vertical lines or planes are privileged as reference frames, whether actual or virtual, as in the sides of a page or map at odd orientations. Acuity is better for horizontal and vertical lines, as is memory, and both perception and memory are distorted toward them (see Howard, 1982 and Tversky, 1992 for reviews). Horizontal and vertical lines are relatively easy for children to copy, but diagonal lines cause difficulties and are drawn toward horizontal and vertical (Ibbotson and Bryant, 1976). The human body, especially one's own, also serves as a natural reference object. The projections of the natural horizontal and vertical axes of the body, head/feet, front/back, and left/right, are a privileged reference frame, with certain of the axes more accessible than others, depending on body posture and viewpoint (e. g., Bryant, Franklin and Tversky, 1992; Bryant, Tversky and Lanca, 1998; Franklin and Tversky, 1990). Regions defined by the axes also vary depending on viewpoint; for example, for self, front is larger than back, and both are larger than left, and right, but not for other.(Franklin, Henkel and Zangas, 1995).

3.1 Schematization and Language of Spatial Relations

Spatial relations, then, are schematized toward reference objects and frames, especially horizontal and vertical planes. Spatial relations are frequently but not always referred to by closed-class terms, prepositions, such as "at," "on," and "in," or "in front of," "on top of," "across," "near," "between," and "parallel to." The schematization of closed-class elements is topological, according to Talmy. It abstracts away the metric properties of shape, size, angle, and distance, distinctions that are normally expressed in lexical elements. Talmy's analyses have been extended by others, especially in the direction of examining the topological constraints underlying prepositions, that is, the expression of spatial relations between a figure and a ground (e. g., Herskovits, 1986, Lakoff, 1986; Landau and Jackendoff, 1993; Vandeloise, 1986). Some languages, however, don't have prepositions. Even in English, which does, open class terms also describe spatial relations, as in "support," "hold," "lean," or "approach."

The scene alone does not determine how it is schematized to spatial relations, though it is often presupposed that the perceptual array is primary (e. g., Carlson-Radvansky and Irwin, 1993; Logan and Sadler, 1996; Hayward and Tarr, 1995). The speaker's perspective, intent, and goals, as well as cultural practices, are some of the influences on schematization. The interpretation of the scene in light of current goals and cultural practices are among the influences on selection of spatial relation terms. As Talmy (1983) noted, we can go "through" or "across" a park, and get "in" or "into" a car. Appropriateness of words like "near" or "approach" depend on the nature of the figure and the ground (Morrow and Clark, 1988). What's more, abstract uses of prepositions depend entirely on functional, not spatial relations, as in "on welfare" or "in a bad mood" (Garrod and Simon, 1989). Even spatial uses have a functional basis. One can say "the pear is in the bowl" where the expression is even though in fact the pear is outside the bowl on top of a pile of fruit. This is because the pear's location is controlled by the location of the bowl (Garrod and Simon, 1989). Although the qualities of schematization of spatial relations in both language and perception are similar, the open-class terms that are used to refer to figures preserve far more detailed spatial information than the terms used to refer to spatial relations. Moreover, although memory for spatial location and orientation is biased toward reference frames and objects, it does not coincide with them. The schematization of the language of spatial relations may be in the same directions as the schematization of 'ception of spatial relations, but it is far more extreme.

4   Motion

Figures in space are not necessarily static, nor are viewers. Perceiving and conceiving of motion are needed from the beginning of life, and, in fact, motion in concert is another clue to figurality (e. g., Spelke, Breinlinger, Macomber, and Jacobson, 1992). Perceiving motion accurately is not a simple matter. For example, generations of paintings of horses galloping have portrayed their legs in impossible configurations. When motion is relatively simple, as in the path of a pendulum or a falling object, people are able to recognize correct and incorrect paths of motion. Yet, some people correctly recognizing paths of motion may nevertheless produce incorrect paths, indicating flawed conceptions of motion (Kaiser, Proffitt, Whelan, and Hecht, 1992). Although motion is continuous, people seem to conceive of it as sequences of natural chunks (Hegarty, 1992). And although motion is continuous, people tend to conceive of it hierarchically (e. g., Newtson, Hairfield, Bloomingdale, and Cutino, 1987; Zacks and Tversky, 1997). As for objects, there seems to be a preferred or basic level, the level of going to a movie (Morris and Murphy, 1990; Rifkin, 1985). Although more can be said about actions and events, we focus here on schematization of motion in 'ception and language.

4.1   Schematization of Motion

Many aspects of motion, such as frequency and causality, are carried by closed-class terms (Talmy, 1975, 1983, 1985, 1988), yet other aspects of motion are referred to by open-class terms, particularly verbs. Verbs vary notoriously within and across languages as to what features they code (e. g., Gentner, 1981; Huttenlocher and Lui, 1981; Talmy, 1975, 1985, 1988). For example, some languages like English regularly encode manner of motion in verbs, as in "swagger," "slink," "slide," and "sway," others primarily encode path in verbs, as in "enter," "exit," and "ascend" (Talmy, 1985). Choice of verb is open to construal. The same perceptual sequence, such as leaving a room may be described in many different ways (Gentner, 1981), such as "went," "raced," "stumbled," "cried," "got chased," "got pushed," or "escaped" out the door. Although activities, like objects, are conceived of hierarchically, descriptors of activities are not necessarily organized hierarchically. Huttenlocher and Lui (1981) have argued that verbs, in contrast to the nouns used to refer to objects, are organized more as matrices than as hierarchies.

166

Like figures, motion can be schematized at various levels of specificity. The simplest way of thinking about motion is the path of an entire figure, a point moving in space. Like objects, paths are perceived in terms of frames of reference and distorted toward them. Just as in locating objects, in perceiving paths of motion, horizontal and vertical coordinates often serve as a reference frame (e. g., Pani, William, and Shippey, 1995; Shiffrar and Shepard, 1991). A more complex level of schematization than a path of motion is a pattern of parts moving in relation to one another. This level is analogous to schematizing an object as a configuration of parts. It is the level of understanding of pulleys (Hegarty, 1992) or gears (Schwartz and Black, 1996) or of distinguishing walking from running, which people readily do from patch-light displays (e. g., Cutting, Proffitt, and Kozlowski, 1978; Johansson, 1975). Yet another level of schematization is manner of motion, as in distinguishing modes of walking, such as swaggering or slinking.

5 Route Directions and Maps

The simplest schematization of motion to a path or route is readily encoded in language (e. g., Denis, 1994; Levelt, 1982; Linde and Labov, 1975; Klein, 1982; Perrig and Kintsch, 1985; Talmy, 1975; Taylor and Tversky, 1992a, 1992b, 1996; Wunderlich and Reinhelt, 1982). Routes are schematized as a point changing direction along a line or a plane, or as a network of nodes and links. Though by no means identical with perceptual or conceptual schematization, route maps can be regarded as schematizations that are closer to externalizations of perceptions than descriptions. Depictions of routes use spatial relations on paper to represent spatial relations in the world. Moreover, they can use iconic representations of entities in the world to represent those entities. Routes, then, can be externally represented as descriptions or depictions. Like route directions, route maps are commonly used to convey how to get from A to B. Which is better seems to depend on the specifics of the navigation task (e. g., Streeter, Vitello, and Wonsiewicz, 1985; Taylor, Naylor, and Chechile, in press; Taylor and Tversky, 1992a). Both route directions and route maps, then, seem adequate to convey information sufficient for arriving at a destination. We were interested in whether descriptions and depictions of routes schematize them similarly.

To get at this question, we approached students outside a campus residence and asked them if they knew how to get to a popular off-campus fast-food restaurant. If they did, we handed them a piece of paper, and asked them to either write down the directions or sketch a map. We obtained a total of 29 maps and 21 directions. Sample descriptions appear in Table 1 and sample maps in Figures 1. Note that route maps differ from other kinds of sketch maps in that they contain only the paths and landmarks relevant to the specific route. Following Denis (1994), we broke down the depictions and descriptions into segments consisting of four elements each: start point, reorientation (direction), path/progression, and end point. As the paths are continuous, the start point for one segment served as the start point for the next. In this situation, the segments corresponded to changes of direction (action) in the route. It would be possible to have segments separated by, say, major intersections or landmarks without changes in direction, but this did not happen in this corpus. Because the sketch maps, unlike street maps, contained very little information about the environment not directly related to the path, it was not difficult to segment the maps. As defined, each segment contains sufficient information to go from node to node. Together, these segments contain the information essential to reach the destination. Two coders coded the maps and descriptions for these categories of information and for categories of supplementary information. They first coded a subset of the protocols, and after reaching agreement on those, coded the rest separately.

Table 1. Examples of Route Directions

DW 9
From Roble parking lot
R onto Santa Theresa
L onto Lagunita (the first stop sign)
L onto Mayfield
L onto Campus drive East
R onto Bowdoin
L onto Stanford Ave.
R onto El Camino
go down few miles. it's on the right.

BD 10

Go down street toward main campus (where most of the buildings are as
opposed to where the fields are) make a right on the first real street
(not an entrance to a dorm or anything else). Then make a left on the
2nd street you come to. There should be some buildings on your right
(Flo Mo) and a parking lot on your left. The street will make a sharp
right. Stay on it. that puts you on Mayfield road. The first
intersection after the turn will be at Campus drive. Turn left and stay
on campus drive until you come to Galvez Street. Turn Right. go down
until you get to El Camino. Turn right (south) and Taco Bell is a
few miles down on the right.

BD 3

Go out St. Theresa
turn Rt.
Follow Campus Dr. way around to Galvez
turn left on Galvez.
turn right on El camino.
Go till you see Taco Bell on your Right

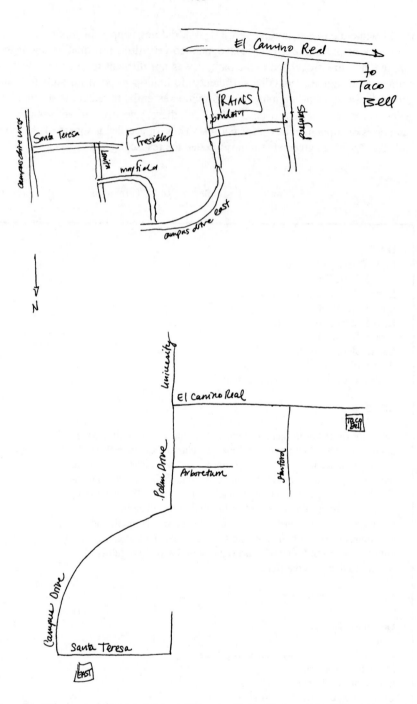

Figure 1. Examples of Route Maps

5.1 Essential Information in Descriptions and Depictions

Not all of the information included in both maps and directions fit into the essential four categories. In fact, 91% of the people giving directions and 90% of the people sketching maps added some information in addition to the start and end points, reorientation and path/progression. The additional information for maps included cardinal directions, arrows, distances, and extra landmarks. That same kind of information was added to directions. In addition, some directions also contained detail describing the landmarks and paths. This information, while not essential, may be important for keeping the traveler confidently on track. It anticipates that travelers may become uneasy when there is a relatively long distance without a change of orientation or distinguishing feature or when there is uncertainty about the identity of a landmark. The descriptions obtained by Denis (1994) and by Gryl (1995) had the same character.

5.2 Schematizing in Descriptions and Depictions

Not only did the same critical and supplementary information constitute the majority of content in route descriptions and depictions but also that information was represented in parallel ways. For both, start points and end points were landmarks, paths, buildings, fields, intersections, and the like. For maps, these were often presented as icons, typically schematized as rough geometric shapes, and often named, such as street or building names. Reorientations or turns were also schematized. In maps, they were typically portrayed as lines that were more or less perpendicular. About half the participants used arrows to explicitly indicate direction. Nearly half used double lines to indicate paths, though single lines predominated. There are at least two ways to interpret the use of double lines to indicate streets. The double lines could be iconic, as streets have width. Alternatively, they could indicate a perspective on the scene, conceiving of paths as planes rather than lines. In directions, there was a limited vocabulary and a limited structure, with slots for actions (verbs), directions, and paths. The common actions were the verbs "turn," "take a," "make a, "and "go." The verb was omitted in some descriptions, especially those that were simply a list of the form: left on X, right on Y. Thus, in both maps and directions, changes of orientation were schematized as turns of unspecified angles. In maps, they were depicted as approximate right angles irrespective of their actual angle. Memory for intersections is also biased toward right angles (e. g., Byrne, 1979; Moar and Bower, 1983; Tversky, 1981). Progressions, too, were schematized. In maps, they appeared either as straight lines or as slightly curved ones, leaving out much of the detail in the real environment. The distinction between straight and curved paths was also made in language. By far, the two most frequent verbs for expressing progressions were "go" and "follow," and they were used differentially for straight and curved paths. "Go" was used 17 times for straight paths and only twice for curved (Chi-square = 9.0, p <.005), whereas "follow" was used only 5 times for straight paths but 20 times for curved ones (Chi-square=11.8, p <.001). Thus, although the actual paths and intersections had many forms, a single category of intersection and two categories of path shape sufficed in schematization, whether verbal or pictorial.

5.3 Sufficiency in Descriptions and Depictions

We found that both descriptions and depictions consist of the same critical and supplementary information schematized in similar ways. Was the information sufficient for conveying the route? That is, did each segment contain all the essential components: start and end points, path, and direction? For maps, the answer was a rousing yes. All of the maps contained all of the essential information. For descriptions, in contrast, the initial answer was no, much information was missing. In fact, 75% of the descriptions were missing a start or an end point and 45% of the descriptions were missing path/progression information. However, many communications contain missing information that can be inferred from context or medium (e. g., Clark and Clark, 1977). In the case of route directions, two simple rules of inference allow recovery of most of the missing information. The first is *continuity*. According to continuity, if a start point is omitted, it is assumed to be the same as the previous end point, or conversely, if an end point is missing, it is assumed to be the same as the subsequent start point. In fact, for depictions, where continuity is inherent in maps, start and end points are not well-defined or distinguishable. The second inference rule is *forward progression*. According to forward progression, the direction of motion is assumed to be forward. The first protocol in Table 1 lacks any end points, yet they can be easily inferred from the subsequent action. This protocol has only one explicit mention of forward progression, at the end ("go down few miles"); rather, the forward progression is implicit. After applying these two rules of inference, 86% of the directions were sufficient. In three of the descriptions, the direction of a turn was missing and could not be recovered.

Although maps and directions schematize routes in similar ways, maps are more complete than directions, which need to be supplemented with inference rules. Another way to put it is that directions are more schematized than maps. This difference, we believe, is inherent in the graphic medium, in the mapping of real space to representing space. Paths in real space are continuous and forward moving given particular start and end points; they are portrayed as such in representing space. Even though the mapping from real space to representing space is schematic rather than strictly iconic, it pragmatically presupposes the two inference rules, continuity and forward progression. The naturalness of the mapping of space to space is further evident in the greater variability of verbal expressions than pictorial expressions for the four elements.

6   Schematization of Space in Language and Cognition

Clearly, there are parallels in the way that cognition and language schematize the spatial world. Language is revealing in this enterprise, not just closed-class terms, but open-class terms as well. Graphic communications, such as route maps, are also schematized, again with similarities to language and cognition. But both language and cognition are rich, and are able to express and encode more or less schematically, depending on the situation and how it is construed.

Figures, spatial relations among them, and paths between them seem to be schematized similarly in language and cognition. Language serves well to convey routes and environments, provided the routes and environments are well-known (e. g., Taylor and Tversky, 1992a, 1992b, 1996). In contrast, language seems to be inadequate at conveying faces, voices, and emotions. We can only speculate on the answer. It is likely that languages develop in small groups of people who know each other and who use language in direct social encounters. Faces and voices are present in those situations so they convey themselves--and emotions--directly, they do not have to be conveyed in words. They are tagged with proper names known to the community so that, if needed, they can be gossiped about in their absence. Proper names, in contrast to category names and closed-class terms, do not convey any spatial information in themselves. Like addresses or GPS coordinates, they point to individuals or locations without giving any other information about them. Routes are described by terms with more general spatial meanings. Unlike faces and voices, they may not be present in the social encounters in which they are discussed. Many cannot even be viewed in entirety from a single vantage point, much less the current one. What's more, individuals often set out alone to forage or hunt, so that developing ways to communicate route directions is useful in communal living. For faces, voices, and routes, then, the Schematization Similarity Conjecture--that language will be successful in communication to the extent that language and 'ception schematize similarly--receives support, along with a speculative explanation.

At the outset we observed that perception inevitably affects language, that at least in part, people develop vocabularies and syntax to communicate about things in the world as they perceive them, and as they need to talk about them. It is equally clear that language influences perception. Language calls attention to particular things and states and qualities and distinctions in the world and ignores others. Over repeated experience, the selective attention encouraged by language can become habitual, so that it seems as if language is no longer involved. Undoubtedly, habitual attention to certain things, states, qualities, and distinctions in space affects the way space is schematized, further intertwining the schematization of language and cognition.

References

1. Biederman, I.: Recognition-by-components: A theory of human image understanding. Psychological Review, 94, (1987) 115-147

2. Bowerman, M.: Semantic and syntactic development: A review of what, when, and how in language acquisition. In: R. L. Schiefelbusch (ed.): Bases of language intervention, Vol. 1. University Park Press, Baltimore. (1978a) 97-189

3. Bowerman, M.: The acquisition of word meaning: An investigation into some current conflicts. In: N. Waterson and C. Snow (eds.): The development of communication: Social and pragmatic factors in language acquisition. Wiley, NY. (1978b) 263-287

4. Brown, R.: How shall a thing be called? Psychological Review, 65, (1958) 14-21

5. Bryant, D. J., Tversky, B., & Franklin, N.: Internal and external spatial frameworks for representing described scenes. Journal of Memory and Language, 31, (1992) 74-98

6. Bryant, D. J., Tversky, B., and Lanca, M.: Retrieving spatial relations from observation and memory. Unpublished manuscript. (1998)

7. Bruce, V. : Recognizing faces. Erlbaum, Hillsdale, NJ (1988)

8. Byrne, R. W.: Memory for urban geography. Quarterly Journal of Experimental Psychology, 31, (1979) 147-154

9. Carlson-Radvansky, L. & Irwin, D. E.: Frames of reference in vision and language: Where is above? Cognition, 46, (1993) 223-244

10. Clark, E. V. What's in a word? On the child's acquisition of semantics in his first language. In: T. E. Moore (ed.): Cognitive development and the acquisition of language. Academic Press, NY. (1973). 65-110

11. Clark, H. H. & Clark, E. V.: Psychology and language. Harcourt Brace Jovanovich, NY. (1977)

12. Clark, H. H. and Clark, E. V.: When nouns surface as verbs. Language, 55, (1979), 420-477

13. Cutting, J. E., Proffitt, D. R., and Kozlowski, L. T.: A biomechanical invariant for gait perception. Journal of Experimental Psychology: Human Perception and Performance (1978) 357-372

14. Denis, M.: La description d'itineraires: Des reperes pour des actions. Notes et Documents LIMSI No. 94-14 Juillet, 1994 (The description of routes: Landmarks for actions) (1994)

15. Diamond, R. and Carey, S.: Why faces are and are not special: An effect of expertise. Journal of Experimental Psychology: General, 115, (1986). 107-1227

16. Franklin, N., Henkel, L. A., and Zangas, T.: Parsing surrounding space into regions. Memory and Cognition, 23, (1995) 397-407

17. Franklin, N. and Tversky, B.: Searching imagined environments. Journal of Experimental Psychology: General, 119, (1990) 63-76

18. Garrod, S. C. and Sanford, A. J.: Discourse models as interfaces between language and the spatial world. Journal of Semantics, 6, (1989) 147-160

19. Gentner, D.: Some interesting differences between verbs and nouns. Cognition and Brain theory, 4, (1981) 161-178

20. Glenberg. A. M., Meyer, M., and Lindem, K.: Mental models contribute to foregrounding during text comprehension. Journal of Memory and Language, 26, (1987) 69-83

21. Gryl, A.: Analyse et modelisation des processus discursifs mis en oeuvre dans la description d'itineraires. Notes et Documents LIMSI No. 95-30. Decembre, 1995 (Analysis and modeling of discourse in the description of routes) (1995)

22. Hagen, M. A. and Perkins, D.: A refutation of the hypothesis of the superfidelity of caricatures relative to photographs. Perception, 12, (1983) 55-61

23. Hayward, W. G. and Tarr, M. J.: Spatial language and spatial representation. Cognition, 55, (1995) 39-84

24. Hegarty, M.: Mental animation: Inferring motion from static displays of mechanical systems. Journal of Experimental Psychology: Learning, Memory, and Cognition, 18, (1992) 1084-1102

25. Herskovits, A.: Language and spatial cognition: An interdisciplinary study of the prepositions in English. Cambridge, MA: Cambridge University Press (1986)

26. Hochberg, J.: Perception. Prentice-Hall, Englewood Cliffs, NJ (1978)

27. Hoffman, D. D. & Richards, W. A.: Parts of recognition. Cognition, 18, (1984) 65-96

28. Howard, I. P.: Human visual orientation. Wiley, NY (1982)

29. Huttenlocher, J. Hedges, L. V., and Duncan, S.: Categories and particulars: Prototype effects in estimating spatial location. Psychological Review, 98, (1991) 352-376

30. Huttenlocher, J. and Lui, F.: The semantic organization of some simple nouns and verbs. Journal of Verbal Learning and Verbal Behavior, 18, (1979) 141-162

31. Ibbotson, A. and Bryant, P. E.: The perpendicular error and the vertical effect. Perception, 5, (1976) 319-326

32. Johansson, G.: Visual motion perception. Scientific American, 232, (1975) 76-88

33. Kaiser, M. K., Proffitt, D. R., Whelan, S. M., and Hecht, H.: Influence of animation on dynamical judgments. Journal of Experimental Psychology: Human Perception and Performance, 18, (1992) 669-690

34. Klein, W.: Local deixis in route directions. In: R. J. Jarvella and W. Klein (eds.): Speech, place and action. Wiley, NY (1982) 161-182

35. Lakoff, G.: Women, fire and dangerous things: What categories reveal about the mind. University of Chicago Press, Chicago (1987)

36. Landau, B. and Jackendoff, R.: "What" and "where" in spatial language and spatial cognition. Behavioral and Brain Sciences, 16:2, (1993) 217-238

37. Landau, B. Smith, L. B. and Jones, S. S.: The importance of shape in early lexical learning. Cognitive Development, 3, (1988) 299-321

38. Landau, B. Smith, L. B. and Jones, S. S.: Syntactic context and the shape bias in children's and adults' lexical learning. Journal of Memory and Language, 31, (1992) 807-825

39. Leibowitz, H. W., Guzy, L. T., Peterson, E., & Blake, P. T.: Quantitative perceptual estimates: Verbal versus nonverbal retrieval techniques. Perception, 22, (1993) 1051-1060

40. Levelt, W. J. M.: Linearization in describing spatial networks. In: S. Peters and E. Saarinen (eds.): Processes, beliefs, and questions. Reidel, Dordrecht (1982) 199-220

41. Linde, C. and Labov, W.: Spatial structures as a site for the study of language and thought. Language, 51, (1975) 924-939

42. Logan, G. D. and Sadler, D. D.: A computational analysis of the apprehension of spatial relations. In: P. Bloom, M. A. Peterson, L. Nadel, and M. Garrett (eds.): Language and Space. MIT Press, Cambridge, MA (1996) 493-529

43. Marr, D.: Vision: A computational investigation into the human representation and processing of visual information. Freeman, NY (1982)

44. Morrow, D. G., Bower, G. H., and Greenspan, S.: Updating situation models during narrative comprehension. Journal of Memory and Language (1989) 292-312

45. Morrow, D. G. and Clark, H. H.: Interpreting words in spatial descriptions. Language and Cognitive Processes, 3, (1988) 275-291

46. McBeath, M. K., Schiano, D. J., and Tversky, B.: Three-dimensional bilateral symmetry bias in judgments of figural identity and orientation. Psychological Science, 8, (1997) 217-223

47. Moar, I. & Bower, G. H.: Inconsistency in spatial knowledge. Memory and Cognition, 11, (1983) 107-113
48. Morris, M. W. & Murphy, G. L.: Converging operations on a basic level in event taxonomies. Memory and Cognition, 18, (1990) 407-418
49. Murphy, G. L. and Smith, E. E.: Basic-level superiority in picture categorization: Journal of Verbal Learning and Verbal Behavior, 21, (1982) 1-20
50. Newtson, D., Hairfield, J., Bloomingdale, J., and Cutino, S.: The structure of action and interaction. Social Cognition, 5, (1987) 191-237
51. Palmer, S., Rosch, E. & Chase, P.: Canonical perspective and the perception of objects. In: J. B. Long & A. D. Baddeley (eds.): Attention and Performance, IX. Erlbaum: Hillsdale, New Jersey (1981)
52. Pani, J. R., William, C. T., and Shippey, G. T.: Determinants of the perception of rotational motion: Orientation of the motion to the object and to the environment. Journal of Experimental Psychology: Human Perception and Performance, 21, (1995)
53. Nelson, T. O. and Chaiklin, S.: Immediate memory for spatial location. Journal of Experimental Psychology: Human Learning and Memory, 5, (1980) 212-228
54. Perrig, W. and Kintsch, W.: Propositional and situational representations of text. Journal of Memory and Language, 24, (1985) 503-518
55. Rifkin, A.: Evidence for a basic level in event taxonomies. Memory and Cognition, 13, (1985) 538-556
56. Rock, I.: The logic of perception. MIT Press, Cambridge (1983)
57. Rosch, E.: Cognitive representations of semantic categories. Journal of Experimental Psychology: General, 104, (1975) 192-233
58. Rosch, E.: Principles of categorization. In: E. Rosch and B. B. Lloyd: Cognition and categorization. Erlbaum, Hillsdale, NJ (1978) 27-48
59. Rosch, E., Mervis, C., Gray, W., Johnson, D. and Boyes-Braem, P.: Basic objects in natural categories. Cognitive Psychology, 6, (1976) 382-439
60. Schooler, J. W. and Engstler-Schooler, T. Y.: Verbal overshadowing of visual memories: Some things are better left unsaid. Cognitive Psychology, 22, (1991) 36-71
61. Schwartz, D. L. & Black, J. B.: Analog imagery in mental model reasoning: Depictive models. Cognitive Psychology, 30, (1996) 154-219
62. Shapiro, P. N. and Penrod, S.: Meta-analysis of facial identification studies. Psychological Bulletin, 100, (1986) 39-156
63. Shiffrar, M. M. & Shepard, R. N.: Comparison of cube rotations around axes inclined relative to the environment or to the cube. Journal of Experimental Psychology: Human Perception and Performance, 17, (1991) 44-54
64. Spelke, E. S., Breinlinger, K., Macomber, J., and Jacobson, K.: Origins of knowledge. Psychological Review, 99, (1992) 605-632
65. Streeter, L. A., Vitello, D. and Wonsiewicz, S. A.: How to tell people where to go: Comparing navigational aids. International Journal of Man-Machine Studies, 22, (1985) 549-562
66. Talmy, L.: Semantics and syntax of motion. In: John P. Kimball (ed.): Syntax and semantics, Vol. 4. Academic Press, NY (1975) 181-238
67. Talmy, L.: How language structures space. In: H. L. Pick, Jr. & L. P. Acredolo (eds.): Spatial orientation: Theory, research and application. Plenum, NY (1983) 225-282
68. Talmy, L.: Lexicalization patterns: Semantic structure in lexical forms. In: Language typology and syntactic description, Vol. 3, T. Shopen (ed.): Grammatical categories and the lexicon. Cambridge University Press, Cambridge (1985) 57-149

69. Talmy, L.: Force dynamics in language and cognition. Cognitive Science, 12, (1988) 49-100

70. Talmy, L.: Fictive motion and change in language and perception. In: P. Bloom, M. A. Peterson, L. Nadel, and M. Garrett (eds.): Language and space. MIT Press, Cambridge (1996) 211-276

71. Taylor, H. A., Naylor, S. J. and Chechile, N. A.: (In press) Goal-specific influences on the representation of spatial perspectives. Memory and Cognition.

72. Taylor, H. A. & Tversky, B.: Descriptions and depictions of environments. Memory and Cognition, 20, (1992a) 483-496

73. Taylor, H. A. & Tversky, B.: Spatial mental models derived from survey and route descriptions. Journal of Memory and Language, 31, (1992b) 261-292

74. Taylor, H. A. and Tversky, B.: Perspective in spatial descriptions. Journal of Memory and Language, 35, (1996) 371-391

75. Taylor, M. M.: Effect of anchoring and distance perception on the reproduction of forms. Perceptual and Motor Skills, 12 (1961) 203-230

76. Tversky, B.: Distortions in memory for maps. Cognitive Psychology, 13, (1981) 407-433

77. Tversky, B.: Categories and parts. In: C. Craig and T. Givon (Eds.): Noun classes and categorization . John Benjamins Publishing Co, Philadelphia (1985) 63-75

78. Tversky, B.: Distortions in memory for maps. Cognitive Psychology, 13, (1981) 407-433

79. Tversky, B.: Distortions in cognitive maps. Geoforum, 23, (1992) 131-138

80. Tversky, B. and Hemenway, K.: Objects, parts, and categories. Journal of Experimental Psychology: General, 113, (1984) 169-193

81. Tversky, B., Taylor, H. A., and Mainwaring, S.: Langage et perspective spatial (Spatial perspectives in language). In: M. Denis (Eds.): Langage et cognition spatiale. Masson, Paris (1997) 25-49

82. Vandeloise, C.: L'espace en francais: Semantique des prepositions spatiales. Editions du Seuil, Paris. (1986)

83. Wunderlich, D. and Reinelt, R.: How to get there from here. In R. Jarvella and W. Klein (eds.): Speech, place, and action. Wiley, Chichester (1982)183-201

84. Yin, R. K.: Looking at upside-down faces. Journal of Experimental Psychology, 81, (1969) 141-145

85. Zacks, J. and Tversky, B.: What's happening? The structure of event perception. Proceedings of the Meetings of the Cognitive Science Society. Erlbaum, Mahwah, NJ (1997)

Shape Nouns and Shape Concepts:
A Geometry for 'Corner'

Carola Eschenbach, Christopher Habel, Lars Kulik, Annette Leßmöllmann[*]

FB Informatik (AB WSV) and Graduiertenkolleg *Kognitionswissenschaft*,
Universität Hamburg, Vogt-Kölln-Str. 30, D-22527 Hamburg, Germany
{eschenbach, habel, kulik, lessmoellmann}@informatik.uni-hamburg.de

Abstract. This paper investigates geometric and ontological aspects of shape concepts underlying the semantics of nouns. Considering the German shape nouns *Ecke* and *Knick* (*corner* and *kink*) we offer a geometric framework to characterize substantial aspects of shape based on features of the object's boundary. Using the axiomatic method, we develop a geometric system, called 'planar shape geometry', enriching the basic inventory of ordering geometry by shape curves. The geometric characterization is not sufficient to decide which are the referents of the nouns *Ecke* and *Knick* among the entities involved in the spatial constellation. Different tests using the German topological prepositions *in* and *an* (*in* and *at*) are employed to bring forth this decision for the case of *Ecke*. Since these tests do not give uniform evidence in favor of one solution, we have to conclude that *Ecke* is flexible in selecting the referent and the characterizations discussed reflect its meaning spectrum.

1 Introduction

1.1 Language and Space: The Role of Conceptual Representations

Human behavior is anchored in space. The processing of spatial information has a central position for human cognition, since it subsumes information about spatial properties of the entities in our environment, about spatial constellations in our surroundings, and about the spatial properties and relations of our bodies with respect to these surroundings. Spatial information is essential for the recognition of objects and events by different sensory channels, i.e., in visual, haptic or auditory perception. Locomotion and body movement are based on such information as well.

[*] The research reported in this paper has been supported by the Deutsche Forschungsgemeinschaft (DFG) in the project 'Axiomatik räumlicher Konzepte' (Ha 1237/7). We are indebted to Christie Manning, Bernhard Nebel, Esther Rinke, Christoph Schlieder, Hedda Rahel Schmidtke, Mark Siebel, and Heike Tappe for their helpful comments. This paper also benefits from the fruitful discussions in the Hamburg Working Group on Spatial Cognition.

Beyond perception and motor action, some higher cognitive capacities such as memory, problem solving, and planning are based on spatial representations (cf. Eilan, McCarthy & Brewer, 1993). Not only communication about space using natural language involves spatial language, but the systematic use of spatial terms in other domains is a general feature of human communication. This suggests that abstract spatial concepts play an important role in non-spatial domains (see Mandler, 1996; Habel & Eschenbach, 1997).

The study of the relationship of language and space is a major field within cognitive science (see, e.g., Bloom et al., 1996). A widely "consensually accepted framework within which the relations between language and space have been considered" has been established (Peterson et al., 1996, p. 553), namely that conceptual representations are an interface between language and spatial cognition. This holds despite terminological variations and different demarcations between the modules involved among distinct branches of the research on language and space (see Peterson et al., 1996; Habel, 1990; Jackendoff, 1991; Landau & Jackendoff, 1993; Bierwisch & Lang, 1989; Bierwisch, 1996).

The human cognitive system has two major ways of gaining information about space: On the one hand, via perception and proprioception, and on the other hand, via communication, especially using natural language. Conceptual representations are fundamental for the relation between language and space since linguistic and spatial conceptual representations constitute the linguistic-spatial interface. Conceptual representations encode meaning independent from any particular language. They "refer not to the real world or to possible worlds, but rather to the world *as we conceptualize it*" (Jackendoff, 1996, p. 5). Leaving aside details of the interface, studying spatial terms leads to further insights about the deeper levels of conceptual representations (see Jackendoff, 1996; Peterson et al., 1996, p. 555).

The units of the conceptual structure are called 'concepts' in the following. This usage of 'concept' is a generalization of the standard usage of 'concept' in psychology, which focuses on the type called 'nominal concept' by Miller (1978). The restriction of the standard view is based on taking concepts as the means for categorization (Smith, 1995, p. 3) and categorization as the placement of objects in classes only. But since humans are able to categorize constellations of objects or object parts and to recognize one object as being the same as an object encountered before, it seems suitable to extend the notion of 'concept'. According to Miller (1978) we assume that the conceptual structure includes different types of concepts, e.g. concepts for relations, properties and objects. In this sense, touching, betweenness or corner are relational concepts, while sphere, round or angular are nominal concepts. In contrast to this, our concepts of the moon or of Ray Jackendoff are object concepts.

Since conceptual structures are independent from individual languages, the correspondence between lexemes and concepts is usually not of the one-to-one type. Hence, it is important to distinguish between lexemes and their conceptual counterparts, even if they are referred to by the same string; we consider this by typographical differences, e.g. the lexeme *corner* vs. the concept corner.

The structure of representations relating language and space (see Fig. 1) reflects the tasks to be performed in the analysis of spatial language and spatial cognition:

Lexemes (1) constitute the starting point of the analysis. Examples are spatial prepositions as *in* or *behind*, verbs as *enter* or nouns as *corner*. A systematic variation of different combinations of lexemes determines their conceptual similarities and differences. In addition, the applicability and interpretability of simple and complex spatial expressions can be tested with regard to different spatial situations.

A core idea of conceptual semantics is that the lexical entry of a spatial term specifies a concept (2) that serves as a representation of some spatial constellation in the external world. The geometric characterization (3) of spatial concepts (2) is the mathematical description of empirical considerations on conceptual semantics. The primary goal of the formal characterization is to determine some candidates for an inventory of basic spatial concepts, which can be seen as building blocks for a system of spatial concepts.

Lexicon (1) $\rightarrow$ Conceptual Structure (2) $\leftarrow$ Spatial (Geometric) Representations (3)
lexemes   concepts   geometric characterizations

Fig. 1. Types of entities used in the analysis of spatial terms of natural language

To develop a mathematical, e.g. topological or geometric, framework we employ the axiomatic method.[1] Instead of defining the basic concepts, an axiomatic system constitutes a system of constraints that determines the properties of these concepts by specifying their interrelations. Hence, axiomatic specifications of spatial properties and relations provide exact characterizations. Different axiomatic systems—of one given set of spatial relations—can be compared as to how restrictive they are.

In section 3 we develop a geometric framework, providing a formal basis to model empirical linguistic findings. In general, the axiomatic systems we propose for groups of spatial concepts are motivated by the analysis of natural language. The generality of the theories developed using the axiomatic method is sometimes seen as a disadvantage, since they cannot be restricted to the one (intended) model the analysis is based on. But in the context of the analysis of natural language expressions, this property turns out to be an advantage again, since the variability of expressions with respect to their domain of application is a well known feature of natural languages (cf. Habel & Eschenbach, 1997).

1.2 Shape Concepts: Axes and Boundaries

The importance of shape information is emphasized in many areas of cognitive science, especially with respect to visual perception and categorization.[2] In spite of its importance, there is a lack in lucidity what shape is. For instance, there is no explicit characterization of shape, but only an informal agreement among scientists in the field

---

[1] Starting in mathematics and physics the axiomatic method has spread to other disciplines (see, Henkin et al., 1959, compare also the detailed discussion by Luce et al., 1990).

[2] Rosch et al. (1976) point out that shape belongs to the key properties that structure 'basic level categories', e.g. those which are named by count nouns like *apple, hat, cup*.

of visual perception. This agreement has been the basis of many approaches to understand visual perception, which are described in Pinker's (1984) overview of central issues in visual cognition. Starting with the Marr-Nishihara theory (1978), the representation of shape is mainly based on spatial symmetry and object axes, which play a central role both on the level of the whole object and that of the components.

In linguistic semantics the approach of Bierwisch and Lang (1989) on dimensional adjectives, e.g. *long* or *high*, also emphasizes the role of axes for shape descriptions. Object schemata, which serve as explanatory basis for the semantics of these terms (see Lang, 1989), are conceptual representations of objects that specify their main axes as well as the proportions between the axes; this information is matched against a corresponding slot in the lexical entry of dimensional adjectives. The interdependence of shape and axes as seen by Bierwisch and Lang is expressed in Bierwisch's specification of shape of objects as "the proportional metrical characteristics of objects and their parts with respect to their conceptually relevant axes or dimensions (3-D models)" (Bierwisch, 1996, p. 47). [3]

The other core notion for characterizing shape is 'boundary'. Perception research lays emphasis on the use of boundaries in decomposing objects into their parts, especially on describing rules for the detection of part boundaries, e.g. based on notions as "concavities" of "concave regions" or the "minima rule" (see Marr & Nishihara, 1978; Biederman, 1987; Hoffman & Richards, 1984; Hoffman & Singh, 1997). Biederman's (1987) recognition-by-component (RBC) theory supplies an inventory of 36 geons (generalized cones) that are derived from properties of edges in a planar image. These properties include curvature, collinearity, symmetry, parallelism and cotermination. The resulting geons can be seen as building blocks for the specification of objects by their components (cf. Biederman, 1995).

The RBC-theory and the idea of geons have been influential in linguistic and psycholinguistic approaches to shape. For example, Landau and her colleagues assume the same basic principles of characterizing 'shape' by the arrangement of object parts as the object recognition theories described above. On this basis, Landau, Smith, and Jones (1988) show that children and adults use similarity in shape much more than similarity in size or texture in generalizing a novel count noun to new objects.

Characteristics of boundaries seem to play an important role also on the conceptual level: Landau et al. (1992) report a series of experiments that emphasize the influence of different types of boundaries (angular vs. curved edges) as cues for the acceptability of shape transformations. These experiments show that, despite differences in the global shape, objects with curved boundaries are more easily grouped together than objects with angular boundaries. They suggest that angular boundaries in contrast to curved boundaries are seen as evidence for rigidity.

In the present paper we focus on German lexemes—namely, *Ecke (corner)* and *Knick (kink)*[4]—that involve properties of the boundary of an object. Starting with an

---

[3] "Dimension" as used by Bierwisch & Lang (1989) is widely identified with the concept of axis, which is taken to be an internal bounded straight line. Hence, it is possible to compare axes with respect to their length and thus to use axes for characterizing proportion.

[4] The analysis reported was carried out with respect to the German lexemes. Note that their English counterparts, which we use in the text for readability, differ in some respects. E.g.,

informal and general characterization of the spatial constellation underlying the uses of these nouns, we present in section 3 a geometric framework that allows us to specify the spatial aspects of their semantics. The spatial constellation specified is neutral with respect to the question of what is the referent of a phrase like *a corner of the carpet*. The referent might be just a point, a part of the boundary of the carpet, or a part of the carpet itself. The details of these possibilities are given in section 4. Which alternative is the most appropriate one will be discussed in section 5 in connection with the more detailed linguistic analysis.

2 A First Glance at Corners and Kinks

The most obvious similarity between corners and kinks is that they possess vertices. Still, the terms *Ecke* and *Knick* are mostly mutually exclusive.

2.1 Dimensionality

One important factor for differentiating corner and kink can be grasped by the idea of 'dimensionality'. Since we are not aiming at a general account on dimensionality, we will leave it at an informal characterization by giving examples of some mathematical entities: points are zero-dimensional, line segments are linear or one-dimensional, squares are planar or two-dimensional and cubes are three-dimensional. This exemplification meets the interpretation of 'dimensionality' mostly used by linguists, such as Bierwisch & Lang (1989) or Jackendoff (1991), who characterize dimensionality via the number of orthogonal axes of the object. This traditional view has been replaced in modern geometry and topology using recursive or inductive definition schemes that explain the dimension of a space (or a subspace) by the dimension of its boundary.[5]

The differences in applicability of *Ecke* and *Knick* correspond to differences in the dimensionality of the objects. *Knick* concerns a property of paths, sticks, or other objects with a linear appearance.

(1) a. Der Weg hat einen Knick.   (The path has a kink.)
    b. Der Metallstab hat einen Knick.   (The metal stick has a kink.)

(2) ?Der Metallstab hat eine Ecke.   (The metal stick has a corner.)

---

the English noun *corner* is used for parts of the mouth or the eye. In contrast, the German noun *Ecke* cannot be used to refer to such parts. Instead *Winkel (angle)* is used. This observation suggests that *corner* is not as clearly restricted to objects with straight edges as *Ecke* is.

[5] See Kline (1972, p. 1161f) on the history of these mathematical ideas. Although Jackendoff (1991, p. 32) also formulates the idea that "a boundary has one dimension fewer than what it bounds", he does not take the step of separating dimensionality and axes. From our point of view, this separation is necessary to analyze boundary-shape concepts independently from axes-shape concepts.

In both cases in (1), it is impossible to replace the noun *Knick* by *Ecke*. A metal stick that exhibits a vertex where two straight parts of the stick meet can be described using (1.b). In contrast, (2) is not appropriate to describe the same situation. Since this is independent from any contextual influence, there seems to be a semantic or conceptual conflict between *Stab* (*stick*) and *Ecke*, which prohibits their combination. However, the contour itself is not sufficient to judge whether *Knick* or *Ecke* is the appropriate lexeme. To clarify this, we consider the example of a metal stick having four kinks such that it bounds an area (Fig. 2). This situation can be described by (3.a), (3.b) and by (3.c), but not using (3.d).

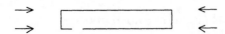

Fig. 2. Depiction of a metal stick with four (highlighted) kinks

(3) a. Der Metallstab hat vier Knicke. (The metal stick has four kinks.)
 b. Die Fläche hat vier Ecken.   (The area has four corners.)
 c. Der Bilderrahmen hat vier Ecken. (The picture frame has four corners.)
 d. ?Der Metallstab hat vier Ecken. (The metal stick has four corners.)

The differences between the linguistic descriptions in (3) reflect the different conceptualizations of the entities in question. While the metal stick is considered as a linear object in (3.a), the area in (3.b) and the picture frame in (3.c) are considered as planar objects. Thus, the conceptualized dimension of an object in combination with the shape properties determines the lexical decision. Since conceptualization, and therefore mental representation, is involved, this difference can be seen as the difference between the concepts kink and corner.

Kink and corner differ with respect to the conceptualized dimensionality of the underlying objects: Kink requires linear objects and corner requires at least planar objects. Although a metal stick can induce a planar region as in Fig. 2, we explicitly have to refer to the induced region and not to the object inducing the region, if we want to refer to its corners, cf. (3.b) vs. (3.d).

Thus, although sticks are extended with respect to three dimensions, and paths are at least planar, their linear conceptualizations are the basis for the constitution of a kink-constellation as expressed in (1). Sticks and paths can be considered as linear objects due to their prominent elongation axes. The characterizations of kink in section 4.2 reflects this observation.

The analysis of *Ecke* in the present paper focuses on planar examples like *corner of a window* or *corner of a carpet*. Corners of three-dimensional objects—e.g. corners of rooms, houses, cupboards, etc.—can be classified depending on whether the object is conceptualized as planar or three-dimensional: *the carpet in the corner of the room* needs only the consideration of the planar outline of the room, while *the spider net in the corner of the room* can geometrically best be described based on three dimensions. The restriction to the planar case is due to our observation that the basic points we want to discuss with respect to the linguistic behavior of *Ecke* and the geometric characterization of corner do not depend on the distinction between two- and three-dimensional corners.

Dimensionality is an important aspect in investigating the ontological essence of corners also in another way. In section 5 we discuss whether corners are vertex points, parts of the boundary (i.e., linear), enclosed regions (i.e., planar), or parts of objects. This line of investigation is in principle independent of the question whether the objects they are corners of are conceptualized as linear or planar, since corners or kinks need not be of the same dimension as the object they belong to.

2.2 Sharp Concepts in Flexible Use

In the remainder of this section we discuss geometric aspects of corner to motivate the geometric analysis given in section 4. The general points hold for kink and other shape nouns as well. The more detailed discussion of the lexeme *corner* in section 5 focuses on aspects of its meaning that cannot be determined by geometric means.

Ideal corners, like the corners of a square or a triangle, exhibit a vertex at which two straight parts of the object's boundary meet. But the applicability of the lexeme *corner* is vague with respect to at least three independent aspects.

- Vagueness in size: Corners can, depending on the context, reach far to the middle of an object, or can be a very small part close to the vertex.
- Idealization of the vertex: Corners of material entities do not have to exhibit a proper vertex at which two straight lines meet, e.g., corners of tables are not sharp-edged in most cases (see Fig. 3.b).
- Idealization of the edges: The two edges meeting at the corner need not be straight as line segments in geometry. The boundary of a material object can be wavy or saw-toothed up to a certain degree, like the boundary of a stamp (see Fig. 3.c, d).

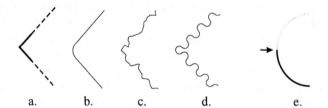

a.　　　b.　　　c.　　　d.　　　e.

Fig. 3. Four depictions of corners in contrast to smoothly meeting boundary parts

The review of these aspects of vagueness leads to the question, whether we have to consider all geometric deviations that real-world referents of the count noun *Ecke* show in the geometric characterization of corner. First of all, the vagueness in size seems intrinsic to the concept of corner. The extent of corners cannot be given sharply with respect to other parts of the object (this is indicated by the dashed lines in Fig. 3.a). Thus, the formal characterization of corner shall reflect this vagueness.

In contrast to this, the vagueness in regard to the sharpness of the vertex and the straightness of the edges concerns the relation between corner-instances in the real world and the concept corner or the lexeme *Ecke*, respectively. It regards the question how well reality and geometry fit together. Akin to Miller & Johnson-Laird (1976), we assume a definite conception for corner and flexible mappings of the real world to the conceptual level that can make sense of the world as it is perceived, i.e., that are able to handle the wide variety of instances of members of a category humans are confronted with in perception.

We want to exemplify this flexibility with the case of the corners of stamps (see Fig. 3.d). In order to assign corners to a stamp, it can be ignored that its border is saw-toothed. The detailed geometric characteristics of the border of the stamp are not relevant for its global shape. It is sufficient to conceive of this border as consisting of straight line segments that are induced from the arrangement of the teeth of the stamp. Thus, the attribution of having corners is based on the possibility to cognitively induce straight edges from global characteristics of the object. Hence, the formal characterization of corner developed in section 4 sticks to straight edges.

Kink on the other hand does not need this requirement: If a kink is ascribed to a metal stick as in (1.b), this does not demand that the adjacent parts of the vertex have to be straight. Instead they may be curved. To reflect this distinction between kink and corner we develop a framework that distinguishes straight from curved lines.

Another essential component of this concept is the vertex. Shapes that do not exhibit a vertex (see Fig. 3.b) may be called *runde Ecke (round corner* or *rounded corner)*. They are constituted by straight line segments smoothly connected by an arc. The line segments can be extended in a straight manner, obeying the Gestalt principle of good continuation, such that they meet in a non-smooth way at a point outside the object's boundary, which can be called a 'virtual vertex' or 'virtual corner point'. This condition of 'non-smooth meeting' is essential, as Fig. 3.e exemplifies: Two arcs that meet smoothly do not constitute a corner. Without the possibility to generate at least a virtual vertex, we cannot ascribe the concept corner.

Summarizing the above considerations, we assume corner to be an idealized concept for which straight lines and vertices play an important role. Therefore the formal characterization we develop in section 4 is done in a geometric framework; by this, we formulate the geometric requirements for the concept specifying the meaning of the lexeme *Ecke*. Nevertheless it is not justified to say that *Ecke* is merely a geometric notion that is part of a specialized mathematical register we learn at school (this might be discussed for lexemes like *parallel* or *angle*). *Ecke* is a natural language term with a common-sense denotation. But probably due to the fact that corner is based on straight lines, vertices, and non-smooth meeting, it is mainly applied to artifacts. Their regular shapes more easily induce straight edges that meet in a non-smooth, sharp manner than the shapes of plants or animals. In addition to *corner* there are a variety of natural language terms for object parts with vertices that are not restricted to straight edges: *apex, point, thorn, horn, tip*, etc.

Giving corner a mainly geometric characterization leads to a view corresponding to Pinkal's (1985) general discussion of 'vagueness and precision': Using the lexeme *Ecke* means to neglect some of the properties objects have and to look for properties that are congruent with the geometry of corner. Geometric notions can be meaningfully used in our scope of experience only if they are used flexibly, i.e. allowing some level of imprecision.

3 The Framework of Planar Shape Geometry

This section presents a formal characterization of a geometric framework that is able to specify the spatial constraints discussed so far. The goal is to identify the underlying structures of the spatial concepts corner and kink. The geometric framework is structured similarly to the system presented by Hilbert (1899), which is divided into different groups of axioms. It is developed employing the axiomatic method: An axiomatic system constitutes a system of constraints that determines the properties of basic terms like 'point', '(shape) curve' and 'region' by specifying their interrelations. Since we aim at identifying the general constraints, we develop a description of contours of planar objects that does not require concepts of differential geometry like differentiability, tangents or real numbers. We thereby show that such concepts are not necessary to describe essential shape features. Therefore, a description of shape curves can forgo the use of coordinates or metrical information.[6] The next section presents proposals for characterizing corner and kink as a basis for further discussions.

The geometric structure introduces five types of entities and two primitive relations. The entities are points (denoted by $P, Q, R, P', P_1, \ldots$), (straight) lines (denoted by $l, l', l_1, \ldots$), half-planes (denoted by $H, H', \ldots$), (shape) curves (denoted by $c, c_1, \ldots, s, s_1, \ldots, a, a_1, \ldots$) and (shape) regions (denoted by $Reg, Reg', \ldots$). Segments and arcs are simple shape curves characterized below. (We use the symbols $s, s_1, \ldots$ and $a, a_1, \ldots$ for reference to simple curves.)[7]

The basic idea is that closed shape curves represent the contours of objects conceptualized as planar, i.e., whose geometric representatives are shape regions. Open shape curves are able to represent linear object conceptualizations as well as trajectories of moving objects (Eisenkolb et al., 1998). Shape curves are constituted by line segments and arcs. Half-planes are introduced in order to distinguish vertices and smooth points of curves. This classification of points will be indispensable in order to characterize the concepts corner and kink.

The primitive relations are the binary relation of incidence (symbolized by ι) and the ternary relation of betweenness for points (symbolized by β). The relation of incidence sets up the relation between points and the other entities and characterizes the fact that a point lies on a line or curve, or in a half-plane or region. The relation of betweenness relates three different points on one line.

---

[6] The geometric framework we present is closely related to the framework defined by Eschenbach & Kulik (1997), on which an investigation of spatial orderings in the plane and the structures contributed by different frames of reference is based.

[7] As can easily be shown, any of these entities can be represented as a set of points. Since points, in turn, could be represented as sets of lines, half-planes, curves, or regions (cf., Laguna, 1922, Vieu, 1993), any type of entity we consider could be taken as the ontological basis. Since there are several ways of how such a representation (or coding) can be done, we do not assume any of such possibilities as preferable to the others.

3.1 Points and Straight Lines, Incidence and Betweenness

Axioms for Incidence of Points and Straight Lines

Four axioms relate points and straight lines using incidence. Axiom (I1) guarantees that for every line there are at least two different points on it. Axiom (I2) states that any two points lie on one common line. And (I3) says that two lines have at most one point in common. The last axiom (I4) ensures that the underlying structure is at least planar, i.e., not all points are incident with one line.

(I1) $\forall l \, \exists P \, \exists Q$ $[P \neq Q \wedge P \, \iota \, l \wedge Q \, \iota \, l]$

(I2) $\forall P \, \forall Q \, \exists l$ $[P \, \iota \, l \wedge Q \, \iota \, l]$

(I3) $\forall P \, \forall Q \, \forall l_1 \, \forall l_2$ $[P \, \iota \, l_1 \wedge Q \, \iota \, l_1 \wedge P \, \iota \, l_2 \wedge Q \, \iota \, l_2 \Rightarrow (P = Q \vee l_1 = l_2)]$

(I4) $\forall l \, \exists P$ $[\neg(P \, \iota \, l)]$

Definition (Collinear)

The definitions of collinearity of three points will be useful in the following. For matters of convenience, we define points to be collinear only if they are different. Three (different) points P, Q and R are *collinear*, if they lie on one line

$$\text{col}(P, Q, R) \quad \Leftrightarrow_{\text{def}} P \neq Q \wedge P \neq R \wedge Q \neq R \wedge \exists l \, [P \, \iota \, l \wedge Q \, \iota \, l \wedge R \, \iota \, l]$$

Axioms for Betweenness of Points

The formula '$\beta(P, Q, R)$' can generally be read as 'Q is between P and R'. We provide seven axioms to specify this relation, six of them describe the order of points on the line. According to axiom (β1), if Q is between P and R, then they are collinear, and consequently distinct. Axiom (β2) expresses the symmetry of betweenness with respect to the first and the third argument: If Q is between P and R, then Q is between R and P. Axiom (β3) ensures that at most one of three points is between the other two. Axiom (β4) states that for three collinear points at least one of them is between the other two, therefore betweenness constitutes a total order. Axiom (β5) secures that if Q is between P and R, and Q' is collinear with them, then Q' is on either side of Q.[8] Axiom (β6) guarantees that lines are unlimited. Axiom (β7) (the axiom of Pasch) specifies an additional constraint on the ordering in the plane according to Fig. 4: If a straight line enters the interior of a triangle, then it also leaves it.

(β1) $\forall P \, \forall Q \, \forall R$ $[\beta(P, Q, R) \Rightarrow \text{col}(P, Q, R)]$

(β2) $\forall P \, \forall Q \, \forall R$ $[\beta(P, Q, R) \Rightarrow \beta(R, Q, P)]$

(β3) $\forall P \, \forall Q \, \forall R$ $[\beta(P, Q, R) \Rightarrow \neg\beta(Q, P, R)]$

(β4) $\forall P \, \forall Q \, \forall R$ $[\text{col}(P, Q, R) \Rightarrow (\beta(P, Q, R) \vee \beta(Q, P, R) \vee \beta(P, R, Q))]$

(β5) $\forall P \, \forall Q \, \forall R \, \forall Q'[\beta(P, Q, R) \wedge \text{col}(Q, Q', P) \Rightarrow (\beta(P, Q, Q') \vee \beta(Q', Q, R))]$

(β6) $\forall P \, \forall Q$ $[P \neq Q \Rightarrow \exists R \, [\beta(P, Q, R)]]$

(β7) $\forall P_1 \forall P_2 \forall P_3 \forall l$ $[\neg(P_1 \, \iota \, l) \wedge \neg(P_2 \, \iota \, l) \wedge \neg(P_3 \, \iota \, l) \wedge \exists Q \, [Q \, \iota \, l \wedge \beta(P_1, Q, P_3)] \Rightarrow$
                            $\exists R \, [R \, \iota \, l \wedge (\beta(P_1, R, P_2) \vee \beta(P_2, R, P_3))]]$

---

[8] Huntington (1924) proves the complete independence of this system of five axioms when restricted to one line only.

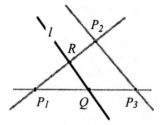

Fig. 4. Illustration of Pasch's axiom (β7)

Pasch's axiom guarantees that the structure we deal with is at most planar. Combined with axiom (I4) this yields that the resulting structure is a plane.

3.2 Half-Planes in Connection to Points and Straight Lines

The axioms for half-planes are formulated with reference to points and straight lines. They make use of the notion of boundary point.

Definition (Boundary Point)

A point is called *boundary point* of a half-plane, if it is in the half-plane and there is a point outside the half-plane such that all points between them are also outside the half-plane: The boundary point is the last point before leaving the half-plane.

$$\text{bdpt}(H, P) \quad \Leftrightarrow_{\text{def}} P \iota H \wedge \exists R \left[\neg (R \iota H) \wedge \forall Q \left[\beta(P, Q, R) \Rightarrow \neg (Q \iota H) \right] \right]$$

Axioms for Half-Planes

Half-planes are bordered by a straight line (H1) and have a point that is not a boundary point (H2). If a point is in a half-plane but not a boundary point, then every point R is in the half-plane, if and only if no point between them is a boundary point (H3). For every straight line and every point there is a half-plane bordered by the line such that the point is in the half-plane (H4). Finally, half-planes can be distinguished by the points that lie in them (H5).

(H1) $\forall H \exists l \forall P \quad [P \iota l \Leftrightarrow \text{bdpt}(H, P)]$

(H2) $\forall H \exists P \quad [P \iota H \wedge \neg \text{bdpt}(H, P)]$

(H3) $\forall H \forall P \quad [P \iota H \wedge \neg \text{bdpt}(H, P) \Rightarrow$
$\qquad \forall R [R \iota H \Leftrightarrow \forall Q [\beta(P, Q, R) \Rightarrow \neg \text{bdpt}(H, Q)]]]$

(H4) $\forall l \forall P \exists H \quad [\forall Q [Q \iota l \Leftrightarrow \text{bdpt}(H, Q)] \wedge P \iota H]$

(H5) $\forall H \forall H' \quad [\forall P [P \iota H \Leftrightarrow P \iota H'] \Rightarrow H = H']$

Since every half-plane is bordered by a line and border-lines are uniquely determined, we use the notion $\text{bl}(H)$ to refer to the border-line of half-plane H. As a consequence, half-planes are uniquely determined by their border-line and an additional point on them, and half-planes are convex.

Definition (Border-Line of a Half-Plane)

$$\text{bl}(H) = l \quad \Leftrightarrow_{\text{def}} \forall P [P \iota l \Leftrightarrow \text{bdpt}(H, P)]$$

3.3 Simple and Complex Shape Curves

Curves are usually not described in a geometric framework. But since we need a basis to describe diverse contours of real-world objects, we cannot limit ourselves to straight lines and line segments. Therefore, we introduce shape curves as geometric entities. According to the current context we confine ourselves to curves that do not branch or intersect themselves like the curve of the figure eight. But it will be obvious, how our characterization of shape curves can be generalized.

Curves and points are related by incidence. Arcs and (line) segments are the simplest curves. We proceed by defining segments and arcs first and then presenting the general axioms for curves. On this basis we give a classification of points relative to curves, including a general notion of vertex.

Definition (Enclosed Point of a Curve, Segment, Endpoint of a Segment)

A point is *enclosed* by a curve, if it lies between two points of the curve.

$$\text{enc}(c, Q) \qquad \Leftrightarrow_{\text{def}} \exists P \,\exists R \,[P \iota c \wedge R \iota c \wedge \beta(P, Q, R)]$$

Segments are connected and bounded parts of lines. Thus, they can be described on the basis of betweenness. Any segment has two points such that a point is incident with the segment, exactly if it lies between them or is identical to one of them.

$$\text{seg}(s) \qquad \Leftrightarrow_{\text{def}} \exists P \,\exists Q \,\forall R \,[R \iota s \Leftrightarrow \beta(P, R, Q) \vee P = R \vee Q = R]$$

It is convenient to use the notion of *endpoint* in specifying segments. Endpoints are those points of segments that do not lie between any other points of the segment.

$$\text{ept}_{\text{seg}}(P, s) \qquad \Leftrightarrow_{\text{def}} P \iota s \wedge \text{seg}(s) \wedge \neg\text{enc}(s, P)$$

The definition for 'segment' guarantees that the points of a segment are incident with one line, that segments are convex, and that segments have (exactly) two endpoints. As a consequence of the axiom that curves can be distinguished by the points on them (C11), we will derive that different segments have different pairs of endpoints.

Definition (Supporting Half-Plane, Outer Smooth Point, Outer Vertex)

The description of arcs is more complex. As a preparation we need a way to classify points on curves. The following definition of the *supporting half-plane* and its border-line is the fundamental step towards the classification we need.

Half-plane H *supports* point P with respect to curve c, symbolized by $\text{sup}(H, P, c)$, if P is incident with c and with the border-line of H, and every point incident with c is incident with the half-plane. Border-lines of supporting half-planes are tangents to the curve in this point.

$$\text{sup}(H, P, c) \qquad \Leftrightarrow_{\text{def}} P \iota c \wedge P \iota \,\text{bl}(H) \wedge \forall Q \,[Q \iota c \Rightarrow Q \iota H]$$

We call a point P an *outer smooth point* of curve c, symbolized by $\text{spt}_o(P, c)$, iff it is incident with c and all supporting half-planes of P with respect to c have the same border-line.

$$\text{spt}_o(P, c) \qquad \Leftrightarrow_{\text{def}} P \iota c \wedge \exists H \,[\text{sup}(H, P, c) \wedge \forall H' \,[\text{sup}(H', P, c) \Rightarrow \text{bl}(H) = \text{bl}(H')]]$$

If, in contrast, several half-planes with different border-lines support one point P with respect to a curve c, then we call P *outer vertex* of c ($\text{vtx}_o(P, c)$; see Fig. 5).

$$\text{vtx}_o(P, c) \qquad \Leftrightarrow_{\text{def}} P \iota c \wedge \exists H \,\exists H' \,[\text{sup}(H, P, c) \wedge \text{sup}(H', P, c) \wedge \text{bl}(H) \neq \text{bl}(H')]$$

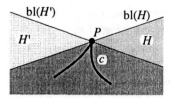

Fig. 5. An outer vertex

In differential geometry uniquely determined tangents ensure that a curve described by real coordinates is differentiable at that point. That we can capture this notion in our more general framework shows that the notion of differentiability and the use of real coordinates is not essential to describe a smooth point.

Since all points of a segment are supported by the half-planes bordered by the line the segment lies on, all points of a segment are outer smooth points or outer vertices and the outer vertices are exactly the endpoints. Line segments are straight. In order to allow curves to be smoothly bent, we introduce arcs. First, we give the complete definition, then we comment the individual clauses of the definition.

Definition (Arc)

An arc is a curve that does not enclose any point on itself, and any point on it is supported by a half-plane. (Therefore, arcs contain only outer smooth points and outer vertices.) An arc has two outer vertices, and one half-plane supports all outer vertices. (This excludes more than two outer vertices.) For any point R that does not lie on the arc and is not enclosed by the arc, there is a segment such that the arc is between R and the segment in the following sense: any point on the arc is between R and some point on the segment and for any point on the segment there is a point on the arc between this point and R. More informally stated: The central projection from R (the central point) maps the arc bijectively onto the segment (see Fig. 6.c). Therefore the denseness and rectifiability of segments are passed on to arcs.

$$\text{arc}(a) \quad \Leftrightarrow_{\text{def}} \forall P \, [P \iota a \Rightarrow \neg\text{enc}(a, P) \wedge \exists H \, [\text{sup}(H, P, a)]] \wedge$$
$$\exists P \, \exists Q \, [P \neq Q \wedge \text{vtx}_0(P, a) \wedge \text{vtx}_0(Q, a)] \wedge$$
$$\exists H \, [\forall P \, [\text{vtx}_0(P, a) \Rightarrow \text{sup}(H, P, a)]] \wedge$$
$$\forall R \, [\neg(R \iota a) \wedge \neg\text{enc}(a, R) \Rightarrow$$
$$\exists s \, [\text{seg}(s) \wedge \forall Q \, [Q \iota s \Rightarrow \exists P \, [P \iota a \wedge \beta(R, P, Q)]]$$
$$\wedge \forall P \, [P \iota a \Rightarrow \exists Q \, [Q \iota s \wedge \beta(R, P, Q)]]]]]$$

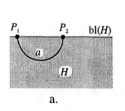

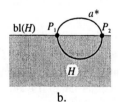

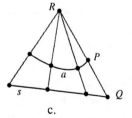

a.   b.   c.

Fig. 6. The definition of arc guarantees a supporting half-plane like H in (a) and excludes that $\dot{a}$ in (b) is an arc. (c) is an illustration of the last clause of the definition of arc

Definition (Part of a Curve, Simple Part of a Curve, Endpoint of a Curve, Curves Meeting at an Endpoint, Thin at P)

A curve c' is *part of* another curve c or a *sub-curve* of c, if all points of c' are incident with c.[9]

$$c' \cdot c \qquad \Leftrightarrow_{\text{def}} \forall P\,[P\,\iota\,c' \Rightarrow P\,\iota\,c]$$

We call curve c' a *simple part* of curve c (in symbols $c'\,\$\cdot\,c$), if c' is a segment or an arc and part of c.

$$c'\,\$\cdot\,c \qquad \Leftrightarrow_{\text{def}} (\text{seg}(c') \vee \text{arc}(c')) \wedge c' \cdot c$$

An endpoint of a curve is on the curve and of any two simple shape curve parts that include it one is part of the other.

$$\text{ept}(P, c) \qquad \Leftrightarrow_{\text{def}} P\,\iota\,c \wedge \forall c_1, c_2\,[c_1\,\$\cdot\,c \wedge c_2\,\$\cdot\,c \wedge P\,\iota\,c_1 \wedge P\,\iota\,c_2 \Rightarrow$$
$$(c_1 \cdot c_2 \vee c_2 \cdot c_1)]$$

Two shape curves c, c' *meet* at endpoint P, symbolized by $\text{meet}(P, c, c')$, if P is a common point and all their common points are endpoints. (This allows two curves to meet at both ends.)

$$\text{meet}(P, c, c') \quad \Leftrightarrow_{\text{def}} P\,\iota\,c \wedge P\,\iota\,c' \wedge \forall Q\,[Q\,\iota\,c \wedge Q\,\iota\,c' \Rightarrow \text{ept}(Q, c) \wedge \text{ept}(Q, c')]$$

A curve c is *thin at point* P, if P is on c and there is a line l, such that P is on l and between P and any other point both on c and l there is a point that is not on c.

$$\text{thin}(c, P) \qquad \Leftrightarrow_{\text{def}} P\,\iota\,c \wedge \exists l\,[P\,\iota\,l \wedge \forall Q\,[Q\,\iota\,l \wedge Q\,\iota\,c \wedge P \neq Q \Rightarrow$$
$$\exists R\,[\beta(P, R, Q) \wedge \neg(R\,\iota\,c)]]]$$

Considering segments and arcs we find that they are thin by definition at all their points, and that the outer vertices of segments are their endpoints also in the general sense defined here.

Axioms for Shape Curves

On this basis we can give the collection of axioms for curves. First, we exclude space-filling curves by stating that curves are thin at all their points (C1). This corresponds to Cantor's definition (cf. Parchomenko 1957).

$$(\text{C1}) \quad \forall c\,\forall P \qquad [P\,\iota\,c \Rightarrow \text{thin}(c, P)]$$

Second, all points of a curve are incident with a simple part of that curve (C2), and no point of a curve is the meeting point of three (or more) simple parts of the curve (C3).

$$(\text{C2}) \quad \forall c\,\forall P \qquad [P\,\iota\,c \Rightarrow \exists c'\,[c'\,\$\cdot\,c \wedge P\,\iota\,c']]$$

$$(\text{C3}) \quad \forall c_1\,\forall c_2\,\forall c_3 \qquad [\exists c\,[c_1\,\$\cdot\,c \wedge c_2\,\$\cdot\,c \wedge c_3\,\$\cdot\,c] \Rightarrow$$
$$\neg \exists P\,[\text{meet}(P, c_1, c_2) \wedge \text{meet}(P, c_1, c_3) \wedge \text{meet}(P, c_2, c_3)]]$$

Curves have at most two endpoints (C4) and, if a curve has one endpoint, then it has another one (C5).

$$(\text{C4}) \quad \forall c\,\forall P\,\forall Q\,\forall R \quad [\text{ept}(P, c) \wedge \text{ept}(Q, c) \wedge \text{ept}(R, c) \Rightarrow (P = Q \vee P = R \vee Q = R)]$$

$$(\text{C5}) \quad \forall c\,\forall P \qquad [\text{ept}(P, c) \Rightarrow \exists Q\,[\text{ept}(Q, c) \wedge P \neq Q]]$$

If two curves meet at one endpoint, then there is a curve that has exactly the points of the two given curves (C6). On the other hand, if a curve c_1 is part of another curve c,

---

[9] The sub-curves of an open curve are related according to interval relations as investigated by Knauff et al. (1998).

then there is an additional sub-curve c_2 of c that meets c_1 (C7). This secures that curves are (path-)connected.

(C6) $\forall c_1 \, \forall c_2$ $[\exists P \, [\text{meet}(P, c_1, c_2)] \Rightarrow \exists c \, \forall Q \, [Q \, \iota \, c \Leftrightarrow (Q \, \iota \, c_1 \vee Q \, \iota \, c_2)]$

(C7) $\forall c \, \forall c_1$ $[c_1 \cdot c \wedge c \neq c_1 \Rightarrow \exists c_2 \, \exists P \, [c_2 \cdot c \wedge \text{meet}(P, c_1, c_2)]]$

Axiom (C8) secures that every curve is constituted by finitely many simple curves. We have to notice that this formulation needs quantification over natural numbers.

(C8) $\forall c \, \exists n \, \exists c_1 \, ... \exists c_n [c_1 \$ \, c \wedge ... \wedge c_n \$ \, c \wedge \forall P \, [P \, \iota \, c \Rightarrow (P \, \iota \, c_1 \vee ... \vee P \, \iota \, c_n)]]$

As an additional axiom we assume that any two points define a segment such that they are its endpoints (C9).

(C9) $\forall P \, \forall Q$ $[P \neq Q \Rightarrow \exists s \, [\text{seg}(s) \wedge \text{ept}_{\text{seg}}(P, s) \wedge \text{ept}_{\text{seg}}(Q, s)]]$

Axiom (C10) states that for every arc a and two points on it there is an arc that has exactly those points of a that lie in a half-plane bordered by a line through the two points. (One simple result is that the outer vertices of arcs are their (only) endpoints.)

(C10) $\forall a \, \forall P \, \forall Q$ $[\text{arc}(a) \wedge P \neq Q \wedge P \, \iota \, a \wedge Q \, \iota \, a \Rightarrow$
$$\exists a' \, \exists H \, [\text{arc}(a') \wedge P \, \iota \, \text{bl}(H) \wedge Q \, \iota \, \text{bl}(H) \wedge$$
$$\forall R \, [R \, \iota \, a' \Leftrightarrow R \, \iota \, a \wedge R \, \iota \, H]]]$$

Finally, curves differ in the points they are incident with (C11). Therefore, curves can be represented as sets of points, although we do not employ such a representation.

(C11) $\forall c \, \forall c'$ $[\forall P \, [P \, \iota \, c \Leftrightarrow P \, \iota \, c'] \Rightarrow c = c']$

Definition (Vertex of a Curve, Smooth Point of a Curve, Turning Point of a Curve, Open Shape Curve, Sum of Two Meeting Curves)

We call a point P a *smooth point* of a curve c, symbolized by $\text{spt}(P, c)$, if it is an outer smooth point of a sub-curve of c.

   $\text{spt}(P, c)$ $\Leftrightarrow_{\text{def}} P \, \iota \, c \wedge \exists c' \, [c' \cdot c \wedge \text{spt}_o(P, c')]$

If a point P is an outer vertex of a sub-curve of curve c and not an endpoint of this sub-curve, then it is an *inner vertex* of c, symbolized by $\text{vtx}(P, c)$ (see Fig. 7).

   $\text{vtx}(P, c)$ $\Leftrightarrow_{\text{def}} P \, \iota \, c \wedge \exists c' \, [c' \cdot c \wedge \neg\text{ept}(P, c') \wedge \text{vtx}_o(P, c')]$

A point P is called *turning point* of c (symbolized as $\text{tpt}(P, c)$), if it is an endpoint of every sub-curve c' such that P is supported by a half-plane with respect to c'.

   $\text{tpt}(P, c)$ $\Leftrightarrow_{\text{def}} P \, \iota \, c \wedge \neg\text{ept}(P, c) \wedge \forall c' \, [c' \cdot c \wedge \exists H \, [\text{sup}(H, P, c')] \Rightarrow \text{ept}(P, c')]$

It is possible to show that every point on a curve that is not an endpoint of this curve belongs to exactly one of the three classes: The point is either a smooth point or a turning point or an inner vertex.

If a shape curve c does not have an endpoint, then we call the shape curve *closed* (in symbols: $\text{cl}(c)$). Otherwise we call the shape curve *open*.

   $\text{cl}(c)$ $\Leftrightarrow_{\text{def}} \neg\exists P \, [\text{ept}(P, c)]$

If two shape curves c_1 and c_2 meet, then we call the curve c that has exactly the points of c_1 and c_2 (see C6 and C11) their *sum* (symbolized as $c = c_1 \, \eth \, c_2$).

   $c = c_1 \, \eth \, c_2$ $\Leftrightarrow_{\text{def}} \forall Q \, [Q \, \iota \, c \Leftrightarrow (Q \, \iota \, c_1 \vee Q \, \iota \, c_2)]$

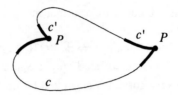

Fig. 7. A curve with two inner vertices

3.4 Shape Regions

The characterization of shape regions we present here is not meant to be a worked out theory of regions in general as, e.g., underlying the calculus investigated by Renz & Nebel (1998). It just needs to fit certain purposes for describing corners. The most important aspect for this purpose is that regions are bounded by closed curves. One consequence is that they are connected and have no holes.

Definition (Part of a Region, Boundary Point of a Region, Convex)

A region *Reg'* is *part of* another region *Reg* or a *sub-region* of *Reg*, if all points of *Reg'* are incident with *Reg*.

$$Reg' \cdot Reg \quad \Leftrightarrow_{def} \forall P\,[P \iota Reg' \Rightarrow P \iota Reg]$$

A point is *a boundary point* of a region *Reg*, if it is in *Reg* and any point lying between it and a point not in *Reg* is also not in *Reg*.

$$\text{bdpt}(Reg, P) \quad \Leftrightarrow_{def} P \iota Reg \wedge \exists R\,[\neg(R \iota Reg) \wedge \forall Q\,[\beta(P, Q, R) \Rightarrow \neg(Q \iota Reg)]]$$

A region is convex, if every point between two points of the region is in the region.

$$\text{cvx}(Reg) \quad \Leftrightarrow_{def} \forall Q\,[\exists P \exists R\,[P \iota Reg \wedge R \iota Reg \wedge \beta(P, Q, R)] \Rightarrow Q \iota Reg]$$

Axioms for (Shape) Regions

The axioms for shape regions state that every region has a boundary that is a closed curve (R1), that it includes a point that is not a boundary point (R2), that between any point (properly) in the region and any point outside the region there is a boundary point of the region (R3), that shape regions do not contain any straight line completely (R4), and that two regions are distinguished by the points in them (R5).

(R1) $\forall Reg \exists c$ $[\text{cl}(c) \wedge \forall P\,[P \iota c \Leftrightarrow \text{bdpt}(Reg, P)]]$

(R2) $\forall Reg \exists P$ $[P \iota Reg \wedge \neg\text{bdpt}(Reg, P)]$

(R3) $\forall Reg \forall P$ $[P \iota Reg \wedge \neg\text{bdpt}(Reg, P) \Rightarrow$
                   $\forall R\,[\neg(R \iota Reg) \Rightarrow \exists Q\,[\beta(P, Q, R) \wedge \text{bdpt}(Reg, Q)]]]$

(R4) $\forall Reg \forall l \exists P$ $[P \iota l \wedge \neg(P \iota Reg)]$

(R5) $\forall Reg \forall Reg'$ $[\forall P\,[P \iota Reg \Leftrightarrow P \iota Reg'] \Rightarrow Reg = Reg']$

Axiom (R4) guarantees that the complement of a region is not a region itself. Since every shape region has a uniquely determined boundary curve, we use the notion $\text{bd}(Reg)$ to refer to the boundary of region *Reg*.

Definition (Boundary of a Region)

$$\mathsf{bd}(Reg) = c \qquad \Leftrightarrow_{\mathrm{def}} \forall P \, [P \iota c \Leftrightarrow \mathsf{bdpt}(Reg, P)]$$

The geometric framework developed in this section is not specialized to the concepts corner and kink that we investigate in the present paper. It can be applied to any contour or planar drawing of a physical object without holes. Since it is formulated without reference to coordinates and differentiability, we have shown that formal representation and characterization of shape features does not presuppose the use of differential geometry. Additionally, the geometric system does not introduce or measure angles. As a consequence, the framework is not strong enough to specify orthogonality, but could be enriched by a group of congruence axioms, if necessary. In the next section we show that this general framework is sufficient to characterize the geometric aspects of corner and kink.

4 Characterizations of Corner and Kink

The framework of planar shape geometry forms the basis for the formal characterizations of corner and kink. Based on the geometric constellation described by these concepts there are several alternatives to select one of the entities involved as the referent of the corresponding nouns. To give an impression of the general spectrum of possibilities and to show the independence of this problem from the geometric specification, we propose a selection of alternatives. The linguistic analysis in the next section discusses these characterizations in order to develop criteria to select among them.

To refer to the mapping of objects to their spatial conceptualization (see section 2), we employ a function named loc. Objects are referred to by O and O'. We use the notion 'linear(loc(O))' to state that the representation of object O is a shape curve and 'planar(loc(O))' to state that it is a shape region.

$$\mathsf{loc: objects} \rightarrow \mathsf{shape\ regions} \cup \mathsf{shape\ curves}$$

4.1 Characterizations of Corner

We give five alternatives for the characterizations of corner. They differ with respect to the entity they directly specify: The referent can be a point, a boundary, a region or an object part. In addition, two characteristic regions are discussed as referents of *Ecke*. They agree in assuming the object in question to be planar and in their specification of the basic geometric structure. All characterizations share the geometric characteristics that an inner vertex and a part of the object's boundary that is constituted by two segments meeting at the vertex are involved.

Corners as Points: Corner_pt

The first characterization (named 'corner_pt') focuses on the vertex of the boundary of the object as the geometric referent of the noun *Ecke*. It says that the corner P of a planar object O is a vertex of O's boundary given by two straight boundary parts (s_1 and s_2) that meet non-smoothly.

$$\text{corner}_{pt}(P, O) \quad \Leftrightarrow_{def} \quad \text{planar}(\text{loc}(O)) \wedge \text{vtx}(P, \text{bd}(\text{loc}(O))) \wedge \exists s_1 \exists s_2 \, [\text{seg}(s_1) \wedge \text{seg}(s_2)$$
$$\wedge \, \text{meet}(P, s_1, s_2) \wedge s_1 \cdot \text{bd}(\text{loc}(O)) \wedge s_2 \cdot \text{bd}(\text{loc}(O))]$$

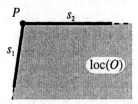

Fig. 8. Depiction of a corner for the characterizations of corner$_{pt}$ and corner$_c$

Corners as Curves: Corner$_c$

The characterization named 'corner$_c$' singles out the boundary part formed by the two meeting segments. It says that a corner c of an object O is a part of the boundary of O that is constituted by two straight boundary parts (s_1 and s_2) that meet in a vertex P. Thus, the geometric referent of the noun *Ecke* is an open curve. This specification is indefinite concerning the extent of the denoted sub-curve. It does not include any assumption concerning the length of the constituting segments. This reflects the indefiniteness of a corner's size discussed in section 2.

$$\text{corner}_c(c, O) \quad \Leftrightarrow_{def} \quad \text{planar}(\text{loc}(O)) \wedge c \cdot \text{bd}(\text{loc}(O)) \wedge \exists P \, [\text{vtx}(P, \text{bd}(\text{loc}(O))) \wedge$$
$$\exists s_1 \exists s_2 \, [\text{seg}(s_1) \wedge \text{seg}(s_2) \wedge \text{meet}(P, s_1, s_2) \wedge c = s_1 \eth s_2]]$$

Corners as Regions: Corner$_{Reg}$

In the third characterization we shift the focus to the region included by the boundary part specified by corner$_c$. It states that the corner Reg of object O is a convex sub-region of the object's region, such that the two regions share a boundary part constituted by straight segments (s_1 and s_2) that meet in a vertex P. This specification is neutral concerning the extent and the exact shape of the region: it can be the triangle or a rectangle enclosed by the boundary part, or some other convex region. (Based on evidence for or against some shape, this characterization could of course be refined.)

$$\text{corner}_{Reg}(Reg, O) \quad \Leftrightarrow_{def} \quad \text{planar}(\text{loc}(O)) \wedge Reg \cdot \text{loc}(O) \wedge \text{cvx}(Reg) \wedge$$
$$\exists P \, [\text{vtx}(P, \text{bd}(\text{loc}(O))) \wedge$$
$$\exists s_1 \exists s_2 \, [\text{seg}(s_1) \wedge \text{seg}(s_2) \wedge \text{meet}(P, s_1, s_2) \wedge$$
$$\forall Q \, [Q \, \imath \, s_1 \eth s_2 \Leftrightarrow \text{bdpt}(\text{loc}(O), Q) \wedge \text{bdpt}(Reg, Q)]]]$$

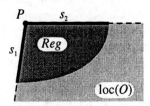

Fig. 9. Depiction of a corner for characterizations of corner$_{Reg}$ and corner$_O$

Corners as Object Parts: Corner$_\text{O}$

The fourth characterization named 'corner$_\text{O}$' is analogous to corner$_\text{Reg}$ except that not the region but an object part occupying the region is singled out. This reflects the intuition that corners are parts of objects. The region of the object part (loc(O')) has to fulfill the same conditions as the region in the characterization of corner$_\text{Reg}$.

$$\text{corner}_\text{O}(O', O) \quad \Leftrightarrow_\text{def} \quad \text{planar}(\text{loc}(O)) \wedge \text{loc}(O') \cdot \text{loc}(O) \wedge \text{cvx}(\text{loc}(O')) \wedge$$
$$\exists P \, [\text{vtx}(P, \text{bd}(\text{loc}(O))) \wedge$$
$$\exists s_1 \, \exists s_2 \, [\text{seg}(s_1) \wedge \text{seg}(s_2) \wedge \text{meet}(P, s_1, s_2) \wedge$$
$$\forall Q \, [Q \, \iota \, s_1 \, \eth \, s_2 \Leftrightarrow \text{bdpt}(\text{loc}(O), Q) \wedge \text{bdpt}(\text{loc}(O'), Q)]]]$$

Corners as Internally Structured Regions: Corner$_\text{sReg}$

In contrast to the characterizations above, corner$_\text{sReg}$ specifies a corner of an object O as an internally structured region Reg (see Fig. 10). It is modeled as a convex region again, though this region is not completely included in the object's region, but intersects with it. Its internal structure includes the vertex P and two segments (s_1 and s_2) that are part of the boundary of the object's region and meet at P. This characterization treats the convex and the concave region parts around the vertex as more symmetric than corner$_\text{Reg}$.

$$\text{corner}_\text{sReg}(Reg, O) \quad \Leftrightarrow_\text{def} \quad \text{planar}(\text{loc}(O)) \wedge \text{cvx}(Reg) \wedge \exists P \, [\text{vtx}(P, \text{bd}(\text{loc}(O))) \wedge$$
$$\exists s_1 \, \exists s_2 \, [\text{seg}(s_1) \wedge \text{seg}(s_2) \wedge \text{meet}(P, s_1, s_2) \wedge$$
$$\forall Q \, [Q \, \iota \, s_1 \, \eth \, s_2 \Leftrightarrow \text{bdpt}(\text{loc}(O), Q) \wedge Q \, \iota \, Reg]]]$$

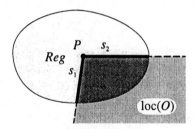

Fig. 10. Depiction of a corner for the characterization of corner$_\text{sReg}$

4.2 Characterizations of Kink

Since kinks are ascribed to linear objects, they can be treated in a manner similar to the first or second characterization of corner. But there are two main differences for the characterization of kink: First of all, the object itself and not its boundary specifies the underlying curve. Secondly, the simple curve parts that form the vertex of a kink need not to be straight. The corresponding sub-curves can be arcs as well.

Kinks as Points: Kink$_\text{pt}$

In the characterization called 'kink$_\text{pt}$', a kink is just a vertex of a linear object.

$$\text{kink}_\text{pt}(P, O) \quad \Leftrightarrow_\text{def} \quad \text{linear}(\text{loc}(O)) \wedge \text{vtx}(P, \text{loc}(O))$$

Kinks as Curve Parts: Kink$_c$

In the second characterization, a kink is a part of the curve including exactly one vertex. The curve is constituted by two simple curves meeting in this vertex.

$$\text{kink}_c(c, O) \qquad \Leftrightarrow_{\text{def}} \quad \text{linear}(\text{loc}(O)) \wedge c \cdot \text{loc}(O) \wedge \exists P \, [\text{vtx}(P, c) \wedge$$
$$\exists c_1 \exists c_2 \, [(\text{seg}(c_1) \vee \text{arc}(c_1)) \wedge (\text{seg}(c_2) \vee \text{arc}(c_2)) \wedge$$
$$\text{meet}(P, c_1, c_2) \wedge c = c_1 \, \eth \, c_2]]$$

The axiomatic characterizations given in this section form the basis for our further discussion of the semantics of *Ecke* as one example of shape nouns. Based on its interaction with spatial prepositions we aim at selecting among the alternatives the proper conceptual counterpart by answering the question, what kind of entities are denoted by *Ecke*?

5 An Analysis of *Ecke*

Having given proposals for the geometric characterization of the concepts corner and kink, we now return to the relation between concepts and lexemes as discussed in section 1 (see Fig. 1). Whereas the axiomatic method is able to yield a variety of conceptual designs for the semantics of shape nouns, the linguistic analysis shall give criteria to choose among them. The aim of discussing the interpretations of *Ecke* in different linguistic contexts is to extract the most appropriate conceptual structure as the semantics of the lexeme.

The notion 'shape noun' reflects the idea that these nouns encode shape features of objects. The referent of the relational noun *Ecke* is established on the basis of the shape of a complex object like a room or a sheet of paper. Shape properties are spatial properties of extended objects, i.e. objects that properly occupy some area. In lexical semantics spatial properties are mainly investigated in relation to prepositions: Topological prepositions like *in, on* or *at* spatially relate two objects and thereby impose requirements on the spatial properties of these objects. This is made explicit by Herskovits (1986), and–for German *in*–by Pribbenow (1993) and Buschbeck-Wolff (1994). Therefore, the analysis of the shape noun *Ecke* in relation to topological prepositions should give us insights concerning the spatial properties of its referents. Our discussion is based on prepositional phrases containing the topological prepositions *in (in)* and *an (at, on)*, and *Ecke* as the head noun of their internal argument. We start by considering the German preposition *in*.

As assumed by Bierwisch (1988), Klein (1991), Herweg (1989) and Wunderlich & Herweg (1991), the semantics of *in* is roughly that an object is localized in the interior of another object; thus *in* yields the relation of local containment. This containment may be realized in different ways; i.e., the localized object being in the interior of a hollow object (*der Stuhl im Zimmer/ the chair in the room*) or being contained in a solid object (*der Splitter im Finger / the splinter in the finger*). Leaving details aside, one object is localized *in* another object, the so-called 'reference object', if and only if the reference object supplies a spatial container for the localized object. Following this line, (4) leads to the demand that *Ecke* has to supply a region in which chairs or spots can be contained. [10]

(4) a. der Stuhl in der Ecke des Zimmers   (the chair in the corner of the room)
  b. der Fleck in der Ecke des Teppichs   (the spot in the corner of the carpet)
  c. der Kratzer in der Ecke des Fensters   (the scratch in the corner of the window)
  d. der Riß in der Ecke des Papiers   (the tear in the corner of the sheet)

In (4.b–d) the localized object is contained in a planar region with a vertex and two boundary parts of the objects as boundaries of the corner. In (4.a), a three-dimensional corner provides the region for the localized three-dimensional object, the chair. [11]

As *in* demands a region, the formal characterizations of *Ecke* to denote a point (corner$_{pt}$) or a curve (corner$_c$) seem to be less plausible. Instead, *Ecke* should be captured either as denoting a region, or an object part that, like other material objects that can occur as reference objects, has to be mapped onto its IN-region, as suggested in Herskovits (1986). But based on (4) we can exclude one of the region-based characterizations of *Ecke*, namely corner$_{sReg}$. *In der Ecke* cannot specify a region outside the underlying object (e.g., the room). Since the region corner$_{sReg}$ spreads outside the object, it does not qualify as the referent of *Ecke*.

Now let us turn to another topological preposition, *an (at)*. Following Herweg (1989), *an* means 'in the region proximal to a reference object'.[12] Obviously, the proximal region that *an* might bring about varies if the corner is a vertex, a curve, an object part or a region. Consider the following examples:

(5) a. Der Kiosk ist an der Ecke.
     (The kiosk is at the corner.)
  b. Die Spinne sitzt an der Ecke des Fensters.
     (The spider is sitting at the corner of the window.)

The interpretation of the examples in (5) allows the localized objects to be inside or outside the corner region as long as they are close enough. In (5.a), the kiosk is nearby a corner, e.g., the corner constituted by a crossing or the corner of a house. It can even be inside the house with the corner. Accordingly, (5.b) does not inform us whether the spider is sitting on the outer edge of the window frame or just *in the corner*, i.e. on the window. Thus, *an der Ecke* allows the localized object to be in the exterior as well as in the interior of the corner with respect to the reference object denoted by *Ecke*.

---

[10] Note that this requirement of supplying a region only holds, if the contained object is two- or three-dimensional. Points and linear objects like kinks can be contained in linearly conceptualized objects like sticks as well (as in *der Knick in dem Stab (the kink in the stick)*).

[11] Examples like *der Stuhl in der Ecke des Teppichs (the chair in the corner of the carpet)* can be seen as evidence that an analysis based on planar conceptualizations would also be sufficient for (4.a), as in this example the relevant containing region is planar.

[12] Herweg (1989) contrasts *an* with *bei (near, by)*, which, according to his analysis, additionally specifies the condition 'not being in contact with the reference object'. Thus, *an* allows contact, but does not require it, whereas *bei* excludes contact. Other authors assume that *an* requires contact and *bei* excludes contact. Note that the difference 'allowing contact' vs. 'requiring contact' for the semantics of *an* is not essential for the following.

In contrast to the results for the preposition *in*, combinations with *an* suggest a more symmetric treatment of the inside and the outside of corners. This prefers the formal characterizations of $corner_{pt}$ and $corner_c$ above $corner_{Reg}$ and $corner_o$.

This observation can have consequences for the semantics of the topological prepositions we employed for analyzing *Ecke*: In connection with nouns denoting concrete objects with clear boundaries, *an* and *in* exclude each other. In other words, if an object is in a reference object then it is not at (*an*) the reference object, and vice versa. We found that this dependence does not exist for *Ecke*. *An der Ecke* and *in der Ecke* may denote overlapping regions. Thus, there is no mutual exclusion of the outside of an object (which is relevant for *an*) and the inside of an object (being relevant for *in*).

One way to handle this may be to analyze *Ecke* on the conceptual level as $corner_c$ and *in* and *an* as mapping this partial contour to regions based on convexity and closeness and contact, respectively. But another consequence might be to have another look at the shape dependencies of *in* and *an* by further analyses of combinations with other shape nouns.

The following examples give further evidence that the choice we planed to make is not clearly decided by the linguistic behavior of *Ecke*. They emphasize the role of the vertex point in the concept of corner. They give evidence that the reference to the vertex can be regarded as essential for the linguistic behavior of *Ecke*.

Considering (6.a), the cupboard has to be localized as close as possible to the wall to make the sentence true. If *genau (exactly)* is omitted, the cupboard might also be standing further away from the wall; *an* thus allows for a proximal region which is not restricted to the closest region. But *genau* restricts the possible ranges of locations for the localized object. *Genau* therefore serves as a test to determine where the exact AN-region is to find, and, in doing so, it indicates which part of the reference object this region has to be close to. In (6.a), it is the surface of the wall.

(6) a. Der Schrank steht genau an der Wand.
      (The cupboard is standing exactly at the wall.)
  b. Die Spinne sitzt genau an der Ecke des Fensters.
      (The spider is sitting exactly at the corner of the window.)

The interpretation of (6.b), is that the spider is sitting as close as possible to the vertex. Hence, we are led to conclude that $corner_{pt}$ is the most plausible characterization of *Ecke* in this context. Example (7) offers additional evidence in this direction.

(7) Der Teppich reicht bis an die Ecke des Zimmers.
    (The carpet reaches the corner of the room.)

In (7), *reichen bis (to reach)* uses the directional version of *an* (obvious from the accusative case of *die Ecke* in contrast to the dative case of *der Ecke* in, e.g., (6.b)). The interpretation we derive is that some edge (or corner) of the carpet is as close as possible to the vertex of the room's corner. Thus, if exactness is emphasized in connection to *Ecke*, closeness (or contact) to the vertex is the preferred interpretation.

Concerning the axiomatic characterization this seems to suggest that *Ecke* refers to the vertex. We even might conclude that $corner_{Reg}$ is inadequate since being in the region is not enough for being exactly at the corner. On the other hand, assuming the vertex to be the referent of *Ecke* leads to, as stated above, more complex explanations for combinations with *in*.

Summarizing the above we find that *Ecke* does not behave as nouns denoting concrete objects do. In connection with *in* and *an* we see that *Ecke* does not clearly distinguish between interior and exterior. In connection with *in*, *Ecke* seems to refer to a region or an object part, and the conceptualization as a vertex is excluded. But the preposition *an* seems to require direct access to the vertex.

We conclude that *Ecke* has a variable denotation within a conceptual spectrum based on a stable geometric characterization. Hence, it behaves like a polysemous word, comparable to institution words like *school, university,* or *government* investigated by Bierwisch (1983). These words offer systematic alternations, including the readings *institution* and *building* (compare *The school bored him to death* and *The school is burning*). These readings correspond to different conceptual variants, comparable to the variants we found for *Ecke*.

In this paper, we have offered five options for conceptually different referents of *Ecke*. We only exclude one of them on the basis of linguistic considerations, namely corner$_{sReg}$, but found evidence for all the others. Thus the spectrum still ranges from a vertex (corner$_{pt}$) via a curve or a region (corner$_c$, corner$_{Reg}$) to an object part (corner$_0$).

6 Conclusion

The present paper offers a formal, geometric approach to describe shape and specify shape concepts, which are part of our spatial knowledge that underlies the semantics of natural language expressions and that diverse cognitive abilities like object recognition and haptic or auditory perception are based on. This characterization is based on features of the contour of an object, rather than its axes.

The formal description is formulated in planar shape geometry developed in this paper as well. Employing the axiomatic method it enriches the basic inventory of planar ordering geometry by shape curves. In this framework we are able to dispense with the use of coordinates and limits, thus showing that a formal description of curves–as needed for the current purpose of describing contours–does not require the notions of differential geometry. Planar shape geometry supplies a general inventory for the description of contours of planar objects without holes. Thus, we offer a tool for future research on contour information, especially concepts that make use of vertices, e.g., concepts related to lexemes like *apex, point, thorn, horn, tip*, etc.

The linguistic analysis of the German nouns *Ecke* and *Knick* sheds light on the underlying concepts corner and kink. It yields the result that the concepts kink and corner primarily differ in dimensionality information. In addition, we found that the behavior of *Ecke* in combination with the topological prepositions *in* and *an* cannot systematically be explained on the basis of only one characterization for corner. Of the five alternative characterizations we offered, we only found evidence to exclude one. This also suggests that the case of analyzing *in* and *an* has to be reopened, especially in combination with shape nouns in addition to nouns denoting concrete material and bounded objects.

References

Biederman, I. (1987). Recognition-by-components: A theory of human image understanding. *Psychological Review, 94*, 115–147.

Biederman, I. (1995). Visual object recognition. In S.M. Kosslyn & D.N. Osherson (eds.), *Visual Cognition—An Invitation to Cognitive Science (2nd ed.) Vol. 2.* (pp. 121–165). Cambridge, MA: MIT.

Bierwisch, M. (1983). Semantische und konzeptuelle Repräsentation lexikalischer Einheiten. In W. Motsch & R. Ruzicka (eds.), *Untersuchungen zur Semantik.* (pp. 61–99). Berlin: Akademie-Verlag.

Bierwisch, M. (1988). On the grammar of local prepositions. In M. Bierwisch, W. Motsch & I. Zimmermann (eds.), *Syntax, Semantik und Lexikon.* (pp. 1–63). Berlin: Akademie-Verlag.

Bierwisch, M. (1996). How much space gets into language? In P. Bloom, M.A. Peterson, L. Nadel & M.F. Garrett (eds.), (pp. 31–76).

Bierwisch, M. & Lang, E. (1989). Somewhat longer—much deeper—further and further. In M. Bierwisch & E. Lang (eds.), *Dimensional Adjectives: Grammatical Structure and Conceptual Interpretation.* (pp. 471–514). Berlin, Heidelberg, New York: Springer.

Buschbeck-Wolff, B. (1994). *Konzeptuelle Interpretation und interlinguabasierte Übersetzung räumlicher Präpositionen* (Working Papers of the Institute for Logic and Linguistics, IBM TR-80.95-015). Stuttgart: IBM.

Bloom, P., Peterson, M.A., Nadel, L. & Garrett, M.F. (eds.) (1996). *Language and Space.* Cambridge, MA: MIT.

Eilan, N., McCarthy, R. & Brewer, B. (eds.) (1993). *Spatial Representations.* Oxford: Blackwell.

Eisenkolb, A., Musto, A., Schill, K. Hernández, D. & Brauer, W. (1998). Representational Levels for the Perception of the Courses of Motion. This volume.

Eschenbach, C. & Kulik, L. (1997). An axiomatic approach to the spatial relations underlying *left–right* and *in front of–behind.* In G. Brewka, C. Habel & B. Nebel (eds.), *KI-97— Advances in Artificial Intelligence.* (pp. 207–218). Berlin: Springer.

Habel, Ch. (1990). Propositional and depictorial representations of spatial knowledge: The case of *path* concepts. In R. Studer (ed.), *Natural Language and Logic.* (pp. 94–117). Berlin: Springer.

Habel, Ch. & Eschenbach, C. (1997). Abstract structures in spatial cognition. In C. Freksa, M. Jantzen & R. Valk (eds.), *Foundations of Computer Science. Potential–Theory–Cognition.* (pp. 369–378). Berlin: Springer.

Henkin, L., Suppes, P. & Tarski, A. (eds.) (1959): *The Axiomatic Method, with Special Reference to Geometry and Physics.* Amsterdam: North-Holland.

Herskovits, A. (1986). *Language and Spatial Cognition.* Cambridge, Eng.: Cambridge University Press.

Herweg, M. (1989). Ansätze zu einer semantischen Beschreibung topologischer Präpositionen. In Ch. Habel, M. Herweg & K. Rehkämper (eds.), *Raumkonzepte in Verstehensprozessen.* (pp. 99–127). Tübingen: Niemeyer.

Hilbert, D. (1899). *Grundlagen der Geometrie.* (8th ed. (1956), with revisions and additions by Paul Bernays.) Stuttgart: Teubner.

Hoffman, D.D. & Richards, W.A. (1984). Parts of recognition. *Cognition, 18*, 65–97.

Hoffman, D.D. & Singh, M. (1997). Salience of visual parts. *Cognition, 63*, 29–78.

Huntington, E.V. (1924). A new set of postulates for betweenness, with proof of complete independence. *Transactions of the American Mathematical Society, 26*, 257–282.

Jackendoff, R. (1991). Parts and boundaries. *Cognition, 41*, 9–45.

Jackendoff, R. (1996). The architecture of the linguistic-spatial interface. In P. Bloom, M.A. Peterson, L. Nadel & M.F. Garrett (eds.), (pp. 1–30).

Klein, W. (1991). Raumausdrücke. *Linguistische Berichte, 132*, 77–114.

Kline, M. (1972). *Mathematical Thought—From Ancient to Modern Times*. New York: Oxford University Press.

Knauff, M., Rauh, R., Schlieder, Ch.. & Strube, G. (1998). Mental Models in Spatial Reasoning. This volume.

Laguna, T. de (1922). Point, line and surface, as sets of solids. *The Journal of Philosophy, 19*, 449–461.

Landau, B. & Jackendoff, R. (1993). "What" and "where" in spatial language and spatial cognition. *Behavioral and Brain Sciences, 16*, 217–238.

Landau, B., Leyton, M., Lynch, E. & Moore, C. (1992). Rigidity, malleability, object kind, and object naming. Paper presented at the Psychonomics Society, St. Louis, MO.

Landau, B., Smith, L., & Jones, S. (1988). The importance of shape in early lexical learning. *Cognitive Development, 3*, 299–321.

Lang, E. (1989). The semantics of dimensional designation of spatial objects. In M. Bierwisch & E. Lang (eds.), *Dimensional Adjectives: Grammatical Structure and Conceptual Interpretation*. (pp. 263–417). Berlin, Heidelberg, New York: Springer.

Luce, R.D., Krantz, D.H., Suppes, P. & Tversky, A. (1990). *Foundations of Measurement. Vol. III. Representation, Axiomatization and Invariance*. San Diego, CA: Academic Press.

Mandler, J. M. (1996). Preverbal representation and language. In P. Bloom, M.A. Peterson, L. Nadel & M.F. Garrett (eds.), (pp. 365–384).

Marr, D. & Nishihara, H.K. (1978). Representation and recognition of the spatial organization of three-dimensional shape. In *Proc. of the Royal Society, Series B, 200*. (pp. 269–294).

Miller, G.A. (1978). Semantic relations among words. In M. Halle, J. Bresnan & G. Miller (eds.), *Linguistic Theory and Psychological Reality*. (pp. 60–117). Cam., MA: MIT.

Miller, G.A. & Johnson-Laird, P. (1976). *Language and Perception*. Cam., MA: Belknap.

Parchomenko, A.S. (1957). *Was ist eine Kurve?* Berlin: Deutscher Verlag der Wissenschaften.

Peterson, M.A., Nadel, L., Bloom, P. & M.F. Garrett (1996). Space and language. In P. Bloom, M.A. Peterson, L. Nadel & M.F. Garrett (eds.), (pp. 553–577).

Pinkal, M. (1985). *Logik und Lexikon: Die Semantik des Unbestimmten*. Berlin: de Gruyter.

Pinker, S. (1984). Visual cognition: An introduction. *Cognition, 18*, 1–63.

Pribbenow, S. (1993). *Räumliche Konzepte in Wissens- und Sprachverarbeitung—Hybride Verarbeitung von Lokalisierung*. Wiesbaden: Deutscher Universitäts-Verlag.

Renz, J. & Nebel, B. (1998). Spatial Reasoning with Topological Information. This Volume.

Rosch, E., Mervis, C., Gray, W., Johnson, D. & Boyes Braem, P. (1976). Basic objects in natural categories. *Cognitive Psychology, 8*, 382–439.

Smith, E.E. (1995). Concepts and categorization. In E.E. Smith & D.H. Osherson (eds.), *Thinking. An Invitation to Cognitive Science (2nd ed.) Vol. 3*. (pp. 3–33). Cambridge, MA: MIT.

Vieu, L. (1993). A logical framework for reasoning about space. In A.U. Frank & I. Campari (eds.), *Spatial Information Theory. A Theoretical Basis for GIS*. (pp. 25–35). Berlin: Springer.

Wunderlich, D. & Herweg, M. (1991). Lokale und Direktionale. In A. von Stechow & D. Wunderlich (eds.), *Semantik*. (pp. 758–785). Berlin, New York: de Gruyter.

Typicality Effects in the Categorization of Spatial Relations

Constanze Vorwerg & Gert Rickheit[1]

Universität Bielefeld, SFB 360 „Situierte Künstliche Kommunikatoren"
Postfach 100131, 33501 Bielefeld, Germany
e-mail: vorwerg@nov1.lili.uni-bielefeld.de, rickheit@lili.uni-bielefeld.de

Abstract. The chapter provides an overview of linguistic, neuropsychological and experimental psychological approaches and findings that support the idea that spatial relation categories are analog, overlapping, internally structured categories based on prototype comparison and with fuzzy boundaries. The main focus is on viewpoint dependent relations (direction relations) in visuospatial cognition. The notion of a frame of reference in spatial cognition is related to the more general concept of a frame of reference in categorization. Categorization constitutes the bridge between spatial vision and spatial language. For visual space, a spatial framework is proposed that is based on perceptually salient directions which act as standard values in relation to which object relations can be judged.

1 Introduction

The question how we link up spatial expressions with our representations of object relations has received particular attention from cognitive scientists in different disciplines in recent years. This great interest is related to the general theoretically important issue of how language and conceptual representational system map onto each other. It is also motivated by the conceptual primacy of space assumed by cognitive linguistics, and new application needs in domains, such as medicine techniques, navigational systems, Geographic Information Systems, human-machine interfaces to CAD and multimedia systems, and robotics. The general trend toward interdisciplinary research in recent years has opened novel ways to the study of spatial relations and locative terms as well as enhanced the attraction of this field, which is predestined for interdisciplinarity.

[1] This work was supported by the Special Research Group „Situated Artificial Communicators" (SFB 360) of the German Science Foundation (DFG), project B1: „Interaction of visual and linguistic information processing".

1.1 Background

Our research interests concerning spatial cognition focus on the relation between spatial language and spatial vision. The integration of linguistic and spatiovisual information processing is necessary in the cooperative solution of an assembly task in a certain situation. What cognitive abilities are needed to solve such a task and how they can be transmitted and implemented in artificial communicators is explored in the Special Research Group „Situated Artificial Communicators" at the University of Bielefeld. Our long-term aim is the development of an integrated system with interacting visual, linguistic, senso-motoric, and other cognitive abilities that can take on the role of the human partner in the accomplishment of assembly tasks. The system is supposed to understand building instructions given by a human and to carry them out. It will be equipped with a stereo camera and has to relate verbal instructions to the observed construction scene. As an empirical basis, a setting has been chosen that involves the cooperative assembly of toy airplanes using a wooden construction kit. In such a setting, communication is situated in a given spatio-temporal context and the processing of visual and linguistically encoded information have to be mapped onto each other. An object's location in space is one of its fundamental attributes and can be used to identify this object. Accordingly, locative specifications are often used in context discriminative object naming (see Herrmann & Grabowski, 1994).

1.2 Spatial Relations

Usually, one object's position is expressed in terms of another one's[2]. That secondary object is used as a reference object. Several factors have been identified to influence the choice of a reference object, as size, mobility, salience, knowledge of speaker and listener, and - in localization sequences - cohesion strategies (see Herrmann & Grabowski, 1994; Herskovits; 1986; Miller & Johnson-Laird, 1976; Talmy, 1983; Vandeloise, 1991). As the localized object and the reference object are usually not interchangeable, spatial relations can be called cognitively asymmetrical. (This can be illustrated by the classic example given by Talmy, 1978. Compare the sentences: THE BIKE IS NEAR THE HOUSE. and: THE HOUSE IS NEAR THE BIKE.)

The visually perceived location is specified by perceived direction and distance (Loomis et al., 1996). Many spatial expressions denote either distance or directional relations. Both types may combine in natural language use (see Schober, 1993). Distance relations are often called *topological relations* borrowing the Piagetian metaphor, because this type of spatial relation are independent of the position of the observer (A IS NEAR TO B). In contrast, directional relations appear more bound to the observer. (An object may be located to the left of the observer now, but to her/his right after s/he does an about-face.) Objects are located with respect to each other in

---

2 In some relations, more than one relatum is used for localizing the intended object, such as between.

terms of their relation to the observer. (A IS TO THE LEFT OF B, or BEHIND B - depending on the position of the observer.) This type of spatial relation moves with the observer (see O'Keefe & Nadel, 1978). Since these are relations in terms of a particular perspective or point of view they are referred to as *projective relations* (Moore, 1976). Understanding projective relations requires taking into account the point of viewing.

1.3 Directional Relations and Reference Frames

Our primary focus here is on viewpoint related localizations. Three elements are needed to establish a projective relation: the intended object, a relatum (or reference object), and a point of view (which determines the *frame of reference*). Relatum and point of view may coincide when the relatum possesses an intrinsic orientation (a *two-point* localization, in the terminology of Herrmann, 1990). In that case, the inherent axes of the reference object determine the frame of reference. A different viewpoint is adopted when objects are located with respect to each other in terms of their relation to either a third object, or the observer/speaker, or the addressee (a *three-point* localization; for a systematic taxonomy of horizontal projective relations, see Herrmann, 1990). An overview of classifications and terminology concerning projective prepositions is given in (Retz-Schmidt, 1988).

A variety of vocabulary describing spatial perspective options is used, including *deictic*, *intrinsic*, and sometimes *extrinsic* perspectives (common in psycholinguistics and linguistics) as well as *viewer-centered*, *object-centered*, and *environment-centered* reference frames (mainly in vision research). It is generally agreed that the interpretation of projective relation terms is only possible taking into account the used frame of reference[3] (see Carlson-Radvansky & Irwin, e.g.). The common usage of the term „frame of reference" in spatial cognition research is similar to its use in physics. Consideration of frames of reference is a key element in phenomena involving velocity and displacement or sameness of place (see Bowden et al., 1992; Brewer & Pears, 1993). Suppose that a pair of glasses is on somebody's nose, as they were an hour ago, but an hour ago that person was in a different room in the house. Are the glasses in the same place as they were an hour ago ? (The example has been taken from Brewer & Pears, 1993). Other examples include a ball rolling toward the back of a train that is traveling forward at a constant speed (Bowden et al., 1992) or a pen falling off a table: the viewer can only know which object moved if one of them has a stationary setting within a reference-frame, with respect to which the displacement of

---

3 In navigation and large-scale space contexts, a common distinction is drawn between egocentric systems of spatial representation in which things are located within spatial frameworks fixed to body parts (the eye, the head, the body, etc.) and an (exo- or) allocentric (maplike) representation that seems to be referenced to the environment. the term. Such a cognitive mapping system enables the organism to move the point of view without actual physical movement in the environment, to view the environment from any vantage point (O'Keefe, 1993; O'Keefe & Nadel, 1978).

the other object receives characterization (Talmy, 1978). There is no way to distinguish uniform motion from rest, neither physically nor in perception. It is purely relative. This „Galilean Principle of Relativity" has been generalized by Einstein to the „Specific Relativity Principle" in 1905. In 1632, Galilei wrote (Galilei, 1632, transl. 1967, p. 116):

> *„Motion, in so far as it is and acts as motion, to that extent exists relatively to things that lack it; and among things which share equally in any motion, it does not act, and is as if it did not exist. Thus the goods with which a ship is laden leaving Venice, pass by Corfu, by Crete, by Cyprus and go to Aleppo. Venice, Corfu, Crete, etc. stand still and do not move with the ship; but as to the sacks, boxes, and bundles with which the boat is laden and with respect to the ship itself, the motion from Venice to Syria is as nothing, and in no way alters the relation among themselves. "*

However, the notion of a frame of reference is by no means restricted to the area of spatial cognition. Following its usage in physics, gestalt psychology used the term to describe the fact that an entity in perception is qualified out of its relation to (preceding and concurrent) elements of the whole situation. Important factors are adaptation level and context as well as (dynamically) memorized or physiologically determined standards. Generally, perception and categorization can be understood as scaling w.r.t. a frame of reference (see Thomas, Lusky, & Morrison, 1992, e.g.); examples are contrast effects, differing temperature sensations depending on prior effects, or the judged height of houses. Both, standards given by memory representation and by the actual situation work together in categorizations. All categorizations require reference frames: a set of values to which each given stimulus can be referred; e.g., focal colors in color vision, adaptation levels in velocity perception, or known size distributions of African elephants.

Given the basic notion of a frame of reference for this quite distinct empirical domain as well as for spatial representation, one interesting question to ask is whether these two concepts have something in common and what the connection is between them.

1.4 Direction Terms as Categories: Use of Cognitive Reference Points

In 1925 Wertheimer combined three of his already published papers in a small book: „Drei Abhandlungen zur Gestaltheorie". One of them (Wertheimer, 1925a) deals with the perception of motion and laid the foundation of a gestalt theory of reference frames; a second one (Wertheimer, 1925b) suggested that among perceptual and abstract categorical entities, there are certain „ideal types" which act as the anchoring points for perception and thinking. This proposal was taken up in a series of studies by Rosch (1975a). She concluded that focal colors act as *cognitive reference points* in relation to which other colors are judged and argued that such reference points form cognitive „prototypes" for the categories (Rosch, 1977). Evidence for being internally structured in a similar way was found for different categories, such as geometrical forms (circle, square, and equilateral triangles), judgements of physical distance, and facial expressions (see Rosch, 1977; Smith & Medin, 1981, for a review of the findings). Following Rosch (1977), natural categories can be characterized in terms of prototypes (clearest cases, best examples) and deviations from prototypes; for many categories, the process of categorization can be treated as an analog function, categories having unclear boundaries.

Different studies have shown the relevance of *typicality* (degree of category membership; goodness of example) for the *processing* of a category (Rosch, 1977), for both general attribute domains and categories of concrete objects. Those studies include verification time experiments (e.g., Rips et al., 1973), priming techniques (Rosch, 1975a), order and probability of item output (Rosch, Simpson & Miller, 1976), and the use of linguistic hedges (Rosch, 1975b). When object attributes are only partially correlated or when attributes are continuous, they are cognitively maintained as distinct by being cognitively coded in terms of prototypes and distance from the prototypes (Rosch, 1977).

However, Rosch stressed that „prototypes only constrain but do not specify representation and process models" (1978, p. 41) and that the „relative typicality of an instance, on her account, could be the result of a variety of structural principles" (Rips, 1989), e.g., contingency relations, cue validity, central tendency or family resemblance. „Prototypes appear to be just those members of a category which most reflect the redundancy structure of the category as a whole." (Rosch, 1977, p. 36). In a somewhat different approach, Reed (1972) has mathematically defined a prototype as the average pattern in a category („that pattern which has for each component x_m the mean value of the mth component of all other patterns in that category"; p. 386f.). According to this definition, a prototype neither needs to be an actual member of the category nor is mathematically equivalent to an average distance model (in which the average distance to the patterns in the category is calculated instead of the distance to the average pattern).

The question of structural principles concerns mainly concrete object categories, formed on the basis of bundles of perceptual and functional attributes. In *attribute categories*, prototypes can be conceptualized as values on a stimulus dimension — physiologically based in some perceptual categories —that serve as reference points in relation to which other stimuli of the domain are judged. We propose that the position of an object relative to a *reference object* or *relatum*, which can be specified in terms of direction, is one of those attribute categories. Direction is similar to categories like color, in that qualitatively differing reference points can be identified (adjectives denoting such attributes are often called „absolute" in linguistics), in contrast to other spatial domains, such as size attributes, (adjectives referring to them called „relative", accordingly). Those reference values constituting direction categories include LEFT, RIGHT, IN-FRONT, BEHIND, ABOVE, and BELOW.[4] The LEFT-RIGHT, FRONT-BEHIND, and ABOVE-BELOW axes and their origin (*origo*)

---

4 Using the geocentric frame of reference would yield to the cardinal directions south, north, west, and east (plus above, and below).

make up the frame of reference; places are individuated by their spatial relations to them.

Linguistic, psychological, and computational considerations suggest that spatial domains are segmented into categories in a manner akin to other categorical structures (Bialystock & Olson, 1987; Hayward & Tarr, 1995; Landau & Jackendoff, 1993; Regier, 1995; Talmy, 1983). There is increasing evidence that spatial categories, such as projective relations, are *not* discrete, mutually exclusive either-or-categories based on critical features with well defined boundaries. Converging experimental and computational results support the idea that projective relation categories are analog, overlapping, internally structured categories based on prototype comparison and with fuzzy boundaries (for an alternative view, see Bialystock & Olson, 1987). In the following sections, we will present some of the evidence that spatial categories, such as projective relations, possess an analog prototype structure and are processed in terms of the prototype and distance from prototype. Furthermore, we will argue that spatial reference frames are a special case of the broader notion of a frame of reference in perception and categorization and that typicality gradients in visuospatial cognition can be put down to the fact that orientation and directional preferences in vision can act as cognitive reference values in relation to which a given spatial relation can be judged. A considerable amount of studies has addressed the question *what* different kinds of reference frames can be employed in spatial cognition; our intention here is to explore *how* one certain frame of reference might be used in categorical judgments on spatial relations. While the intended object is located exactly at one of axes of the reference frame in most studies on the choice of reference frames, we will focus on research on the question how all the deviating positions can be categorized. In these studies, the type of reference frame is usually held constant. We think, verbal object localization will result from an *interaction* of both processes: choice of reference frame and deviation-from-reference-axis computation in many situations (a reference frame might be chosen exactly because of a small prototype deviation).

2 Linguistic Analyses

Talmy (1983) has provided an extensive cross-linguistic study of spatial expressions, aimed „beyond pure description of spatial categories to an account of their common fundamental character and place within larger linguistic-cognitive systems" (p. 225). His analysis shows that *schematization*, a process involving the systematic selection of particular aspects of a spatial configuration while ignoring the remaining aspects, plays a fundamental role in linguistic space descriptions. Following Talmy's analysis, the cognitive processes attending schematization involve decision-making (concerning alternatives of schematization and degree of specificity) on the part of the speaker, and image-constructing (depending on this selection) on the part of the listener. Rather than a exhaustive, contiguous array of specific references partitioning a semantic domain, language provides under-specific general terms to refer to different spatial configurations. There is a small number of references in a scattered distribution over a semantic domain, such as spatial relations, providing a *representative* scattering rather than a comprehensive classification. „The particular schematic abstractions that are represented by individual spatial expressions, such as English prepositions, can be called *schemas*" (Talmy, 1983, p. 258).

The basic properties of schematization include *abstraction, idealization,* and a *topological* kind of plasticity. Abstraction and idealization are complementary properties. While idealization refers to the process of conceptually mapping a spatial entity to a schema applied to it (e.g., a pencil or a skyscraper is idealized as a line, when used with the preposition ALONG), abstraction includes disregarding the rest of it (e.g., for the use of ACROSS, it is irrelevant whether an object has or lacks side boundaries, as in the cases of a bed or a river, respectively). According to Talmy, the ACROSS schema can be characterized as a trajectory from one side to another on a level surface bounded by two relatively long parallel edges, forming a right angle to both edges. The term „topology" is here used to refer to a sort of further abstraction away from any metric specificity as to shape or magnitude of idealized physical objects, also to angles or distances between them. For example, the two edges required for the ACROSS schema, need not be veridical parallel lines; such, ACROSS can be used referring to a path of motion on a lake, where the opposite sides are without uniformity.

The processes of idealization and topology require a cognitive capacity for abstraction and allow for a great flexibility of language. The same spatial configuration can be conceptualized according to alternative schemata; on the other hand, a whole range of spatial configurations has to be captured by the same spatial expression. All possible spatial configurations are to be represented by a small set of expressable schemata. The speaker has to choose the closest available schema in order to linguistically encode a spatial relation. Talmy's investigation has shown that specific terms are well-distributed over semantic space; they usually do *not* have neighbors of equal specificity: spatial references are not partitioning spatial domains in a contiguous and exhaustive way, but rather are representative of them. It can be concluded from this that the naming of a spatial relation requires a comparison between a given spatial configuration and available schemata (or prototypes; see Rosch, 1977) and that a spatial configuration can deviate from a schema to a variable degree and be more or less typical or representative for a given spatial category (see also, Hayward & Tarr, 1995).

3   Orientation and Directional Preferences in Vision

Given that spatial categories are represented by schemata or prototypes, an important question is what factors determine the positioning of them in semantic space. Talmy (1983) argues that — besides some factors, such as frequency of occurrence or cultural significance — their location must be to a great extent arbitrary, an assumption based on the observed enormous amount of non-correspondence between specific morphemes of different languages. Contrary to this viewpoint, Rosch (1977, e.g.) holds that categories are *not* arbitrary[5] and that the psychological principles underlying the formation of categories are subject to investigation. Categories of concrete objects reflect the high correlational structure objects of the world are perceived to possess, determined by many factors, in particular functional needs (Rosch, 1978) with a variety of structural principles accounting for their formation. However, some attribute categories probably have a physiological basis (Rosch, 1977) and an intrinsic qualitative distinctiveness (Bornstein, 1987). For colors (Berlin & Kay, 1969; Kay & McDaniel, 1978; Kay, Berlin, & Merrifield, 1991), forms (see Rosch, 1977), and tastes (Steiner, 1977), there is evidence that categories originate in and are constrained by perceptually salient stimuli. Hence, an interesting question is whether certain orientations or directions might act as perceptually — possibly physiologically — based cognitive reference points. And indeed, a number of experimental results show evidence for a preference of particular orientations and directions in perception.

Orientation effects in visual perception. Ogilvie and Taylor (1958, 1959) showed for several types of test that visibility of a fine wire is better in a horizontal or vertical orientation than in oblique orientations. Lashley (1938) found that rats could learn to discriminate a horizontal pattern from a vertical one much more easily than a right-oblique from a left-oblique (although the two patterns were separated by 180 degrees in both cases). For both adults and children, similarity judgments have been shown to be easier to make for horizontal and vertical line segments than for diagonal line segments (Arnheim, 1974; Palmer, 1977). Orientation also has a strong influence on the perception of shape; vertically or horizontally oriented areas are preferentially perceived as objects. Symmetry tends to be recognized more easily if related to a vertical axis (Rock, 1973). The described orientation effects show the existence of preferred (i.e., perceptually salient) orientations/directions in the visual system, which might in principal provide reference values for spatial relations within a viewer-centered frame of reference.

Neural mechanisms and representations. The above examples of perceptual orientation saliencies can be found again in neural representation. The visual system provides information on the visual horizontal, the visual vertical, and orientation of lines on several processing levels. The retinotopic mapping is preserved in the visual cortex. The *vertical* meridian of the visual field is represented at the boundary between area 17 (Visual Area I) and area 18 (Visual Area II) of the contralateral hemisphere. This representation of the vertical meridian has been shown to be the boundary between a medial (Visual Area I) and a lateral (Visual Area II) retinotopic

---

5 It should be noted here that arbitrariness in the sense of conventionalization is very likely to contribute to the formation of categories, lexical items, and language usage norms in a culture (as exemplified by the preposition use for the spatial relation of a passenger to a bus: in English one says on the bus, whereas in German in the bus is used. The English usage of the platform schema instead of the enclosure schema can historically or diachronically be explained in that it was originally applied to topless carts and stages (Talmy, 1983).

representation of the visual half-field in the cat and the monkey; the vertical meridian being the only retinal region that projects to only one region on the visual cortex. The fovea is represented at the occipital pole. The *horizontal* meridian is represented (in the medial map) at the Calcarine Sulcus with the superior half of the visual field being represented ventrally, and the inferior dorsally, with respect to the calcarine sulcus. Besides the retino-geniculo-cortical projection, there is also a projection of the visual field to the superior colliculi, which are involved in the control of visual orienting reflexes. A retinotopic mapping can be found, in which the *vertical* and the *horizontal* meridians of the visual field cross at the fixation point. Taken together, the perceptual and neural saliency of the visual vertical and the horizontal meridian might account for the intrinsic qualitative distinctiveness of spatial relations reflected in spatial language compared to other spatial attribute domains such as size, length, or distance, in which only quantitative distinctions can be drawn.

Neural selectivity. Contrary to the lateral geniculate body, there is a profound reorganization of the incoming messages in the cortex. One of the features of cortical neurons discovered by Hubel & Wiesel (1959) is their *orientation selectivity*. Whereas some simple cells depend in their reaction on an optimal stimulus (long narrow dark or light rectangles, or edges) of particular orientation and retinal position, complex cells respond to the appropriate axis orientation of an elongated stimulus *irrespective of its exact shape and position*. In a similar way, some visual neurons (ganglions cells as well as simple and complex cells in the visual cortex and visually responsive cells outside area 17) exhibit *directional selectivity* — a preference for stimuli moving in a particular direction. These findings demonstrate the ability of the visual system to abstract spatial information from visual input and the special role of direction/orientation in neural representation.

Spatial encoding. A Euclidean frame of reference for spatially oriented perception and cognition is provided by the three mutual orthogonal canal planes of the (gaze dependent) vestibular system, which is relevant for the perception of the vertical, maintaining an upright posture and visual orientation constancy (Berthoz, 1991). Aligned with these planes of the semi-circular canals are the preferred directions of activation found in a neural network of several parallel pathways that is specialized in the processing of visual motion (Berthoz, 1991; Cohen & Henn, 1988). The vestibular system is one of the factors guiding oculomotor activity, as exemplified by the vestibulo-ocular reflex. The geometry of the canals is paralleled by the directions of the oculomotor system (Berthoz, 1991), which in turn plays an important role in human binocular spatial vision. Rotations of the globe of the eye can occur about the visual axis, the vertical axis and the horizontal axis. Directing the visual axis straight ahead through the crossing point of vertical and horizontal meridian is called the *primary position*, whereas merely vertical or horizontal movements from primary position place the eyes in *secondary position* (all remaining positions are called *tertiary*). Direction of gaze (requiring eye-head coordination) plays an important role in visual localization (Haustein, 1992) and accounts for the obvious capability of fairly accurate judgments of visual directions (see Loomis et al., 1996) - as opposed to distances.

Conclusion. Based on the reviewed evidence, it can be assumed that the reference points constituting direction categories include LEFT, RIGHT, IN-FRONT, BEHIND, ABOVE, and BELOW are related to physiologically anchored preferred directions in perception. The particular LEFT-RIGHT, FRONT-BEHIND, and ABOVE-BELOW axes and their origin (*origo*) establishing a deictic frame of reference depend on the point of view superimposed.

4 Reports from Memory

Huttenlocher, Hedges, & Duncan (1991) proposed a model of category effects on reports of particulars and applied this model to the estimation of spatial location. When memory is inexact, schemata may be used as retrieval and reconstruction aid (Bartlett, 1932; Alba & Hasher, 1983). The model proposed by Huttenlocher et al. includes estimation processes combining the (inexactly) remembered stimulus value with category information: boundary values in truncation processes and a prototype value in weighting. They found that people when reporting the location of a dot in a circle spontaneously impose horizontal and vertical lines that divide the circle into quadrants. These lines serve as reference points in that dots are systematically misplaced away from them with distance to them being a strong predictor of bias (being strongest near them). Huttenlocher et al. argue that these reference directions (half-axes from the center to the circumference line) constitute the boundaries of four spatial categories with values near the angular center of each quadrant (neutral towards the horizontal and the vertical axes) at the point of zero bias being their prototypes. It is a somewhat surprising result that the prototypes should lie along the obliques. An alternative account, in our view, might consider the horizontal and vertical half-axes as prototypes (instead of boundaries) and *deviations* from prototypes could be encoded resulting in a cognitive enhancement of these deviations by contrast. This account would attribute the obtained prototype effects rather to encoding than to retrieval processes, contrary to the model proposed by Huttenlocher et al. (1991). A the time being, we are conducting an experiment to explore this hypothesis. Irrespective of what are the precise processing principles producing the obtained categorization data, these data clearly show prototype effects in memorizing angular directions and the use of the vertical and the horizontal axes as reference directions.

For space surrounding oneself, Franklin, Henkel, & Zangas (1995) have shown that spatial categories, named by projective expressions, such as adpositions or adverbs, have fuzzy boundaries, overlap each other, and seem to be defined with respect to their corresponding canonical pole (0°, 90°, 180°, and 270°). They found FRONT — the perceptually and functionally most salient region of surrounding space (see Clark, 1973) — to be largest and to be recalled with the greatest precision. Not only was the absolute error of reported location significantly smaller for FRONT than for other regions but absolute error also increased as a function of distance in degree from the FRONT pole. Furthermore, errors were biased away from the FRONT and BEHIND pole toward LEFT and RIGHT. In terms of typicality effects, the last two findings are the most interesting ones. In an egocentric trunk-centered (subjects were allowed to move head and shoulders) frame of reference, the FRONT/BEHIND axis seems to constitute the most important reference point and to predict error and bias patterns.

5 Linguistic Hedges

Natural languages possess means for expressing gradients of category membership. The term „hedges" has been coined by Lakoff to refer to those qualifying expressions, such as „almost", „virtually", or „exactly" (see Rosch, 1977). Use of hedges was one of the linguistic variables studied in another experiment by Franklin et al. (1995), in which subjects described the directions of objects placed in various *positions around themselves*. Nonspatial qualifiers used by the subjects in their descriptions were rated according to their „emphasis" on a particular direction (e.g., „directly" would be rated higher than „almost", a low value would be assigned to „slightly" or „kind of"). It was found that emphasis decreases as a function of deviation in degrees from the nearest pole.

We obtained similar results for the categorization of spatial relations in *visual* space (Vorwerg & Rickheit, in prep.). German native speakers were asked to name the spatial position of an intended object with respect to a reference object with no intrinsic orientation. Hedges were used to qualify the degree of direction category membership. According to the experimental setting, there were six distances between located object and reference object. Examples of frequently used hedges dependent on the experimentally varied proximity between the object's location and the vertical/horizontal axis are given in table 1.

Table 1. Hedges used to qualify the degree of direction category membership

German hedge	English translation
genau/exakt; direkt	exactly; directly
fast; fast genau; nicht ganz	almost; almost exactly; not quite
sehr leicht; (ein) bißchen; ein Stück	very slightly; a little bit; a bit
leicht; etwas	slightly; somewhat
versetzt	shifted
schräg	oblique

6 Use of Direction Terms

Another linguistic measure of typicality of a given direction for a direction category is the proportion of use of direction terms associated with that direction category. Typicality (degree of category membership) should be indicated by a gradient of frequency of use of spatial expressions. Hayward & Tarr (1995) studied the use of vertically and horizontally oriented prepositions in verbal descriptions of a configuration consisting of a reference object (with intrinsic TOP/DOWN axis oriented coaxially to the geocentric and egocentric axes) and an object located at one of 48 systematically varied positions. The authors found a tendency by subjects to use single prepositions (i.e., to assign just one direction category) only in positions that were directly aligned vertically or horizontally with the reference object. An analysis of the proportion use of the respective prepositions generated first in a description revealed a maximum use at locations along the vertical and the horizontal axes, respectively, and a gradual decrease with growing distance from the particular axis. (The absolute level of use as primary descriptor was found to be clearly higher for vertically oriented prepositions.)

We used a similar design to investigate the apprehension of *horizontal* direction categories (front, behind, left, right) in 3D space and did find a gradient of proportion of use, too. In an experimental evaluation of a spatial computational model developed in our research group (Fuhr et al., 1995), subjects were asked to describe spatial configurations consisting of a reference object with no intrinsic sides and another object located in one of 72 positions around the reference object (Vorwerg et al., 1997). Four different orientations (0°, 45°, 90°, 135°) of the (elongated) reference object were studied. A graded use of single directional terms was found for all direction categories in each orientation of the reference object (see Fig. 1). Distance from the nearest reference direction was found to predict the relative frequency of use.

From both experiments follows that spatial relations are not categorized in an all-or-non fashion. Rather, proximity or similarity between a given direction and a cognitively used reference direction is determined. These results hold true for 2D (Hayward & Tarr) as well as for 3D (Vorwerg et al., 1997) space. Comparable gradients were found for the use of direction terms in the vertical (Hayward & Tarr) and in the horizontal (Vorwerg et al., 1997) plane. Additionally, we were able to show that frequency gradients *within* areas adjacent to one side of the reference object. The cognitive reference direction indicated by the best agreement between subjects seems to result from an interaction between point of view and nearest point (edge, corner) of the intended object w.r.t. the reference object (Vorwerg et al., 1997).

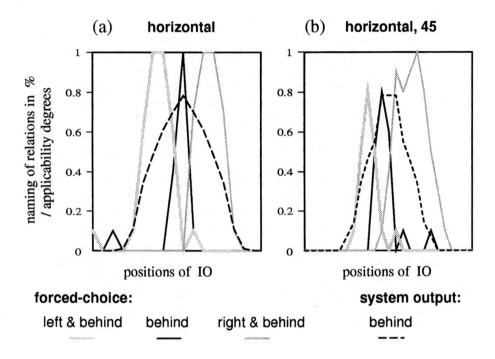

Fig. 1. Acceptability judgments (Vorwerg et al., 1997).

7 Acceptability Judgments

Typicality can be described in terms of similarity to representative category members and dissimilarity to representatives of contrasting categories. Several studies have addressed the question of how good projectivec terms are judged to describe a particular spatial relation.

Using the same displays as described in section 5, Hayward and Tarr (1995) asked subjects to provide a rating of goodness of a spatial preposition to describe the relationship between the located object and the reference object. They found predominant regions of acceptability along the horizontal and the vertical axes. Ratings were determined by the distance from the axis and in part by the angle between reference and located object. Following Hayward & Tarr, the predominant regions along the orienting axes can be interpreted as *prototypical regions* for a given spatial term; the applicability of a term at a particular position seems to vary with the distance from the prototypical axis and with the absolute and angular distance to the reference object.

Vorwerg et al. (1997) studied the rated acceptability of horizontal projective terms generated by a system implementation of a spatial computational model (Fuhr et al.,

1995). The spatial configurations used were the same as described in section 5. The use of elongated reference objects (bars) allowed to compare the influence of distance from the reference object and distance from the horizontal and sagittal axes, as distances between orientational axis and edges differ according to the orientation of the bar. Locations nearer to the prototypical axis proved to be rated higher in acceptability even if they have the same absolute or angular difference from the edges of the reference object, although equal ratings were predicted by the spatial model.

Gapp (1997) used a radial array of locations around one of differently sized quadratic or rectangular reference objects. His data show a linear decrease of applicability with increase of angular deviation from the respective axis. A projective relation was fully applicable if there was no angular deviation. Ratings depended on the extensions of the reference object. Further, results from this study suggest that it may be proximal angular deviation (to the edge of the reference object) rather than the angular deviation in relation to the center of mass, that underlies projective term acceptability ratings. This parameter had not been varied by Hayward & Tarr, but contradicts the above mentioned finding of Vorwerg et al., that proximity to the axis significantly contributes to the acceptability of a projective term even for locations with the same deviation to the edge of the reference object. Our results suggest that both factors, position of a proximal point of the reference object and proximity to view-point and center-of-mass defined axes, contribute to the applicability of a direction term.

8 Chronometric Methods

Several categorization studies have used tasks that required subjects to verify category membership of a given particular. Responses have found to be faster for items that had been rated more typical (see Rosch, 1977). Those findings „demonstrate that the internal structure of categories has an effect on cognition" (Rosch, 1977, p. 23). Spatial attention is necessary to map conceptual representations onto perceptual representations of spatial relations (Logan, 1994, 1995). Hence, assuming that prototype effects reflect in some way the conceptual representation of directional categories, reaction times in direction related verification tasks can be predicted to vary with distance from prototypical values.

A reaction time paradigm was applied in an experiment studying the assignment of horizontal projective terms (Vorwerg, 1997). An array of 120 systematically varied positions (11 by 11 with exception of those positions covered by the respective reference object) was used. In each trial, the reference object could have one of four different orientations (0°, 22,5°, 45°, 90°). Objects with no intrinsic front were chosen to avoid the use of conflicting intrinsic and deictic reference frames. The located object was a ring in order to prevent problems of axial alignment with the reference object. A bar was used as a reference object in order to be able to investigate orientation effects (these are not of concern here and will be presented elsewhere). The bar was located at the center of the array. Therefore, five equidistant locations exist at both sides of the reference object. The reference object's width was approximately the diameter of the ring, its length corresponded five adjacent positions in the array, so that six positions in line with reference object remained (three at each side) on the one hand and five positions existed within the area adjacent to one long side of the bar on the other hand.

Subjects were asked to decide for each possible configuration of one located and one reference object either if the ring is left to the bar or right or neither (LEFT-RIGHT-condition), or if the ring is in front of the bar or behind or neither (FRONT-BEHIND-condition). This task was chosen instead of a simple YES-NO reaction, in order to prevent subjects from simply dividing the whole scene by drawing an imaginary horizontal or sagittal line. One of the results obtained is a significant increase of reaction times with growing distance from the nearest axis (see Fig. 2, 3 showing the overall relations between reaction time and distance for all configurations). It should be noted that reaction time gradients can even be found *within* regions adjacent to a side of the reference object (contrary to what is sometimes suggested in the literature).

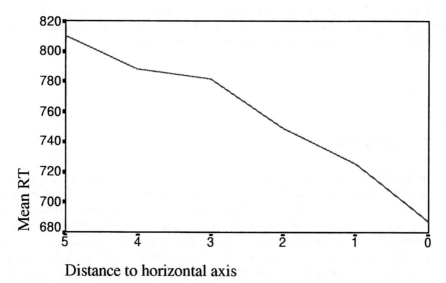

Distance to horizontal axis

Fig. 2. Verification of left and right. The dependence of reaction time on the distance to the horizontal axis is shown. A value of 0 indicates location at the axis, a value of 5 indicates the furthest location of the intended object

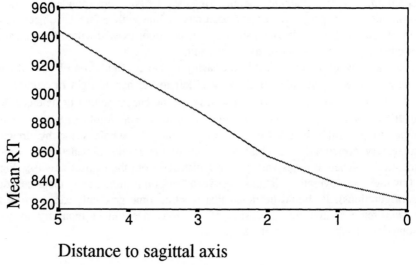

Fig. 3. Verification of front and behind. The dependence of reaction time on the distance to the vertical axis is shown. A value of 0 indicates location at the axis, a value of 5 indicates the furthest location of the intended object

9 Conclusion

In this chapter, we have presented some of the evidence that spatial categories, such as projective relations, possess a graded structure. Several kinds of empirical data suggest an analog structure around perceptually salient prototypes, similar to other (non-binary) attribute domains. A prototype view of spatial categorization is also supported by the prototype effects obtained in a connectionist model (Regier, 1996), although this behavior was not explicitly trained. The model's graded responses to two-object configurations for spatial terms correspond well with empirically found response gradients. Graded applicability of spatial relation terms has found its way into several computational spatial models (e.g., Abella & Kender, 1993; Fuhr et al., 1995; Gapp, 1997).

The perceptually salient prototypes are presumably based on direction/orientation preferences within the visual system (the visual vertical, the horizontal meridian, additionally the obliques might play a smaller role). Factors influencing the membership degree of a given spatial relation to a spatial relation category include the distance from point-of-view and center-of-mass determined axes and the distance from prominent or proximal parts (edges, corners, etc.) of the reference object. Furthermore, an interaction between thus determined deviations from prototypes and the choice of a certain frame of reference. Categorization of spatial relations seems to be guided by the same principles as other kinds of categorization phenomena. Reference objects, in our opinion, do not simply partition space into regions with clear-cut boundaries. Instead, the categorization of a spatial relation can be regarded as a process involving different kinds of information.

It should be noted that categorization is always context-dependent (for instability in category representation, see Barsalou, 1985). The context of confusable alternatives, the functional relation between objects (Carlson-Radvansky & Radvansky, 1996; Coventry, Carmichael, & Garrod, 1994), the purpose of localization (compare identification and exact description) can have an effect on the categorization of spatial relations. The relative importance of reference points may depend not only on the particular task, but also on the point of view determining the frame of reference; such, for some frames of reference, the binocular visual direction (with a cyclopean eye; Mansfield & Legge, 1997) will probably play an important role in determining the FRONT/BEHIND direction (see also, Bühler, 1934). All these factors contribute to the flexible and adaptive nature of categorization as do the changeability of category boundaries and the variability of category assignments of particulars provided by actual comparison processes with situationally, perceptually, and conceptually given points of reference (representatives of cognitive categories).

Our intention in this chapter was to give an overview of empirical results on the question *how* people use frames of reference in visuospatial cognition. A typical finding are gradients of category membership similar to results in other categorization domains. It is concluded that the apprehension of spatial relations can be regarded as a categorization phenomenon. Therefore known categorization effects such as contrast effects can be expected and have been found to occur with spatial cognition. Direction categories are particular in that they seem to be based on visual preferences and show qualitative distinctiveness. The notion of a frame of reference for spatial representation is related to the broader concept of reference frames categorization. For direction relations, the frame of reference is determined by point of view and reference object. It provides the *cognitive reference points* in relation to which object relations are judged.

References

Abella, A. & Kender, J. R. (1993). Qualitatively describing objects using spatial prepositions. *Proceedings of AAAI-93*, 536-540.

Alba, J. W. & Hasher, L. (1983). Is memory schematic? *Psychological Bulletin, 93*, 203-231.

Arnheim, R. (1974). *Art and visual perception: A psychology of the creative eye.* Berkeley: University of California Press.

Barsalou, L. W. (1985). Ideals, central tendency, and frequency of instantiation as determinants of graded structure in categories. *Journal of Experimental Psychology: Learning, Memory and Cognition, 11*, 629-654.

Bartlett, F. C. (1932). *Remembering.* Cambridge: Cambridge University Press.

Berlin, B. & Kay, P. (1969). *Basic color terms. Their universality and evolution.* Berkeley: University of California Press.

Berthoz, A. (1991). Reference frames for the perception and control of movement. In J. Paillard (Ed.), *Brain and space* (pp. 81-111). Oxford: Oxford University Press.

Bornstein, M. H. (1987). Perceptual categories in vision and audition. In S. Harnad (Ed.), *Categorical perception. The groundwork of cognition* (pp. 287-300). Cambridge: Cambridge University Press.

Bowden, J., Dall'Alba, G., Martin, E., Laurillard, D., Marton, F., Masters, G., Ramsden, P., Stephanou, A. & Walsh, E. (1992). Displacement, velocity, and frames of reference: Phenomenographic studies of students' understanding and some implications for teaching and assessment. *American Journal of Physics, 60*, 262-269.

Brewer, B. & Pears, J. (1993). Introduction: Frames of reference. In N. Eilan, R. McCarthy & B. Brewer (Eds.), *Spatial representation* (pp. 25-30). Oxford: Blackwell.

Bühler, K. (1934). *Sprachtheorie*. Jena: Gustav Fischer.

Carlson-Radvansky, L. A. & Radvansky, G. A. (1996). The influence of functional relations on spatial term selection. *Psychological Science, 7*, 56-60.

Clark, H. (1973). Space, time, semantics, and the child. In T. Moore (Ed.), *Cognitive development and the acquisition of language* (pp. 27-63) Academic Press.

Cohen, B. & Henn, V. (eds.). (1988). *Representation of three-dimensional space in the vistibular, oculomotor, and visual systems*. New York: The New York Academy of Sciences.

Coventry, K. R., Carmichael, R. & Garrod, S. C. (1994). Spatial prepositions, object-specific function, and task requirements. *Journal of Semantics, 11*, 289-309.

Franklin, N., Henkel, L. A. & Zangas, T. (1995). Parsing surrounding space into regions. *Memory and Cognition, 23*, 397-407.

Fuhr, T., Socher, G., Scheering, C. & Sagerer, G. (1995). A three-dimensional spatial model for the interpretation of image data. In P. L. Olivier (Ed.), *IJCAI-95 Workshop on Representation and Processing of Spatial Expressions* (pp. 93-102), Montreal.

Galilei, G. (1632). *Dialogue concerning the two chief world systems, Ptolemaic and Copernican*. Berkeley: University of California (Transl., 1967).

Gapp, K. (1997). *Objektlokalisation. Ein System zur sprachlichen Raumbeschreibung*. Wiesbaden: Deutscher Universitätsverlag.

Haustein, W. (1992). Head-centric visual localization with lateral body tilt. *Vision Research, 32*, 669-673.

Hayward, W. G. & Tarr, M. J. (1995). Spatial language and spatial representation. *Cognition 55*, 39-84.

Herrmann, T. (1990). Vor, hinter, rechts und links: das 6H-Modell. *Zeitschrift für Literaturwissenschaft und Linguistik, 78*, 117-140.

Herrmann, T. & Grabowski, J. (1994). *Sprechen. Psychologie der Sprachproduktion*. Heidelberg: Spektrum Akademischer Verlag.

Herskovits, A. (1986). *Language and spatial cognition: An interdisciplinary study of the prepositions in English*. Cambridge: Cambridge University Press.

Hubel, D. H. & Wiesel, T. N. (1959). Receptive fields of single neurons in the cat's striate cortex. *Journal of Physiology, 148*, 574-591.

Huttenlocher, J., Hedges, L. & Duncan, S. (1991). Categories and particulars: Prototype effects in estimating spatial location. *Psychological Review, 98*, 352-376.

Kay, P., Berlin, B. & Merrifield, W. (1991). Biocultural implications of systems of color naming. *Journal of Linguistic Anthropology, 1*, 12-25.

Kay, P. & McDaniel, C. K. (1978). The linguistic significance of the meanings of basic color terms. *Language, 54*, 610-646.

O'Keefe, J. (1993). Kant and the sea-horse: An essay in the neurophilosophy of space. In N. Eilan, R. McCarthy & B. Brewer (Eds.), *Spatial representation* (pp. 43-64). Oxford: Blackwell.

O'Keefe, J. & Nadel, L. (1978). *The hippocampus as a cognitive map.* Oxford: Oxford University Press.

Landau, B. & Jackendoff, R. (1993). "What" and "where" in spatial language and spatial cognition. *Behavioral and Brain Sciences, 16*, 217-265.

Lashley, K. S. (1938). The mechanism of vision: XV. Preliminary studies of the rats' capacity for detailed vision. *Journal of General Psychology, 18*, 123-193.

Logan, G. D. (1994). Spatial attention and the apprehension of spatial relations. *Journal of Experimental Psychology: Human Perception and Performance, 5*, 1015-1036.

Logan, G. D. (1995). Linguistic and conceptual control of visual spatial attention. *Cognitive Psychology, 28*, 103-174.

Loomis, J. M., Da Silva, J. A., Philbeck, J. W. & Fukusima, S. S. (1996). Visual perception of location and distance. *Current Directions in Psychological Science, 3*, 72-77.

Mansfield, J. S. & Legge, G. E. (1997). Binocular visual direction: Reply to Banks, van Ee & Backus (1997). *Vision Research, 37*, 1610-1613.

Miller, G. & Johnson-Laird, P. N. (1976). *Language and perception.* Cambridge: Cambrige University Press.

Moore, G. T. (1976). Theory and research on the development of environmental knowing. In G. T. Moore & R. G. Golledge (Eds.), *Environmental knowing* (pp. 138-164). Stroudsburg, Penn.: Dowden, Hutchinson & Ross.

Ogilvie, J. C. & Taylor, M. M. (1958). Effects of orientation of the visibility of a fine line. *Journal of the Optical Society of America, 48*, 628-629.

Ogilvie, J. C. & Taylor, M.M. (1959). Effect of length on the visibility of a fine line. *Journal of the Optical Society of America, 49*, 898-900.

Palmer, S. E. (1977). Hierarchical structure in perceptual representation. *Cognitive Psychology, 9*, 441-474.

Reed, S. K. (1972). Pattern recognition and categorization. *Cognitive Psychology, 3*, 382-407.

Regier, T. (1995). A model of the human capacity for categorizing spatial relations. *Cognitive linguistics, 6*, 63-88.

Regier, T. (1996). *The human semantic potential. Spatial language and constrained connectionism.* Cambridge, MA: MIT Press.

Retz-Schmidt, G. (1988). Various views on spatial prepositions. *AI magazine, 9*, 95-105.

Rips, L. J. (1989). Similarity, typicality, and categorization. In S. Vosniadou & A. Ortony (Eds.), *Similarity and analogical reasoning.* Cambridge: Cambridge University Press.

Rips, L. J., Shoben, E. J. & Smith, E. E. (1973). Semantic distance and the verification of semantic relations. *Journal of Verbal Learning and Verbal Behaior, 14*, 665-681.

Rock, I. (1973). *Orientation and form.* New York: Academic Press.

Rosch, E. (1975a). The nature of mental codes for color categories. *Journal of Experimental Psychology: Human Perception and Performance, 1*, 303-322.

Rosch, E. (1975b). Cognitive reference points. *Cognitive Psychology, 7*, 532-547.

Rosch, E. (1977). Human categorization. In N. Warren (Ed.), *Studies in cross-cultural psychology* (pp. 1-49). London: Academic Press.

Rosch, E. (1978). Principles of categorization. In E. Rosch & B. B. Lloyd (Eds.), *Cognition and categorization* (pp. 27-48). Hillsdale, N.J.: Lawrence Erlbaum Associates.

Rosch, E., Simpson, C. & Miller, R. S. (1976). Structural bases of typicality effects. *Journal of Experimental Psychology: Human Perception and Performance, 2*, 491-502.

Schober, M. F. (1993). Spatial perspective taking in conversation. *Cognition, 47*, 1-24.

Smith, E. & Medin, D. L. (1981). *Categories and concepts*. Cambridge, MA: Harvard University Press.

Steiner, J. E. (1977). Facial expressions of the neonate infant indicating the hedonics of food-related chemical stimuli. In J. M. Weiffenbach (Ed.), *Taste and development*. Bethesda, MD: U.S. Department of Health, Education, and Welfare.

Talmy, L. (1978). Figure and ground in complex sentences. In J. H. Greenberg (Ed.), *Universals of human language* (pp. 625-649). Stanford/CA: Stanford University Press.

Talmy, L. (1983). How language structures space. In H. Pick & L. Acredolo (Eds.), *Spatial orientation: Theory, research and application* (pp. 225-282). Stanford: Stanford University Press.

Thomas, D. R., Lusky, M. & Morrison, S. (1992). A comparison of generalization functions and frame reference effects in different training paradigms. *Perception & Psychophysics, 51*, 529-540.

Vandeloise, C. (1991). *Spatial propositions: A case study from French*. Chicago: Chicago University Press.

Vorwerg, C. (1997). Kategorisierung von Richtungsrelationen. In E. van der Meer et al. (Eds.), *Experimentelle Psychologie. Abstracts der 39. Tagung experimentell arbeitender Psychologen* (p. 225). Lengerich: Pabst Science Publishers.

Vorwerg, C. & Rickheit, G. (in prep.). Richtungsausdrücke und Heckenbildung beim sprachlichen Lokalisieren von Objekten im visuellen Raum. (Direction terms and hedges in verbal localisations of objects in visual space)

Vorwerg, C., Socher, G., Fuhr, T., Sagerer, G. & Rickheit, G. (1997). Projective relations for 3D space: Computational model, application, and psychological evaluation. *Proceedings of AAAI-97* (pp. 159-164). Menlo Park, CA: AAAI Press / The MIT Press.

Wertheimer, M. (1925a). Experimentelle Studien über das Sehen von Bewegung. In M. Wertheimer, *Drei Abhandlungen zur Gestalttheorie* (pp. 1-105). Erlangen: Verlag der Philosophischen Akademie.

Wertheimer, M. (1925b). Über das Denken der Naturvölker, Zahlen und Zahlgebilde. In M. Wertheimer, *Drei Abhandlungen zur Gestalttheorie* (pp. 106-163). Erlangen: Verlag der Philosophischen Akademie.

The Use of Locative Expressions in Dependence of the Spatial Relation between Target and Reference Object in Two-Dimensional Layouts[*]

Hubert D. Zimmer[1], Harry R. Speiser[1], Jörg Baus[2],
Anselm Blocher[3], Eva Stopp[3]

[1] Department of Psychology, University of Saarland
PO Box 151150, D- 66041 Saarbrücken
huzimmer@rz.uni-sb.de; h.speiser@rz.uni-sb.de
[2] SFB 378, University of Saarland
PO Box 151150, D- 66041 Saarbrücken
baus@cs.uni-sb.de
[3] Department of Artificial Intelligence, University of Saarland
PO Box 151150, D- 66041 Saarbrücken
blocher@cs.uni-sb.de; stopp@cs.uni-sb.de

Abstract. In two experiments we investigated the use of German locative expressions as a function of the spatial relation between a reference object (RO) and a to-be-located object (LO). In the experiments, a speaker described to another participant, by locative expressions, where LO can be found in relation to RO. LO (a blue dot) was presented at different positions around RO (a red dot). The listener saw RO only, and her or his task was to find LO by moving a small window over the screen using the computer mouse. The positions of LO were circularly arranged around RO and their angular relations were varied in steps of 15 degrees. In Experiment 1, only the four canonical expressions (left/right, above/below) and their single composites were allowed. In Experiment 2, no constraints were made. Both experiments yielded comparable results. The canonical expressions were used nearly exclusively for prototypical relations, and their production latencies were the shortest. Composites were used for all non-prototypical relations. There was only a small spatial area next to the canonical directions in which two different locative expressions were used, and in these areas of competition the longest production times were observed. Thus, canonical expressions were used in a much smaller area around the prototypical axes than predicted by selection rules based on applicability ratings obtained in meta-linguistic judgments.

---

[*] This research was supported by a grant from the German Research Foundation in the Special Research Division SFB 378 'Resource-Adaptive Cognitive Processes'.

1 Introduction

Imagine a situation in which someone is standing in front of a city map which is mounted at the wall of a tourist office. She or he is searching for a target position on this map, e.g., the TV tower. The actual position is highlighted as a red spot which can be easily found. However, dependent on the density of the depicted information, it might be hard to find the intended location. The visitor might ask another person for help, and this person might answer: "The TV tower is to the right of the red spot, which indicates your present position". This is the task we wanted to simulate in the experiments.

We are interested in the use of projective locative expressions. More precisely speaking, we are interested in the use of such expressions depending on the angular spatial relation between the to-be-located object and a reference. We investigated the use of the four canonical locative expressions in German - 'unten' [below], 'oben' [above], 'links' [on the left], 'rechts' [on the right] - and their single composites - 'links oben', [on the left and above], 'rechts oben' [on the right and above], 'links unten' [on the left and below] and 'rechts unten' [on the right and below].

Projective locative expressions are comprised of three components: a to-be-located object, a reference object, and a directional spatial expression, e.g., 'on the left of', 'above', etc. (Herskovits 1985, Logan 1995). In the case of a person wanting to describe a location using verbal means, the speaker must explicitly encode spatial relations into verbal expressions. We assume that people have a non-verbal spatial representation of a perceived layout, and that they map this non-verbal representation onto a verbal one. During the verbal encoding process, two basic problems have to be solved: a reference system has to be specified, and a locative expression has to be selected to indicate the target position within this reference system. We assume, that for this purpose, the speaker projects a dimensional coordinate system onto the spatial layout, and dependent on the angular relation between RO and LO a verbal expression is then selected which specifies the intended region within these coordinates.

Furthermore, we assume that the origin of the coordinate system is located in the reference point, and that the orientation of the coordinates is given by the perspective taken by the speaker and adopted by the listener (cf., Herrmann, 1996). The point of view can either be suggested by the reference object itself (the intrinsic case) or by an external point of observation (the perspective of the speaker or the listener) (cf., Grabowski, 1996; Levelt 1989; Retz-Schmidt 1988). In the example, the reference point will most likely be the highlighted position (the red spot), and the speaker and the listener share their point of view, i.e. they behave as if they were standing side by side facing the map.

The use of a specific coordinate system depends on the cultural habits (Klein, 1994). Members of the Western culture most likely use a frame of reference with three canonical axes: up-down, left-right, and front-back (e.g., Bryant, Tversky & Franklin, 1992). Under certain conditions, one of these axes can be ignored, so that the system is reduced to two dimensions. In the example, it is reduced to a vertical plane. Projective locative expressions label spatial areas in the direction of the four base axes, e.g., 'to the right of' or 'above'. However, these expressions are ambiguous in respect to the spatial area which they indicate. For instance, 'to the right of' can correspond (a) to the right half plane, (b) to a more or less extended sector to the right of the reference object, and (c) to a position that is on the right axis or in a narrow spatial area next to it (cf. Herskovits, 1985, p. 181). A position on the right diagonal in a horizontal plane, for example, is therefore equally acceptable as being 'to the right' as well as being 'in front of' the reference position (cf. Franklin, Henkel & Zangas, 1995). In other words, the spatial expressions are not mutually exclusive. Consequently, there is no one to one correspondence between a spatial location and an expression, forcing the speaker to select among a set of spatial expressions which are correct in a logical sense.

In order to solve this problem, we have to define a selection mechanism or a rule to select the expression. For this purpose, we can use the construct of *applicability*. The applicability of an expression is the degree to which a real spatial relation fits the ideal or prototypical condition which would represent the meaning of that expression in the best way. The ideal meaning of a projective expression is defined by a focal area which holds a specific spatial angular relation to the reference object. For example, "'to the right' is an instance of a graded concept with the right axis as its focal region" Herskovits (1986, p. 184). By this definition for each spatial expression, a field of decreasing applicability is defined with it's maximum at the prototypical spatial area. The speaker might then select the expression which has the highest applicability for the spatial region, and if more than one expression is left, the most specific one might be used (cf. Levelt (1989) for a similar general conception of the selection of a lexical entry).

Based on these theoretical assumptions, the applicability of a projective spatial expression (and also its selection) should primarily be dependent on the deviation of the angular direction in reality from the prototypical angular direction. Empirical support for this relationship was reported by Gapp (1995). He presented his subjects with configurations of two simple geometric objects with different angular relations between a to-be-located object and a reference object, and he asked his subjects to judge the applicability of different expressions from 0 to 1 (from not applicable to perfectly applicable). He observed that the ratings decreased with increasing angular deviations from the prototypical positions. The same result was reported by Hayward & Tarr (1995) who also investigated the applicability of projective spatial expressions (left, right, above, below) dependent on the angular direction between RO and LO. Due to these results Gapp (1997) used the applicability as a means to select the best fitting spatial expression in a verbal localization task, in which a computer automatically generated spatial descriptions. Similarly, Logan & Sadler (1996) suggested a spatial template rule which was also based on the idea of a varying degree of applicability. These authors subdivided the spatial area in three fields of applicability to describe the use of the four locative expressions. The prototypical direction (good), the 'not acceptable' area (a deviation of 90° and more) and the acceptable region (the area between the prototypical and the not acceptable area).

However, although these studies revealed very consistent data, they have the weakness of being based on metalinguistic judgments regarding language use, and this behavior does not necessarily correspond to a real production situation. Subjects rated the applicability of a *given* spatial expression to describe a specific angular relation between LO and RO. This is a listener's perspective. The evaluation process, from this perspective might differ from the conditions in which the real language output is generated. Metalinguistic judgments might therefore not reflect the true selection conditions. It is also possible that the ratings reflect rational arguments which give the presented examples a post-hoc plausibility. In the real production situation the observed language output might have been shown for a different reason. There exits empirical data supporting this assumption. Franklin, Henkel & Zangas (1995), for example, observed that subjects used more often 'left' and 'right' than 'in front of' which had been expected from the metalinguistic judgments, and Hayward & Tarr (1995) reported that the use of 'left and 'right' was restricted to the area of prototypical use, although these expressions were accepted as applicable in a much wider area.

These data provide evidence that it is necessary to investigate the use of these spatial expressions, not only by judgments, but also in overt production. In order to do this, we simulated, in the laboratory, a production situation in which a speaker had to describe a spatial relation between LO and RO to a listener who only knew the position of RO. We introduced the listener to make the production conditions more real and to set up a situation in which locative expressions had a real pragmatic function. In addition, we systematically varied the spatial relation between RO and LO. We did this the same way it was done in the study from Gapp (1995) in which the applicability was rated.

2 Experiment 1

In the experiment, the production of locative expressions was investigated under conditions which simulated everyday situations. A speaker informed another person, using locative expressions about the position of an object which was being searched for. Therefore, the subjects took part in the experiment in pairs. One subject served as speaker, and the other as listener.

Subjects produced spatial descriptions to a visual display presenting two small dots, one served as RO and one as LO. LO was placed in different locations circularly arranged around RO. The speaker was instructed to describe where LO could be found by using one of eight spatial expressions: the four canonical expressions ('links', 'rechts', 'oben' 'unten') or one of the four single composites, e.g., 'links oben'. We tape-recorded the produced descriptions so that we could count the frequencies of usage, and we compared it with the frequencies which were expected due to the selection rules based on the applicability ratings of the same eight expressions. In addition, we measured the production latencies (speech-onset-time), by a voice key, as an indicator of the difficulties of selection. The task of the recipient was to find the hidden LO on the computer screen. Using a computer mouse, the listener moved a small window over the display. The LO was then made visible as soon as the window hit the target.

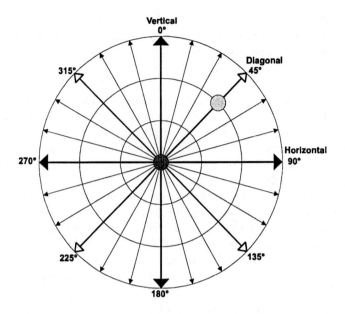

Fig. 1. An illustration of the angular relations obtained in the experiments. LO (the light gray dot) could be presented in each of the 24 directions around RO (the dark gray dot). The lines are for illustration. During the experiment only RO and LO were visible

As an independent variable, we varied the angular relations between LO and RO (cf. Figure 1). We expected that the observed frequencies of expressions and the production latencies were a function of the angular relations between LO and RO. If the use of a spatial expression depends on the deviation of the actual angular relation between LO and RO from the prototypical relation of the expression, we should find that the production frequencies decrease with an increase of deviation from the ideal area of use. In addition, if the selection rules which were based on metalinguistic judgments, can be used for productions (Gapp, 1997), the frequency of using a spatial expression should follow the rating of applicability. Gapp computed the applicability of composite relations in such a way that their applicability decreased faster with angular deviation than the applicability of the canonical locative expressions. Consequently, the sector, in which composite expressions are used, should be smaller than the one for canonical expressions. Composite expressions should therefore be less frequent than canonical expressions.

For the production latencies similar assumptions can be made. It is plausible to predict that the shortest latencies should occur when the angular relation has the best fit to template. The latencies should become longer the more the conditions deviate from the optimal angular relation. The spatial expressions for canonical directions

(0°, 90°, 180°, 270°) should therefore be produced very quickly. If corresponding spatial templates for composite expressions exist, these expressions should behave in the same manner. The diagonal directions (45°, 135°, 225°, 315°, cf. Figure 1) should be named quickly. All other relations which are non-prototypical should be processed more slowly.

However, regarding the production latencies we have to consider a further influence. The selection model discussed so far, is based on the assumption that the 'mental space' is as homogeneous as the physical space. All things being equal, comparing the different directions no anisotropy should exist, and as a consequence, the only factor which influences the selection of a spatial expression is the deviation from the prototypical direction. Contrary to this, data from psychological experiments suggest that the accessibility of spatial information varies with orientation. If subjects mentally accessed an object in a specific location, the above-below direction was easier to evaluate than the ahead-behind direction which was again easier than the left-right direction (Franklin & Tversky, 1990; Logan, 1995). It is therefore possible that the production latencies are influenced by two factors. One is the deviation from the prototypical direction, and the other is the absolute direction in which LO is located.

In addition to these production data, we also measured the search time of the listener from the initiation of the mouse movement up to the moment when the window touched the dot. This time was taken as an indicator for the appropriateness of the used expression. The better the locative expression indicated a specific position, the easier it should have been for the listener to find the dot, therefore shortening the search time.

2.1 Method

Subjects. Thirty students of the University of Saarland took part in the experiment. All subjects were native German speakers and were paid for their participation. All factors were manipulated within subjects.

Material and Design. The experiments were controlled by two HP workstations with a PC connection. The configurations for the speaker and the listener were presented on the workstations' 17'' computer screens. The PC measured the verbal response latencies using a soundcard which was utilized as a voice key.

As 'objects' a blue and a red dot were presented, which were solid circles with a diameter of 1 cm. Circles were chosen because they do not favor a specific orientation of the coordinate system. The red dot was defined as reference object and the blue dot was defined as the to-be-located object.

The reference object was randomly displayed at the corner of an invisible square which was centered in the middle of the screen. This variation should help avoid that subjects use absolute screen positions instead of relative positions, and it should make plausible to use projective expressions.

The to-be-located object was presented on 72 different positions which laid on 24 directions around the RO. Figure 1 illustrates the possible positions. The angles between these lines were 15°. The set of relations consisted of the four canonical directions (0°, 90°, 180°, 270°), the four diagonal directions (45°, 135°, 225°, 315°), and the non-prototypical directions in between. The directions which were prototypical for a spatial expression are highlighted by bold lines in Figure 1. On each line three different positions were defined to allow for variable distance between RO and LO (3.50 cm, 5.25 cm, and 7.00 cm).

Procedure. Subjects were run in pairs. One subject served as speaker, the second subject, the listener, had to find, as quickly as possible, the hidden target on the screen. After the completion of one phase, i.e., all 72 relations were presented, the roles were exchanged. The 72 trials, i.e., different angular relations, were presented in random order.

Subjects were seated in front of their computers. Their stations were separated by a folding screen so that they could talk, but not see each other. At the beginning, both subjects could practice in both roles to familiarize themselves with the complete task. All subjects reported to be experienced in using a computer mouse.

Each trial had the following structure. A short warning signal (a beep) was presented. One second later, the speaker saw RO (the red dot), and 250 ms after stimulus onset LO (the blue dot) was presented. The speaker was instructed to say aloud a locative expression that indicated the spatial position of the target, as quickly as possible. As spatial expressions 'unten' [below], 'oben' [above], 'links' [on the left]', 'rechts' [on the right] and their single composites 'links oben' [upper left], 'rechts oben' [upper right], 'links unten' [lower left] and 'rechts unten' [lower right] were allowed. Subjects were free to choose which way they sequentially arranged the locative expressions within the composites. They could use, for example, 'rechts oben' or 'oben rechts'. – A later analysis revealed that in nearly all cases, subjects started with the horizontal dimension first. – The speaker was instructed to verbalize only the spatial expressions and not to formulate whole sentences. We tape recorded the verbalizations for a later analysis of the used expressions. This procedure should have minimized the contribution of additional sentence planning processes to the production latencies which were outside the scope of our research. Subjects were neither allowed to verbalize absolute positions (e.g., in the top right-hand corner of the screen) nor to specify the distances between RO and LO.

The speaker's production triggered a voice key. This impulse released the presentation of the listener's reference object surrounded by a square the size of the dot. This square served as a search window. It was attached to the mouse pointer, so that the move of the window corresponded the movement of the mouse in speed and direction. The search window was first bound to the position of RO, and it could be unlocked by a mouse click. As soon as the window touched the invisible target, the dot was displayed. A further mouse click finished the current trial and started the next one.

2.2 Results

The frequencies of the used expressions, the production latencies, the comprehension times, and the search times were analyzed. Overall there were only a few invalid trials, 3.4 % were caused by the speaker, and 2.8 % by the recipient. The production latency was defined as the time from the presentation of RO up to the speech onset which triggered the voice key. The comprehension time was estimated using the time from the initiation of the speech up to the first mouse click by the listener. However, because the comprehension times strongly correlated with the production latencies we have not reported these data because we could not exclude the possibility that the comprehension times were influenced by the time the speaker needed to utter the expressions. The search time was measured using the time between this mouse click and the contact of the search window with the invisible dot. The order of runs, i.e. whether the subject first had the role of a speaker or of a recipient, did not influence the results. We therefore excluded this factor from all analyses.

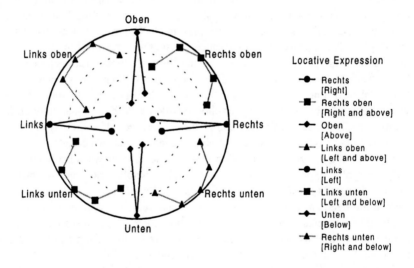

Fig. 2. Frequencies of use of the different spatial expressions (in percentages) as a function of the angular relation between the to-be-located object and the reference object in Experiment 1. The radius of the circle indicates the maximum possible value (0% to 100 %); the concentric dotted circles indicate steps of 25 %. The spatial relations are the same as in Figure 1

First, we analyzed the used spatial expressions. For each expression, the average frequency of its use was computed dependent on the spatial relation between RO and

LO. In Figure 2 these results are illustrated as percentages of use dependent of the spatial relation between LO and RO. Since subjects did not produce wrong descriptions, only the proportions of prototypical and composite expressions are given. Frequencies less than 10% are not shown.

Clearly, subjects used the canonical locative expressions only in the exact prototypical case, i.e. the four main axes. All diagonal and nearly all non-prototypical relations were labeled with composite expressions, and the data were symmetric around these axes. Therefore, for further analyses, the naming frequencies were combined in three classes of expressions each of which corresponded to one prototypical area: the vertical direction (above/below), the horizontal direction (left/right) and the two diagonal directions (single composites). Geometrically, this procedure is equivalent to a projection of all conditions into the upper right-hand quadrant. Figure 3 illustrates the average frequencies of the three types of expressions dependent on the angular deviation from the prototypical area.

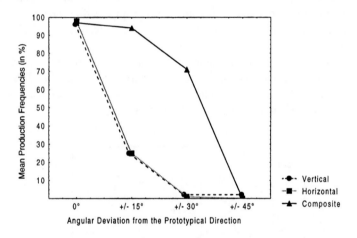

Fig. 3. Frequencies of use of vertical, horizontal and composite expressions (in percentages) as a function of the angular deviation of the to-be-located object from the prototypical direction

As expected, the frequencies of production decreased with the increase of the angular deviation from the prototypical direction. However, the shape of this function is different for the three classes of expressions. A comparison of the vertical and horizontal dimension showed no difference, ($\chi^2(3) = 4.3$, $p > .20$), but both were different from the diagonal dimension, ($\chi^2(3) = 71.1$, $p < .001$). This is reflected by the different slopes of the three frequency curves over angular deviation. At its prototypical area, each of the three 'correct' expressions was used nearly exclusively. However, if the spatial relation deviates only by 15° from this area, for canonical expressions, the curve declined abruptly to zero, whereas the decrease for the composite expressions started at stronger deviations. At 15° deviation, composites were produced as frequently (94 %) as at the prototypical area (97 %), whereas at the same deviation the canonical relations already dropped to a very low frequency of usage (25 %).

Similarly to the frequencies, the production latencies were analyzed. A difficulty with voice key measures is that the point at which the stopwatch is triggered, depends on the amplitude of the speech signal which is partially dependent on the phoneme at the word's beginning (Pechmann, Reetz & Zerbst, 1989). The latencies of different locative expressions which start with different phonemes, e.g., 'links' and 'rechts', are therefore not necessarily comparable. Based on this information, we ran a control experiment to test whether such differences would influence our production latencies. In this experiment, subjects had only to read aloud the visually presented locative expressions, and we measured the production latencies. No significant differences between the two sets of locative expressions starting with 'links' [left] and 'rechts' [right] were found ($F(1,9) < 1$). However, the vertical canonical descriptions ('oben'/'unten') [above/below] were produced about 100 ms more quickly than all other expressions, $F(1,9) = 19.6$, $p < .01$. We should therefore be careful when interpreting effects of this dimension in the production latencies of the main experiment. More important, however, is that the speech onset of all other expressions were comparable.

Due to these results in the control experiment we next analyzed the production latencies as a function of the angular relation between LO and RO. In a first step, the mean latencies for the production were calculated independent of the used expressions. These data are depicted in Figure 4.

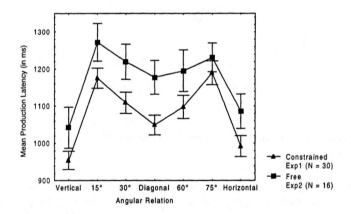

Fig. 4. Average production latencies of the spatial expressions (in ms) as a function of the angular relation between the to-be-located object and the reference point in Experiment 1 (set of locative expressions was pre-specified) and 2 (free production). All relations are projected into the upper right quadrant of the complete circular layout. The error bars indicate plus minus one standard error

In a first analysis, we compared the latencies for the three classes of expressions (vertical, horizontal, and diagonal) in the prototypical directions, i.e., when only the typical expressions were used. Latencies increased from vertical to horizontal to diagonal dimensions, $F(2,58) = 7.5$; $p < .01$, $MSE = 9748$ (946 ms, 994 ms, and 1044 ms, respectively).

In a second analysis, the latencies for the diagonal directions were analyzed with the additional factor angular deviation ($\pm0°$, $\pm15°$). The main effect of angular deviation was significant, $F(1,29) = 6.2$; $p < .02$, $MSE = 5448$. The latencies increased from the exact direction (1045 ms) to the $\pm15°$ condition (1097 ms). In other words, with a stronger deviation from the prototypical direction, longer production latencies were observed. A more extreme deviation ($\pm30°$) produced even longer production latencies (1189 ms). However, this effect cannot be accepted as support of the angular deviation hypothesis. The longer production latencies are probably not due to the stronger angular deviation from the prototypical area, but are more likely dependent on the fact that subjects had to select between two different expressions. This is shown in the following analysis.

In an over all analysis, independent of the used expression, the spatial relations which were next to the canonical dimensions showed the longest production latencies of all conditions, $F(1,29) = 104.9$, $MSE = 8276$, $p < .001$. It should be again noted, that this was the only condition in which different expressions were used. It is therefore likely that, in this case, two expressions are in competition to each other. This assumption is supported by a final analysis in which we compared the production latencies for canonical expressions and composites, both at their prototypical angular relations and at the conditions with competition (cf. Fig. 5).

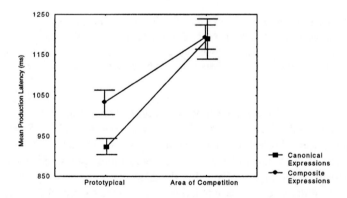

Fig. 5. Production latencies for canonical and composite locative expressions in their prototypical positions and in the area of competition in Experiment 1. The error bars indicate plus minus one standard error

This analysis was based on only 19 subjects because all subjects were excluded who did not use different kinds of locative expressions at the area of competition. The analysis revealed that in the prototypical case composites were produced more slowly than canonical expression, although both were equally adequate. In addition, canonical and composite expressions did not differ from each other in the conditions with competition, and in this case both expressions were produced more slowly than in the prototypical ones, $F(1,18) = 78.6$; $p <.001$, $MSE = 10948$. This does not support the assumption that the production times increase with the degree of angular deviation, because in the competition area canonical expressions only had an angular deviation of 15° from their prototypical area, whereas composite expressions had a deviation of 30° from their prototypical area.

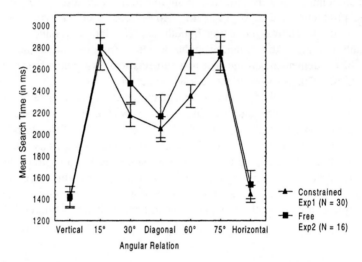

Fig. 6. Average search times (in ms) as a function of the angular relation between the to-be-located object and the reference point in Experiment 1 (set of expressions was pre-specified) and 2 (free production). All relations are projected into the upper right quadrant of the complete circular layout. The error bars indicate plus minus one standard error

Finally, we analyzed the search times (cf. Fig. 6). These times also varied over the angular relations between LO and RO, $F(6,174) = 50.4$, $p < .001$, $MSE = 174992$. The search times were shortest with the canonical relations. Objects in diagonal directions and at ±15° on both sides of it were discovered more slowly, but there was only a small increase from the prototypical to the ±15° condition. The longest search times were observed in those cases where LO was next to the canonical dimensions, i.e., the conditions with competition. Once more, we separated the two cases in which canonical expressions and composites were used. We also observed that the search times for the canonical directions (1411 ms) were faster than that of the prototypical diagonal directions (1977 ms). Subjects who had to locate an object in the competition area, found the targets faster if a canonical expression was used (2354 ms) than if a composite expression was used (3119 ms). For the listener the canonical expression was more useful than the composite expression, although the speaker used the canonical expression less frequently than the composite.

3 Experiment 2

Before we discuss these results, we want to report a second experiment. The data of Experiment 1 were clear-cut, however, they had one restriction. Subjects were instructed as to which set of locative expressions they should select their expression from. This was a necessary restriction in the experiment, due to our interest in the production conditions for the eight locative expressions. However, in order to be able to generalize the data to everyday production conditions, which do not have this restriction, it would be nice to have data from an experiment in which subjects were free to choose any locative expression.

In such an experiment we expected a higher variation of the used spatial expressions. This would make it impossible to analyze the complete data as a function of specific expressions and angular relations. However, if our data were at least partially replicated, we should observe the same results as in Experiment 1 for those subjects who used the locative expressions. We therefore expected fast productions of canonical expressions for prototypical spatial relations. In these cases, in Experiment 1, canonical expressions were used almost exclusively, and this should remain so in a free production condition. With angular deviations from these axes, the variability of expressions might increase but if the results are comparable, with small angular deviations from the canonical axes again composites or other types of qualified expressions should be used. In addition, if different expressions are in competition to each other in the areas, on both sides of the canonical axes, the longest production times should be found. Furthermore, if subjects produce adequate descriptions of the spatial relations, the pattern of the search times over angular relations should not differ from those of Experiment 1.

3.1 Method

Subjects. Thirty students from the University of Saarland who were paid for their participation took part in the experiment. All subjects were native German speakers.

Material and Procedure. The material and the general procedure were the same as in Experiment 1. The only change in the procedure was that now subjects were free in their usage of locative expressions. During the instructions, the subjects were not directed as to which expression should or should not be used. The only remaining restriction was that only verbal descriptions were allowed, e.g., no pointing, etc.

3.2 Results

In this experiment the number of errors was again low. The speaker caused 6.1 % of errors, and the recipient 4.3 %.

As we expected, the variability of the used spatial descriptions increased. Half of the subjects used other reference systems rather than locative expressions. Ten used a reference system analogous to an analog clock, e.g., 'at two o'clock' or 'five to twelve'. Two subjects used circular degrees, two used a geographical reference system (north, east, south-west, etc.), and two subjects mixed different systems. The other 16 subjects used locative expressions. However, for these subjects the variability of expressions increased. In those cases, in Experiment 1 composite expressions were used, more complex qualifications were now more frequently given. Examples are: 'rechts und etwas oben' [to the right and a bit above], 'rechts diagonal oben' [to the right and diagonally above]', etc.

Due to this variability of the used complex locative expressions, we only reported the production and search latencies in dependence with the angular relation between LO and RO independent of the used expressions. Additionally, only those sixteen subjects, who used locative expressions, were further analyzed. For prototypical relations, these data reflect the latencies for canonical locative expressions, because these were nearly exclusively used with these directions. For all other angular relations, the latencies were averaged over all types of expressions. Therefore, these data integrated latencies over different types of expression specifying a canonical locative term. These data are depicted together with those of Experiment 1 in Figure 4 and 6.

These data reveal that the production latencies as well as the search times showed a pattern that was similar to Experiment 1. The production latencies strongly varied with the angular relation between LO and RO $F(6,90) = 11.5$; $p < .001$, $MSE = 9389$. Canonical expressions, with prototypical positions were produced faster than all expressions at other spatial relations, $F(1,15) = 29.4$; $p < .001$, $MSE = 18786$. The longest production latencies were again observed at the area of competition, i.e. ±15 beside the canonical directions, $F(1,15) = 33,7$; $p < .001$, $MSE = 7856$. The differences between the exact diagonal and the intermediate spatial relations, however, were even smaller than in Experiment 1, and these latencies did not significantly differ from each other, although the advantage of the diagonal relation was numerically present.

If one compares the production latencies in the two experiments in a two way analysis of variance with Experiment as an additional factor, and Spatial relation (canonical or diagonal) as the second factor, the two main effects were significant. The production latencies increased from Experiment 1, constrained production: 1011 ms, to Experiment 2, unconstrained production: 1121 ms, $F(1,44) = 6.2$; $p < .02$, $MSE = 40407$. Furthermore, locative expressions were produced more quickly for canonical (1005 ms) than for diagonal relations (1094 ms), $F(1,44) = 23.5$; $p < .001$, $MSE = 7988$. Furthermore, locative expressions were produced more quickly for canonical (1005 ms) than for diagonal relations (1094 ms), $F(1,44) = 23.5$, $MSE = 7988$, $p < .001$. Finally, the general pattern of the production latencies, as a function of the angular relation between LO and RO, was the same in both experiments ($F < 1$).

The same is true for the search times (cf. Fig. 6). The search times were shorter for the canonical directions than for all other spatial relations, $F(1,15) = 148.9$; $p < .001$, $MSE = 2848586$, and the times in the area of conflict were the longest, $F(1,15) =$

49.1; $p < .001$, $MSE = 3506595$. The search times, in the diagonal direction, were longer than in the canonical direction but shorter than in the area of conflict. If one compares the search times in the two experiments neither the factor Experiment nor the interaction was significant.

4 General Discussion

The experiment should reveal how the use of locative expressions depends on the spatial relation between LO and RO, and how it depends on the deviation of LO's direction from the prototypical axes of the specific locative expressions. In addition, we wanted to test whether the actual usage of the locative expressions corresponded to the frequencies which were predicted on the basis of the applicability ratings. We got a clear answer to both questions. The use of locative expressions clearly depended on the angular relation between LO and RO but the frequencies were different from those which we expected from the applicability ratings.

In the prototypical regions of all expressions, we observed that in these ideal cases the 'correct' expression was used exclusively. However, the two classes of expressions showed different usage frequency patterns dependent on angular deviation when outside of the prototypical region. The canonical expressions were only used in their prototypical area. In all other directions composite relations were used which was even the case if LO was displaced from the prototypical axis by only 15°. Canonical relations therefore had a very small area of use, whereas composite relations had a wide area of application. From the selection rule, which was used by Gapp (1997), we had expected that composite expressions would have a narrow area of use, and canonical ones a large area of use, which was the opposite of the result.

Two possibilities exist to adjust the model to the observed data. One possibility is to find a different selection rule to modulate the observed data using the existent functions. The alternative is to use different applicability functions, possibly, in addition, with another selection rule. The applicability ratings, as metalinguistic judgments, might differ from the applicability values which are actually used in a production situation. This possibility is supported by the fact that also others reported deviations between the usage of locative expressions and the given ratings (e.g., Franklin, Henkel & Zangas, 1995; Hayward & Tarr, 1995). Perhaps this change could even be observed when subjects were required to judge the 'probability of use' and not the 'applicability of the expression'. Such a difference between the listener's view and the speaker's (functional) view is plausible, if one considers that 'what is acceptable' might differ from those things 'which are suitable or which are optimal to reach the communicative goal', i.e., a quick search for LO. Since locative expressions serve as means to reach the latter goal, one should consider this functional perspective as an alternative to the metalinguistic perspective when modulating the selection of locative expressions.

The production latencies followed a similar pattern as the frequencies. The productions of expressions at canonical directions were the fastest. It took longer to produce the composite expressions to diagonal relations, although these spatial relations matched the ideal direction as well as the horizontal and vertical relations matched the ideal spatial relations for canonical expressions. The longest production times were observed in the area of competition, in which both types of relations, i.e. composite and canonical expressions, were used. However, the production latencies did not directly follow the angular deviations. It looked more like an ordinal relationship. The verbalizations of the canonical relations were the fastest, followed by the composites which in a large sector around the diagonal direction were produced nearly at the same speed. In the area of competition, the longest production times were observed. This pattern of results is in good correspondence to the idea of a spatial template (cf. Logan & Sadler, 1996), if one adds the assumption that composite locative expressions are used outside of the prototypical area of the canonical locative expressions.

Based on these results, we can speculate which cognitive processes caused these data. We assume, that two groups of processes are operating. One set specifies the canonical locative expressions. These processes are quickly finished as soon as the coordinate system is adjusted to RO. With the frame of reference, the prototypical directions for the four canonical locative expressions are given. For example, if LO is located "in a strip with axis the right base axis, and abutting on the reference object" (Herskovits, 1986, p.184) the locative expression 'to the right of' is immediately accessed. The same is true for the other half axes.

If the object is clearly not located in these positions, but in the diagonal sector, other processes are operating. Here we can think of two possibilities. One alternative is, combining two locative expressions that label the axes of the corresponding quadrant of LO. The two expressions should already be activated, because the application of a reference system specifies the applicability of canonical directions, i.e., the half planes bounded by a coordinate axis. These expressions are then combined to a composite expression. In this case, no additional evaluation processes for the applicability of diagonal relations are necessary, but only an additional process which combines the canonical terms.

Another alternative is to treat composites as primary expressions. In this case, a further process might be started which evaluates the exact diagonal direction as the prototypical relation for composites. In support of the latter assumption, the small advantage of the exact diagonal compared to the intermediate positions can be mentioned. However, this might not be a general effect. It might only occur in constrained situations, as in Experiment 1. There, the composite expressions were explicitly pre-specified as the alternatives to be used. In Experiment 2, no expressions were mentioned. In this unconstrained condition, the advantage of the exact diagonal direction was clearly reduced. The latter effect speaks in favor of the first alternative.

Finally, we have to assume the existence of a third component: an additional mutual inhibition of two possible expressions, followed by an active selection. These processes are necessary if two expressions are available, which compete with each other. In our experiments, this is the case at directions near to but not on the canonical axes. In this condition, one expression has to be selected, and vice versa the other one has to be inhibited. This additional processing slows down the production latencies by about 200 ms. Such delays were also observed in other verbalization tasks in which two words compete with each other, and one has to be inhibited, whereas the other one has to be selected (e.g. Glaser & Glaser, 1989). It is therefore very likely that a comparable selection process is effective at positions on either side of the canonical directions (±15°). These directions are near enough to the area of application for canonical terms, so that these terms are activated but at the same time LO is also located within the 'diagonal area' that favors the use of a composite term. This exactly defines the competition.

A final word has to be said on the search times. Here too, the canonical conditions were the fastest. This could be expected because the areas of application for canonical expressions were small, and in correspondence, they clearly indicated a small area where the target could very likely be found. In contrast, the composite expressions had a wide area of use, and as a consequence they were less informative about the location. Subjects therefore had to search in a large area and which caused longer search times. The longest times were observed in the regions next to the canonical axes because they were last to be touched if subjects started to search from the diagonal which is the prototypical area for composite expressions. Due to this asymmetry of the areas of competition, the canonical expressions were more useful in terms of the search times, although they were used less often.

In summary, our data clearly support the idea that the use of locative expressions are a function of the deviation of the actual direction of LO from the prototypical directions of the respective expressions. Canonical expressions were used nearly exclusively for canonical relations, and they were uttered very quickly. In all other cases, composites were used which required more time. In addition, the areas of application were nearly non-overlapping, so that there was a clear correspondence between a spatial direction and the used expression. Only with relations next to the canonical direction was it observed that two different locative expressions were used, and these are conditions of competition which were associated with the longest production times. However, the observed data on the actual use of the different expressions also suggest that one should be cautious in using applicability ratings for the purpose of selection. Ratings are given under another perspective, and therefore they might measure a different aspect than the one which is effective during production. Although this is not necessarily the case, the data from Franklin, Henkel & Zangas (1995) as well as our data compared with those from Gapp (1997) demonstrated this difference.

References

Bryant, D. J., Tversky, B. & Franklin, N. (1992). Internal and external spatial frameworks for representing described scenes. *Journal of Memory and Language, 31*, 74-98.

Franklin, N., Henkel, L. A. & Zangas, T. (1995). Parsing surrounding space into regions. *Memory and Cognition, 23*, 397-407.

Franklin, N. & Tversky, B. (1990). Searching imagined environments. *Journal of Experimental Psychology: General, 119*, 63-76.

Gapp, K.-P. (1995). An empirically validated model for computing spatial relations. In I. Wachsmuth, C.-R. Rollinger, and W. Brauer (Ed.), *KI-95: Advances in artifical intelligence. Proceedings of the 19th Annual German Conference on Artifical Intelligence* (pp. 245-256). Berlin, Heidelberg: Springer.

Gapp, K.-P. (1997). *Objektlokalisation: Ein System zur sprachlichen Raumbeschreibung.* [The localization of objects: A system for verbal description of locations.] Wiesbaden: Deutscher Universitätsverlag.

Glaser, W. R. & Glaser, M. O. (1989). Context effects in Stroop-like word and picture processing. *Journal of Experimental Psychology: General, 118,* 13-42.

Grabowski, J., & Weiß, P. (1996). Determinanten der Interpretation dimensionaler Lokalisationsäußerungen: Experimente in fünf Sprachen. [Factors that determine the interpretation of dimensional spatial expressions: experiments in five languages.] *Sprache & Kognition, 15,* 234-250.

Hayward, W. G. & Tarr, M. J. (1995). Spatial language and spatial representation. *Cognition, 55,* 39-84.

Herrmann, T. (1996). Blickpunkte und Blickpunktsequenzen. [Points of view and point of view sequences.] *Sprache & Kognition, 15,* 159-177.

Herskovits, A. (1985). Sematics and pragmatics of locative expressions. *Cognitive Science, 9,* 341-378.

Herskovits, A. (1986). *Language and Spatial Cognition. An interdisciplinary Study of the Preposition in English.* Cambridge, London: Cambridge University Press.

Klein, W. (1994). Keine Känguruhs zur Linken. [No kangaroos at the left.] In H. J. Kornadt, J. Grabowski & R. Mangold-Allwinn (Eds.), *Sprache und Kognition: Perspektiven moderner Sprachpsychologie* (pp. 163-182). Heidelberg: Spektrum.

Levelt, W. J. M. (1989). *Speaking: From intention to articulation.* Cambridge: MIT Press.

Logan, G. D. (1995). Linguistic and conceptual control of visual spatial attention. *Cognitive Pychology, 28,* 103-174.

Logan, G. D. & Sadler, D. D. (1996). A computational analysis of the apprehension of spatial relations. In P. Bloom, M. A. Peterson, L. Nadel & M. F. Garrett (Eds.), *Language and space. Language, speech, and communication* (pp. 493-529). Cambridge, MA: MIT Press.

Pechmann, T, Reetz, H., & Zerbst, D, (1989). Kritik einer Meßmethode: Zur Ungenauigkeit von voicekey Messungen. [The unreliability of voice key measurements.] *Sprache & Kognition, 8,* 65-71.

Retz-Schmidt, G. (1988). Various views on spatial prepositions. *AI Magazine, 9 (2),* 95-105.

Reference Frames for Spatial Inference in Text Understanding

Berry Claus, Klaus Eyferth, Carsten Gips, Robin Hörnig, Ute Schmid, Sylvia Wiebrock and Fritz Wysotzki*

Methods of Artificial Intelligence
Department of Computer Science, Technical University Berlin
email: sppraum@cs.tu-berlin.de

Abstract. We present an approach to spatial reasoning that is based on homogenous coordinate systems and their transformations. In contrast to qualitative approaches, spatial relations are not represented by symbolic expressions only but additionally by parameters with constraints, which are subsets of real numbers. Our work is based on the notion of mental models in text understanding introduced in cognitive science. That is, we model the understanding of descriptions of spatial configurations by constructing a representation of the class of situations that are compatible with the description. Within our approach a spatial relation between two objects is represented by constraining the position of one of the objects with respect to the reference frame of the other one. That is, inferences of spatial relations not given explicitly in a text depend on the frame of reference which is presumed by the (human or computer) system. In this paper we describe a general framework for spatial reasoning based on the notion of mental models and we present some empirical results concerning the influence of the selected reference frame on the inference process.

1   Introduction

To reason about space is a fundamental human ability pervading our everyday life. In this paper we focus on a special aspect of spatial reasoning: the understanding of natural language descriptions of spatial scenes. Without this ability we could not make use of route descriptions in navigation ("to reach the post office, you have to turn left at the next crossing"), of instructions for installing or using technical equipment ("insert the disc in the slot under the console and press the right button") or for locating things ("the book is in the left part of the shelf in my room"). As the examples illustrate, understanding information of this kind can be crucial for performing purposeful actions. This is not only true for humans but also for computer systems and robots: It is a long standing goal

---

* Fritz Wysotzki is the coordinator of the project. Authors are listed in alphabetical order. This research was supported by the Deutsche Forschungsgemeinschaft (DFG) in the project "Modellierung von Inferenzen in Mentalen Modellen" (Wy 20/2–1) within the priority program on spatial cognition ("Raumkognition").

in computer science to control graphic user-interfaces or robot actions by means of a (quasi) natural language input. For this purpose there is need for a higher programming language making use of the operational semantics of spatial relations (like "left" or "in front") and taking into account physical constraints such as occupied space (cf. Ambler and Popplestone 1975; Lenzmann and Wachsmuth 1996).

Our aim is to construct an algorithm which is able "to understand" descriptions of spatial scenes. A common view in cognitive science is that human text understanding is to be construed as building up a so called *mental model* (cf. Garnham 1985, p. 6). That is, to understand a given text, the understanding (human or computer) system has to construct an internal representation of the situation described in the text. This idea guides our work on the understanding of spatial descriptions. A mental model of a description of a spatial configuration has to represent the objects introduced in the text and the spatial relations between them in a way which is "structurally identical" (cf. Johnson-Laird 1983, p. 419) to the described situation. Accordingly, understanding does not end up in determining the propositions (qualitative representation) capturing the "meaning" of the text (i.e. the intensional semantics). Instead, a representation formalism is needed that captures the class of all possible spatial configurations which are consistent with the description (i.e. a kind of model-theoretic semantics). Furthermore, this "spatial representation" has to be constructed incrementally from a sequence of propositions given in a text. The representation has to be updated with respect to new incoming information, and relations between objects not given explicitly should be inferrable if needed.

As a representation formalism, realizing the characteristics of mental models given above, we are using a directed graph with objects as nodes and spatial relations between pairs of objects as arcs. Each object node is associated with a 3-dimensional coordinate system with the origin determining the object position. Orientation of objects is represented by the axes of this coordinate system. A spatial relation between objects A and B is represented by a transformation matrix specifying the translation and rotation operations that map the coordinate system of object A onto that of object B. The idea that information about object positions is represented by homogeneous transformation matrices has biological plausibility, as was shown by O'Keefe (1989). We use parameters associated with constraints to represent the class of possible situations compatible with the given description. That is, the parameters represent subsets of real numbers. Our system is provided with an (extendable) set of predefined spatial relations (like "right of", "in front") associated with constraints on the transformation matrices representing the operational semantics of these spatial prepositions. To infer spatial relations not explicitly given in a text transformation matrices are multiplied. Inference proceeds along the arcs of the labelled graph representing a described configuration. If, for example, the information "B is *right-of* A" and "C is *on* B" is given, the relation "C is *right-of* A" can be inferred: The matrices annotated at the arcs $A - B$ and $B - C$ are multiplied and a new arc representing the relation between A and C is introduced. Note that inference is done not on the level of linguistic variables (qualitative expression), but on the level of possible positions and orientations of objects (as subsets of real numbers). Therefore, it is always possible to infer a spatial relation regardless of whether there exists a corresponding linguistic label or not.

The crucial characteristic of our approach is that it relies on "spatial representations" (cf. Glasgow and Papadias 1992, p. 356). It can be considered as an intermediating alternative to "language oriented", logic-based models and to "perception-oriented" depictional models. Logic-based models (e.g. Guesgen 1989, Hernández 1991, Mittal and Mukerjee 1995, Renz and Nebel 1997) operate on qualitative information, i.e. symbolic expressions (like *left-of(x,y)*). Inference of relations between objects ("qualitative reasoning") is performed by syntactical methods. A variant of "purely" logic approaches is to represent conceptual knowledge about spatial configurations by semantical networks (cf. Anderson and Bower 1973). As in our graph-based model, objects are represented by unique nodes. But objects and relations are represented by linguistic labels (symbolic expressions) only. In our approach objects and geometric relations are represented by coordinate systems and real valued parameters which can additionally be labelled by spatial concepts. We believe that representing quantitative aspects of spatial relations can be a fruitful approach to spatial (or other domains of) reasoning (cf. Guesgen 1997). Working on the model of a described configuration instead of a syntactical representation can have methodological advantages for constructing inference algorithms. The usefulness of constructive approaches to reasoning has already been demonstrated in the research on type theory in computer science (cf. Thompson 1991) and is also discussed in the area of spatial reasoning (see Schlieder 1998). Furthermore, model construction can be regarded as "semantic reasoning" which mimics human information processing (Johnson-Laird and Byrne 1991).

In depictional models (e.g. Waltz and Boggess 1979, Paivio 1971, Pribbenow 1993) object positions are represented by absolute values with respect to a global coordinate system. There is no need for inference of spatial relations. Instead information can be retrieved from the picture by a scanning process. In our approach, object relations are represented not absolutely but relatively. Because the information about a spatial configuration given in a text is underspecified with respect to object positions and orientations, our representation denotes the class of all possible depictions of the described situation (Johnson-Laird 1983). In contrast to qualitative approaches, where underspecification of symbolic descriptions is an inherent feature, the class of possible depictions is represented explicitly by the parameters of the transformation matrices (and their constraints). That is, our model could be extended to verify whether a picture corresponds to a verbal description, and to generate prototypical depictions.

We believe that this approach to spatial reasoning can be fruitful for different applications in the area of natural language interaction with computer systems and robots. The introduction of spatial inference by means of the manipulation of coordinate systems offers a more "natural" framework for spatial reasoning than the more prominent qualitative syntactic approaches. Additionally, our model corresponds to the approach of mental models in text understanding and reasoning. This enables us to mutually profit from work in cognitive psychology and artificial intelligence. On the one hand, we profit from insights gained in experimental cognitive psychology, and on the other hand, the model can be seen as a formal and more precise version of the cognitive approach of mental models in text understanding. In the following, we first give some general ideas about reference frames, arguing that the choice of reference frames is crucial for inferring spatial relations. Afterwards our model is described in more detail. In Sect. 4 psychological experiments regarding the problem of online inferences are presented. Finally we will give an outlook on further work to be done.

2   Frames of Reference

We investigate texts that contain sentences of the form "B is standing in spatial relation r to A.", such as (1):

(1)   Der Kühlschrank steht links von Torsten.
      The refrigerator stands on the left of Torsten.

The expression "links von" in (1) is itself ambiguous in that it allows for an intrinsic interpretation as well as a deictic one. The difference between the two readings relies crucially on the reference frame that is critical in linguistically localizing object B (the refrigerator in (1), we call it the *referent*) with respect to A (Torsten in (1), called the *relatum*). The intrinsic reading calls for a reference frame in terms of the intrinsic properties of the relatum itself, with respect to which the object has to be localized. Given an intrinsic reading of "left of" in (1), the refrigerator is located with respect to the left-hand side of Torsten. A deictic reading demands a reference frame not explicitly mentioned in the sentence, which is distinct from the one of the relatum. The most prominent candidate for the reference frame in question is the intrinsic one of the speaker as oriented towards the relatum (see Miller and Johnson-Laird 1976). Now, with respect to semantics, to know which of the two readings is the intended one is to know which of the candidate reference frames is the relevant one in order to interpret the sentence. But with respect to a mental model this is not the whole story: taking the intrinsic reading of the expression as the intended one in (1), the reader not only has to represent the objects mentioned in the sentence, but moreover, the represented relatum actually has to provide a spatial reference frame within which the referent may then be localized.

One way to think about the reference frame of the relatum is to attach a three-dimensional coordinate system to the represented object, derived from its intrinsic properties. We are then able to define the x-axis as the left-right axis of Torsten's body and take the negative region of the x-axis as the area on Torsten's left, i.e. the region where to localize the refrigerator in the mental model. This is exactly the way in which the formalism we chose in our project serves our needs. The relational expression is used to constrain the possible positions of the referent relative to the reference frame of the relatum. With such a conception in mind, we can speak of a *minimal mental model* just in case that all spatial relations explicitly mentioned in the text, and only these, lead to the addition of further edges (and eventually to the addition of a further node, if a mentioned object lacks a representing node). A minimal mental model does not include any inferred spatial relation between objects. As a valid inference we consider the following one: If sentence (1) is followed by sentence (2), it would be valid to infer that the bowl is on the left of Torsten.

(2) Die Schüssel steht auf dem Kühlschrank.
       The bowl is standing on the refrigerator.

Such an inference would yield the positions of the referents of both sentences (refrigerator and bowl) relative to one and the same reference frame (the one of Torsten). We consider such a mental model in a sense as a more integrated spatial representation.

2.1 Is there any overall organizing principle?

A text describing an object arrangement might include a number of spatial relational expressions calling for an intrinsic reading involving a whole range of different relata (e.g., a text comprised of sentences (1) and (2)). The corresponding minimal mental model lacks an overall organizing principle, in that every object position is specified only with respect to the reference frame of the object to which it was related in the description, i.e. its relatum. Since a minimal mental model might thus be far from being an integrated spatial representation, one may reasonably doubt that a reader ends up in constructing a minimal mental model.

Within our formal approach we may restate the problem as follows: Is there any global or most dominating reference frame within which every object in a given described situation has to be localized at the time of encoding? If this is the case we might postulate a need for an inference in case of an intrinsic relational expression whenever the reference frame of the object that serves as a relatum in the text, differs from the global one. If we were to take the coordinate system of Torsten as the global one with respect to sentences (1) and (2), indeed we would have to infer the position of the bowl within the reference frame of Torsten at encoding time, as a consequence of the overall organizing principle. In terms of the psychology of text comprehension, an overall organizing principle may be taken as a coherence principle, defined under the auspices of a spatial mental model. Thus to infer a spatial relation not explicitly stated in the text as a consequence of the overall organizing principle may be characterized as a *bridging inference*, i.e. an inference which is drawn at encoding time in order to establish coherence (Garnham 1985, p. 157f).

In the example above, to qualify the reference frame of Torsten as global may seem rather arbitrary. Beyond the scope of text comprehension two structural principles for spatial information are of primary interest: the difference as well as the interplay of an egocentric reference frame and an allocentric reference frame. Both are discussed as organizing principles in structuring space within given contexts, especially in navigating in space using route knowledge as well as survey knowledge, often tacitly connecting route knowledge with an egocentric reference frame and survey knowledge with an allocentric reference frame (see Werner et al. 1997).

2.2 Egocentric reference frame

In analysing deictic expressions Bühler (1982) introduced the term "origo" as the name for the origin of the egocentric reference frame. But the origo does not necessarily coincide with the actual perceptible 'Here', as Bühler discusses under the heading of "deixis at phantasma". As his second major case of transposition Bühler (1982, p. 34) considers the phenomenon when "one is transported in the imagination to the geographical place of what is being imagined, one has this present before the mind's eye from a determinate point of view, which may be given and at which is found the self's own position in the imagined scene." In the spirit of Bühler's note on transposition we claim that a person reading a text about objects surrounding a protagonist has the opportunity to transpose herself into the described situation, the determinate point of view being the one of the protagonist. In this case the reader takes the protagonist's perspective[1], thus constructing an egocentric mental model. That is to say: Given the protagonist's perspective, in comprehending (1), the reader may localize the refrigerator in the egocentric mental model with respect to her own reference frame, which is - by definition of the protagonist's perspective - in congruence with the one of the protagonist (Torsten).

If we take the egocentric reference frame as an overall organizing principle as introduced above, it follows that a reader taking the protagonist's perspective localizes every object mentioned in the description within her own egocentric reference frame. As a consequence, for every object not directly linguistically localized within the protagonist's reference frame - which would give the position within the egocentric reference frame at the same time because of the protagonist's perspective - the position within the egocentric reference frame would have to be inferred at the time of encoding. This is the case whenever the relatum differs from the protagonist, e.g., for the localization of the bowl in (2), which is linguistically localized with respect to the reference frame of the refrigerator. Such an egocentrically organized mental model exhibits the interesting property that the representation of the object arrangement changes as a result of every movement (e.g., rotation) of the protagonist. Given (1), if the protagonist Torsten now turns to his left, the refrigerator will afterwards be in front of him. The same holds for the position of the refrigerator within the egocentric mental model, given a protagonist's perspective.

---

[1] We take the protagonist's perspective as a special case of what we call "internal perspective". With "internal perspective" we indicate any case in which the reader transposes herself into the described scene regardless of the imagined viewpoint.

2.3 Allocentric reference frame

There is an intuitive understanding of an object arrangement according to which the arrangement remains the same when the protagonist turns to his left, keeping his position constant. This intuitive understanding is reflected by the deictic reading of (1), which is not sensitive to the orientation of the relatum Torsten[2], but instead would only be sensitive to the orientation of a reference frame whose origin is marked as origo; let it be that of the speaker. Given a deictic reading, the refrigerator remains on the left of Torsten regardless of his orientation, as long as the speaker's reference frame remains constant. But in this special case of a deictic reading it seems obvious that it is not the object arrangement that changes if the speaker's reference frame changes with respect to the object arrangement. Instead, this indicates a change in the perspective of the speaker, thus altering the conditions for description (see Hörnig, Claus, and Eyferth 1997 on the identity of object positions in dependence on the reference frame and further arguments in favor of an allocentric reference frame.)

In constructing a mental model that captures the identity of the object arrangement when the orientation of the protagonist changes, we may localize every object within an allocentric reference frame. An allocentric reference frame as an overall organizing principle has to be external to the object arrangement and stipulated as constant. We propose a room-reference system for this purpose. Sentences (1) and (2) are part of a text which begins by introducing Torsten as standing in the middle of his kitchen. Now, what remains constant in the described situation, despite any re-orientation of Torsten, are the positions of all of the objects with respect to the kitchen (including that of the protagonist, since our texts did not mention any translation of the protagonist, thus keeping his position constant). The question, which reference frame to select as the allocentric one, can only be answered in dependence on the identity conditions of the represented object arrangement: If one would want to represent a re-orientation of the kitchen, one surely would need an allocentric reference frame external to the one of the kitchen (e.g., the object being originally in the *south* of the kitchen now being in the *west*, if the object position with respect to the kitchen remains constant.)

In order to construct a mental model with an allocentric room-reference frame, e.g., the reference frame of the kitchen, the room-node has to provide a coordinate system. It does not seem in itself problematic to provide an abstract coordinate system for the kitchen. But since the kitchen does not have intrinsic properties, linguistic expressions like "left of" would not allow for a straightforward interpretation based on the reference frame of the kitchen. For

---

[2] In terms of our formal approach, identity of two object arrangements with respect to a single global reference frame is dependent only on those parts of the transformation matrices that encode translation, but independent of rotation. For the identity of object arrangements it does not matter if the localized objects change orientation. In contrast, a re-orientation of the global reference frame would affect those parts of the matrices that encode translation as well, as is the case with the egocentric mental model.

this reason we suggest that the x-axis of the kitchen does not by itself admit an interpretation as the left-right-axis. If such a convention is needed, the allocentric reference frame must be supplemented by an extrinsic orientation as can be given by imposing an overview perspective on the reference frame of the kitchen (occasionally it has been suggested that in case of an overview perspective front/behind has to be mapped onto below/above, respectively).

Even if the allocentric reference frame is tied to an overview perspective, it remains essentially different from an egocentric reference frame as discussed above in several respects. While both cases clearly operate with an internal perspective, i.e. with a determinate point of view, the overview perspective is not egocentric. *Ego* is not surrounded by objects (*internal spatial viewpoint* in Bryant et al. 1992) but the whole object arrangement is in front of *ego* (*external spatial viewpoint* in Bryant et al. 1992). Especially, *ego* is definitely not part of the object arrangement. Finally, at least when tied to an allocentric reference frame, the overview perspective is required to be always in accordance with the allocentric reference frame, which itself is presupposed to be inherently constant.

The question of an overall organizing principle leads directly to the topic of inference. In the next section we will describe *how* spatial relations can be inferred within our formal approach to mental models. In Sect. 4 we will take a closer look at the issue of *when* and *what* inferences humans draw, and propose a possible interplay of the egocentric reference frame with an allocentric one.

3 Modelling Spatial Inferences by Transformation Matrices and Constraints

As discussed above, the understanding of texts about spatial scenes can be described as the incremental construction of a *mental model*. Mental models are characterized as internal representations of the described situation which is structurally identical to it in the sense that not only the relations explicitly given in the text are represented, but also inferences of relations holding in the situation can be made. To capture the characteristics we construct a directed labelled graph. Nodes in this graph represent the objects mentioned in the text, whereas the relations are represented by labelled arcs. We pursue an approach which was developed for robotics applications (cf. Ambler and Popplestone 1975). In this setting objects are associated with a coordinate system with distinguished axes for the left/right, front/back, and up/down directions. The spatial relation between two objects A and B can then be described by a transformation matrix T that maps the coordinate system of A onto the coordinate system of B. T consists of a matrix R for the rotation of the coordinate system, and a translation vector Δ. Transformation matrices have been used for qualitative reasoning as well (Mukerjee and Mittal 1995), but we use symbolic variables that can be instantiated by real numbers. For the simple case where only rotations about

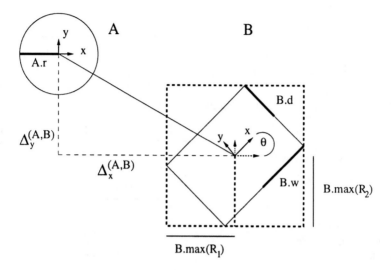

Fig. 1. Relation with relatum A and localized object B

the vertical z-axis are considered, the matrices have the form

$$
T = \begin{pmatrix} & R & & \Delta_x^{(A,B)} \\ & & & \Delta_y^{(A,B)} \\ & & & \Delta_z^{(A,B)} \\ 0 & 0 & 0 & 1 \end{pmatrix} \qquad R = \begin{pmatrix} \cos\theta & -\sin\theta & 0 \\ \sin\theta & \cos\theta & 0 \\ 0 & 0 & 1 \end{pmatrix}
$$

In (Mukerjee and Mittal 1995), all matrix entries are elements of $\{-1, 0, 1\}$. This means that distances cannot be represented. Furthermore, there are only nine possible sectors for spatial relations in two dimensions, and the extension of objects (non-overlapping) cannot even be formulated. Therefore, we think metric information is necessary. The transformation is shown in Fig. 1. R_1 and R_2 are the first and second row vector of R, respectively, and are used to compute the size of the smallest bounding box containing B that is parallel to A. This bounding box is used to simplify computations and is shown with dotted lines. The thickest lines are used for the size parameters $A.r$, $B.w$, and $B.d$. A relation is then defined by constraints on the transformation matrix. Below we will describe our assumptions about the model, the representation of objects, the definition of relations and the inference process.

3.1 Constructing the graph

The texts we consider describe spatial scenes in a room. Therefore the graph initially contains the nodes for the *room* and its parts (objects are discussed below). We represent the background knowledge about the relations between the subparts in the arcs from the *room*-node to the nodes for the parts. Relations between two walls, e.g., can be computed. The constraints with respect to the

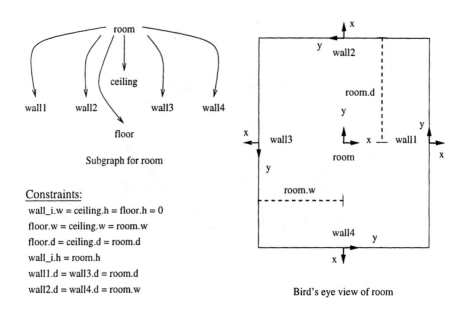

Fig. 2. Default constellation for *room* object(s)

extensions of the subparts are stored in a constraint-table. The default layout of the room with the location of the coordinate systems, the constraints, and the initial graph are shown in Fig. 2. The extensions of the objects are denoted by *w*idth, *d*epth and *h*eight, respectively. Spatial relations are given by propositions like *left_of(Torsten, refrigerator)* for sentence (1). To insert this arc into the graph, we have to check whether there are already nodes for *Torsten* and *refrigerator*. If not, the descriptions of the objects have to be looked up (see below) and the nodes must be created. Then the definition of the relation *left_of* is looked up, the constraints for *left_of(Torsten, refrigerator)* are computed and are stored in the constraint-table. Now the directed arc from *refrigerator* to *Torsten* is added to the graph and is labelled with *left* and the transformation matrix. If a new node is introduced, the background knowledge that all objects are *in* the room, is represented by an arc from the object to the *room* node.

Objects. Objects are represented by *cuboids* and *cylinders*. Simple symmetrical bodies are commonly used to build up objects in programs for spatial inference (Waltz and Boggess 1979) and for the description of office environments (Jörding and Wachsmuth 1996). As we don't get any visual input, we feel we don't need more complicated objects. We don't use default values for the extensions of objects, but it is possible to give upper and lower bounds for the dimensions. The coordinate system is located in the center of the object. The extension is always measured from the origin of the coordinate system to the border of the object (see Fig. 2). The variables for the extensions of objects are *w*idth, *d*epth and *h*eight for cuboids, and *r*adius and *h*eight for cylinders. At present the only object which can be decomposed into its building parts, is the *room*. In Fig. 2, the arcs from the *room* node are labelled with the special relation *has part* (for lack of space, the labels are not shown in the picture) and the corresponding transformation matrix. The relation *has part* is used to ensure that for the mentioned *room* object, all sub-objects are included in the graph. The transformations are fully specified by the extension of the *room*. As a wall has no intrinsic axes for front/back and left/right, the orientation of these axes is arbitrary. We have chosen these coordinate systems for the walls so that the definition for the relation *at_the_wall* is the same for all walls. The constraints on the matrices and on the variables for the extensions of the (sub)objects ensure that the parts "fit together". A more detailed description of objects may become necessary when we want to construct depictions for a given model.

As stated in Sect. 2, to build a mental model from a text, it is necessary to know whether an intrinsic or a deictic interpretation of a relation is intended. Therefore it is essential to know which readings are possible. We provide the information which of the axes of the coordinate system correspond to intrinsic axes of the object. Currently we assume an intrinsic use of spatial propositions. In the future we hope to get results which allow to decide whether an allocentric or egocentric reference frame must be used for the interpretation of a proposition.

Relations. The definition of a relation consists of a name and the constraints for the transformation matrix. These constraints can restrict the rotation of the localized object with respect to the relatum and describe a region of R^3. In Fig. 3, a possible definition for the relation *right* is shown. In this case B must be fully inside the defined region. Because of the variables in the transformation matrix and the unknown extensions of objects we cannot use collision detection procedures to decide whether B is in the depicted region *right* of A. We simplify the problem by considering the smallest cuboid that is parallel to A and contains B. The size of this cuboid can be computed with the function max which takes an object and a direction vector as arguments. The result is still a symbolic expression containing the rotation θ.[3]

The constraints for the above definition of *right* are

$$\Delta_x^{(A,B)} \geq A.\max(1,0,0) + B.\max(R_1)$$
$$\Delta_y^{(A,B)} \leq \ (\Delta_x^{(A,B)} - A.\max(1,0,0) - B.\max(R_1))$$
$$+ (A.\max(0,1,0) - B.\max(R_2))$$
$$\Delta_y^{(A,B)} \geq -[(\Delta_x^{(A,B)} - A.\max(1,0,0) - B.\max(R_1))$$
$$+ (A.\max(0,1,0) - B.\max(R_2))]$$

The function max gets a vector w.r.t. A's coordinate system as argument and computes the maximum distance from the object's origin to the border of the bounding box in this direction. For A, $\max(1,0,0) = A.w$ for a cuboid, and

---

[3] Sometimes we may be able to restrict the possible orientations with defaults or background knowledge.

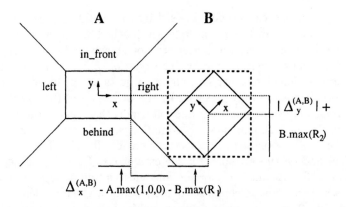

Fig. 3. A possible definition for *right*

$\max(1,0,0) = \max(0,1,0) = A.r$ for a cylinder. As can be seen in the picture, we don't get an exhaustive set of relations when we consider extended objects and require that the whole object is positioned within the region. This means we will often have situations where the relation (i.e. a transformation matrix) between two objects can be computed, but the corresponding arc cannot be labelled with the name of a defined relation. An example is shown in Fig. 4. For these specific positions of the objects and the above definitions, B is *right* of A, and C is *left* of B, but the relation between A and C cannot be named, because C overlaps the sector for *right* and the sector for *behind*.

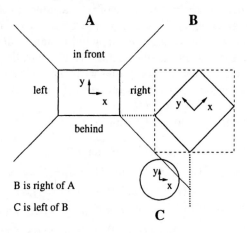

Fig. 4. Bird's eye view of a constellation of three objects

Finding linguistically justified constraints for spatial relations is not a goal of our work. We are experimenting with different sets of relations. In principle it is no problem to include results from psychological experiments that report overlapping, nonsymmetric regions of different sizes for the four basic directions

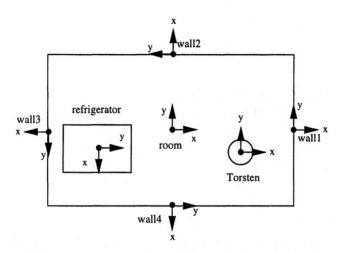

Fig. 5. Possible situation for "The refrigerator stands on the left of Torsten"

(Franklin, Henkel, and Zangas 1995; Vorwerg 1997; Vorwerg and Rickheit 1998). Whether the models that are possible for these regions are psychologically justified for text understanding tasks is an open problem. The constraints for the transformation matrix can be equations or inequations containing the elements of the matrix and the extensions of the two related objects. A possible topic of research is the introduction of functional roles for objects which would enable us to include context dependent information.

For our example sentence (1), one situation is depicted in Fig. 5. We have used the following assumptions: a) the protagonist faces a wall. As walls are so far undistinguished, we have set the coordinate system of Torsten parallel to that of the room. b) The refrigerator is oriented with its front side (positive y-axis) towards Torsten. We get the following transformation matrices (abbreviations: *To*rsten, *re*frigerator, *w*idth, *d*epth, *h*eight, and *r*adius).

$$
T^{(room, \ To)} = \begin{pmatrix} 1 & 0 & 0 & \Delta_x^{(room, To)} \\ 0 & 1 & 0 & \Delta_y^{(room, To)} \\ 0 & 0 & 1 & To.h - room.h \\ 0 & 0 & 0 & 1 \end{pmatrix}
$$

with the constraints

$$
-(room.w - To.r) \leq \Delta_x^{(room, To)} \leq room.w - To.r
$$
$$
-(room.d - To.r) \leq \Delta_y^{(room, To)} \leq room.d - To.r
$$

$$
T^{(To, \ re)} = \begin{pmatrix} 0 & 1 & 0 & \Delta_x^{(To, re)} \\ -1 & 0 & 0 & \Delta_y^{(To, re)} \\ 0 & 0 & 1 & re.h - To.h \\ 0 & 0 & 0 & 1 \end{pmatrix}
$$

with the constraints

$$-(\Delta_x^{(\text{room,To})} + \text{room.w} - \text{To.r} - 2\text{re.d}) \leq \Delta_x^{(\text{To,re})} \leq -(\text{To.r} + \text{re.d})$$

$$(\Delta_x^{(\text{To,re})} + \text{To.r} + \text{re.d}) - (\text{To.r} - \text{re.w}) =$$
$$(\Delta_x^{(\text{To,re})} + \text{re.d} + \text{re.w}) \leq \Delta_y^{(\text{To,re})} \leq -(\Delta_x^{(\text{To,re})} + \text{re.d} + \text{re.w})$$

and
$$T^{(room,\ re)} = \begin{pmatrix} 0 & 1 & 0 & \Delta_x^{(\text{room,re})} \\ -1 & 0 & 0 & \Delta_y^{(\text{room,re})} \\ 0 & 0 & 1 & \text{re.h} - \text{room.h} \\ 0 & 0 & 0 & 1 \end{pmatrix}$$

with the constraints

$$- (\text{room.w} - \text{re.d}) \leq \Delta_x^{(\text{room,re})} \leq \Delta_x^{(\text{room,To})} - \text{To.r} - \text{re.d}$$
$$-(\text{room.d} - \text{re.w}) \leq \Delta_y^{(\text{room,re})} \leq (\text{room.d} - \text{re.w})$$

3.2 Inferences

To infer a relation between objects B and relatum A not explicitly given in a text, we have to find a directed path $B = n_0 - n_1 - \ldots - n_k = A$ in the graph. We can then multiply the transformation matrices to get

$$T^{(A,\ B)} = T^{(A, n_{k-1})} \times \cdots \times T^{(n_1, B)}$$

Constraint propagation yields the constraints for the computed matrix which have to be matched against the predefined relations if we are interested in an interpretation (a proposition label, e.g. *right*) for the arc.

It is a topic of research when and how (along which paths) relations between objects in the model are inferred. As stated in Sect. 2, this also depends on the reference frame chosen for the interpretation of a spatial proposition. If a global reference frame P exists (e.g. the *room* or the protagonist typically introduced in the text), it would be possible to construct a tree where only the relations between the objects and P are represented by arcs (and inferred at encoding time) whereas no relation between any two other objects might be represented, even if explicitly given in the text. The other extreme would be to build a complete graph with an arc between every pair of objects, even if most arcs cannot be labelled with the name of a relation (i.e. the transformation matrix does not satisfy the constraints of any predefined relation). At present we have implemented a medium strategy. The graph consists of those arcs for which relations are explicitly given, the arcs for the background knowledge (any object is *in* the room), and arcs that are inferred after a query to the program. The existence of an arc from an object to the *room* node can be regarded as choosing the *room* object as a global reference frame. From the graph in Fig. 6 below, it should be obvious that we have more than one fixed reference frame, because otherwise the arcs labelled *left* and *on*, respectively, would not be included. The

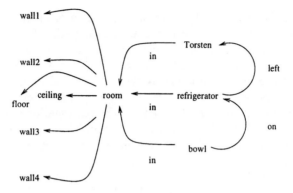

Fig. 6. The graph after two sentences

goal of this work is the implementation of several strategies dependent on the results of the psychological experiments described below.

If we continue our example with sentence (2), we have to add the arcs *bowl* $\longrightarrow$ *refrigerator* and *bowl* $\longrightarrow$ *room*. A *bowl* is modelled as a cylinder without intrinsic x- and y-axes. Therefore the orientation of its coordinate system is of no importance. For sake of convenience, we choose an orientation identical with that of the refrigerator. The graph is shown in Fig. 6.

For the arc *bowl* $\longrightarrow$ *refrigerator*, the rotation matrix is the identity matrix. The constraints are:

$$-(\text{re.w} - \text{bowl.r}) \leq \Delta_x^{(\text{re,bowl})} \leq (\text{re.w} - \text{bowl.r}),$$
$$-(\text{re.d} - \text{bowl.r}) \leq \Delta_y^{(\text{re,bowl})} \leq (\text{re.d} - \text{bowl.r}), \text{ and}$$
$$\Delta_z^{(\text{re,bowl})} = \text{re.h} + \text{bowl.h}.$$

These constraints are propagated, so that we get

$$T^{(room,\ bowl)} = \begin{pmatrix} 0 & 1 & 0 & \Delta_x^{(\text{room,bowl})} \\ -1 & 0 & 0 & \Delta_y^{(\text{room,bowl})} \\ 0 & 0 & 1 & 2\text{re.h} + \text{bowl.h} - \text{room.h} \\ 0 & 0 & 0 & 1 \end{pmatrix}$$

with the constraints

$$\Delta_x^{(\text{room,re})} - (\text{re.d} - \text{bowl.r}) \leq \Delta_x^{(\text{room,bowl})} \leq \Delta_x^{(\text{room,re})} + (\text{re.d} - \text{bowl.r})$$
$$\Delta_y^{(\text{room,re})} - (\text{re.w} - \text{bowl.r}) \leq \Delta_y^{(\text{room,bowl})} \leq \Delta_y^{(\text{room,re})} + \text{re.w} - \text{bowl.r})$$

To infer the relation between the bowl and Torsten (the relatum), we have to find a directed path from *bowl* to *Torsten*. In the figure there is no such path. Now, an advantage of the transformation matrix approach over propositional inference calculi is the fact that for every relation, the reverse relation can be computed by inverting the matrix[4].

---

[4] In general, the inverse matrix need not correspond to any predefined relation. For instance, in Fig. 5, *wall1* is left of *wall3*, and *wall3* is left of *wall1* for the relations

When we are allowed to follow arcs in both directions (the inverse arc is not inserted in the graph), we find the following four paths: *bowl – refrigerator – Torsten, bowl – room – Torsten*$^{-1}$, *bowl – refrigerator – room – Torsten*$^{-1}$, and *bowl – room – refrigerator*$^{-1}$ *– Torsten*, where the raised $^{-1}$ denotes inverting the preceding arc. Since the graph is always connected, there is at least one path between any two nodes. The search strategy for finding a path depends heavily on the organization of the mental model and our representing graph. In this first implementation, we can always take the shortcut via the *room* node, i.e., the path *bowl–room–Torsten*$^{-1}$. Now we can compute the matrix

$$T^{(\text{Torsten,bowl})}$$

$$= T^{(\text{room,Torsten})^{-1}} \times T^{(\text{room,bowl})}$$

$$= \begin{pmatrix} 1 & 0 & 0 & -\Delta_x^{(\text{room,To})} \\ 0 & 1 & 0 & -\Delta_y^{(\text{room,To})} \\ 0 & 0 & 1 & -(\text{To.h} - \text{room.h}) \\ 0 & 0 & 0 & 1 \end{pmatrix}$$

$$\times \begin{pmatrix} 0 & 1 & 0 & \Delta_x^{(\text{room,bowl})} \\ -1 & 0 & 0 & \Delta_y^{(\text{room,bowl})} \\ 0 & 0 & 1 & 2\text{re.h} + \text{bowl.h} - \text{room.h} \\ 0 & 0 & 0 & 1 \end{pmatrix}$$

$$= \begin{pmatrix} 0 & 1 & 0 & \Delta_x^{(\text{room,bowl})} - \Delta_x^{(\text{room,To})} \\ -1 & 0 & 0 & \Delta_y^{(\text{room,bowl})} - \Delta_y^{(\text{room,To})} \\ 0 & 0 & 1 & 2\text{re.h} + \text{bowl.h} - \text{room.h} - (\text{To.h} - \text{room.h}) \\ 0 & 0 & 0 & 1 \end{pmatrix}$$

In the next step, we have to match the elements of the matrix against the constraints for all predefined relations (unless the query asks whether a particular relation holds). For the matrix above, it is easy to prove that *left_of(Torsten, bowl)* holds although the inference would be easier for the obvious path *bowl – refrigerator – Torsten*. Constraint solving is inherently difficult, because in the general case, there are inequations involving trigonometric functions. As shown above, a possible simplification of the problem is the use of defaults. Currently, we restrict the orientations of objects to multiples of 90° around the *z*-axis. This means that the coordinate systems of all objects become parallel or orthogonal. This makes the inferences easier without too much loss of expressive power. We are at present working on the problem of finding heuristics to prove trigonometric (in)equations for arbitrary rotations around the *z*-axis. We are also investigating the application of machine learning algorithms for constraintsolving (see Gips et al. 1998).

---

in Fig. 4, because the coordinate system of *wall3* is rotated 180° with respect to the coordinate system of *wall1*, and the relations are defined w.r.t. the intrinsic axes of the relatum.

In the above example, all possible situations are captured within one model. Underdeterminate descriptions may be handled by introducing disjunctive constraints, which no longer yield a unique model. Wysotzki et al. (1997) present an example of several pairwise incompatible models (all compatible with the description), which may eventually be disambiguated by information given later in the text. In the area of spatial reasoning, a similar problem arises if the premises do not license a straightforward conclusion. In this latter case there is evidence that human problem solvers prefer some conclusions over others (see Knauff et al. 1998). Schlieder (1996) and Berendt (1996) formulate cognitive principles in terms of the construction of *preferred mental models* in order to explain such preferences. In our domain we investigate if human text understanders prefer to infer object positions with respect to some reference frame as compared with some other. For this purpose we proposed two alternative organizational principles (see Sect. 2) that would lead to the construction of quite different graphs. Which of these graphs corresponds to the preferred mental model has to be verified empirically. The next section discusses the hypotheses that follow from these organizational principles, as well as one of our experiments and the results obtained so far.

4   An Empirical Approach to Online Inferences in Spatial Reference Frames

In this section we restrict our considerations to egocentric mental models, i.e. mental models constructed by a reader taking the protagonist's perspective. We primarily focus on the question whether the egocentric reference frame serves as an overall organizing principle, or if an egocentric mental model is provided with an allocentric reference frame. This section relies on the work done by Barbara Tversky and her colleagues on egocentric spatial mental models (Franklin and Tversky 1990, Franklin, Tversky, and Coon 1992, Bryant and Tversky 1992).

4.1   Spatial framework

In a series of experiments, Tversky and her colleagues demonstrated that object access in spatial mental models depends on the direction in which the object is located. In the experiments of interest here, the texts described objects located either around the reader (e.g., "object A is behind you") or around a protagonist. We will consider those instances where the reader took the protagonist's perspective. Along the horizontal plane, objects located in front of *ego* were accessed fastest. Throughout, objects located behind *ego* could be accessed faster than objects beside *ego*[5]. Tversky et al. explained these results by characterizing egocentric mental models as (internal) spatial framework models. A spatial

---

[5] While we examine only object arrangements along the horizontal plane, Tversky et al. additionally compared those objects with objects above and below *ego*. Objects above and below *ego* proved to be accessed fastest, as long as the recipient imagined herself as standing upright. Again, we limit our considerations to this last condition.

framework is a three-dimensional reference frame imposed on the surrounding space which is derived from an observer's body-axes. Differences in access latencies are explained in terms of dominance relations between the axes and their symmetry properties. While the above/below axis renders the primary orientation, the left/right axis is dominated by the front/back axis from which it is derived. Therefore, objects located on the front/back axis are accessed faster than on the left/right axis. Moreover, the front-back axis is asymmetric by keeping objects in front perceptually as well as functionally more salient, thus causing access to objects in front to be faster.

Access latencies are explained by properties of the egocentric reference frame, i.e. in terms of representation. Consequently, access latencies should exhibit the spatial framework pattern whenever the objects are localized directly within the egocentric reference frame. If access latencies deviate from the spatial framework pattern this indicates that the examined objects are not represented directly within an egocentric reference frame. With this argument at hand it is possible to empirically test the hypothesis that an egocentric mental model sets up an overall organizing principle as described above. As is stated by this principle, objects introduced in a text relative to a relatum distinct from the protagonist should nevertheless be located directly within the egocentric reference frame, already at the time of encoding by virtue of online inference. Consequently, all objects should be accessed as predicted by the spatial framework model[6]. Before we go into more detail, we will discuss an alternative model, whose predictions for access latencies deviate from the spatial framework pattern: the mental transformation model.

4.2 Mental transformation

In presenting the spatial framework model, Franklin and Tversky (1990) contrasted it to the mental transformation model. According to the mental transformation model, object access proceeds in much the same way as a perceptual search within a surrounding environment. Only objects in front are immediatly accessible. In order to access an object located someplace else one has to mentally re-orient (i.e. rotate) until the object lies in front and therefore becomes accessible. According to this model, access latencies increase as a function of the angle by which one has to re-orient in order to enable immediate access to the object. Fastest access along the horizontal plane is predicted for objects located in front of *ego*, as was predicted by the spatial framework model. With respect to objects behind and beside *ego*, the mental transformation model contradicts the spatial framework model. In order to access an object behind *ego*, one has to re-orient by 180°, which takes longer than to access an object on either side, which only requires a re-orientation by 90°.

---

[6] As far as we know, in the texts used by Tversky et al., the objects were always linguistically located directly within the egocentric reference frame. Therefore, the results do not call for an overall organizing principle in order to postulate immediate access within the egocentric reference frame.

It seems worthwhile to take a closer look at the presumptions of this idea of a mental transformation. As is evident from the findings by Tversky et al., objects located linguistically within the egocentric reference frame are immediately accessible without any need for a mental transformation. In terms of reference frames, a mental transformation has to be stated as a mental rotation of the egocentric reference frame against the surrounding space, i.e. an allocentric reference frame. So, it is only for objects not already represented directly within the egocentric reference frame that one may reasonably assume an access mediated by a mental transformation. As long as we rely on the egocentric reference frame to establish an overall organizing principle, there is no need for a mental transformation. On the other hand, if we do suggest such a need, we must admit for egocentric mental models in which objects are not always directly located within the egocentric reference frame but only within an allocentric one. This assumption concerning a twofold representation is then supplemented by a procedural principle. If object access within the egocentric reference frame fails, it has to proceed within the allocentric reference frame. This case demands an indirect access, maybe governed by a procedural principle such as mental transformation. In summary, we acknowledge the possibility of two concurrent organizing principles in case of an egocentric mental model: an egocentric reference frame connected with an allocentric reference frame.

If there are two concurrent organizing principles for egocentric mental models, how do they interrelate? We assume this as the question of how a reader gains orientation within a spatial mental model. As long as one considers solely the egocentric reference frame, the question of orientation does not arise, because *left* is necessarily on the left-hand side of *ego*. It is with respect to an allocentric reference frame that the question becomes a sensitive one. At the time the reader encodes sentence (1), given the protagonist's perspective, egocentric *left* coincides with the location of the refrigerator[7]. We claim that an object like the refrigerator, which may be assumed to be localized invariantly within the kitchen, enables the reader to establish her orientation within the spatial mental model by anchoring her egocentric reference frame within the allocentric room reference frame. An object serving this purpose we call an *anchoring* object (anchoring objects play roughly the role of local landmarks in navigation tasks; see Werner et al. (1997)). The location of an anchoring object in an egocentric mental model is always represented in at least two ways. It is specified with respect to the egocentric reference frame as well as with respect to the allocentric reference frame, thus interconnecting the two. As a consequence, anchoring objects are always represented directly within the egocentric reference frame and therefore directly accessible. Because of this, access latencies for anchoring objects should always exhibit the spatial framework pattern. On the other hand, there may be objects in the mental model which are not located with respect to the egocentric

---

[7] This point becomes clear when the protagonist re-orients: in this case *left* is still left within the egocentric reference frame, but egocentrical *left* changes relative to the refrigerator and its constant position within (the allocentric reference frame of) the kitchen (see Hörnig, Claus, and Eyferth 1997).

reference frame, but merely with respect to the allocentric reference frame (we call such objects *critical objects*). If subjects have to access critical objects, latencies may deviate from the spatial framework pattern. Rather, critical objects may be accessed as predicted by the mental transformation model.

Our formal approach raises a severe argument against the egocentric reference frame as an overall organizing principle. In case of a re-orientation of the protagonist, the reader taking the protagonist's perspective has to execute the re-orientation within the egocentric mental model. As a consequence, every edge from an object node in the graph linked to the *ego*-node has to be updated online, since the mental model represents the actual egocentric situation, which changes in case of a re-orientation. If our approach to represent mental models is valid, a re-orientation in an egocentric mental model will be more complex - and, in terms of computation, more costly - as more edges have to be updated. This makes it highly implausible that a reader infers every object location within the egocentric reference frame online, leading to a maximal number of edges to be updated. On the other hand, if object access within the allocentric reference frame requires a mental transformation, which in turn demands an update of all of the edges from the anchoring objects (whether continuously or in discrete steps), it turns out to be a highly complex process too. Mental transformation as an access principle is even more complex in that the reader has to keep her actual orientation in mind. Moreover, if the reader has to generate the current location of a given critical object by means of a mental transformation, she has to determine the object location in the allocentric reference frame given her current orientation (before she initiates mental transformation) by "subtracting" the mental transformation she has performed in order to access the object in front of her (after having finished the mental transformation).

4.3 An empirical investigation

To illustrate our empirical approach we report one of our latest experiments. Our primary goal is to compare access to anchoring objects with critical objects in egocentric mental models. If the egocentric reference frame sets up an overall organizing principle, both types of objects are represented directly within the egocentric reference frame at the time of testing. Therefore, both types of objects are immediately accessible and should exhibit the same latency pattern.

Material. We constructed eight experimental texts which described a protagonist standing in the middle of a room. Along the horizontal plane there were four anchoring objects placed in all four directions of the protagonist (front, behind, left, right). Anchoring objects were not introduced as in (1), taking the protagonist as the relatum. Instead we used sentences like "An der Wand links von Torsten steht der Kühlschrank." ("At the wall to Torsten's left, there is the refrigerator."). Thus, in order for the so-called anchoring object to be localized directly within the egocentric reference frame, it is already necessary for the reader to infer the spatial relation between the refrigerator and Torsten, i.e. that the refrigerator is positioned on the left of Torsten. In addition, four critical objects were introduced, each of them combined with one of the anchoring objects as in (2). For every text there were four versions to systematically vary the direction, in which the object combinations (anchoring and critical object) were placed. The order of mentioning objects was held constant across the versions. The four walls were described one after another, by introducing the anchoring object, followed by the critical object. Subjects read the texts on a PC screen. Reading times were self-paced.

Procedure. In most of the cases, Tversky et al. presented direction probes to their subjects who had to remember the object located in that direction. Of course, we could not adapt this paradigm, because direction probes would be ambiguous in that there are always two objects located in each direction: an anchoring object and a critical object. We therefore adopted the procedure used by Bryant and Tversky (1992), who presented their subjects object probes and subjects had to remember the location of the named object. Our subjects were presented an object name of either an anchoring object or a critical object on the PC screen and had to respond in two successive steps. As soon as they knew the location of the object, they had to press the enter-button (RT1). Afterwards they had to specify the remembered direction by pressing one of four keys indicating front, behind, left and right (RT2). For both steps response times were measured. In the subsequent presentation of the results, we report the latencies for the first step (RT1).

For every text, subjects were probed three times. The first probe followed the description of the object arrangement, with a filler sentence between them. After the first and second probe, the text was continued by mentioning a re-orientation of the protagonist either to his left, to his right or towards his back (in two of all cases the protagonist did not re-orient). Subjects were instructed to answer in accordance with the current orientation of the protagonist and were told that taking the protagonist's perspective would aid in this task. The object names chosen for probes as well as the order in which they were presented were the same for every text and subject. Test items differed only in the direction which had to be remembered, dependent on the text version. For every subject 24 answers were recorded: for both object types all of the four directions were probed three times.

Hypotheses. We predict different patterns of response times for anchoring objects and critical objects. With regard to anchoring objects we suppose that object positions with respect to the egocentric reference frame are inferred at encoding time. Anchoring objects are therefore represented directly within the egocentric reference frame. As was demonstrated by Tversky and colleagues, objects directly located within the egocentric reference frame exhibit access latencies according to the spatial framework pattern. Especially, objects behind should be accessed faster than objects beside (left or right). In contrast we assume critical objects not to be located directly within the egocentric reference

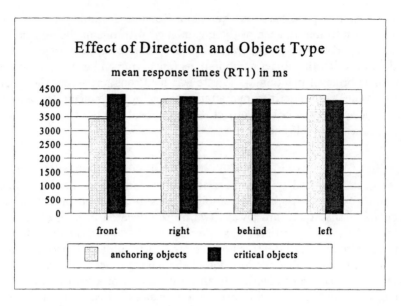

Fig. 7. Mean response times (RT1) for anchoring objects and critical objects as a function of direction

frame, i.e. the location of a critical object with respect to the egocentric reference frame is not inferred online. Critical objects are solely represented within the allocentric reference frame and are not accessible directly within the egocentric reference frame. Our general prediction is that access latencies for critical objects deviate from the spatial framework pattern. Specifically, we suggest that access latencies in this case follow the predictions of the mental transformation pattern, i.e. objects beside are accessed faster than objects behind. For both object types, objects in front should be accessed fastest.

A direct comparison of mean response times for anchoring objects and critical objects should reveal faster responses in case of the former than in the case of the latter. The task of the subjects is to indicate the direction in which an object is currently located. In the case of an anchoring object, the object location is claimed to be inferred at encoding time and updated online when reading about a re-orientation of the protagonist. So, it has only to be read off the mental model. In contrast, the location of a critical object is supposed not to be inferred at encoding time but rather has to be inferred at test time.

Results. Analysis of response times was based on the correct responses (71%) of 28 subjects and limited to RT1. Mean response times were significantly shorter for anchoring objects (3838 ms) than for critical objects (4200 ms) $[F(1, 27) = 10.36; p < .01]$. While object direction yielded no overall main effect $[F(3, 81) = 1.98; p > .10]$, it interacted with object type $[F(3, 81) = 2.69; p = .05]$. Separate analyses for both object types revealed an influence of object direction only for anchoring objects $[F(3, 81) = 4.51; p < .01]$ but not for critical objects $[F(3,

81) < 1]. As is illustrated in Fig. 7, for critical objects there was no difference at all in response times as a function of object direction. In contrast, anchoring objects were responded to faster if located in front (3426 ms) or behind (3496 ms) - front being only slightly faster than behind - than if located on the left (4287 ms) or right (4141 ms).

Discussion. Our results clearly indicate different access times for anchoring objects and critical objects. Latencies for anchoring objects mainly exhibit the spatial framework pattern in that objects located on the front/back axis are responded to faster than on the left/right axis. However, objects behind are judged as fast as are objects in front, a finding not conforming to the expected pattern. While the dominance of the front/back axis over the left/right axis did affect latencies, the asymmetry of the front/back axis yielded no influence. This lack of influence in the last case might be due to the higher complexity of the mental models in our case (eight objects in four directions) as compared with those of Tversky et al. (five objects in five directions). Yet the effectiveness of the dominance of the front/back axis speaks in favor of egocentric mental models.

With critical objects, object access did not exhibit the slightest tendency of a spatial framework pattern. As we argued above, access to critical objects does not conform to the predictions of the spatial framework model because critical objects are not directly represented within the egocentric reference frame, and therefore are not immediately accessible. We take this as a strong argument against the egocentric reference frame as an overall organizing principle. The relative positions of the critical objects with respect to the egocentric reference frame are not inferred online. This finding agrees with our assumption that the positions of critical objects are determined with respect to an allocentric reference frame, even in egocentric mental models. Nevertheless, critical objects were obviously not accessed in a manner of a mental transformation. Because for critical objects response times were not influenced by their location at all, we may not ultimately decide if these objects are really accessed within an allocentric reference frame.

On average, anchoring objects are responded to faster than critical objects, as predicted. But since mean response times for critical objects are as high as those obtained for anchoring objects located on the left/right axis, it is difficult to explain the overall disadvantage for the critical objects in terms of an inference necessary at test time. Subjects might have been immediately able to discriminate critical objects from anchoring objects, hence being able to circumvent to try to access them within the egocentric reference frame. But with respect to our initial hypothesis this explantion is not satisfactory. The task of our subjects was to respond with the egocentrically defined location of the probed object. To access the object is not sufficient for this purpose if the object location in the mental model is not egocentrically defined, as we assume in the case of critical objects. Since we expected a time-consuming inference necessary for critical objects, this would mean that they are accessed faster than anchoring objects on the left/right axis, because their positions are assumed to be egocentrically defined.

In summary, we have evidence that for anchoring objects located on the front/back axis their location is easiest to determine. At least for those anchoring objects we maintain that they are immediately accessible within the egocentric reference frame and that their location may be directly read off the mental model, since it is inferred or updated online. We leave open the possibility that the egocentric reference frame is anchored within the allocentric reference frame solely by the objects located on the front/back axis but that there are no anchoring objects on the left/right axis.

5   Conclusion and Further Work

First results of our attempts to experimentally explore and to model mechanisms that allow for a representation of spatial configurations are implemented in the model prototype SPACE/0 (cf. Wiebrock et al. 1998). Since model construction with SPACE/0 does not put any constraints on the selection of arcs that built up the graph, the resulting model may realize multiple perspectives simultaneously. That is, we may represent a spatial scene in a way that cannot possibly be *perceived* under any perspective in the real world. The perceptible part of a spatial configuration is restricted by the viewpoint of an observer. In extending our approach to the construction of depictions (planned as a next step), principles for constructing graphs for a selected visual perspective tied to a given frame of reference have to be specified. Likewise, spatial mental models constructed in text understanding are not restricted to a specific viewpoint in the same way as depictions. Instead - even under an egocentric perspective - the mental model can represent the space surrounding *ego* simultaneously. But the whole surrounding space cannot possibly be visible from an egocentric viewpoint.

Generating depictions from a mental model is closely connected to the problem of inferring spatial relations. In graph representations that allow a "direct" transformation into depictions, all object positions have to be represented with respect to a unique - viewpoint dependent - reference frame. That is, the graph has to contain a unique node (representing the viewpoint), which is directly connected to every other node (representing the objects of a spatial configuration). Such graph representations are more restrictive than the class of mental models constructed by SPACE/0 and the class of egocentric mental models constructed by human text understanders. To transform a mental model into a "depictable" graph, some relations represented explicitly in the mental model have to be deleted and others have to be inferred.

In expanding our model by generating depictions, we hope to gain more precise guidelines for the assessment of mental models concerning spatial inferences as well as for techniques of visualization guided by descriptive terms for several fields of application.

References

Ambler, A. P. and R. J. Popplestone (1975). Inferring the Positions of Bodies from Specified Spatial Relationships. *Artificial Intelligence 6*, 157–174.

Anderson, J. and G. Bower (1973). *Human Associative Memory*. Washington, D.C.: Winston.

Berendt, B. (1996). The utility of mental images: How to construct stable mental models in an unstable image medium. In U. Schmid, J. Krems, and F. Wysotzki (Hrsg.), *Proceedings of the First European Workshop on Cognitive Modeling*, Forschungsberichte des Fachbereichs Informatik, Report No. 96-39, S. 97–103. TU Berlin.

Bryant, D. J. and B. Tversky (1992). Assessing spatial frameworks with object and direction probes. *Bulletin of the Psychonomic Society 30*, 29–32.

Bryant, D. J., B. Tversky, and N. Franklin (1992). Internal and external spatial frameworks for representing described scenes. *Journal of Memory and Language 31*, 74–98.

Bühler, K. (1982). The deictic field of language and deictic words. In R. J. Jarvella and W. Klein (Hrsg.), *Speech, place, and action. Studies in deixis and related topics*. Chichester: John Wiley & Sons Ltd. English reprint. Originally published in 1934.

Franklin, N., L. A. Henkel, and T. Zangas (1995). Parsing surrounding space into regions. *Memory and Cognition 23*(4), 397–407.

Franklin, N. and B. Tversky (1990). Searching imagined environments. *Journal of Experimental Psychology: General 119*, 63–76.

Franklin, N., B. Tversky, and V. Coon (1992). Switching points of view in spatial mental models. *Memory and Cognition 20*(5), 507–518.

Garnham, A. (1985). *Psycholinguistics: Central Topics*. Routledge.

Gips, C., P. Geibel, S. Wiebrock, and F. Wysotzki (1998). Learning Spatial Relations with CAL5 and TRITOP. In preparation.

Glasgow, J. and D. Papadias (1992). Computational Imagery. *Cognitive Science 16*, 355–394.

Guesgen, H. W. (1989). Spatial reasoning based on Allen's temporal logic. Technical Report TR-89-049, ICSI, Berkeley, Cal.

Guesgen, H. W. (1997). Towards hybrid spatial reasoning. In T. P. Martin and A. L. Ralescu (Hrsg.), *Fuzzy Logic in Artificial Intelligence. Towards Intelligent Systems*, LNAI 1188, S. 197–206. Springer. IJCAI'95 Workshop, Montreal CDN, Selected Papers.

Hernández, D. (1991). Relative representation of spatial knowledge: the 2-D case. In D. M. Merk and A. Frank (Hrsg.), *Cognitive and Linguistic Aspects of Geographic Space*, S. 373–385. Dordrecht: Kluwer.

Hörnig, R., B. Claus, and K. Eyferth (1997). Objektzugriff in Mentalen Modellen: Eine Frage der Perspektive. In C. Umbach, M. Grabski, and R. Hörnig (Hrsg.), *Perspektive in Sprache und Raum*. Deutscher Universitätsverlag, Wiesbaden.

Johnson-Laird, P. and R. Byrne (1991). *Deduction*. Hillsdale, NL: Lawrence Erlbaum.

Johnson-Laird, P. N. (1983). *Mental Models: Towards a Cognitive Science of Language, Inference and Consciousness*. Cambridge: Cambridge University Press.

Jörding, T. and I. Wachsmuth (1996). An Antropomorphic Agent for the Use of Spatial Language. In *Proceedings of ECAI'96-Workshop on Representation and Processing of Spatial Expressions*, S. 41–53.

Knauff, M., C. Schlieder, R. Rauh, and G. Strube (1998). Mental Models in Spatial Reasoning. This volume.

Lenzmann, B. and I. Wachsmuth (1996). A user–adaptive interface agency for interaction with a virtual environment. In G. Weiss and S. Sen (Hrsg.), *Adaptation and Learning in Multi–Agent Systems*, LNAI 1042, S. 140–151. Springer Verlag.

Miller, G. A. and P. N. Johnson-Laird (1976). *Language and perception*. Cambridge, MA: Cambridge University Press.

Mittal, N. and A. Mukerjee (1995). Qualitative Subdivision Algebra: moving towards the Quantitative. Technical Report TR 95-23, UT Austin, CS Dept.

Mukerjee, A. and N. Mittal (1995). A qualitative representation of frame-transformation motions in 3-dimensional space. In *Proceedings of the IEEE Conference on Robotics & Automation*.

O'Keefe, J. (1989). Computations the hippocampus might perform. In *Neural connections, mental computation*, S. 225–284. Cambrdige, MA: MIT Press.

Paivio, A. (1971). *Imagery and Verbal Processes*. New York: Holt, Rinehart and Winston.

Pribbenow, S. (1993, Dezember). Räumliche Inferenzen und Bilder. *KI 4*, 7–13.

Renz, J. and B. Nebel (1997). On the Complexity of Qualitative Spatial Reasoning: A Maximal Tractable Fragment of RCC-8. In *Proceedings of the 11th International Workshop on Qualitative Reasoning QR 97*.

Schlieder, C. (1996). A Computational Account of Preferences in Mental Model Construction. In U. Schmid, J. Krems, and F. Wysotzki (Hrsg.), *Proceedings of the First European Workshop on Cognitive Modeling*, Forschungsberichte des Fachbereichs Informatik, Report No. 96-39, S. 90–96. TU Berlin.

Schlieder, C. (1998). Two Approaches to Qualitative Spatial Reasoning: Constraint Satisfaction and Model Construction. Forschungsberichte des Fachbereichs Informatik, Report No. 98-2. Technische Universität Berlin. Results of the Colloquium in Berlin, june, 26.-27. 1997. http://ki.tu-berlin.de/~sppraum/juni-paper.ps.

Thompson, S. (1991). *Type Theory and Functional Programming*. Reading, Massachusetts: Addison Wesley.

Vorwerg, C. (1997). Kategorisierung von Richtungsrelationen. In E. van der Meer et al. (Ed.), *Experimentelle Psychologie. 39. Tagung experimentell arbeitender Psychologen (TeaP), HU Berlin, 24. - 27.3.97*, Lengerich. Pabst Science Publishers.

Vorwerg, C. and G. Rickheit (1998). Typicality effects in the Categorization of Spatial Relations. This volume.

Waltz, D. and L. Boggess (1979). Visual analog representations for natural language understanding. In *Proc. of the 6th IJCAI*, S. 926–934.

Werner, S., B. Krieg-Brückner, H. A. Mallot, K. Schweizer, and C. Freksa (1997). Spatial cognition: the role of landmark, route, and survey knowledge in human and robots. In *Jahrestagung der Gesellschaft für Informatik, LNCS*.

Wiebrock, S., C. Gips, U. Schmid, and F. Wysotzki (1998). SPACE/0 – Documentation of a Program for Spatial Inference. In preparation.

Wysotzki, F., U. Schmid, and E. Heymann (1997). Modellierung räumlicher Inferenzen durch Graphen mit symbolischen und numerischen Constraints. In C. Umbach, M. Grabski, and R. Hörnig (Hrsg.), *Perspektive in Sprache und Raum*. Deutscher Universitätsverlag, Wiesbaden.

Mental Models in Spatial Reasoning

Markus Knauff[1], Reinhold Rauh[1], Christoph Schlieder[2], & Gerhard Strube[1]

[1]University of Freiburg
Center for Cognitive Science
{knauff,reinhold,strube}@cognition.iig.uni-freiburg.de
[2]Technical University of Munich
Department of Computer Science
csc@tiki.informatik.tu-muenchen.de

Abstract. This chapter gives an overview of our ongoing experimental research in the *MeMoSpace* project, concerning the cognitive processes underlying human spatial reasoning. Our theoretical background is mental model theory, which conceives reasoning as a process in which mental models of the given information are constructed and inspected to solve a reasoning task. We first report some findings of our previous work and then two new experiments on spatial relational inference, which were conducted to investigate well-known effects from relational and syllogistic reasoning. (1) Continuity effect: n-term-series problems with continuous *(W r_1 X, X r_2 Y, Y r_3 Z)* and semi-continuous *(X r_2 Y, Y r_3 Z, W r_1 X)* premise order are easier than tasks with discontinuous order *(Y r_3 Z, W r_1 X, X r_2 Y)*. (2) Figural bias: the order of terms in the premises *(X r Y, Y r Z or Y r X, Z r Y)* effects the order of terms in the conclusion *(X r Z or Z r X)*. In the first experiment subjects made more errors and took more time to process the premises when in discontinuous order. In the second experiment subjects showed the general preference for the term order *Z r X* in the generated conclusions, modulated by a "figural bias": subjects used *X r Z* more often if the premise term order was *X r Y, Y r Z*, whereas *Z r X* was used most often for the premise term order *Y r X, Z r Y*. Results are discussed in the framework of mental model theory with special reference to computational models of spatial reasoning.

1 Introduction

In a large number of everyday contexts, people make extensive use of binary spatial relations which locate one object (LO = located object) with respect to another (RO = reference object). Examples of such relations are *"lies to the left"*, *"lies to the right"*, *"is in front of"* and so on. Furthermore, we are able to use such relations for making inferences, that is, we are able to infer relations not explicitly given from the ones we already know. If we know, for instance, that object X lies to the left of object Y and that Y is to the left of object Z, it is very easy for us to infer that X must be to the left of Z.

Such inferences based on binary spatial relations have long been studied in the psychology of thinking and have recently received increased attention in the literature on mental model theory of human reasoning (Johnson-Laird, 1983; Byrne & Johnson-Laird, 1989; Johnson-Laird & Byrne, 1991; Evans, Newstead & Byrne, 1993).

The general scheme of an important class of tasks studied in the psychology of reasoning are the so called *n-term-series problems*, in which subjects have to find a conclusion on the basis of given premises. In the special case of a spatial *three-term series problem (3ts-problem)*, two spatial relational terms $X\,r_1\,Y$ and $Y\,r_2\,Z$ are given as premises (Johnson-Laird, 1972). The goal is to find a conclusion $X\,r_3\,Z$ that is consistent with the premises. In a *four-term series problem (4ts-problem)*, three premises $W\,r_1\,X$, $X\,r_2\,Y$ and $Y\,r_3\,Z$, are given, and three relations not explicitly given, namely $W\,r_4Y$, $X\,r_5\,Z$ and $W\,r_6\,Z$, can be inferred. However, there are slight differences depending on which inference paradigm is being applied. Two such paradigms are commonly found in the literature on reasoning. The first we will call an *inference verification task*. The second, the *active inference task*, can be broken down into two different cases, *active general inference* and *active particular inference*. To make the difference explicit, we introduce the notation $\{\varphi_1,\varphi_2\} \rhd \varphi_3$ to denote the fact that the conclusion φ_3 is consistent with the premises φ_1 and φ_2. The two paradigms can be written as follows:

(1)   *inference verification:* does $\{\varphi_1,\varphi_2\} \rhd \varphi_3$ hold?

(2a)   *active general inference:* find <u>all</u> φ_3 such that $\{\varphi_1,\varphi_2\} \rhd \varphi_3$.

(2b)   *active particular inference:* find <u>some</u> φ_3 such that $\{\varphi_1,\varphi_2\} \rhd \varphi_3$.

There are two main theories that attempt to explain the underlying mental processes of such inferences. The first is called *theory of mental proof* and goes back to the idea that the human mind contains something like a mental logic consisting of formal inference rules (Rips, 1994). According to this theory, language-like and context-independent formal rules of inference are represented in the human mind, and inference tasks are solved by applying these rules to the given premises. Rips (1994) characterized the main idea as follows:

> "... *reasoning consists in the application of mental inference rules to the premises and conclusion of an argument. The sequence of applied rules forms a mental proof or derivation of the conclusion from the premises, where these implicit proofs are analogous to the explicit proofs of elementary logic*". (Rips, 1994, p. 40)

The key idea of the theory is clearly a repertoire of inference rules represented in human long-term-memory (LTM). These rules can be used to solve inference tasks by transferring them into working memory (WM) and applying them to the given premises, which are also represented there. For this reason, the given premises must be kept separate in the mind throughout the whole reasoning process, which means that no integrated or unified representation of the given information is generated. The premises as well as the inference rules are represented as separate entities in WM.

Here lies the main difference between this and the second approach, *mental model theory*. Since this second approach seems to be empirically more successful, and is the theoretical background of our project, we will briefly review the essential points of mental model theory of spatial relational inference in the following section. However, it is important for the reader to keep in mind that although we restrict ourselves here to the spatial domain, the theory also accounts for other types of reasoning.

1.1 Spatial Inference According to Mental Model Theory

In general, the key idea of *mental model theory* is that people translate an external situation of the real world into a *mental model* and use this representation to solve given inference tasks. In other words, a mental model is a representation of objects and relations ("structure") in working memory that constitutes a model (in the usual logical sense) of the premises given in the reasoning task. According to this view, spatial reasoning does not rely primarily on syntactic operations like in the rule-based approaches, but rather on the construction and manipulation of mental models (Johnson-Laird & Byrne, 1991). The common denominator of all mental model accounts is the conception of reasoning as a process in which, at first, unified mental representations of the given premises are generated and then, due to the fact that this information can be ambiguous, alternative models of the premises are sequentially generated and inspected. This process can be broken down into three separate phases, which are often called the comprehension, description and validation phases (Johnson-Laird & Byrne, 1991). In our project, we use the terms *construction, inspection* and *variation phase* in order to clarify the character and function of these phases.

In the *construction phase* reasoners use their general knowledge and knowledge about the semantics of spatial expressions to construct an internal model of the *"state of affairs"* that the premises describe. This is the stage of the reasoning process in which the given premises are integrated into an unified mental model. According to the theory, only this mental model needs to be kept in memory, i.e. the premises may be forgotten. It is important to point out that numerous spatial descriptions are vague and that often we can find more than one possible model that is consistent with the given premises. For this reason, we have to distinguish *determinate* tasks in which only a single model can be constructed from *indeterminate* tasks that make multiple models possible. The influence of this difference on the difficulty of reasoning tasks is one of our main research topics and we will return to this point later.

In the *inspection phase*, a parsimonious description of the mental model is constructed, including a preliminary conclusion. In other words, the mental model is inspected to find out relations which are not explicitly given. This phase was called the description phase by Johnson-Laird and Byrne (1991) because they conceived the preliminary conclusion as a kind of description of the model: *"This description should assert something new that is not explicitly stated in the premises"* (Johnson-Laird & Byrne, 1991, p. 35).

In the *variation phase*, people try to find alternative models of the premises in which the conclusion is false. If they cannot find such a model, the conclusion must be true. If they find a contradiction, they return to the first stage – and so on until all possible models are tested (Johnson-Laird & Byrne, 1991). For this reason the variation phase could be viewed as an iteration of the first two phases in which alternative models are generated and inspected in turn.

Also characteristic of the mental model theory is the concentration on particular questions, namely (1) why some inference tasks are harder to solve than others, and (2) how the inference process starts and which of the possible models reasoners generate first.

Mental model theory answers the first question quite simply: mental models are entities represented in working memory, which has a very limited capacity. Due to this fact, an inference task becomes more difficult with an increasing number of possible models that the reasoner has to keep in mind. According to this hypothesis, inference tasks with multiple models must be harder than those with single models, which means that they will take longer to solve and will be more likely to result in errors.

The second question is discussed in the literature under the keywords "initial" or "preferred models" and was investigated in our project extensively. In general, we can say that in multiple model cases, the sequence of generated models is not random. On the contrary, the construction of a first mental model seems to be a general cognitive process that works the same way for most people. We will later discuss this question in more detail.

Before we summarize our most relevant empirical results together with some findings reported in the literature, and report two new experiments, we must say a few words about the material used in our experiments.

2 The Material: Spatial Relations with an Unambiguous Semantic

When investigating spatial relational inference, cognitive psychologists usually present natural language expressions (such as *"right-of"*, *"left-of"*, *"in-front-of"*, *"behind"*) in spatial descriptions to their subjects (Byrne & Johnson-Laird, 1989; Ehrlich & Johnson-Laird, 1982; Mani & Johnson-Laird, 1982). However, as illustrated in Fig. 1, the semantics of these natural language expressions and their underlying relational concepts are far from being clear. As illustrated in the lefthand picture, numerous empirical results (in particular those concerning ratings of acceptability) have shown that the semantics of spatial descriptions such as *"lies to the right"* are very ambiguous and fuzzy (see for example Gapp, 1997; Knauff, 1997). It is not clear how to deal with such problems when investigating spatial inferences involving these natural language descriptions. The problem is more clearly demonstrated in the righthand picture. It is easy to see in this picture that the relation of the natural language expressions to the diagram is far from being clear and makes an assessment of the mental models generated by the subjects very difficult.

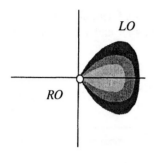
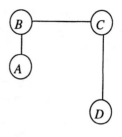

| *Ratings of acceptability* | *Is this configuration a model* |
| *for "LO lies to the right of RO"* | *of "D is before A"?* |

Fig. 1. The two examples illustrate problems with spatial expressions. As shown in the lefthand figure, the semantics of "right" is vague and fuzzy. The righthand figure demonstrates such a problem in the context of spatial inference tasks.

To avoid these difficulties in our experiments, we use the set of thirteen qualitative relations introduced by Allen (1983). In AI research on spatial reasoning the interval relations are commonly used as a representational device and found their way into several applications, e.g. in geographic information systems. There is one main reason for preferring this set of relations over natural language expressions in our experiments: one can formulate a model-theoretic semantic of the relations that allows the exact determination of what counts as a model and what does not (e.g., Nebel & Bürckert, 1995; Schlieder, 1995). This is because the relations are *jointly exhaustive* and *pairwise disjoint* (JEPD), i.e. exactly one relation holds between any two intervals. It is easy to see that these 13 relations can be used to express any qualitative relationship that can hold between two (one-dimensional) objects in an inference task.

Later it should be possible to generalize the results from our experiments in order to understand inferences on the bases of a hitherto unknown natural set of relations. In this sense, Allen's set of relations can be used as a means to study the properties of the inference processes in the spatial domain.

The set consists of the following 13 relations: *before (<)* and its converse *after (>)*, *meets (m)* and *met by (mi)*, *overlaps (o)* and *overlapped by (oi)*, *finishes (f)* and *finished by (fi)*, *during (d)* and *contains (di)*, *starts (s)* and *started by (si)*, and *equal (=)*, which has no converse. Table 1 gives pictorial examples for these relations, natural language expressions for the spatial domain, and the ordering of startpoints and endpoints as the basis for the model-theoretical foundation. In the following, we will refer to a specific point ordering as an "ordinal model", because it ignores the metrical properties of the spatial description and is based solely on ordering information of the startpoints and endpoints of the intervals.

Table 1. The 13 qualitative interval relations, associated natural language expressions, one graphical realization, and ordering of startpoints and endpoints (adapted and augmented according to Allen, 1983).

Symbol	Natural language description	Graphical realization	Point ordering (s=start, e=endpoint)
$X < Y$	X lies to the left of Y		$s_X < e_X < s_Y < e_Y$
$X\,m\,Y$	X touches Y at the left		$s_X < e_X = s_Y < e_Y$
$X\,o\,Y$	X overlaps Y from the left		$s_X < s_Y < e_X < e_Y$
$X\,s\,Y$	X lies left-justified in Y		$s_Y = s_X < e_X < e_Y$
$X\,d\,Y$	X is completely in Y		$s_Y < s_X < e_X < e_Y$
$X\,f\,Y$	X lies right-justified in Y		$s_Y < s_X < e_X = e_Y$
$X = Y$	X equals Y		$s_X = s_Y < e_Y = e_X$
$X\,fi\,Y$	X contains Y right-justified		$s_X < s_Y < e_Y = e_X$
$X\,di\,Y$	X surrounds Y		$s_X < s_Y < e_Y < e_X$
$X\,si\,Y$	X contains Y left-justified		$s_X = s_Y < e_Y < e_X$
$X\,oi\,Y$	X overlaps Y from the right		$s_Y < s_X < e_Y < e_X$
$X\,mi\,Y$	X touches Y at the right		$s_Y < e_Y = s_X < e_X$
$X > Y$	X lies to the right of Y		$s_Y < e_Y < s_X < e_X$

Combining two relations r_i and r_j gives the composition $c(r_i, r_j)$ that specifies the possible relationships between an interval X and Z given the qualitative relationship between X and Y, and Y and Z. For instance, given that *X meets Y* and *Y is during Z* then the following relations between *X* and *Z* are possible: *X overlaps Z* or *X is during Z* or *X starts with Z*. Since Allen's theory contains thirteen relations, there are 144 compositions $c\ (r_1\ r_2)$, when omitting the trivial "=" relation. They are presented in Table 2. As can be seen in the example mentioned above, there are compositions (exactly 72 of the 144) that have multiple solutions. They are presented in the table as shaded cells, whereas the white cells are single model cases. From these compositions it is easy to construct inference tasks that are known in the psychology of reasoning as 3ts-problems (e.g. Johnson-Laird, 1972).

Table 2. Composition table for the 12 qualitative relations (omitting the trivial relation "=") introduced by Allen (1983). (shaded cells = multiple model cases; white cells = single model cases).

	<	m	o	fi	di	si	s	d	f	oi	mi	>
<	<	<	<	<	<	<	<	<, m, o, s, d	<, m, o, s, d	<, m, o, s, d	<, m, o, s, d	<, ... >*
m	<	<	<	<	<	m	m	o, s, d	o, s, d	o, s, d	fi, =, f	di, si, oi, mi, >
o	<	<	<, m, o	<, m, o	<, m, o, fi, di	o, fi, di	o	o, s, d	o, s, d	o, fi, di, si, =, s, d, f, oi	di, si, oi	di, si, oi, mi, >
fi	<	m	o	fi	di	di	o	o, s, d	fi, =, f	di, si, oi	di, si, oi	di, si, oi, mi, >
di	<, m, o, fi, di	o, fi, di	o, fi, di	di	di	di	o, fi, di	o, fi, di, si, =, s, d, f, oi	di, si, oi	di, si, oi	di, si, oi	di, si, oi, mi, >
si	<, m, o, fi, di	o, fi, di	o, fi, di	di	di	si	si, =, s	d, f, oi	oi	oi	mi	>
s	<	<	<, m, o	<, m, o	<, m, o, fi, di	si, =, s	s	d	d	d, f, oi	mi	>
d	<	<	<, m, o, s, d	<, m, o, s, d	<, ... >	d, f, oi, mi, >	d	d	d	d, f, oi, mi, >	>	>
f	<	m	o, s, d	fi, =, f	di, si, oi, mi, >	oi, mi, >	d	d	f	oi, mi, >	>	>
oi	<, m, o, fi, di	o, fi, di	o, fi, di, si, =, s, d, f, oi	di, si, oi	di, si, oi, mi, >	oi, mi, >	d, f, oi	d, f, oi	oi	oi, mi, >	>	>
mi	<, m, o, fi, di	si, =, s	d, f, oi	mi	>	>	d, f, oi	d, f, oi	mi	>	>	>
>	<, ... >†	d, f, oi, mi, >	d, f, oi, mi, >	>	>	>	d, f, oi, mi, >	d, f, oi, mi, >	>	>	>	>

*. All 13 relations are possible.

3 Empirical Evidence and Previous Work

In the following section we will give a brief overview of our previous work, which used Allen's calculus as a basis, and discuss some empirical results reported in the literature of mental model theory. The sections about the "order of premises" and the "figural effect" describe two effects that are very often found in experiments concerned with other kinds of deductive reasoning and which can be largely accounted for by the assumptions of mental model theory. The question as to whether we could also find such effects in the spatial domain motivated us to perform the two experiments reported afterwards in detail.

3.1 Premise Integration in Spatial Relational Inference

As outlined above, rule theories assume that inferences are drawn on the basis of the linguistic-semantic representation of the premises, whereas mental model theory states that the premises are integrated into a unified representation–the mental model. Evidence for an integrated representation was gained (i) indirectly and (ii) mainly in the field of transitive inference (Maybery, Bain & Halford, 1986; Johnson-Laird & Byrne, 1991).

Therefore, we conducted an experiment using the *active particular inference paradigm* (2b in section 1) that was aimed to test for premise integration in spatial relational reasoning tasks in a direct manner (for details see Rauh & Schlieder, 1997). Subjects read referentially continuous, indeterminate 4ts-problems and were asked to construct one possible relationship between each implicit pair of intervals separately, namely between the first and the third, the second and the fourth, and the first and the fourth interval. Giving subjects the opportunity to provide answers to the three implicit relationships separately, a distinction could be made between correct answer triples and model-consistent answer triples. Since the latter make up a subclass of the former, they have the additional property that the three answers together with the premises were consistent with an integrated representation of the premises, whereas the correct, but non model-consistent answer triples needed to be consistent only with the premises. Materials were selected to minimize the ratio of model-consistent answers to correct answer triples, in order to result in a strong test for premise integration. The main result of this experiment was that nearly all correct answer triples were also model-consistent, which is strong evidence that people constructed one integrated representation of the premises and scanned it for the implicit informative relationships.

3.2 Model Construction: Preferred Mental Models

Within the mental model framework with its three phases of inference, we concentrated further on the phase of construction of an initial model of the premises. As a general theory of human reasoning, mental model theory ought to explain the construction of mental models from the premises as a serial process that always produces the same first mental model.

In a further experiment, also using the *active particular inference paradigm* (2b in section 1; see: Knauff, Rauh & Schlieder, 1995 for details) we tested the assumption of the existence of generally preferred mental models: subjects had to read 3ts-problems and give one possible relationship between the first and the third interval. We were able to show that a significant majority of subjects were in agreement with respect to the given answer for all of the 72 indeterminate (multiple models) problems. This suggests that the construction of an initial mental model is a general cognitive process that seems to work the same way for most people. The preferred models, with respect to the composition table (shown in Table 2), are presented in Table 3.

Table 3. Empirical model preferences (Knauff, Rauh & Schlieder 1995).

	<	m	o	fi	di	si	s	d	f	oi	mi	>
<	<	<	<	<	<	<	<	<d	oi	o	o	=
m	<	<	<	<	<	m	m	o	o	o	=	oi
o	<	<	<	<	m	o	o	o	o,d	=	oi	>
fi	<	m	o	fi	di	di	o	d	=f	oi	oi	>
di	<	o	o	di	di	di	o	=	oi	oi	oi	>
si	<	o	o	di	di	si	=	d	oi	oi	mi	>
s	<	<	o	o	fi	si	s	d	d	oi	mi	>
d	<	<	o	o	=	oi	d	d	d	oi	>	>
f	<	m	o	fi	di,oi	oi	d	d	f	oi	>	>
oi	<	o	=	oi	mi	mi	d	oi	oi	>	>	>
mi	<	si	oi	mi	>	>	oi	oi	mi	>	>	>
>	=	oi	oi	>	>	>	oi	>	>	>	>	>

3.3 On the Causal Influence of Preferred Mental Models

Based on the fact that the initially constructed mental model is the first one that is available in working memory, it follows that this will favor certain inferences before others. We tested this prediction in an experiment using the *verification task paradigm* (1 in section 1), where subjects first read referentially continuous indeterminate 3ts-problems and then had to verify a presented relationship between the first and the third interval (see Rauh, Schlieder & Knauff, 1997, for details). As shown in Fig. 2 the results corroborated our prediction in two ways: relationships that conformed to the preferred mental model in the study of Knauff et al. (1995) were (1) verified faster than other possible relationships, and (2) they were also more often correctly verified than other possible relationships.

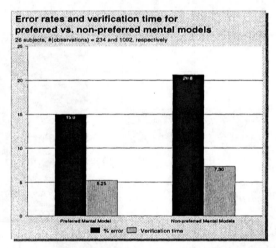

Fig. 2. Percentage of errors and verification latencies [in sec.] for 3ts-problems combined with relations that correspond to the preferred mental models as opposed to the same problems combined with other valid relations.

3.4 Symmetry Properties of Preferred Mental Models

Investigating the process of model construction in more detail, we aimed at characterizing abstract properties of this process for a cognitively adequate algorithmic reconstruction. With recourse to formal studies of Allen's calculus (Ligozat, 1990), it is possible to investigate (i) whether the model construction process works in the same manner from left-to-right and right-to-left, and (ii) whether processing an interval relation is dependent on what has been already processed (context-dependent processing of spatial relationships). Rauh and Schlieder (1997) devised an experiment in which related pairs of referentially continuous indeterminate 4ts-problem were presented in a generation experiment (2b in section 1). For each original 4ts-problem there was a twin problem that differed only with respect to orientation and an additional twin problem that differed from the original only with respect to the transposition (see Fig. 3). The main conclusion of this study was that the model construction process works in the same manner from left-to-right or right-to-left, but is context-sensitive, i.e., the processing of spatial relationships is dependent on what already has been processed. These two general properties, *symmetry of reorientation* and *asymmetry of transposition*, rule out whole classes of possible cognitive modelings and provide restrictions that a cognitively adequate algorithmic reconstruction must take into account.

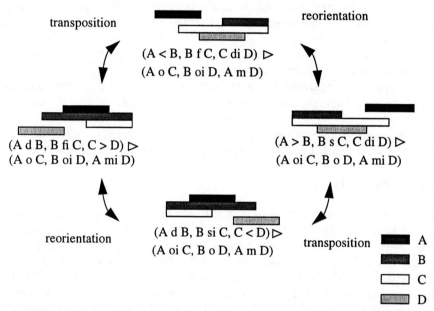

Fig. 3. Symmetry transformations on Allen-based 4ts-problems constituting an *orbit* of 4 inference tasks.

In summary, we obtained empirical evidence that (1) people construct an integrated representation of the premises, (2) that they generally come up with the same integrated representation–the preferred mental model, (3) that these preferred mental models have causal effects, because they facilitate certain inferences and suppress others, and (4) that the model construction process has the two properties of working in the same manner from left to right and right to left and of being context-sensitive.

3.5 The Order of Premises Effect

Further evidence that people construct integrated representations of the given premises in the sense of a mental model has been found through the investigation of premise order. The reported result is often called in the literature "continuity effect" or "order of premises effect" (Evans, Newstead & Byrne, 1993).

Ehrlich and Johnson-Laird (1982), for instance, gave subjects relational 4ts-problems and the three premises $W\ r_1\ X,\ X\ r_2\ Y,\ Y\ r_3\ Z$ were presented in continuous *(W r_1 X, X r_2 Y, Y r_3 Z)*, semi-continuous *(X r_2 Y, Y r_3 Z, W r_1 X)*, and discontinuous order *(Y r_3 Z, W r_1 X, X r_2 Y)*. Subjects had to infer only the conclusion $X\ r_4\ Z$. The dependent measures were the error rates (and premise processing times) for each kind of premise order.

The results support the prediction of mental model theory that continuous order (37% errors) is easier than discontinuous order (60% errors) and there is no significant difference between continuous and semi-continuous (39% errors) tasks.

Mental model theory explains these results as an effect of the difficulty of integrating the information from the premises. In the continuous and semi-continuous orders, it is possible to integrate the information of the first two premises into one model at the outset, whereas when they are presented with the discontinuous order subjects must wait for the third premise in order to integrate the information in the premises into a unified representation. Before they get this information they have to temporarily store the information from the first and second premise separately, making the task much harder.

Experiment 1 below was conducted to investigate the effect of premise order in spatial relational inference through the application of Allen's interval relations.

3.6 The Figural Effect

When investigating the effect of premise order an obvious question is whether there is a similar effect for the order of objects (terms) inside the premises. This has been done extensively in the area of syllogistic reasoning and researchers have come up with an extremely reliable and very robust effect that is called the "figural effect" or "figural bias" (Hunter, 1957; De Soto, London & Handel, 1965; Trabasso, Riley & Wilson, 1975). We explain this effect according to an experiment on relational inference by Johnson-Laird and Bara (1984). They asked subjects for a possible conclusion (1 in section 1) for the following types of problems:

Type 1:	*Type 2:*
X is related to Y	*Y is related to X*
Y is related to Z	*Z is related to Y*

The result was, that in problems of Type 1 subjects tend to spontaneously generate more conclusions in the form *"X is related to Z"* than the other correct conclusion *"Z is related to X"*, whereas they tend to generate more conclusions in the form *"Z is related to X"* for problems of Type 2. According to the rule-based, mental proof theory, the surface features of the premises determine the figural effect (Rips, 1994). However, Johnson-Laird and Bara (1984) explained the "figural effect" according to mental model theory. They assumed that the integration of the premises in working memory is more difficult in Type 2 problems because of the need to bring the *Y* term into the middle. According to this view, the construction of a mental model is easier for premises that have the repeated term as first term in the next premise. In this case, the information of the given premises can be integrated immediately and no cognitive resources are needed for mental operations that bring the middle term into the middle.

There are many good reasons to be sceptical as to whether a figural effect can also be found in spatial relational inference tasks. In particular, the syntactic structure of spatial tasks - without quantifiers - are very similar to each other and there are only a few different surface features.

Experiment 2 below was conducted to find out whether a figural effect can be found in the spatial domain as well.

4 Experiment 1: Order of Premises

As in the experiments reported above, this computer-aided experiment was separated into three blocks: a *definition*, a *learning*, and an *inference phase*. The reasons for the procedure are discussed extensively in Knauff, Rauh and Schlieder (1995). The main idea was to distinguish between conceptual and inferential aspects of Allen's calculus and to refer the obtained results to the pure inference process, holding constant the conceptual aspects.

Subjects

Thirtysix paid students (18 female, 18 male) of the University of Freiburg, ranging in age from 21 to 33 years.

Method and Procedure

In the *definition phase*, subjects read descriptions of the locations of a red and a blue interval using the 13 qualitative relations (in German). Each verbal description was presented with a short commentary about the location of the beginnings and endings of the two intervals and a picture with a red and blue interval that matched the description.

The *learning phase* consisted of blocks of trials, where subjects were presented with the one-sentence description of the red and blue interval. They then had to determine the startpoints and endpoints of a red and a blue interval using mouse clicks. After confirmation of her/his final choices, the subject was told whether her/his choices were correct or false. If they were false, additional information about the correct answer was given. Trials were presented in blocks of all 13 relations in randomized order. The learning criterion for one relation was accomplished if the subject gave correct answers in 3 consecutive blocks of the corresponding relation. The learning phase stopped as soon as the last remaining relation reached the learning criterion. Subjects needed 15 to 30 minutes to accomplish the learning phase.

In the *inference phase*, subjects had to solve 12 spatial *4ts-problems* in the *active particular inference paradigm* (2b in section 1), and the premises $W\, r_1\, X$, $X\, r_2\, Y$, and $Y\, r_3\, Z$ were presented in continuous ($W\, r_1\, X$, $X\, r_2\, Y$, $Y\, r_3\, Z$), semi-continuous ($X\, r_2\, Y$, $Y\, r_3\, Z$, $W\, r_1\, X$) and discontinuous ($Y\, r_3\, Z$, $W\, r_1\, X$, $X\, r_2\, Y$) order. They were selected on the basis of our first 4ts-experiment reported above, thus the number of correct answers given by the subjects were relatively high and each of the 12 relations were presented in the first premise exactly once. According to the separated-stages paradigm (Potts & Scholz, 1975), premises were presented successively in a self-paced manner, each on an extra screen.

Afterwards, subjects had to specify the three conclusions, namely the implicit relations $W\ r_4\ Y$, $X\ r_5\ Z$ and $W\ r_6\ Z$, each on an extra screen, by choosing the startpoints and endpoints of the intervals in lightly colored rectangular regions, as they had done in the learning phase. To avoid the effects of presentation order we systematically varied the color of the intervals and the order of conclusions asked for on the separate screens. This made the tasks relatively difficult, since subjects not only had to specify the relations but also to remember the combination of colors in each premise.

The three instances of each of the 12 4ts-problems (12x3=36 tasks) were compiled in different blocks, and there was also one practice block in the beginning consisting of 6 other simple 4ts-problems. The sequence of experimental blocks was counterbalanced across subjects according to a sequentially counterbalanced Latin square. The experiment took approximately 1.5 hours.

4.1 Results

All 36 subjects successfully passed the learning phase, and all data collected in the inference phase could be further analyzed. Individual performance showed considerable variation, ranging from 44% to 95% correct answers.

As shown in Fig. 4, the results corroborated our prediction in two ways: (1) there was no significant difference in the percent of errors between continuous (39.7%) and semi-continuous (40.1%) premise order, but (2) both were significantly easier than the discontinuous order which lead to 50.0% errors on average [$\chi^2_{(1)} = 9.643$, p < .001; $\chi^2_{(1)} = 8.864$, p < .002.].

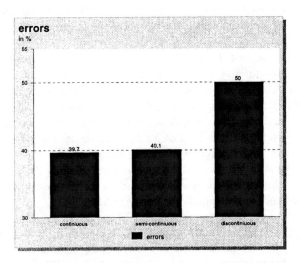

Fig. 4. Error rates for continuous, semi-continuous and discontinuous premise order in the 4ts-problems using Allen's interval relations.

Another important finding is reported in Table 4: the data on premise processing times support the assumption of mental model theory that a discontinuous premise order will increase the processing time for the third premise, because information from all premises must be integrated at this point.

Table 4. Premise processing times for the first, second and third premise in the tasks with continuous, semi-continuous and discontinuous premise order.

	Premise processing time in sec.		
premise order	premise 1	premise 2	premise 3
continuous	13.0	11.2	10.9
semi-cont.	13.6	11.0	14.4
discontinuous	12.4	13.9	19.5

Reliable differences can be found in the processing times of the third premise between continuous and semi-continuous order [$F(1,35) = 37.61$, $p < .001$], semi-continuous and discontinuous order [$F(1,35) = 40.44$, $p < .001$], and continuous and discontinuous order [$F(1,35) = 74.87$, $p < .001$]. For the second premise the differences between continuous and discontinuous order [$F(1,35) = 17.63$, $p < .001$], and semi-continuous and discontinuous order [$F(1,35) = 22.89$, $p < .001$] are significant. All other differences, in particular in the first premise, and the difference between continuous and semi-continuous order in the second premise, are not reliable.

4.2 Discussion

The experiment was conducted to investigate the continuity effect in the spatial domain with the aid of the interval relations. The error rates as well as the premise processing times showed a strong continuity effect. Subjects made more errors in tasks with discontinuous premise order than in continuous and semi-continuous order and it took more time to process the third premise in the discontinuous condition. These results can be seen as evidence for the most important assumption of mental model theory, namely that the information of the premises is integrated in a unified representation–the mental model. With this background, the result can be explained as an effect of the difficulty of integrating the information from the premises. Only in the continuous and semi-continuous order, is it possible to integrate the premises immediately into one unified representation, whereas in the discontinuous order the information from the first and second premise must be kept temporarily separated (may be in a language-like propositional form or as separate models) in working memory until the third premise is given.

This assumption is supported by the premise processing times as well, which have shown that it took much more time to process the third premise in the discontinuous order. Again, these results are compatible with the assumption that subjects build an integrated representation of the given premises. In fact, the processing time for the third premise in the discontinuous premise order must be longer, because at this point in the model construction process subjects get the first opportunity to integrate the first two premises.

5 Experiment 2: Order of Terms

As mentioned above, the figural effect is a very robust finding in the area of syllogistic reasoning (Johnson-Laird & Bara, 1984; Johnson-Laird & Steedman, 1978) that was also found in relational reasoning (Johnson-Laird & Bara, 1984, Exp. 2). For the spatial domain, however, the figural effect has not yet been investigated systematically. For this reason, the following experiment is an explorative one designed to determine whether there is the same figural effect in spatial relational inference, and whether the order of terms effects the preferred mental model.

Subjects

Twentyfour paid students (12 female, 12 male) of the University of Freiburg, ranging in age from 20 to 33 years.

Material and Procedure

The computer-aided experiment was again separated into the three phases. The definition phase and the learning phase were conducted as in Experiment I. In the *inference phase* subjects had to solve spatial *3ts-problems* (plus 10 practice trials) in the *active particular inference paradigm* (2b in section 1).

Of the 144 possible 3ts-problems, we selected 32 indeterminate task (i.e., multiple model problems) that showed the highest degree of preference from our preferred mental models experiment reported in Knauff, Rauh and Schlieder (1995). For each task we constructed "twin" tasks, which use the inverse relation but describe the same spatial relation between the three intervals. For example, the spatial arrangement of *"X lies to the left of Y"* and *"Y lies to the left of Z"* is identical to *"Y lies to the right of X"* and *"Z lies to the right of Y"*.

As shown in Table 5, based on the location of the terms, we constructed tasks of four different types (4x32=128 3ts-problems). The complement lines in the table denote the fact that the inverse relation was used. With respect to the terminology of research on syllogistic reasoning the "types" can also be called "figures". In all four types *Y* is the middle term, which occurs in both premises of the problem but on different locations. The conclusions connect the two end terms *X* and *Z*, which occur in the premises at different locations as well.

Table 5. The 3ts-problems of experiment 2 were constructed in four different types, by changing the term orders and using the inverse interval relations.

type	premise 1	premise 2	possible conclusions
1	$X\,r_1\,Y$	$Y\,r_2\,Z$	
2	$Y\,\bar{r}_1\,X$	$Y\,r_2\,Z$	$X\,r_3\,Z$ or $Z\,\bar{r}_3\,X$
3	$X\,r_1\,Y$	$Z\,\bar{r}_2\,Y$	
4	$Y\,\bar{r}_1\,X$	$Z\,\bar{r}_2\,Y$	

In each trial, after reading the premises, subjects first had to decide which interval to use to begin the one-sentence description of the conclusion (in German). This was done by pressing associated keys on the keyboard, namely <B> for *"The blue interval ..."*, <R> for *"The red interval ..."* and <G> for *"The green interval ..."*. Afterwards a new screen was shown, where this phrase and the second part of the sentence was displayed automatically. This was possible because the middle term could not be used in the conclusion. If, for example, the green interval was the middle term of the task, and the subject had pressed the key <R> initially, the two phrases *"The red interval ..."* and *"... the blue interval"* were displayed. Between these, a list of all 13 interval relations were displayed (in randomized order), and the subject could choose one of them with the cursor.

5.1 Results

As in our previous experiment, all 24 subjects successfully passed the learning phase. Individual performance in this (easier) experiment ranged from 43% to 98% correct answers.

The most important result is concerned with the term orders chosen in the conclusions. It can be predicted from the results of Johnson-Laird and Bara (1984) that subjects tend to choose the order $X\,r_3\,Z$ (abbreviated in the following as "$X - Z$") in the conclusion more often than the reverse order $Z\,\bar{r}_3\,X$ (abbreviated as "$Z - X$"). This assumption is not supported by our results: 62.8% of all conclusions given by the subjects were in the order $Z - X$.

Independently from this result, we analyzed how often the conclusions $X - Z$ and $Z - X$ were used for the four types of premise term orders. As shown in Fig. 5, the term order $X - Z$ was used for $X - Y$, $Y - Z$ (44.3%) more often than for $Y - X$, $Y - Z$ (38.7%), $X - Y$, $Z - Y$ (39.6%), and $Y - X$, $Z - Y$ (31.1%). The difference between $Y - X$, $Y - Z$ and $X - Y$, $Z - Y$ is not reliable [$\chi^2_{(1)} = 0.134$, p > .71], whereas the other differences are significant and show a figural bias: The conclusion $X - Z$ was used more often for the premise order $X - Y$, $Y - Z$ than for $Y - X$, $Y - Z$ [$\chi^2_{(1)} = 4.959$, p < .015], $X - Y$, $Z - Y$ [$\chi^2_{(1)} = 3.465$, p < .035] and $Y - X$, $Z - Y$ [$\chi^2_{(1)} = 28.259$, p < .001] and for the premise order $Y - X$, $Y - Z$ more often than for $Y - X$, $Z - Y$ [$\chi^2_{(1)} = 9.640$, p < .001]. The difference between $Y - X$, $Z - Y$ and $X - Y$, $Z - Y$ is also reliable [$\chi^2_{(1)} = 12.036$, p < .001].

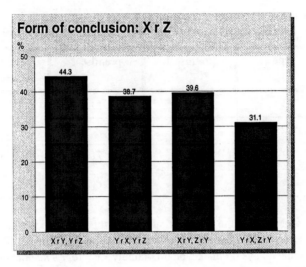

Fig. 5. The effect of term order in the premises on the form of conclusions. The figure shows the distribution of X r Z conclusions [in %] for 32 problems in each of the four types.

The next question is which type of problem is more difficult. In general, with an relatively low error rate of 16.4% the subjects performed the tasks surprisingly well. In Fig. 6 the error rates for the four term orders are depicted. Only the difference between $Y - X, Y - Z$ and $X - Y, Z - Y$ is reliable [$\chi^2_{(1)} = 7.107$, p < .05]. We have also analyzed the premise processing times, but found no significant differences.

As mentioned above, in an earlier experiment we have found preferred mental models for problems with multiple solutions (see Table 3). We now look at the solutions of our 32 indeterminate problems and compare the conclusions with these preferences. Two results are important: (1) the preferences we found in the experiment were independent from the order of terms. Only in one of the 32 tasks did the preferences differ in the four term orders; (2) in all cases we found strong preferences, the majority of which (24 out of 32, or 75%) were identical to those found in our previous investigation (Knauff & al., 1995).

Finally we computed an item- and subject analyses and found slight differences for both factors. First, the subjects' tendency to prefer the conclusion term order $Z - X$ was not equally distributed [$\chi^2_{(23)} = 363.440$, p < .001]. In fact $X - Z$ conclusions were chosen by five of our 24 subjects in more than 50% of the problems. Second, the item analyses also have shown differences for the tasks with respect to the distribution of $X - Z$ and $Z - X$ conclusions [$\chi^2_{(31)} = 54.842$, p < .005].

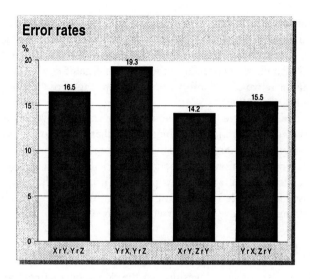

Fig. 6. Error rates [in %] for the four types of term order.

5.2 Discussion

As mentioned above, Experiment 2 were aimed more at proving the existence of some effects found in other domains of reasoning, than on testing predictions from mental model theory of spatial relational inference. We found (i) contrary to the results of Johnson-Laird and Bara (1984) no general bias towards $X - Z$ conclusions, and (ii) in accordance with Johnson-Laird and Bara a figural effect, i.e. the figure $X - Y$, $Y - Z$ favored $X - Z$ conclusions, whereas the figure $Y - X$, $Z - Y$ favored $Z - X$ conclusions. Additionally, (iii) we found significant differences for the $X - Z$ preference between subjects, and (iv) minor, but still reliable differences for the $X - Z$ preference between different tasks.

The first contradictory result of the $Z - X$ preference could be attributed either to the spatial domain or to the property of the used relation in the Johnson-Laird and Bara study; they used the relation "is related to" denoting kinship, a relation that has the property of symmetry in contrast to the normally used material in relational reasoning studies ("better than", "taller than", ...) and also in contrast to our qualitative relations that do not have the property of symmetry. The effect of abstract properties of relations like symmetry, asymmetry, or anti-symmetry, and the effect of the domain of relations (spatial v. non-spatial) on the preference of $X - Z$ conclusions has to be determined in future experiments. At least, the overall effect of a $Z - X$ preference can be explained by a cognitive process that inspects the mental model by means of a spatial focus and is sensitive with respect to the outcome of the model construction. This explanation is sketched in the general discussion below.

The next result was that the type of preferred mental model seems to be independent of the order of terms in the premises, since in nearly all the *3ts problems* the same relation between the end terms was chosen for all four figures per task. Also, the stability of preferred mental models determined in the study of Knauff et al. (1995) was not perfect, but within the range of variability found in a replication of the latter experiment in Kuß, Rauh and Strube (1996).

That some subjects differed in the overall preference for $Z - X$ conclusions and that some tasks favored $Z - X$ conclusions more than others may have to do with different inspection strategies (for example re-focussing before inspection of the mental model or not) and preferred direction of inspection of mental models (for example preferred inspection direction from left-to-right as practiced heavily in the everyday activity of reading). Subsequent detailed analysis of tasks and further experiments investigating this factor systematically have to be carried out in the future.

6 General Discussion

We reported two experiments investigating the "continuity effect" and the "figural bias" in spatial relational inference tasks. Taken together, our findings support an account of the inference process following mental model theory. In the first experiment we found clear evidence for premise integration and showed that discontinuous premise order is much harder than semi-continuous and continuous order. The result is easy to explain on the basis of the difficulty of integrating the information of the premises. This explanation is clearly supported by the premise processing times. In general the outcome of the experiment, together with numerous similar findings in other areas of inference, leads to a homogeneous image of what makes one inference task more difficult than others.

On the other hand, the results of the second experiment leave us with some open questions. The main idea of the experiment was to investigate another factor possibly effecting spatial relational inference in a similar fashion to the way that premise order does. The results of this experiment were surprising, particularly in one point. Contrary to the results of Johnson-Laird and Bara (1984), we found no general bias towards $X - Z$ conclusions. In fact, subjects tend to generate conclusions of the form $Z - X$. At first glance, this result seems to be counterintuitive, but may have a plausible explanation: the overall effect of a $Z - X$ preference can be explained–in agreement with the assumptions of mental model theory–by a cognitive process that inspects the mental model by means of a spatial focus (see for example: De Vooght & Vandierendonck, in press). In this case, after model construction the focus should be positioned on the last end term of the *3ts-problem,* namely Z. If this is the starting point of the scanning process it is plausible that the first term in the conclusion is Z and not X, because for $X - Z$ conclusions the focus must be shifted back to the term X before the scanning process starts. In contrast, for $Z - X$ conclusions these time consuming re-focussing processes are unnecessary. This seems to be a plausible explanation for the preference of $Z - X$ conclusions in our experiment. In addition to this, the cognitive modeling reported in the next section, gives support to this explanation.

However, the most important result of the experiment is that the model construction process seems to be independent of the term order in the premises. In 31 of the 32 problems the same (preferred) mental model was chosen for all four types of term order.

Taken together, both experiments give us important hints into the processes of model construction and model inspection. The model construction process seems to be sensitive with respect to the premise order and widely independent of the term order. In particular the difficulty of premise integration under discontinuous circumstances seems to be a strong argument for an incremental model construction process and supports the assumptions of mental model theory. The importance of term order emerges when the constructed model is inspected. The model inspection process can be explained as a scanning process, in which a focus is shifted over the mental model. This process seems to be sensitive to the last position of the focus, which is a result of the model construction process (see below). However, further investigations are needed to decide whether model construction and model inspection interact with respect to the positioning of the spatial focus.

7 A Computational Approach to Spatial Reasoning with Allen's Interval Relations

Schlieder (1995) proposed a computational theory of the processes we investigated. The relevant data for which the computational theory claims to account come in particular from the experiments on preferred mental models and symmetry transformations described briefly above and in detail in Knauff, Rauh, and Schlieder (1995) and Rauh and Schlieder (1997).

In the following section, we offer a brief look at the main idea of this computational model. In addition, we hope that this sketch gives an impression why the preference of $Z - X$ conclusions in experiment 2 is plausible. However, the major motivation for formulating a computational theory is the expectation that some general principles guide the model construction process and thereby explain why the preferred mental models arise. In addition, this approach helps to decide what type of spatial information is relevant for the model construction process.

For the tasks investigated in our project, the model construction process could involve either metrical information or ordering information. The former is the more constraining (i.e., stronger) and the latter the less constraining (i.e., weaker) type of spatial information. Following the principle of representational parsimony, our account is based on ordering information alone: the model of two premises $X r_1 Y$ and $Y r_2 Z$ is an (ordinal) *point ordering representation* that represents only the linear order of the *startpoints* and *endpoints* of the intervals. From the many different ways in which a linear order of a startpoint and endpoint can be represented, we choose the most parsimonious, namely to represent only the direct succession and the identification of points.

The modeling also accounts for the asymmetries of our experimental data by making the outcome of the model construction process dependent on the status in which the model was left by the previous premise integration. This status consists of the described point ordering plus a focus position. The model of *X m Y* for instance is represented as shown in Fig. 7.

$$e_{min} \; < \; s_X \; < \; e_X \; = \; s_Y \; < \; e_Y \; < \; s_{max}$$

$$\text{focus}$$

Fig. 7. A model of X *m* Y.

In this account, the inference is modeled as a scanning process over this representation that is realized as a shift of the focus. A further assumption is that the scanning process is directed in one of two possible directions and that the *model construction* should require only a minimum number of changes in scanning direction. Our idea of explaining the $Z - X$ preference comes in exactly at this point. Our modeling is mostly concerned with the *model construction* process, but it is also plausible to postulate a similar shift of focus in the *model inspection* phase. Recapping: the focus should be positioned on the last end term Z after the model construction process. If the shift of focus for the *model inspection* requires as well only a minimum number of changes in scanning direction, the avoidance of a re-focussing to the first term X after the model construction is not surprising. For model inspection, it is much easier to use this point as starting point of the scanning process. In this case the scanning process starts with the end term Z and not with X and we get the term order $Z - X$ in the generated conclusion.

However, for a brief assessment of the computational theory, we will return to the empirical results of our earlier experiments, as the modeling only claims to account for the data from the experiments on preferred mental models and symmetry transformations. We compared the empirical model preferences with the predictions of the computational theory. To sum up, assuming that the model construction process essentially works in the same manner in a left-to-right direction as in a right-to-left direction, only 2 out of the 72 preferences cannot be reproduced by the computational theory. To our mind, the cognitive modeling can therefore be considered as descriptively adequate. However, the experimental results with respect to the figural effect–which have not yet been taken into account–may help us to come up with an increasingly adequate cognitive modeling of spatial relational inference. Apart from this it may help us to compare our approach to a second account in explaining our preferences, which was developed in another group of the DFG special program. This approach goes back to the idea of metrical prototypes as the basis of model construction and comes up with some other predictions (Berendt, 1996).

8 The Conceptual Adequacy of Allen's Interval Relations

So far we have only reported experimental results which are concerned with the inferential aspects of Allen's calculus, and the reader will remember that our experiments all started with a learning phase to control the conceptual aspects of the calculus. However, one might ask whether Allen's interval relations can also be used to describe a part of human conceptual knowledge. In a previous paper on the cognitive adequacy of Allen's calculus we addressed this question by differentiating between two kinds of cognitive adequacy, namely *conceptual* and *inferential* cognitive adequacy (Knauff, Rauh, & Schlieder, 1995). Our understanding of *inferential cognitive adequacy* can be claimed if and only if the reasoning mechanism of the calculus is structurally similar to the way people reason about space. This question was the main topic of this paper. Independently from this, *conceptual cognitive adequacy* can be claimed if and only if empirical evidence supports the assumption that a system of relations is a model of people's conceptual knowledge of spatial relationships.

To answer this question for Allen's interval relations, a number of experiments were conducted and are reported in Knauff (1997; for a short overview see: Knauff, Rauh, Schlieder & Strube, 1997). In addition to several experiments in which subjects gave acceptability ratings for spatial arrangements with respect to Allen's relations or generated spatial arrangements for themselves, two memory experiments were conducted.

In a recognition experiment, for instance, subjects learned spatial arrangements with respect to Allen's relations (targets) and were later presented with them together with instances of other relations as distractors. The task was to find out the earlier learned arrangements (target) and the question was whether or not subjects are able to differentiate all 13 possible relations. We also predicted that the task must become more difficult with an increase in spatial similarity between target and distractor. In a recall experiment, subjects had to generate (draw) the previously learned spatial relations after getting a cue-stimulus.

Taken together, the most important result of these experiments was that subjects could remember the learned arrangements very well (recognition: 92%; recall: 60%) and are able to distinguish all 13 relations without learning anything about them before. To our mind, this result is a strong argument that some parts of human conceptual knowledge about ordinal spatial arrangements can be described with respect to Allen's interval relations. For our project this has the consequence that we are not working with absolutely artificial material but can say that it seems to be conceptually adequate in at least some important aspects (for details see: Knauff, 1997).

To conclude this section we wish to point out that in a new paper we report our research in cooperation with another group from the DFG special program that is concerned with a very similar question translated to the domain of topological knowledge (Knauff, Rauh & Renz, 1997; see also the article of Renz and Nebel in this book).

9 Conclusions and Future Work

Together with the conclusions of our earlier experiments the reported results show a good explanatory coherence within the framework of mental model theory, and challenge other theories of reasoning like approaches based on formal rules of inference. In particular the findings of Experiment 1 support the assumption that spatial relational inference is based on the construction of unified representations. However, further evidence will be needed before a detailed modeling of the inference process is possible. Experiment 2 leaves it an open question as to why the preferred term order in the conclusions is modulated by the term order of the premises. A series of ensuing experiments will be concerned with this point and the interaction of model construction and model inspection. Furthermore, we will investigate whether mental models represent only ordinal spatial information as assumed in the presented computational model, or contain metrical information as well. At present, we are performing dual-tasks experiments, in which the second task is "visual" or "spatial" as distinguished for example in Logie (1995) or Kosslyn (1994). The results will probably give some hints as to the kind of information represented in spatial mental models.

10 Acknowledgments

The reported research was supported by grant *Str 301/5-1* (MeMoSpace) in the research program "Spatial cognition" of the National Research Foundation (Deutsche Forschungsgemeinschaft, DFG). We would like to thank Karin Banholzer, Thomas Kuß and Kornél Markó for their assistance in several parts of the investigations and Bernhard Nebel, Nick Ketley, Judas Robinson, Karl-Friedrich Wender and an anonymous reviewer for helpful comments on an earlier draft of this paper.

References

Allen, J. F. (1983). Maintaining knowledge about temporal intervals. *Communications of the ACM, 26*, 832-843.

Berendt, B. (1996). Explaining preferred mental models in Allen inferences with a metrical model of imagery. *Proceedings of the Eighteenth Annual Conference of the Cognitive Science Society*. (pp. 489-494). Mahwah, NJ: Lawrence Erlbaum Associates.

Byrne, R. M. J., & Johnson-Laird, P. N. (1989). Spatial reasoning. *Journal of Memory and Language, 28*, 564-575.

De Soto, L. B., London, M. & Handel, L. S. (1965). Social reasoning and spatial para-logic. *Journal of Personality and Social Psychology, 2*, 513-521.

De Vooght, G., & Vandierendonck, A. (in press). Spatial mental models in linear reasoning. *Kognitionswissenschaft (Special issue "Räumliche mentale Modelle / Spatial mental models")*.

Ehrlich, K., & Johnson-Laird, P.N. (1982). Spatial descriptions and referential continuity. *Journal of Verbal Learning and Verbal Behavior, 21*, 296-306.

Evans, J. St. B. T., Newstead, S. E., & Byrne, R. M. J. (1993). *Human reasoning. The psychology of deduction*. Hove (UK): Lawrence Erlbaum Associates.

Freksa, C. (1991). Qualitative spatial reasoning. In D. M. Mark & A. U. Frank (Eds.), *Cognitive and linguistic aspects of geographic space* (pp. 361-372). Dordrecht: Kluwer Academic Publishers.

Gapp, K.-P. (1997). *Objektlokalisation [Object Localisation]*. Wiesbaden: Deutscher Universitäts-Verlag.

Hunter, I. M. L. (1957). The solving of three-term series problems. *British Journal of Psychology, 48*, 286-298.

Johnson-Laird, P. N. (1972). The three-term series problem. *Cognition, 1*, 58-82.

Johnson-Laird, P. N. (1983). *Mental models. Towards a cognitive science of language, inference, and consciousness*. Cambridge, MA: Harvard University Press.

Johnson-Laird, P. N., & Bara, B. G. (1984). Syllogistic reasoning. *Cognition,* 16, 1-62.

Johnson-Laird, P. N., & Byrne, R.M.J. (1991). *Deduction.* Hove: Lawrence Erlbaum Associates.

Johnson-Laird, P. N., & Steedman, M. (1978). The psychology of syllogisms. *Cognitive Psychology,* 10, 64-99.

Knauff, M. (1997). *Räumliches Wissen und Gedächtnis [Spatial knowledge and memory].* Wiesbaden: Deutscher Universitäts-Verlag

Knauff, M., Rauh, R., & Renz, J. (1997). A cognitive assessment of topological spatial relations: Results from an empirical investigation. In S. C. Hirtle & A. U. Frank (Eds.), *Spatial information theory. A theoretical basis for GIS.* Proceedings of COSIT '97 (pp.193-206). New York: Springer.

Knauff, M., Rauh, R., & Schlieder, C. (1995). Preferred mental models in qualitative spatial reasoning: A cognitive assessment of Allen's calculus. In *Proceedings of the Seventeenth Annual Conference of the Cognitive Science Society* (pp. 200-205). Mahwah, NJ: Lawrence Erlbaum Associates.

Knauff, M. Rauh, R., Schlieder, C., & Strube, G. (1997). Analogizität und Perspektive in räumlichen mentalen Modellen [Analog representation and perspective in spatial mental models]. In C. Umbach, M. Grabski & R. Hörnig (Hrsg.). *Perspektive in Sprache und Raum* (pp. 35-60). Wiesbaden: Deutscher Universitäts-Verlag.

Knauff, M., Rauh, R., Schlieder, C., & Strube, G. (1998). Continuity effect and figural bias in spatial relational inference. *Submitted to the Twentieth Annual Conference of the Cognitive Science Society,* Madison-Wisconsin.

Kosslyn, S. M. (1994). *Image and brain.* Cambridge, MA: MIT Press.

Ligozat, G. (1990). Weak representations of interval algebras. *Proceedings of the Eighth National Conference on Artificial Intelligence* (Vol. 2, pp. 715-720). Menlo Park, CA: AAAI Press / MIT Press.

Logie, R. H. (1995). *Visuo-spatial working memory.* Hove: Lawrence Erlbaum.

Maybery, M.T., Bain, J.D., & Halford, G. S. (1986). Information-processing demands of transitive inference. *Journal of Experimental Psychology: Learning, Memory, and Cognition,* 12, 600-613.

Nebel, B., & Bürckert, H.J. (1995). Reasoning about temporal relations: A maximal tractable subclass of Allen's interval algebra. *Communication of the ACM,* 42, 43-66.

Potts, G. R., & Scholz, K.W. (1975). The internal representation of a three-term series problem. *Journal of Verbal Learning and Verbal Behavior,* 14, 439-452.

Rauh, R., & Schlieder, C. (1997). Symmetries of model construction in spatial relational inference. In *Proceedings of the Nineteenth Annual Conference of the Cognitive Science Society* (pp. 638-643). Mahwah, NJ: Lawrence Erlbaum Associates.

Rauh, R., Schlieder, C., & Knauff, M. (1997). Präferierte mentale Modelle beim räumlich-relationalen Schließen: Empirie und kognitive Modellierung [Preferred mental models in spatial relational inference: Empirical results and cognitive modeling]. *Kognitionswissenschaft,* 6, 21-34.

Rips, L. J. (1994). *The psychology of proof.* Cambridge, MA: MIT Press.

Schlieder, C. (1995). *The construction of preferred mental models in reasoning with the interval relations* (Tech. Rep. 5/95). Freiburg: Institut für Informatik und Gesellschaft der Albert-Ludwigs-Universität Freiburg (appears also in: C. Habel & al. (Eds.), Mental models in discourse comprehension and reasoning).

Trabasso, T., Riley, C. A., & Wilson, E. G. (1975). The representation of linear orders and spatial strategies in reasoning. A developmenal study. In R. J. Falmagne (Ed.), *Reasoning: Representation and Processes.* New York: Wiley.

Formal Models for Cognition -
Taxonomy of Spatial Location Description and
Frames of Reference

Andrew U. Frank

Department of Geoinformation
Technical University Vienna
A-1040 Vienna, Austria
frank@geoinfo.tuwien.ac.at

Abstract. Language uses location description with respect to spatial frames of reference. For the transformation from a visual perception to the relative expression the reference frames must fix three parameters:
- origin (e.g., the speaker, an object, another person),
- orientation (e.g., the axial frame of the speakers, of the addressee, of another object),
- handedness of the coordinate system (same as a person's or inverse).

These parameters characterize a reference frame. The paper describes the methods used in the English language and proposes exact definitions of egocentric, intrinsic or retinal relative reference frames, and egocentric or allocentric cardinal relative reference frames. Invariants of descriptions with respect to classes of reference frames are discussed and some hints for the pragmatic preference of one or the other reference frame suggested.

The paper demonstrates two alternative computational methods for Levelt's perspective taking, which deduces another person's egocentric perspective from the speaker's egocentric (perceptive) perspective. One method is assuming imagistic (analog) representations and the other method works with a propositional (qualitative) representation. Precise hypotheses can be formulated in the formalized framework to construct human subject tests to differentiate between these alternatives.

1   Introduction

Terminology and implied assumptions often confuse the discussion of cognitive abilities of human beings. An example for terminological problems is the extensive discussion of deictic versus intrinsic frames of reference to describe spatial locations, which is ongoing for many years without conclusive results. It has been pointed out that this was - at least partially - due to varying definitions of terminology: "The analysis of spatial terms in familiar European languages remains deeply confused, and those in other languages almost entirely unexplored." (Levinson 1996, p.134)

The standard solution for clarification of terminology is formalization: defining a few base concepts - preferably not from the discipline to be investigated, thus less likely to lead to confusions - and deriving other concepts from these by formal definition (Tarski 1946). This is often attempted with complex computer programs, simulating human performance (Gopal and Woodstock 1994). However, the resulting models are so complex, contain so much detail and are so difficult to read even for their authors that they contribute little to the clarification and formalization of the discussion.

The current trend towards functional descriptions in cognitive science (Bierwisch 1996) naturally leads to models formulated in a functional language (Bird and Wadler 1988). Formalizations in a functional language can be read as definitions, but can be executed as programs to observe their effects. This allows to understand the interaction of definitions and to test our intuition. It becomes also possible to compare the defined behavior with observed behavior (Montello and Frank 1996).

A detailed description is achieved through the construction of a formal model to produce the sentences to describe the spatial relations between objects as they are typically found in the literature (e.g., (Talmy 1988; Levinson 1996): "The bicycle is in front of the church", "The ball is left of the chair", etc.). The description presented covers the simple case, just sufficient to capture the ongoing discussion (e.g., (Levinson 1996)). This approach is based on the conviction that an explanation must list all the inputs necessary to produce the observed outputs; computational, formal models are the best way to check that all necessary details are considered. Applying Occam's razor whenever possible, I strive for the simplest model and posit complications only when justified by observable behavior.

A taxonomy of reference frames must be sufficient to define mechanisms which are sufficient to produce and differentiate the different verbal descriptions of spatial location using different reference frames. The formalization shows that the current taxonomy is not sufficient for this purpose. Levinson differentiates for English 3 cases: intrinsic, absolute and relative (1996). Levelt combines this with a differentiation between egocentric and allocentric, to give 6 different cases; he uses the term *deictic* to mean *relative egocentric*, intrinsic to mean *intrinsic allocentric* and absolute for *absolute allocentric* (1996). (A detailed discussion how these terms are used by the authors is delayed till section 11, when the geometric foundation has been laid). The current taxonomy is not sufficient to differentiate the use of left/right for another person or a stage (both are *intrinsic* in the standard terminology) (figure 1 left) or with respect to a person or an object (both are *deictic*) (figure 1 right):

> Der Ball ist links von Peter. (The ball is to the left of Peter)
>   -> from the observer's perspective right of Peter,
>
> Der Ball ist auf der linken Seite der Buehne. (The ball is on the left side of the stage)
>   -> from the observer's perspective on the left side.
>
>
> Der Ball ist links von mir. (The ball is to my left)
> Der Ball ist links vom Baum. (The ball is to the left of the tree)

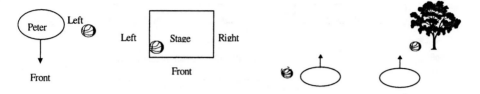

Fig. 1. left) Two different uses of 'left' (right-handed and left-handed coordinate system) right) Two different forms of deictic reference frame (egocentric and retinal)

The formal model presented here leads to the classification of reference frames by 3 values:

- The reference object (*ground*): the speaker, an observer or another object;
- The *orientation* of the reference frame (which implies the selection of the frame type): speaker, ground, and direction from speaker to ground, or one of an externally fixed system (cardinal directions, up/downhill etc.);
- The *handedness* of the reference frame (right- or left-handed).

For English, 4 different situations using the 'front, left, back, right' terminology (in subsection 10.1.1) and 2 using the cardinal direction terms 'north, west, south, east' (in subsection 10.1.2) are characterized. The classification scheme is deduced from a formal (mathematical) model. The formalization can be used as code, which is sufficient to produce acceptable English descriptions for all cases. The classification can be used for other languages as well. The model is not intended to explain the pragmatics of the selection of one or the other perspective, the use of default values and ellipsis in general, etc.

Three points seem novel in this formal approach presented here:

- The model includes the world and the observing cognizant subject and gives a formal description of the observation operation. This can lead to quantitative modeling of errors in human performance (Frank 1998).
- The model combines distance and directions in a single framework following the discussion in (Frank 1991; Frank 1991; Frank 1992; Frank 1996).
- The model includes an imagistic and a propositional alternative of spatial reasoning. Specific hypotheses to differentiate between the two can now be formulated and tested.

2 Overview of the Overall Cognitive Model Used

There seems to be sufficient agreement on the overall content of a cognitive model to situate the discussion. For example, Jackendoff gives a 'coarse sketch of the relation between language and vision' (1996, p.3 Figure 1.3), which can be summarized as Figure 2.

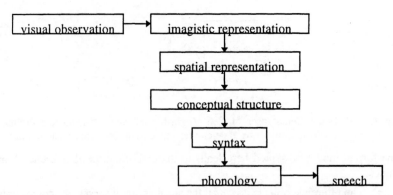

Fig. 2. Relation between language and vision (after (Jackendoff 1996), p.3)

In an attempt to achieve a formal discussion, a simpler model (figure 3) is proposed, for which each step can be formally defined. It merges transformations Jackendoff separates into single units if a distinction seems not to be definable, but adds a transformation from the propositional representation of the environment to a perspective representation, following Levelt (1996 p.96, figure 3.11). It merges syntax and phonology, which are of less interest here. The perspective transformations - which are at the core of this paper - and the respective representations are then given a formally defined meaning in this paper. This fixes the intermediate representations and the code for the transformation. 'Frame of Reference' gets in this context a well defined, operational meaning: it describes the geometric principles that are used to transform from a primary geometric representation of the environment to a perspective representation, which contains all the elements to translate to a verbal expression by a linguistic (non-geometric) transformation.

Unlike Jackendoff's model, the model in figure 3 includes a representation of the world. The representation of the world is not part of the 'cognitive model' *sensu strictu,* and is created and manipulated by special commands. The human observers or cognizant beings, in the sequel called EGO, exist in this world. The EGOs have the facility to observe the world - for example, by a visual channel - and to build an imagistic representation. From this imagistic representation a propositional (qualitative) spatial representation can be deduced by 'internal inspection'. It would be possible that the subjects in the model (humans, animals) interact with the world and change it, but this is not modeled here (it would lead to models as used in Artificial Life research).

Such a simple model is rich enough to posit a number of interesting and testable hypotheses. It can be formalized and provides the framework for other questions, a number of which are sketched at the end of this paper. The novel step is to include the world into the model; this permits to formalize the observation process (perception and cognition in one - here undifferentiated - transformation, following Talmy's 'ception' (1996, with previous literature) in the same model.

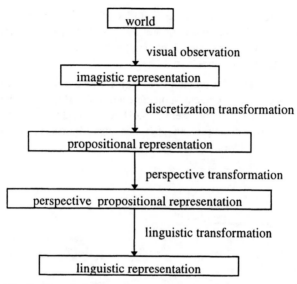

Fig. 3. The relation between language and vision used here

In such a model the question of *intermodal* transfers - of which the first part of Bloom's introduction is mostly concerned - can be discussed (Bloom, et al. 1994): what are the transformation functions, which translate from one to the other modality. One can posit a haptic observation channel from the world and then state precisely which kind of knowledge is acquired and how it is integrated with the knowledge gained through the visual channel. Is it integrated with the imagistic representation or with the propositional one (or partially one, partially the other)? The question can be posed much more specifically, because the formalization forces to define the observations that belong to a specific modality and the representation it builds.

3 Formalization of the Model

I use here a functional programming language, specifically Gofer (Jones 1991), a derivative of the Haskell language (Hudak, et al. 1992). This is a language in the ML tradition, strictly typed with type inference (Milner 1978), which assures that all operations are called with arguments of the correct type. The language is referentially transparent, meaning that substitution is generally permitted as all variables retain the value they are initially assigned; side effects are not allowed. Code can therefore be read as high-school algebra, the left-hand side of an equation is defined as the right-hand side. The code can be obtained from the author.

4 Representation of the World

The representation of the world is not part of the 'cognitive model' *sensu strictu*, but it must be included in order to permit to formalize a model of the observation function. The representation of the world - and in this case a spatial world - does not matter from a cognitive point of view; it is not part of the cognitive model, but is the model of the object, the cognizant beings are observing. The model of the world is not the model of the concept of the world the EGO has. Only the observation function must be cognitively sound. The representation of the world must be sufficient to contain the facts that are necessary to produce the human knowledge about the environment that is compatible with human behavior.

The world consists of a collection of objects that can be identified by a name. Some of the objects are cognizant subjects (persons), which can observe the world. The world is constructed at the beginning, but it could also be manipulated with special operations (for example, an object is moved). As a running example, the configuration in figure 4 will be used.

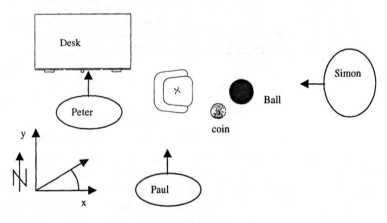

Fig. 4. The example world

The objects in the world are seen as points in a two dimensional plane without extension. We follow Levinson (1996, p. 135) in confining the discussion to the horizontal plane, as this is sufficient to reconstruct and clarify the current discussion in the literature. The objects are 'axial' (Landau 1996) and for each object, the location is given with 2 coordinates (location of the centroid) and an orientation. The orientation for things that have a natural orientation (for example, persons) is the azimuth for the 'front direction' (i.e. the angle with the positive x coordinate axis, measured clockwise) (see figure 5). Objects that do not have a natural coordinate frame are marked with *OmniDir*.

The extension and shape of the objects are not considered; this is sufficient to reconstruct the discussion by Levinson (1996), but would need extension to capture the work of Eschenbach (this volume). The model of the world could be more complex, for example, space could be. This is left out here, however, to present the core of the modeling concept without additional complications.

5   Visual Perception and Imagistic Representation

The complete visual observation process is captured here - for simplicity - in a single process. It observes the world from the perspective of the EGO and produces a single - ego-centered - representation of the object positions. The visual observation is captured in a distance and direction to each object in the world on a ratio scale and in the relative orientation of this object with respect to the observer (see figure 5). At this stage, the arbitrary orientation of the world coordinate system, in which the position of the object is represented in the world model, is removed and replaced by the egocentric coordinate system.

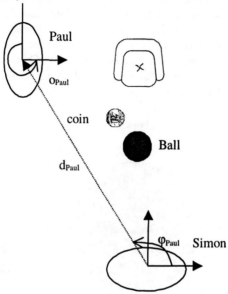

Fig. 5. The construction of the ego-centered representation

The (minimalist) EGO consists of its own name, its orientation in space (which will be used to deduce an absolute representation - see section 8) and the observation of the world (as seen from its perspective). This 'view' consists of a list of objects with their name and the vector (distance and direction) from the observer (same as (O'Keefe 1996)) to these and their orientation (with respect to the observer's orientation). One could introduce limitation to what is visible from the world, but this is not done here.

It is assumed here that the observer builds first a representation in which he uses himself and the organization of space as it emanates from himself as ground and represents the other objects on this ground (egocentric view). Since Piaget, there is an extensive discussion how allocentric representation of space is built from such an egocentric representation, but the observations reported in the literature are not sufficient that a specific, more elaborate formal description can be justified possible. For present purposes, an egocentric representation is sufficient.

6 Propositional Representation for Direct ('Egocentric') View

Following the model of Jackendoff (1996) an abstract propositional representation is deduced from the imagistic representation. An observation function accesses the imagistic representation and deduces relative positions for each object in a propositional representation. Little is known about the encoding of this propositional representation. Levelt and Levinson report experiments, which seem to come to contradictory conclusions (Levelt 1996; Levinson 1996).

We assume here qualitative representation in an egocentric system (figure 6), which differentiates 4 distance relations, where each successive range reaches twice as far as the previous one and 8 equidistant directions, following the schema proposed in (Frank 1992; Hong, et al. 1995). This system seems ecologically plausible; reasoning with directions, human performance gives approximately the same level of errors as a model with 8 direction cones (Montello and Frank 1996). Distance is encoded in zones: the zone up to 1 unit is *here*, between 1 and 2 units is *near*, 2 to 4 units is *far* and further is *very far* (see figure 9).

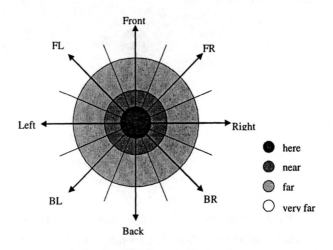

Fig. 6. The qualitative distances and directions

The transformation discretizes for each object the distance and the direction value in 4 levels for distances and 8 values for the direction and replaces the quantitative representation in the imagistic representation by a qualitative representation ("Ball" 3.2 45) becomes ("Ball" Far FrontRight). The propositional representation of the world by the EGO is a propositional, qualitative encoding of the imagistic one. This means that the vectors are discretized (i.e. distances encoded as far, near, etc. and directions as front, left, etc., directions by 8 cardinal direction values). Jackendoff's model assumes that in the propositional representation sufficient information is available for the production of the linguistic code. Sentences like

Simon says: Paul steht links vor mir. (Paul is to my front left)
  Der Sessel steht gerade vor mir. (The chair is in front of me)

etc. can be produced. This produces the most direct representation of a spatial situation. It is often called intrinsic, but Levinson shows the difficulty of this label. We propose the term *egocentric* for this. In Levinson's characteristic, it is described as an intrinsic coordinate system, with origin at the speaker and using as relatum (ground) the speaker.

7 Perception from Other Perspectives

It is generally observed that people are capable of transforming their perception of space into the perception from a different point - Levelt's 'perspective taking' (Levelt 1996). This ability is fundamental to understand other people's description of space from their perspective. This is often called the intrinsic or deictic description of the relative position of objects: the position of an object (figure) is described not with respect to the speaker but with respect to another object (ground) as if this were the observer. Very often the addressee is used as ground, as in:

Peter speaking to Simon: Der Ball liegt vor Dir. (The ball is in front of you)

but any person can be used as a ground (relatum):

Peter speaking to Paul: Der Ball liegt vor Simon. (The ball is in front of Simon)

This section concentrates on the deduction of the relative position of an object with respect to another one. Perspective taking consists of three steps:

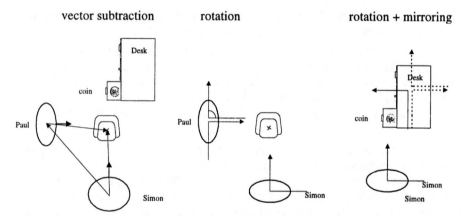

Fig. 7. vector subtraction: $V_{ground \to object} = W_{observer \to object} - U_{observer \to ground}$, rotation and mirroring

1. 'Path completion' (Frank 1992; Hong, et al. 1995) or vector subtraction, computes the vector from the ground to the figure given the vector from the observer to the ground, and the vector from the observer to the figure, (figure 7 left). Vector subtraction gives (in the coordinate system of the observer):

   (vector Simon - chair(w)) - (vector Simon Paul(u))= (vector Paul - chair(v))

2. To achieve Paul's view one has to rotate the result into Paul's orientation (figure 7 middle).

3. The handedness of the resulting coordinate system has to be selected, which decides how the reference axis of the 'new' coordinate system is oriented (figure 7 right). This differentiates the two cases:

   Der Ball ist links von Dir. (The ball is to your left)

   Die Muenze ist in der linken Schublade des Tisches. (The coin is in the left drawer of the table)

Two computational paths could be used to deduce this information: The relative position between two objects observed is either deduced from the imagistic representation or from the discrete representation - using the same formulae. One must assume that in the imagistic representation, some analogue computation similar to vector addition and subtraction can be performed. In the discrete representation, vector operations are carried out as table look up, and the combination for all inputs is stored in tables.

7.1 Construction of an Imagistic Representation from Another Perspective

The person constructs the imagistic (2D top view or distance-direction vectors) representation as it would be deduced from the other perspective. This is, mathematically, simply a transformation of origin and orientation of the imagistic representation gained directly from one's own perspective. For the distance-direction representation, it amounts to the subtraction of vectors, rotation and mirroring. The results are exactly the same as if the other person had observed the world.

From this imagistic view from another person's perspective the discrete representation can be deduced by the same discretization as described in section 6.

7.2 Deduction of the Propositional Representation for Another Perspective from the Propositional Ego-Centered Representation

This requires a qualitative reasoning about the combination of distance, directions and orientations. Simon reasons (figure 7): if the chair is in front of me and far, and Paul is front-left of me and very far, then the chair is (in Simon's orientation system) left and near from Paul. The transformation to Paul's orientation gives: if the chair is in Simon's orientation system left and near from Paul, and Paul is facing left (in Simon's system), then the chair is in front (and near) from Paul.

The deduction of qualitative relations is known as relation combination and traditionally written with the ';' operator (Schroeder in (Bird and Moor 1997)). In the literature several proposals for qualitative relation combination calculations have been made. They are mostly influenced by Allen's calculus for time relations (Allen 1983) and follow a mathematical tradition, to give as a result the conjunction of all possible results. If a combination of two relations can more than one relation, the disjunction of these is given as a result (written as a list of terms connected by *or*).

This was applied to direction reasoning by Freksa (Freksa 1991; Hernandez 1993) and to reasoning with topological relations (Egenhofer, et al. 1994). I have proposed an approximate mode for the composition operation, which gives always a single relation for the combination of two relations - the most plausible or most likely one (Frank 1992;, Egenhofer et al. 1995; Frank 1996). This is cognitively more plausible, because there are no indications that humans would handle disjunctions of relations. It agrees with the observed 'preferred model' tendency of human subjects (Knauff, this volume).

The method used here is based on 4 distances and 8 directions, which results in 32 different positions. It is possible to devise methods using symmetry, such that not the full 32 square matrix for all combination must be present in memory, but only about 120 entries; thus a propositional table look-up seems mentally possible (comparable to the memorized multiplication tables with 100 entries).

7.3 Comparison of an Imagistic or a Qualitative Transformation to Another Perspective

In principle, perspective taking should produce the same result whether applied to an imagistic or a propositional, qualitative representation:

$$discretize\ (a\text{-}b) = discretize\ (a) - discretize\ (b)$$

Due to the errors introduced by the discretization process, the deduction in the discretized (qualiative) representation is only an approximation. This difference can be used to perform experiments, which help to decide which process humans use.

Given the data reported in the literature, it cannot be decided if the perspective transformation is performed before or after translation to a qualitative model. We have shown here that a qualitative model is possible and can be formalized. Experiments with human subjects will be necessary to answer this hypothesis; the model here presented is the necessary formal framework to formulate a precise hypothesis and design the corresponding experiments.

8 Construction of an Absolute Frame of Reference

Humans have a potential to refer to the location of objects in an *absolute frame*, in English mostly the cardinal directions, but also up/down in a valley, towards the sea are used. These are particular cases of perspective taking, different from the cases discussed in section 7. There only the relative position of figure, ground and observer were relevant. If the objects in a relative frame are rotated together, the same expression results. In an absolute frame of reference the directions are invariant under individual rotation of the speaker, the observer or the ground, but if the whole configuration rotates, then the expression changes. If the speaker, ground and figure all together are rotated by a quarter turn, then what was north becomes west (figure 8).

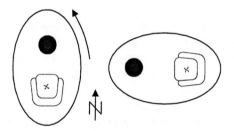

Fig. 8. After rotation 'the ball is north of the chair' becomes
'the ball is west of the chair'

The difference in what is and is not affected by changes is likely to affect the pragmatics of selecting an absolute or a relative reference frame: Absolute directions, are invariant under individual rotation of the speaker, the observer or the ground object. What is to my north is also to your north, independent of our relative position and orientation (provided we are relatively close together with respect to the figure). Absolute directions are therefore mostly used in geographic setting, where both distance relations are given and rotation of the configuration is not possible. The invariance under rotation of the individual objects and the (relative) invariance to the small changes in the positions of ground may explain why rural people pragmatically prefer absolute frames. Direction expressions related to the body frame are invariant if the full configuration (speaker, figure, and ground) are rotated together; this compensates for the uncertainty of the position in an absolute frame typically experienced indoors. This would explain the preference of urban people for expressions related to the body frame (Pederson 1993).

To construct and understand this kind of spatial relation, the EGO needs to know its orientation in this absolute frame. Then all the orientations can be translated from the egocentric reference frame to the absolute reference frame with a single subtraction of angles. This operation can be performed - similar to subsection 7.1 and 7.2 - in the imagistic or in a qualitative, propositional representation.

It is worth noting that absolute directions depend also on the ground object; only the orientation of the reference frame is given by the cardinal direction, or the up/down valley direction etc. This is often not noted, because absolute directions are in English mostly used to express positions in geographic space where the speaker and the listener are relatively close. This advantage is lost when used in proximity:

Simon says:	The ball is to the west.
Peter says:	The ball is to the east.

9   Definition of a Frame of Reference

The computation for 'perspective taking' has as input a view and produces another view - for example, the view, as the observer would have it. Section 7 offered an imagistic or a propositional method for the computation in detail. In both cases and for a relative or an absolute reference frame (section 7 and section 8), the computational model shows that a new frame of reference must be specified with three characteristics:

- the origin of the coordinate system (independent if one assumes the conventional orthogonal coordinate system or not);
- the orientation of the coordinate system, given by the direction of its primary axis (even if the secondary axis is not orthogonal to it);
- the handedness of the coordinate system (i.e. the relations of the axes).

Any description of a spatial relation is thus a relation between: the figure F and the ground G as set up by the reference frame defined by an Origin (which is the ground), an Orientation and a Handedness. All of these can be left away in a verbal description and replaced by default values (for example, ground is speaker, the origin and orientation of the reference frame is to be taken from the speaker and the handedness is conventional right-handed). The abundant use of default values in verbal description and the fact that certain combinations do not or even cannot occur, has certainly led to the confusion in the characterization of spatial frames of reference.

9.1   Formalization

Perspective taking consists of 3 steps:

1. The origin (or ground) indicates the new point with coordinates 0/0. A translation with the vector from the ego to the new origin gives the new coordinate values (operation *translate*).

2. The orientation gives the rotation between the new coordinate system and the coordinate system of the ego (operation *rotate*). (In the literature, the term *Origo* is sometimes used for the orientation only and ground for the origin).

3. The selection of the orientation of the coordinate system. In most cases, the handedness of the coordinate system is the same as that of the observer (by convention called right-handed). If one observes a person, then the person's left seems right to me (this is often characterized by *intrinsic* and *deictic*). For some objects like desks, which have a front, but their left side is to the left of the observer, left and right are exchanged compared to the coordinate system in comparison to the observers. This we call a 'left-handed' coordinate system.

> Peter stands in front of the desk and opens the left drawer.

The handedness of the coordinate system differentiates the system of a person, where anti-clockwise follow: front, left, back, right, front, from the system used for objects like desks, where again anti-clockwise follow: front, right, back, left, front. Mathematicians call a coordinate system right-handed if the three primary axes (front, left, up) are in the same situation as the thumb, index and middle finger of the right hand (otherwise it is called left-handed).

In principle, these three steps are performed for all perspective taking. To understand a verbal description, the parameters for all three steps must be indicated. Default values, customary choices, but also vocabulary are used to indicate these choices. In English, different terminology is used to describe absolute references (cardinal direction terms) or relative references (terms like front, left etc.). Some situations remain ambiguous: English allows to call an armrest of a chair its 'left' or 'right' one, depending on the conceptualization (it may be influenced if the chair is occupied or not, which influences if the right-handed coordinate system of the person sitting in the chair is used or if the left-handed coordinate system of the chair is used).

9.2 Overview of Transformation Steps

In this model, the outside world is first transformed in an imagistic (egocentric perspective) representation. This imagistic representation can be discretized and transformed to a propositional, qualitative representation. The propositional, qualitative representation contains all data for the transformation in a verbal expression.

In order to produce and understand verbal descriptions from other than the egocentric perspective, specific perspective taking transformations are necessary (for which the parameters were described in the previous sub-section). These perspective taking transformations can be performed either on the imagistic representation (with a discretization following) or on the qualitative representation. The available experimental data is not sufficient to differentiate between these two alternatives.

For the verbal expression, the following language 'instructions' are necessary:

- the expressions to identify ground, orientation and handedness,
- the lexical terms used for the direction,
- the distance terms, and
- the position of the observer.

Pragmatic rules must explain when ellipsis is possible and which default values can be assumed and which expressions are to be preferred. I assume here that these rules can be applied after perspective taking has been performed (Levelt's results are not conclusive to this point (Levelt 1996)).

10 Descriptions of Spatial Reference Frames

Every spatial reference frame can be characterized by three values: the selection of ground, orientation and handedness of the coordinate system. Here descriptions of the 4 constructions used in English in conjunction with body-centered terms (front, left, back, right) and then the 2 constructions used with external frames are given. The following section compares these characterizations with the descriptions found in the literature. The characterizations listed here are sufficient to produce the corresponding expressions.

For each method of perspective taking, the object that determines the origin of the coordinate system (ground), the object that gives the orientation of the coordinate system (orientation), and the handedness is given as three parameter values.

10.1   English Allows the Following Constructions:

10.1.1 Using the Body-Centered Direction Terms (Front, Left, Back, Right):

The following 4 relations between point-like objects are used:

- egocentric
  ground = speaker, orientation = speaker, right-handed
  > Peter says: The desk is before me, the chair is to my right
- intrinsic - right-handed (mostly used for persons, but also animals, vehicles)
  Ground = person, orientation = person, right-handed
  > The ball is in front of Simon
- intrinsic - left handed (objects like a stage, desk etc.)
  ground = oriented-object, orientation = object, left-handed
  > The left drawer of the desk
- retinal
  ground = object, orientation = towards the observer, left-handed
  (note: the observer is not necessarily the speaker)
  > Paul says: The ball is to the right of the chair.
  > For Peter, the ball is behind the chair.

For path, the English allows:

ground = current-position, orientation = direction of movement, right-handed

10.1.2 Using Cardinal Directions (North, West, South, East)

Cardinal directions are only used for geographic space in English.

- egocentric
  ground = speaker, orientation = cardinal, right-handed
  > Paul says: The chair is to the north, Simon is north-east

- allocentric

ground = object, orientation = cardinal directions, right handed

> Acity is north of Btown.

10.1.3 Using Other Direction Terms

ground = speaker, orientation = hill-side, right-handed -- only used in geographic space

> The tree is above the house.

10.2 Examples from Other Languages

German allows, but only for geographic space in valley:

ground = speaker, orientation = valley, right-handed (down is the primary axis)

> Aastadt is oben, Bedorf unten (im Tal).
>
> Bedorf ist auf der rechten Seite.

ground = object, orientation = valley, right-handed

> Aastadt ist weiter oben als Bedorf.

This seems to be similar to the system used by Tzeltal (Levinson 1996).

10.3 Converseness and Transitivity

The properties identified by Levelt for the combination of reference frames follow from the above description (Levelt 1996). Converseness $(R\ (a,b) = R^{-1}\ (b,a))$ and transitivity $(R\ (a,b)\ and\ R\ (b,c) = R\ (a,c))$ hold for all cases when the orientation is the same, but not when the orientation (or the handedness) for two expressions differs (which is the case for his 'intrinsic' case: ground = person orientation = person). One can therefore construct more transitive relations for an intrinsic case: if persons Simon, Peter and Paul all face me, then Simon is left of Peter and Peter is left of Paul allows the deduction Simon is left of Paul.

11 Compare to Usual Terminology

The usual terminology for English sentences can be characterized in these terms and we propose some (minor) restrictions to make the regular terminology more precise.

11.1 Absolute Reference Systems

All systems where the orientation is given from the outside (cardinal direction, valley up-down, monsoon, inland/seaward, direction of a landmark, local landmark (Pederson 1993) etc.) are called absolute. For all methods of absolute orientation, the individual must know its own orientation relative to the fixed reference frame (but it need not know the speaker's orientation for understanding a description). Absolute systems are invariant under rotation of ground, but not invariant under rotation of the whole configuration (figure and ground).

We differentiate between egocentric and allocentric absolute systems:

- egocentric absolute: ground = speaker, orientation = an absolute reference frame, RH

> Simon says: The ball is to the west;

- allocentric absolute: ground = object, orientation = an absolute reference
  frame, RH

  Paul says: The ball is to the west of Simon.

11.2   Egocentric

Egocentric is used for a frame of reference that is centered in the speaker and with the speaker's orientation (or the assumed observers).

Ground = observer, orientation = observer

  Paul says: The chair is before me (= in front of me).

This is obviously also a case of intrinsic reference frame (in Levinson's definition), but a further differentiation, as proposed by Levelt, is appropriate.

11.3   Intrinsic

This term is used to describe spatial reference frames where the orientation is taken from an object. We propose to exclude the case where the reference frame is the speaker serving as the ground (we call it egocentric).

Ground = an oriented object (not the speaker), orientation = from the ground object, RH or LH

  Paul says: The ball is in front of the chair.

This agrees with Levinson's characterization that intrinsic frames of reference are invariant under rotation of the viewer, but not under rotation of the ground object.

11.4   Retinal

Retinal means the relative position of an object with respect to another, as it presents itself in the eye of the speaker (this is often described as deictic):

  ground = object, orientation = arrow from observer to ground, RH.

  Simon says: The coin is left of the ball.
  Paul says: The coin is in front of the ball.

Levinson would characterize this among his group of relative frames, which are not invariant under rotation of the speaker. This is not completely accurate, as the orientation of the reference frame is taken from the arrow from observer to ground, not the observer's orientation (Levinson 1996, p.149, figure 4.10).

11.5   Deictic

This is used for relations where ground and observer are not the same and the orientation is taken from the observer.

  Ground = object, orientation = observer

Precisely this definition it is not often used in English. There may be situations - for example, in military situations - where two observers discuss the relative place-ment of objects in a landscape; facing the same direction, the observers will use their (nearly common) reference frame. Usually the retinal frame of reference is meant when a frame of reference is described as deictic (e.g., in Levelt).

A description in a deictic frame of reference would be invariant under rotation (and translation) of the viewer, but not under rotation of the ground (this is Levinson's characteristics of his class of intrinsic frames of reference).

11.6 Relative vs. Absolute

We propose to use the term relative for all 'body-centered' (or generally object part related) reference frames - these are the ones which in English are based on the 'front/back, left/right' lexical terms. All these are invariant under translation or rotation of the full configuration (speaker, ground, and figure), but not under rotation or translation of either speaker or ground. This is different from Levinson's usage, where 'relative frames' are invariant under rotation of the ground.

The term absolute should be used for frames of reference, where descriptions are invariant under rotation of the ground or speaker, but not invariant under rotation of the complete situation (figure, ground and observer).

		egocentric ground=speaker	allocentric ground=object
relative	orientation= speaker related	egocentric	retinal (orientation= speaker-object)
	orientation=object		intrinsic (RH or LH)
absolute	cardinal	egocentric cardinal	allocentric cardinal
	up/down-hill	...	...

Fig. 9. Summary of Frames of Reference used in English

The table (figure 9) classifies the frames of reference in the English language. It differentiates first between relative and absolute reference frames, and then differentiates the relative in those where the orientation is taken from the speaker and those where it is taken from the object. Absolute frames of reference can be separated in those using cardinal directions and those using other, for example, up/down-hill, as reference frames. It further differentiates each case in egocentric (speaker serves as ground) or allocentric (another object serves as ground). In English, it seems not to occur that the orientation is taken from another object, but the speaker serves as ground (Haussa seems to use such a reference frame). Other languages may use other frames of references, which are best classified using the characteristics of ground, orientation and handedness of the coordinate system.

12 Conclusions

A formalization of the operations necessary to produce from a visual observation of the environment the propositional descriptions to verbalize spatial relations between objects has identified the necessary parameters for the description of the spatial reference frames used. A reference frame should give all geometric properties for this transformation and is therefore characterized by:

*   its origin (ground),

- the orientation (orientation of the observer, orientation of ground or externally fixed), and
- the handedness of the coordinate system.

With these parameters, all spatial frames of reference for all languages can be characterized. Pragmatics when to use which frame of reference are not considered here, but it is noted that some parameter combinations are meaningless or not used. The English language uses only some combinations, which are best characterized as:

- relative (front/back, left/right) with the reference frames: egocentric, intrinsic and retinal, and
- absolute (cardinal, up/downhill, etc.) with egocentric and allocentric reference frames.

Precise definitions for these terms were given with respect to the characterization of reference frames.

Other languages use different spatial frames of reference and the current terminology is not adequate for their description. The characterization with the parameters for origin, orientation and handedness, however, capture easily all examples given in the literature. The formalization has also helped to understand assumptions built in this classification schema and may lead to discoveries of mechanisms in languages that are not covered.

The formalization and the computational model produced show that:

- All forms of relative spatial descriptions can be deduced from a single observation of the world, which results in a representation of distance and directions (in the egocentric reference frame) to the other objects in the world.
- A propositional, qualitative representation can be deduced from an imagistic representation by discretization (for example, qualitative distance-directions expressions for each object).
- The translation of the ego's representation to a representation as seen from another person either requires the construction of a translated and rotated (imagistic) representation, using the subtraction of vectors in 2D space, or composition operations for the propositional representation. Observation of error characteristics of human performance may allow identifying one of these possibilities as the one used by humans.
- For objects (at least for persons), the internal representation must include the orientation of the object. This is required for the translation of one's perspective to the perspective of others or to understand another person's description of a scene.
- Knowledge of the ego's orientation with respect to an absolute frame is only required to produce spatial relations in an absolute reference frame.

All absolute frames are invariant under rotation of ground or observer; thus the relative position of the parties involved in an exchange does not affect the understanding of the spatial relations. For geographic space, if the observers are relatively close to each other and the objects described are further apart, the description is even invariant under translation of the observers. These properties apply to cardinal directions as well as to up/down-hill, up/down-wind etc.

All relative frames are invariant under rotation or translation of the full configuration (speaker, ground, and figure). They are thus easier to use inside buildings or towns, where the absolute orientation is not so easily detected.

A number of questions remain open: An expression, using the speaker as ground, and a way to express relations using another person as ground, with an absolute orientation or the body orientation of the person serving as ground, is likely a universal (or at least that if a language has the first, it has also the second: one is necessary to understand the other, spoken by another person).

It is an open question, how different languages indicate which perspective is taken; many acceptable ellipses further complicate the situation. Both in English (Levelt 1996) and in Tamil (Pederson 1993) misunderstandings are common, based on errors in the perspective taken to produce the verbal expression. Lexicon seems to separate the body-related (relative) and the absolute perspectives.

It is further interesting to analyze the preferences in certain languages for certain expressions. Pederson gives an example where two Tamil-speaking populations are differentiated by one preferring a cardinal orientation (and the corresponding vocabulary) and the other body axes (relative) expressions (Pederson 1993). Is this correctly related to the invariants of absolute and relative frames of reference, which make absolute reference systems better suited to rural and relative reference frames easier to use in urban situations?

Acknowledgments

Numerous discussions with Dan Montello increased my awareness of experimental results. Extensive discussions with Carola Eschenbach helped me to frame the question precisely and her comments led to this hopefully improved presentation. Many valuable discussions with David Mark are gratefully acknowledged. Roswitha Markwart carefully edited the text and Hartmut Schachinger drew the figures.

References

Allen, J.F. (1983). Maintaining Knowledge about Temporal Intervals." *ACM Comm.* 832-843.

Bierwisch, M. (1996). How much space gets into language? In: *Language and Space.* P. Bloom, et al. (eds.). Cambridge MA, MIT Press: 31-76.

Bird, R. and O.de Moor (1997). *Algebra of Programming.* London, Prentice Hall.

Bird, R. and P. Wadler (1988). *Introduction to Functional Programming.* Hemel Hempstead, UK, Prentice Hall International.

Bloom, P., et al, (eds.). (1994). *Language and Space.* Cambridge, MA, MIT Press.

Egenhofer, M.J., et al. (1994). *Evaluating inconsistencies among multiple representations.* Proceedings of SDH'94, Edinburgh, UK.

Frank, A.U. (1991). *Qualitative Spatial Reasoning about Cardinal Directions.* Proceedings of Auto-Carto 10, ACSM-ASPRS, Baltimore, MD.

Frank, A.U. (1991). *Qualitative Spatial Reasoning with Cardinal Directions.* Proceedings of 7th Austrian Conference on Artificial Intelligence, Vienna. Berlin Heidelberg, Springer.

Frank, A.U. (1992). Qualitative Spatial Reasoning about Distances and Directions in Geographic Space. *Journal of Visual Languages and Computing* 1992(3): 343-371.

Frank, A.U. (1996). Qualitative spatial reasoning: cardinal directions as an example. *IJGIS* 10(3): 269-290.

Frank, A.U. (1998). Metamodels for Data Quality Description. In: *Data Quality in Geographic Information: from Error to Uncertainty*. R. Jeansoulin and M. Goodchild (eds.). Paris, Editions Hermès.

Freksa, C. (1991). Qualitative Spatial Reasoning. In: *Cognitive and Linguistic Aspects of Geographic Space*. D.M. Mark and A.U. Frank (eds.). Dordrecht, Kluwer: 361-372.

Gopal, S. and C. Woodstock (1994). Theory and Methods for Accuracy Assessment of Thematic Maps Using Fuzzy Sets. *Photogrammetric Engineering & Remote Sensing* 60(2): 181-188.

Hernandez, D. (1993). Maintaining Qualitative Spatial Knowledge. In: *Spatial Information Theory: A Theoretical Basis for GIS*. A.U. Frank and I. Campari (eds.). Lecture Notes in Computer Science, Vol. 716. Berlin Heidelberg, Springer: 36-53.

Hong, J.H., et al. (1995). *On the Robustness of Qualitative Distance- and Direction-Reasoning*. Proceedings of Auto-Carto 12, Charlotte, NC.

Hudak, P., et al. (1992). "Report on the functional programming language Haskell, Version 1.2." *SIGPLAN Notices* 27.

Jackendoff, R. (1996). The architecture of the linguistic-spatial interface. In: *Language and Space*. P. Bloom, et al. (eds.). Cambridge, MA, MIT Press: 1-30.

Jones, M.P. (1991). An Introduction to Gofer, Ph.D. thesis, Yale University.

Landau, B. (1996). Multiple Geometric Representations of Objects. In: *Language and Space*. P. Bloom, et al. (eds.). Cambridge, MA, MIT Press: 317-363.

Levelt, W.J.M. (1996). Perspective Taking and Ellipsis in Spatial Descriptions. In: *Language and Space*. P. Bloom, et al. (eds.). Cambridge MA, MIT Press: 77-108.

Levinson, S.C. (1996). Frames of Reference and Molyneux's Question: Crosslinguistic Evidence. In: *Language and Space*. P. Bloom, et al. Cambridge, MA, MIT Press: 109-170.

Milner, R. (1978). "A Theory of Type Polymorphism in Programming." *Journal of Computer and System Sciences* 17: 348-375.

Montello, D.R. and A.U. Frank (1996). Modeling directional knowledge and reasoning in environmental space: testing qualitative metrics. In: *The Construction of Cognitive Maps*. J. Portugali. Dordrecht, Kluwer Academic Publishers: 321-344.

O'Keefe, J. (1996). The spatial prepositions in English, vector grammar, and the cognitive map. In: *Language and Space*. P. Bloom, et al. (eds.). Cambridge, MA, MIT Press: 277-316.

Pederson, E. (1993). Geographic and Manipulable Space in Two Tamil Linguistic Systems. In: *Spatial Information Theory: Theoretical Basis for GIS*. A.U. Frank and I. Campari. Lecture Notes in Computer Science Vol. 716. Berlin Heidelberg, Springer: 294-311.

Talmy, L. (1988). *How Language Structures Space*. In: Cognitive and Linguistic Aspects of Geographic Space, Report on a Workshop, NCGIA, Santa Barbara, CA.

Talmy, L. (1996). Fictive Motion in language and "ception". In: *Language and Space*. P. Bloom, et al. (eds.). Cambridge, MA, MIT Press: 211-276.

Tarski, A. (1946). *Introduction to logic and to the methodology of deductive sciences*. New York: Oxford University Press.

Spatial Representation with Aspect Maps[1]

Bettina Berendt, Thomas Barkowsky,
Christian Freksa, and Stephanie Kelter

Universität Hamburg, Fachbereich Informatik &
Graduiertenkolleg Kognitionswissenschaft
Vogt-Kölln-Str. 30, D-22527 Hamburg, Germany
{berendt, barkowsky, freksa, kelter}@informatik.uni-hamburg.de

Abstract. This paper describes the *aspect map approach* to model the processing of geographic maps. Geographic maps are described as spatial representation media which play an important role in many processes of human spatial cognition. We focus on the aspectuality of representation and therefore deal with *aspect maps*: spatial organization structures in which one or more aspects of geographic entities are represented. The aspect map architecture is presented, an AI model of processing geographic maps. Two processes contained in this model are investigated in more detail. The first is the transformation of one aspect map into another aspect map which only retains selected entities and aspects (extraction). The second process is the combination of two aspect maps, in order to obtain a third aspect map. The results of an empirical study show that the formal approach can describe and distinguish the ways in which people solve this task.

1 Introduction: Maps

From the viewpoint of spatial cognition and knowledge representation, geographic maps are worthwhile to investigate for at least two reasons. First, maps are a common means of depicting geographic knowledge which have proven their usefulness over centuries. From the viewpoint of representation theory it is of interest to investigate how geographic knowledge is represented in maps and how it can be extracted. Second, and related thereto, from a cognitive point of view, investigating the structure of maps can provide a promising access to the way humans perceive, represent, and interact with their spatial environment. The existent de facto standard of mapping techniques has gradually evolved over a long time in a purposeful way; thus, the way people construct and interact with geographic maps has to be regarded as a valuable clue to the properties of the underlying mental structures and processes for spatial cognition.

---

[1] This work is supported by the Deutsche Forschungsgemeinschaft (DFG) in the framework of the priority program on spatial cognition (grant Fr 806/8).

1.1 Maps as Spatial Representation Media

Maps are spatial media used for representing spatial, or more specifically: geographic knowledge. Considering spatial representation with respect to geographic maps in the first place means examining a representation *medium*. This medium as a form of representation imposes certain constraints on the content encoded in it. The concepts of spatial knowledge relevant for structuring our surrounding world (e.g. distance, shape, orientation, etc.) therefore occur in the representational context at hand as highly dependent on the way they can be depicted.

Maps: Properties and Types. We use the term 'map' to characterize a type of pictorial representations which have in common that they (1) have two-dimensional spatial extent, (2) make use of symbolic depictions which stand for geographic entities in the world, and (3) convey information about spatial relations in the world by the localization of these symbols. In a figurative sense, we will also use the term 'map' to denote cognitive or internal representations of external maps.

These criteria given for maps may apply to a large variety of representations of spatial information. In particular, two types of geographic maps fall into this class of representation: topographic maps and thematic maps. The former are designed to show the primary geographic characteristics of an environment (e.g. shapes of land and water areas, elevations, etc.) as precisely as possible with respect to the given resolution. Thematic maps - which are based on topographic maps - are designed to show data that are attached to the topographic basis (e.g. demographic information or soil data).

General purpose city maps can be viewed as thematic maps which relax the strict location requirements of topographic maps. In city maps the main emphasis lies in depicting and locating labels and other symbols used for orienting in and communicating about urban environments. City maps are highly structured by symbolic information (e.g. names of streets and public locations). The topographic exactness of such special-purpose maps usually is of lesser importance. More typical thematic maps abstract even more stringently from topographic requirements.

Schematic maps (e.g. public transportation maps or tourist maps) are representations of spatial features which omit most of the topographic information. Schematic depictions of urban underground networks, for example, often only depict the sequence of stations on the lines and their coarse direction, thus ignoring exact distances and localizations.

Finally, map sketches are designed to serve certain specific ad hoc purposes; therefore, they represent only few spatial characteristics of the environment they depict, and are highly imprecise with respect to the features they are not intended to show, even omitting them when they are not needed to induce coherence between the pieces of information of interest.

Maps as Integrated Spatio-Symbolic Representation Media. Maps as representational media carry information in two ways: pictorially and by means of symbols. Maps are pictorial representations as they result from an analogical mapping of spatial characteristics onto a spatial (planar) medium [Sloman 1971, 1975]. This means that they form a natural or physical homomorphism in the sense of knowledge representation theory [Palmer 1978]. On the other hand, in every map certain classes of entities are represented symbolically.

So cartographic signs represent geographic entities either pictorially or symbolically. The meaning of these signs, i.e. the correspondence between the signs on the maps and the geographic entities denoted by them, is either explained explicitly in the map's legend or is given implicitly by reference to the map reader's general cartographic understanding.

Every map embodies both kinds of representations. Depending on the map's type and scale either pictorial or symbolic information is predominant. For example, when a survey map at a smaller scale is generated on the basis of a detailed reference map, geographic features often have to be grouped and replaced by a symbol. As a result of this transformation, information about details is lost, and geographic information formerly represented pictorially now is represented symbolically, where the symbols correspond to classes of pictorial representations.

Representation of Aspects. This observation is related to our notion of *aspects* of a representation. As the representational space of a map of a given scale is strictly limited, every map - however detailed it may be - only represents certain features or aspects of geographic entities and omits others as a result of the planar pictorial structure of the medium used. For example, a survey road map represents all roads belonging to a certain class together with the connections between them; however, it does not show smaller roads or environmental features that are presumed irrelevant for the map's intended purpose.

So from a representational point of view, to use a given map correctly, it is necessary to distinguish (1) which aspects are represented pictorially, (2) which aspects are represented symbolically, and (3) which aspects are not represented in the map. In other words, to look at it from the processing point of view, adequate operations have to be used to interpret the map's content correctly. This means that the information contained in a map has to be adapted appropriately to deal with spatial information in general. Map reading can be understood as the process of relating the entities on the map to appropriate entities in the geographic world by employing an adequate body of spatial knowledge.

1.2 Maps and Spatial Cognition

External representations like maps are created to be used by people. The choice of a particular representation medium raises the question of its particular properties and of its advantages and disadvantages in cognitive processes. We will discuss one particular property of maps in this section, namely the depiction of one kind of space in another kind of space. This allows the map user to perform operations that cannot be performed directly in the represented space. A different property may arise from similarities between maps and internal representations of space: since maps allow operations which are also routinely performed on internal representations, these may be 'natural' and easy to accomplish. Vice versa, the fact that external maps have proven useful in solving tasks may provide a clue about the properties of internal representations of geographical spaces. In which respects are internal representations of space 'map-like'? A critical discussion of this question will lead us back to the importance and pervasiveness of the aspectual nature of spatial representations, internal or external.

Maps and Cognitive Spaces. The notion of 'space' in research on cognition encompasses a wide range of objects and relations between them. The objects of a space can be very small, like elements of a picture or objects on a table, and they can be very large, like cities or continents. While mathematical descriptions of such spaces may vary only by a scaling factor, our way of perceiving and interacting with them varies greatly. For cognitive research, one important difference is whether a space can be perceived from a single vantage point (e.g. a picture, or the inside of a room), or whether one has to move around and integrate different perceptions to gain an integrated representation of the space (e.g. a building, a city, a country).[2]

As spaces of different scale form a different cognitive reality, it is likely that our ways of representing and reasoning about these spaces differ too. Many authors propose to distinguish between spaces based on differences in perception, interaction, and the possible necessity of locomotion (for a survey, see [Montello 1993]). Montello distinguishes between figural space, which can be subdivided into pictorial and object space (small flat spaces and small 3-D spaces, respectively), vista space (single rooms, town squares, small valleys, horizons), environmental space (buildings, neighborhoods, cities), and geographical space (large cities, countries, continents). The difference between environmental and geographical spaces is that the former can be apprehended through direct experience (locomotion), while the latter "cannot be apprehended directly through locomotion; rather, [they] must be learned via ... representations such as maps or models that *essentially reduce the geographical space to figural space.*" ([Montello 1993], p. 315; emphasis added). In the following, we will use the term 'geographic space' to refer to Montello's 'geographical space'.

What are the advantages of such reduction? Operations that cannot be performed in the original geographic space can be performed on its representation in figural space. For example, the relative length of a shortcut in comparison to a detour can be seen directly on the map, while it cannot be seen in the real environment. More generally speaking, ways of interacting in one space are utilized to extend ways of interacting in another space. In this sense, a map can be regarded as a "cognitive tool" for conveying geographic relations, or, the map representations are "external components of a cognitive system with which the internal components interact during a dynamic and task-oriented process of development" - 'cognitive externalism' [Peterson 1996, p. 8].

---

[2] Strictly speaking, the perception of a space from a single vantage point also involves movement (head and eye movements). However, this does not affect our central argument and will not be considered here.

Exactly what these interactions are is one of the two cognitive science questions addressed by the aspect map approach. In line with Montello's argument, the two spaces we shall be concerned with are geographic and pictorial spaces.

Internal vs. External Maps. If geographic maps are external components of a cognitive system that help a person answer questions about geographic space and solve tasks concerning that space, one may ask what the related internal components look like. One may suspect that they represent a geographic space in ways that are similar to external map representations. Indeed, the map metaphor has a long tradition in psychology: *Cognitive map* is the term most commonly used to denote various kinds of knowledge about environmental and geographic spaces in humans, animals, and robots (for a survey, see e.g. [Hirtle & Heidorn 1993]). Knowledge about environmental and geographic spaces may be knowledge about routes, or it may be survey knowledge (e.g. [Schweizer et al., this volume; Krieg-Brückner et al., this volume; Rothkegel et al., this volume]). Maps usually convey survey knowledge.

However, the map metaphor is slightly unfortunate, since most research has concentrated on (and found) properties of internal representations of environmental and geographic spaces that are not map-like: Unlike good maps, these internal representations often do not represent the space exhaustively, and they are distorted or even inconsistent. Tversky [1993] therefore proposes to replace the map metaphor by two other ones: It is likely that one only has 'collage-like' knowledge about complex, less familiar spaces (*cognitive collages*). At the same time, one may have an integrated *spatial mental model* of a simple, well-known space. A mental model can be considered weaker than a map in that it need not be metrically specified and stronger than a map in that it may be 3-dimensional.

Also, internal representations seem to be dominated by knowledge about qualitative relations like hierarchical containment (e.g. [Stevens & Coupe 1978; Hirtle & Jonides 1985; Gehrke & Hommel, this volume], rather than by positions as in a map. However, non-exhaustiveness, distortions, and emphasis on (certain kinds of) relations can also be regarded in a different light: Just like external maps (or indeed any kind of representation), internal maps only represent some *aspects* of the represented space (see section 1.1.3). Exactly what these aspects are, and how internally and externally represented aspects interact, is the second of the two cognitive science questions addressed by the aspect map approach. However, this will not be the subject of the present paper (see the outlook in section 6).

2 Modeling Geographic Knowledge

There are numerous research areas dealing with the representation of geographic knowledge in different disciplines. We briefly present some relevant work related to geographic knowledge representation to motivate our approach.

2.1 Existing Approaches

In computer science and geography, Geographic Information Systems (GIS) are designed to store geographic information, operate on them, and present information for human use. Though cognitive science implications of GIS are of growing interest, GIS in the usual sense are merely technical tools for handling geographic information. The questions involved in GIS development address issues of database management, geometric error handling, interface design, etc. Although a GIS can be regarded as a collection of digital maps (together with the functionality for dealing with them) the term 'map' in the GIS context should be taken as a metaphor for digital representations of geographic entities and / or environments.

Hierarchical models of knowledge processing lead to more cognitively motivated forms of representing geographic knowledge. Obviously, mental representations of relations between objects and the relations of parts within objects can be thought of as organized hierarchically [Kosslyn 1987, Kosslyn et al. 1990]. Accordingly, in their computational imagery paradigm, Glasgow and Papadias [1992] represent spatial relationships between objects by so-called symbolic arrays. Spatial relations can be read out when required using the relational properties of the array cells containing the objects. An important property of a symbolic array is that its cells themselves may contain symbolic arrays, thus leading to a recursive information structure. In this way inter-object and intra-object relationships can be represented at different levels of granularity. Although quite complex knowledge structures can be organized within such a system, the hierarchical structure encoded therein is fixed. A map, however, allows perceptual restructuring depending on the task to be solved.

When we view maps as pictorial media, their benefits are due to the geometric properties of the planar map space. As part of Artificial Intelligence, the representational and operational properties of diagrammatic or pictorial representations are explored in the area of diagrammatic reasoning (for a collection of research papers see [Glasgow et al. 1995]). The advantages of diagrams in numerous problem solving contexts are mainly based on (a) the localization of objects in the diagram [Larkin & Simon 1987], (b) the specific constraints that are inherent in pictorial structures [Haugeland 1985], and (c) psychological and computational emergent properties that support specific processes [Koedinger 1994]. The problems discussed in diagrammatic reasoning refer to pictorial properties in general and not to the specific application in geographic knowledge representation.

The relationships between diagrammatic depictions and the respective geographic entities represented in depictions are investigated in *formal map semantics* [Pratt 1993]. The main goal of this research area lies in determining suitable criteria for deciding on the truth or falsehood of map-like diagrammatic representations. It shows that a map as a whole - due to the restrictions of its representational medium - can only be approximately true; numerous types of distortions occur in the relationship between the depicted entities and the real world objects they stand for. A presupposition for deciding upon the truth of map-like-representation is the identification of objects in the map, i.e. to distinguish which items are object entities and which form the geographic background on which the former are localized. The focus in map semantics, however, is rather on logical questions than on cognitive issues.

The investigations in formal map semantics are motivated by attempts to explain map making and map reading in the context of semiotics. Semiotics is a general theory of representation. It can be defined as a theory of signs and signification that examines human communication by means of organized signification systems [Eco 1976]. Regarding maps, we can identify the two components of a semiotic sign, the expression and the content of the sign, as the map element and the geographic concept the map maker and reader relate to it, respectively [Head 1991]. To use maps as communication media, both components are linked by the semiotic code shared by the persons that generate and use the map. Related to semiotic explanations of maps and mapping are investigations that examine the role of the graphic appearance of the signs within the communication process [Bertin 1981, Kosslyn 1994]. Obviously, the communicative intention of graphic representations can be more or less met by the sign's expression. The gradual geographic characteristic 'elevation', for example, is better depicted by gradually varied colors than by arbitrary textures.

As mentioned earlier, there are attempts to take the results of cognitive science into account in the development of Geographic Information Systems. Such an integrative view of geographical representation is expressed in the theses of Naive Geography [Egenhofer & Mark 1995]. Based on these, formal models of common sense geographic understanding shall be developed to facilitate the use of future Geographic Information Systems. Several facets specific to human-map interaction and geographic understanding are claimed by Naive Geography. Some of the most important in the present context concern the incompleteness of geographic knowledge and the diversity of underlying spatial conceptualizations. Furthermore, different levels of detail and the lack of unique object identifications have to be handled. In particular, the observation that information depicted in maps is considered more truthful than the direct experience of the corresponding geographic environment underlines the cognitive importance of geographic maps for conveying knowledge about large-scale spaces.

2.2 Our Research Questions

In Section 1, we have introduced three defining characteristics of maps: (1) maps represent some aspects pictorially; (2) maps represent some aspects symbolically; and (3) maps represent only selected aspects of the underlying geographic space. We have also pointed out that to solve a given task, a cognitive system uses both internal and external representations.

We are particularly interested in the following questions:

Most external maps are ambiguous with respect to the represented aspects, and they represent many aspects. This ambiguity is due to the purpose the map has to serve in combination with the restrictions of the map's space as representation medium. So, in constructing maps, it must be decided which aspects are to be included and how they are represented in the map. How are aspects selected before they are represented in the map? Which constraints are imposed on aspects competing for the same map space? How are aspects mapped, i.e. how are they depicted in the medium? These questions will be discussed in Section 4, demonstrating how different aspect maps can be constructed from the same geographic information.

When reading a map, a cognitive system must choose and disambiguate aspects. Which aspects are identified? Can criteria be described that guide this identification process? These questions can only be addressed with reference to specific tasks. An example of such a task will be used in Section 5 to illustrate general principles.

2.3 Goals

To address these questions, we aim at describing parts of the processes involved in map reading and drawing inferences in maps. Human abilities to deal with knowledge represented in maps are highly developed. We aim at investigating these abilities to understand their strengths and weaknesses. The resulting observations shall be analyzed with respect to their underlying representational and processing principles thus leading to a description of encoding, representing, and decoding spatial information by means of maps.

This description will be given in a state formal enough to allow the development of an AI architecture. In line with the cognitive science questions posed here, the emphasis is on AI as a tool for cognitive modeling, rather than on AI as engineering. Nevertheless, principles developed in this context can be employed in future applications dealing with geographic knowledge processing tasks. The main advantage of the resulting model architecture will be its flexibility with respect to modeling alternative realizations. Therefore, the framework will allow for critical explorations of the model at hand.

3 The Aspect Map Approach

Our basic assumption is that the processing of geographic maps to solve a given task involves a number of external and/or internal spatial representations which can be described as maps, or, more precisely, *aspect maps*.

An *aspect map* is a spatial organization structure that represents one or more aspects of geographic entities. *Aspects* are properties of and relations between geographic entities (in the following, we shall use the term 'relation' also to denote unary relations/properties). Examples are the extent of a forest area, the name of a city, or the geographic orientation of one place with respect to another place. An aspect can be predominantly geographic (e.g. a geographic category: water area, built-up area) or predominantly spatial (e.g. connectedness of subway stations or distance between locations). In this sense, any cartographic representation can be considered an aspect map. In the present context, we are particularly interested in those aspect maps that represent only few aspects and which restrict interpretations by the specification of processes.

We assume that low-level perception is finished and entities are identified, e.g. we assume that the question of what is an entity is resolved by the context in which the map is examined. (Cf. [Pratt 1993; MacEachren 1995] for a discussion of figure-ground problems concerning regions contained within one another, such as land, lakes, and islands.)

The restrictions on the interpretation are described by meta-level knowledge about the aspect map, which we call the *meta knowledge*. The meta knowledge describes how the information contained in the map is read (a map, like other pictures and unlike symbolic representations, always contains concrete values for positions of all entities and other relations derived from these positions). The *correct* meta knowledge describes which information *may* legitimately be read off the map. However, there is also the possibility that a map is misinterpreted. The meta knowledge applied in this case is not the correct meta knowledge: It does not describe which information *may* be read off, but which information *is* read off. For example, while a public transportation network map contains concrete values for directions and distances between stations, these cannot be legitimately interpreted as directions and distances in the represented domain (see the example in section 5 below).[3]

We distinguish operations on aspect maps according to the number of input aspect maps. *Extraction* evaluates one aspect map with respect to relevant aspects, and *combination* joins information from several aspect maps. In each case, the output is a new aspect map. Extraction, whose output is a new aspect map, needs to be distinguished from *selection*, whose output is relevant entities and/or relations. Usually, a combination also involves a selection.

The current focus of the aspect map approach is on the cognitive processing of geographic maps. Aspect maps may be external representations. The meta knowledge associated with an aspect map is usually the knowledge of the cartographer or of the user, i.e. it may be an internal representation. This would also make the legitimate information content an internal representation. In the cases we are considering at the moment, the aspect map resulting from the operations extraction and combination is also an external map. Insofar as the aspect map approach is a cognitive model, it aims at modeling (or externalizing) both the meta knowledge and the legitimate information content. Insofar as it is an AI model, it carries this basic assumption further and assumes that the operations can be chained in a wide variety of ways and in chains of arbitrary lengths. The AI question then becomes: What are the consequences of specific assumptions about representations, operations, and sequences of operations?

In the aspect map architecture, these concepts are formalized as follows: An aspect map AM is described as a triple $<E,R,M>$, where the entities E are identifiers of cartographic objects, and the relations R are relations on this set E as specified by the

---

[3] This selective evaluation of pictures by meta knowledge combines ideas about explicit 'annotations' associated with images which limit the possibilities for interpreting these images [Habel, in press] with ideas about a 'sketch interpretation' of images [Berendt 1996a, 1996b]. The latter means that graphic properties of the image are interpreted as a communicative intention: "If the map contains only straight lines, this straightness should not be taken literally", i.e. the course of these lines is to be coarsened in interpretation.

map as a picture (e.g. attributes, position, distances and angles). M is the meta knowledge associated with the map. It is a function which assigns to the cartographic relations R the aspect relations R_{asp}, which constitute the 'legitimate information content' of the map (and correspond to the aspects).

To understand this structure, consider a miniature example. On a schematic public transportation network map, the connection between STATION-A and STATION-B is depicted as in Figure 1. The aspect map contains the entities $E = \{A,B\}$. Assume the cartographer has produced the map such that the direction left-right corresponds roughly to the direction west-east. In this case, STATION-B is EAST-OF STATION-A. The meta knowledge reflects this kind of mapping. First, the map is inspected: Scan from A to B and note the change in x and y coordinates. In the example, the process shows that B has a greater x value than A, and the same y value: *exactly-right-of(B,A)* $\in R$. The meta knowledge assigns to the map type 'public transportation network map' and the relation *exactly-right-of* the aspect relation *roughly-east-of*. This knowledge can be combined with the knowledge from other aspect maps in order to make inferences (see below). Finally, an interpretation function assigns to this aspect relation its meaning 'in the world', i.e. the aspect EAST-OF and its geographic meaning.

Fig. 1. A simple aspect map

Operations on aspect maps are extraction and combination. Both involve a sequence composed of the following processes:
- *Evaluation*: The transformation of the picture into its legitimate information content, or rather, those parts of its legitimate information content which are relevant for the task.
- *Inference*: Inference geared towards making those relations explicit that the task demands. These may be of different complexity. In the simplest case, extraction only reproduces a part of the input aspect map (e.g. only one line of a public transportation network map). Inference may also be a generalization of one piece of knowledge (e.g. the information that one landmark lies *northwest of* another landmark in a park is filtered to yield that both are contained in this park). Inference may also be a rather complex combination of several pieces of knowledge. This may happen in extraction (e.g. a transitive inference) or in combination (e.g. joining complementary information).
- *Visualization*: the inferences drawn are in turn transformed into an aspect map. This may involve some choices concerning the depiction of aspect relations which add information (e.g. if both landmarks are *contained in* the park, they must still be assigned concrete positions in order to be visualized, and the newly constructed meta knowledge must reflect that these positions should not be interpreted).

The new aspect map therefore consists of a subset of the entities of the old aspect map(s), relations which are results of inferences based on the given relations, and a combination of the meta knowledge of the input maps which reflects the inferences and the visualization. In particular, when there are ambiguities, the new meta knowledge has to assign an aspect relation to a relation, where this assignment depends on which of the input aspect maps the relation was taken from. Also, consequences of visualization have to be considered in the new aspect map.

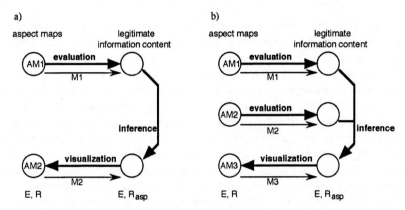

Fig. 2. Operations on aspect maps: (a) extraction, (b) combination

Figure 2 shows schematically how extraction and combination work. Note that all aspect maps need to have a common semantics in order to allow for combination. In a model-theoretic framework, this is linked to the legitimate information content by the process of *interpretation* (see Figure 3), which we will not discuss further here. Examples of inferences with aspect maps are discussed in sections 4 and 5.

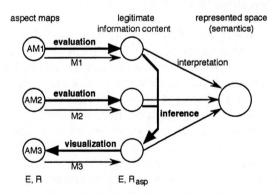

Fig. 3. The aspect map architecture

This kind of depiction also makes it possible to relate the aspect map architecture to other knowledge representation formalisms dealing with the combination of information from more than one representation format (e.g. diagrammatic reasoning formalisms like that of Myers and Konolige [1995]).

The following two sections will discuss examples of extraction and combination of aspect maps, respectively.

4 Extraction and Construction of Aspect Maps

In this section, we show how aspect maps can be constructed from given geographical information as it may be obtained by evaluating a given aspect map. According to our definition in Chapter 3 this involves the questions, (1) which entities shall be depicted, (2) which aspects the aspect map is to convey, i.e. which properties of and relations between the depicted entities shall be represented, and (3) how these aspects are realized by the pictorial environment that is used as representation medium.

We illustrate the involved components and processes using a familiar type of external aspect map: schematic public transportation maps. Figure 4 shows a part of the Hamburg public transportation network map which will serve as a running example in the following sections. This public transportation network map displays underground and city train lines together with their stations and transfer stations in an abstract and schematic way. We can assume that this schematic map has been constructed using a more or less topographically correct city map as a starting point and that some topographic aspects are coarsely kept in this map. Therefore the example we will discuss here is the construction of an aspect map similar to the part of the Hamburg public transportation network map from the corresponding part of the Hamburg city map.

4.1 Overview: Operations Involved

According to our aspect map model the construction of a schematic public transportation map from a topographic city map (both of them regarded as aspect maps) will be described as the process of aspect map extraction. Thus the extraction consists of three steps (cf. Figure 2). First, the entities and aspects contained in the first aspect map (the city map) need to be *evaluated* using the meta knowledge assigned to it, thus gaining the legitimate information content $R_{asp}1$ of the first map with respect to the entities and relations we are interested in according to our task. Second, the process of *inference*, carried out with the information evaluated by the first step, has to construct the legitimate information content $R_{asp}2$ of the new aspect map. In this way the aspects to be contained in the schematic public transportation map are derived from the aspects taken from the city map. Third, based on the newly created set of aspects $R_{asp}2$, the resulting aspect map AM2 can be constructed by *visualizing* the legitimate information content. Thus, the visualization comprises both, the transformation of $R_{asp}2$ to the relations R of the new aspect map AM2 and the construction of the adequate meta knowledge for this aspect map that determines how the legitimate information content $R_{asp}2$ may be reconstructed from the aspect map AM2 when evaluating it.

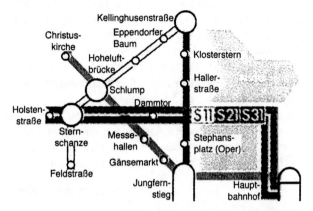

Fig. 4. Part of the Hamburg public transportation network map

In the following we will presuppose that the aspects of the city map are given, and we will concentrate on the processes of inferring and visualizing new aspects. According to the purpose of the type of schematic map under consideration only a few classes of entities will be contained and only a few aspects will be represented in the map.

4.2 Analyzing Entities and Aspects

The schematic public transportation map contains as entities underground and city train stations, transfer stations, and the lines that connect the stations with one another[4]. From the city map we obtain as aspects the exact positions of the stations and the exact course of the lines[5]. When depicting this information we obtain the representation in Figure 5 (due to the visualization, circles are chosen as symbols for stations and the lines are marked by different gray values).

However, the aspect map depicted in Figure 4 above obviously does not contain information that is as detailed as in Figure 5. In the following we will discuss some of the aspects contained in Figure 5, and show whether and how they can be realized in the new aspect map to be constructed.

---

[4] In this context, we will abstract from the labels of the station names also depicted in the map. We will, however, refer to the signs for the stations (black circles) by the names of the stations.

[5] Of course, the city map underlies the same restrictions as every map with regards to the exactness of geographic representation; nevertheless, for reasons of simplicity we do not consider the related problems here.

Station Related Aspects.

- *Exact locations*: The location of stations in the city is of major importance as it facilitates both finding a station near a location that is known and identifying the location of a station with some place in the city. Therefore, positional information should be contained in the schematic public transportation map.
- *Orientation between stations*: The same argument applies to the orientation between stations (in the sense of cardinal directions). The correct depiction of the positions of stations with respect to each other facilitates the imagination of the course of underground and city train lines in the city.
- *Distance between stations (as the crow flies)*: When the exact positions of the stations are given, distances between stations are represented as well. However, both aspects may be separated, as it is possible to modify positions and nevertheless correctly represent the distances in a straight line.

Fig. 5. Aspect map immediately derived from the city map.

Line Related Aspects.

- *Location of start/end point of line segments*: Start and end points should coincide with the stations they are related to. However trivial this statement seems, it will be of importance when generating the visualization (see next section).
- *Ordering information*: The lines connecting the stations impose a segmentation of the geographic plane (as well as of the map) that may be used as an orientation hint, especially when lines are aboveground. Therefore, ordering information with respect to (visible) lines should be contained in the map.
- *Exact courses*: The exact course of a line is regarded as of minor importance. The schematic map does not contain enough information to use the shape of the lines and the course of underground lines cannot be recognized at all. For reasons of map readability this aspect shall be omitted.

- *Length of line segments*: The length of line segments might be represented independently of the exact shape of the lines' courses. This information however, i.e. the distance the train travels on its way from one station to the other, is usually not important for passengers.

So the aspects that shall be contained in the map cover: the *positions of stations* (thus encompassing the *distance* in a straight line), the *orientation* between stations, and *ordering* of stations with respect to the lines. The following considerations show how schematic public transportation maps can realize these aspects in different ways. Existing public transportation maps show that indeed similar principles seem to be applied when designing such maps.

4.3 Visualization

Taking into account the above aspects we obtain a map very similar to the map in Figure 5. However this map shows straight lines connecting the stations with each other instead of the exact courses of the lines, which are omitted (see Figure 6).

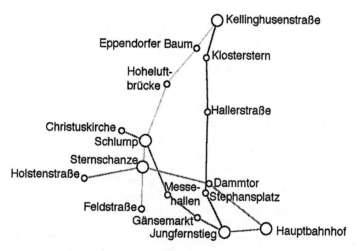

Fig. 6. Aspect map with exact line courses omitted.

Up to now we have selected aspects from the topographic map and we have inferred a new aspect, distance, depicted by straight lines connecting the stations. In schematic maps, however, aspects are often not represented as precisely as they may be read off a topographic map. The Hamburg schematic public transportation map (Figure 4) shows that underground and city train lines are depicted only by horizontal, vertical and diagonal lines. Thus, although orientation information obviously is kept in this map, it is only depicted in a coarse way.

This observation leads us to two claims concerning inference and visualization of aspects. First, aspects may be realized at different levels of precision. These levels may be ordered hierarchically. The aspect of orientation, for example, may be depicted by precise angles or it might be based on a more or less coarse sector model with 8, 4 or even as few as 2 categories to distinguish. Second, and related thereto, the possibilities of depicting several aspects depend on each other. Due to the restrictions imposed by the planar, pictorial representation medium, representing one aspect in the intended way may block the planned depiction of another one. So, in the visualization process one has to decide which of the two aspects shall be depicted correctly and which one only approximately. We therefore propose to order the aspects to be depicted in an aspect map in the form of a depictional precedence hierarchy [Barkowsky & Freksa 1997]. This hierarchy describes in which order the diverse aspects have to be realized in the aspect map, and therefore determines which aspects have to be depicted in a less precise way in case of collision.

When, for example, the orientation between stations is to be represented according to a sector model of 8 cardinal directions (as depicted in the Hamburg public transportation map), the depiction of distance and location, among other things, will be affected. Figure 7 shows an aspect map in which coarse orientations (according to 8 cardinal directions) are visualized by exact horizontal, vertical, and diagonal straight lines. This allows only an approximate depiction of the other aspects discussed.

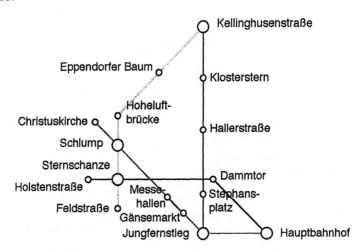

Fig. 7. Aspect map emphasizing coarse classification of orientation between stations

When, in turn, the emphasis lies on the aspect of distance, i.e. if straight-line distances are to be correctly depicted, 8-sector orientation can only be visualized approximately, i.e. horizontal, vertical, and diagonal angles must be relaxed to guarantee that the line segments start and end at the related station symbols (Figure 8).

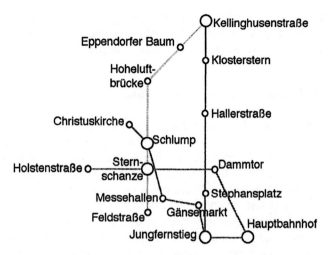

Fig. 8. Aspect map with priority of distance over orientation

5 Combination of Aspect Maps

This section deals with the evaluation and combination of aspect maps for the following kind of problem: What kind of location information can be extracted from a public transportation network map when this map is related to a city map?

When a map is read, the information obtained from the picture must be evaluated to obtain the map's legitimate information content. In some cases, this is un-problematic, because the aspect of interest is known to be represented in the map. For example, one may consult a public transportation network map to determine the name of the next station on a given line: It is clear that the aspect *being neighboring stations on a city train line* is represented by the next dot on the line of the respective color.

Sometimes, however, evaluation is not as straightforward as this. There are cases in which it may not be clear whether the aspect(s) of interest is/are represented in the aspect map. Also, often the visualization that has been used to create this map is an abstraction, i.e. it cannot be inverted (at least not in a strict mathematical sense). For most of the visualization's output values, one can only specify a *range* of input values. As an example, consider the situation that from a public transportation network map, one wants to find out something about the location of a station in the city. This may happen if we have a city map of part of the city that does not show the station we want to walk to. The city map does show directions and distances along with street names, all of which we can use to orient ourselves. Inferring direction and distance to our goal station, if this is possible, allows us to locate that station with respect to the city map, so we can plan a walking route in the right direction (and maybe decide whether walking is worth the effort, depending on the distance). A question is whether direction and distance information is represented in the transportation map and if so, how.

This section can be regarded as complementing section 4, in that in section 4, locations in a city map were transformed into locations on a public transportation network map. Here, the reverse transformation is necessary.

5.1 An Example: Locating a City Train Station

This will be demonstrated using an example which involves two real maps. The first is a city map which shows the Hamburg university campus and some adjoining streets, as well as one city train station (*Hallerstraße*), another city train station (*Dammtor*), and the city train line going through the latter (see Figure 9 (a)).[6] In the following, this map will be called "campus map". An important city train station within the Hamburg public transportation network, *Schlump*, is close to the depicted area, but outside it. The second map is a part of the Hamburg public transportation network map, which shows all three stations (*Hallerstraße*, *Dammtor*, and *Schlump*), the city train line going through *Dammtor*, as well as all other city train lines and stations in that region (Figure 9 (b)). The task is to locate *Schlump* with respect to the campus map, i.e. the task is to describe the output of the combination of the two maps in Figure 9. This output is an aspect map that looks like the city map, but in addition has a location for *Schlump* in it.

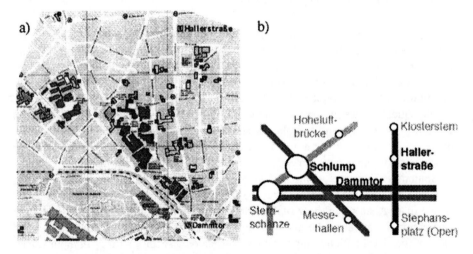

Fig. 9. Example: (a) Campus map, (b) public transportation network map

---

[6] Only this city train line in the area is shown on the campus map. This is because it is overground, while the other lines are underground.

5.2 A Formal Approach

On the basis of the hierarchy of aspects mentioned in section 4 and a general classification of spatial knowledge used in Spatial Reasoning (e.g. [Schlieder 1996]), we distinguish the following five possibilities of evaluation:[7]

(1) *topological*: The public transportation network map only contains information about the connections between stations. *Schlump* could be anywhere with respect to the campus map (hatched region in Figure 10 (a)).

(2) *orientation with respect to axes specified within the map:* The public transportation network map only contains orientation information which locates points (stations) with respect to axes contained in the public transportation network map (lines). Therefore, all that can be inferred from it is that *Schlump* lies on the same side of the city train line as *Hallerstraße*. Assuming a straight continuation of the city train line outside of the campus map, *Schlump* could be anywhere in the hatched region of Figure 10 (b).

(3) *orientation with respect to sectors or axes specified outside the map*: This assumes that the public transportation network map gives approximate information about cardinal directions. For example, what is *left of* and *above* a reference object in the public transportation network map is *northwest* of it on the campus map. Using a sector model for cardinal directions like that proposed by Frank [1992], we see the following: With respect to *Dammtor*, *Schlump* is in the *northwest* or *west* sector (Figure 10 (c)). This means that *Schlump* could be anywhere in the hatched region of Figure 10 (d).

(4) *orientation with respect to axes specified within and outside the map:* This is a combination of (2) and (3) and yields the hatched region in Figure 10 (e).

(5) *metrical*: This assumes that the public transportation network map gives exact information about directions and distances. Therefore, the triangle formed by *Hallerstraße*, *Dammtor*, and *Schlump* on the public transportation network map can be scaled and superimposed on the campus map. This means that *Schlump* must be at the hatched location shown in Figure 10 (f).

5.3 An Empirical Study

The preceding discussion has outlined the logically possible ways of specifying a location for *Schlump*. But how do people tackle this question? In an empirical study [Berendt, Rauh & Barkowsky in press], 26 subjects were given the two maps and the information about them (given above in the description of Figure 9). They were asked to indicate their estimates of the location of the train station *Schlump* on the partial city map. After that, subjects had to give a verbal report on how they arrived at their estimate. The verbal data were classified into the five theoretical categories of evaluation possibilities, according to which objects were explicitly mentioned as reference

[7] From a formal point of view, this list is not exhaustive, but it describes a plausible selection of possibilities based on everyday notions of what a simple schematic map could inform us about: connection, rough cardinal direction, or distances and directions.

objects, and whether angles or distances or the process "projection" were explicitly mentioned. Three subjects were excluded because they knew Hamburg or because they judged themselves as unable to read maps and said they had guessed the position.

According to the analysis in section 5.2, evaluations (2), (3) and (4) are correct, with (4) specifying the smallest region. Evaluation (5) is not correct; it is an over-interpretation of the public transportation network map.

Results were as follows: Four subgroups of subjects could be identified. Each described an evaluation possibility and located *Schlump* in the corresponding region. The two largest subgroups were subgroups corresponding to evaluation possibilities (4) and (5) (34.8% and 39.1%, respectively). The mean locations given by these two subgroups were significantly different.

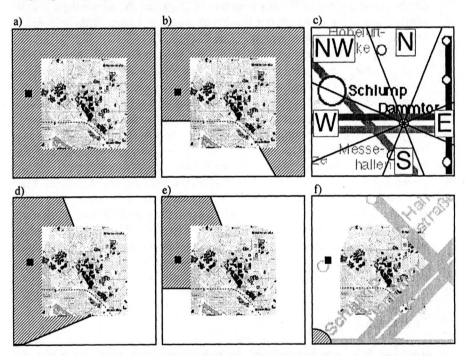

Fig. 10. Positions of *Schlump* for different evaluation possibilities: (a) topological, (b) orientation I, (c) 8-sector interpretation of the public transportation network map, (d) orientation II, (e) orientation I+II, (f) angles + distances (public transportation network map superimposed on campus map). All maps contain a square at the correct position of *Schlump*.

Most subjects read the public transportation network map as containing some rough information about cardinal directions. This corresponds to the evaluation possi-bilities presented in the theoretical analysis above and also to the correct evaluation of the current public transportation network map, as the correct position of *Schlump* shows. Interestingly, no subject described evaluation possibility (2). In other words, if the public transportation network map was taken to contain the aspect 'orientation', this was taken to be 'orientation with respect to axes within and outside the map'.

6 Conclusions and Outlook

We have described the aspect map approach to model the processing of geographic maps. An aspect map is a spatial organization structure in which one or more aspects of geographic entities are represented. Aspects are properties of and relations between geographic entities.

The centerpiece of the aspect map approach is the aspect map architecture, an AI model of the processing of geographic maps. This is a general framework that describes the processing of aspect maps in terms of two fundamental operations: extraction and combination. Both operations transform one (or several) aspect map(s) into a new one. Meta knowledge, which specifies which information may legitimately be read off the map, has been discussed as a central component in this architecture. The description and management of meta knowledge, its correct application, and the derivation of the meta knowledge assigned to a newly produced aspect map are therefore essential components of the aspect map architecture. Further work will further investigate and formalize the management of meta knowledge.

An implementation of the aspect map architecture is being developed. Currently, it models and visualizes operations like those described in section 5. The implementation will be developed further to accommodate different types of maps and more complex sequences of operations.

Two processes contained in the aspect map architecture were investigated in more detail.

In section 4 we have shown how different aspects can be inferred from given legitimate geographic information, and how they can be visualized in different ways. In particular how visualizations of different aspects interact with each other is of major interest for the construction of aspect maps. Further investigations shall deal with the automatization of visualization processes according to predefined aspect hierarchies. According to our realization of aspect maps as related to meta knowledge, which determines the way maps are to be read, the construction of adequate meta knowledge is of central concern. The related meta knowledge must be constructed together with the visualization of newly inferred aspects. Thus, future work will investigate the interrelationships between visualization facilities and the description of related meta knowledge.

In section 5, an example of the combination of two aspect maps to obtain a third aspect map was described. We have shown that this involves, as an essential step, the evaluation of the input map whose depicted metrical relations may not be evaluated. A formal approach to the problem has distinguished five evaluation possibilities (applications of meta knowledge) depending on which aspects are taken to be represented in the input map. These were based on a general classification of spatial knowledge used in Qualitative Spatial Reasoning: topological, orientation, and metrical knowledge. It was shown that some evaluation possibilities are correct, while others are incorrect. The description of evaluation possibilities will be further refined in the framework of a system of aspects. We have also described the results of an empirical study, which shows that the formal analysis can describe and distinguish the ways in which people deal with this task. Further studies involving different types of maps and different types of reasoning problems will allow the investigation of the correspondence between the formally identified possibilities of map processing and the way people deal with geographic maps. Do people use maps correctly? Do they obtain too little information? Do they obtain too much or even wrong information? Further investigations are also planned for determining relevant processes of extraction of regions, of reasoning from the relations between them, and of estimation of points within selected regions that necessarily entail the point that is asked for.

In section 3, we have outlined some possible relationships between external and internal representations within the aspect map approach. The assumption so far has been that aspect maps may be external representations, whereas meta knowledge and legitimate information content may be internal representations. In addition, there may also be internal aspect maps. Since the knowledge that becomes effective in operations (and therefore in thinking) is the legitimate information content or the product of an aspect map and the associated meta knowledge, we cannot make statements about these in isolation; all we can observe from behavior are the aspect map together with its meta knowledge. About this structure, we can ask empirical questions like: Are several aspects stored separately or in one internal aspect map?, Are certain combinations of aspects possible or not?, Do certain extractions and combinations lead to specific distortions?

Acknowledgments

We thank Bernhard Hommel, Markus Klann, Alexander Klippel, Bernd Krieg-Brückner, Christie Manning, Mark Siebel, and an anonymous reviewer for fruitful comments to previous drafts of this paper. We also acknowledge the cooperation of our partner projects in the spatial cognition priority program.

References

Barkowsky, T., & Freksa, C. (1997). Cognitive requirements on making and interpreting maps. In S. Hirtle & A. Frank (Eds.), *Spatial information theory: A theoretical basis for GIS* (pp. 347-361). Berlin: Springer.

Berendt, B. (1996a). Explaining preferred mental models in Allen inferences with a metrical model of imagery. In Cognitive Science Society (Ed.), *Proceedings of the 18th annual conference of the Cognitive Science Society* (pp. 489-494). Mahwah, NJ: Lawrence Erlbaum.

Berendt, B. (1996b). The utility of mental images: How to create stable mental models in an unstable image medium. In *Proceedings of the First European Workshop on Cognitive*

Modeling (pp. 97-103). Berlin, Technische Universität Berlin, Fachbereich Informatik, Report No. 96-39.

Berendt, B., Rauh, R., & Barkowsky, T. (in press). Spatial thinking with geographic maps: an empirical study. In H. Czap, P. Ohly, & S. Pribbenow (Eds.), *Wissensorganisation mit multimedialen Techniken. Fortschritte in der Wissensorganisation, Band 5*. Würzburg: ERGON-Verlag.

Bertin, J. (1981). *Graphics and graphic information-processing*. Berlin: de Gruyter.

Eco, U. (1976). *A theory of semiotics*. Bloomington, Ind.: Indiana University Press.

Egenhofer, M. J., & Mark, D. M. (1995). Naive geography. In A. U. Frank & W. Kuhn (Eds.), *Spatial information theory. A theoretical basis for GIS. LNCS 988* (pp. 1-15). Berlin: Springer.

Frank, A. U. (1992). Qualitative spatial reasoning about distances and directions in geographic space. *Journal of Visual Languages and Computing, 3*, 343-371.

Gehrke, J., & Hommel, B. (this volume). The impact of exogenous factors on spatial coding in perception and memory.

Glasgow, J., Narayanan, H., & Chandrasekaran, B. (Eds.) (1995). *Diagrammatic reasoning: Computational and cognitive perspectives*. Cambridge, MA: MIT-Press.

Glasgow, J., & Papadias, D. (1992). Computational imagery. *Cognitive Science, 16*, 355-394.

Habel, C. (in press). Piktorielle Repräsentationen als unterbestimmte räumliche Modelle. To appear in *Kognitionswissenschaft, 7*.

Haugeland, J. (1985). *Artificial Intelligence - The very idea*. Cambridge, MA: MIT-Press.

Head, C. G. (1991). Mapping as language or semiotic system: Review and comment. In D. M. Mark & A. U. Frank (Eds.), *Cognitive and linguistic aspects of geographic space* (pp. 237-262). Dordrecht, Boston, London: Kluwer Academic Publishers.

Hirtle, S. C., & Heidorn, P. B. (1993). The structure of cognitive maps: Representations and processes. In T. Gärling & R. G. Golledge (Eds.), *Behavior and environment: Psychological and geographical approaches* (pp. 170-192). Amsterdam: North-Holland.

Hirtle, S. C., & Jonides J. (1985). Evidence of hierarchies in cognitive maps. *Memory & Cognition, 13*(3), 208-217.

Koedinger, K. R. (1994). Emergent properties and structural constraints: advantages of diagrammatic representations for reasoning and learning. In B. Chandrasekaran & H. Simon (Eds.), *Reasoning with diagrammatic representations* (pp. 151-156). Menlo Park, CA: AAAI Press.

Kosslyn, S. M. (1987). Seeing and imagining in the cerebral hemispheres: a computational approach. *Psychological Review, 94*, 148-175.

Kosslyn, S. M. (1994). *Elements of graph design*. New York: Freeman.

Kosslyn, S. M., Flynn, R. A., Amsterdam, J. B., & Wang, G. (1990). Components of high-level vision: A cognitive neuroscience analysis and accounts of neurological syndromes. *Cognition, 34*, 203-277.

Krieg-Brückner, B., Röfer, T., Carmesin, H.-O., & Müller, R. (this volume). A taxonomy of spatial knowledge for navigation and its application to the Bremen autonomous wheelchair.

Larkin, J. H., & Simon, H. A. (1987). Why a diagram is (sometimes) worth ten thousand words. *Cognitive Science, 11*, 65-99.

MacEachren, A. M. (1995). *How maps work: representation, visualization, and design*. New York, London: The Guilford Press.

Montello, D. R. (1993). Scale and multiple psychologies of space. In A. U. Frank & I. Campari (Eds.), *Spatial Information Theory: A theoretical basis for GIS (Proc. COSIT'93)* (pp. 312-321). Berlin etc.: Springer.

Myers, K., & Konolige, K. (1995). Reasoning with analogical representations. In J. Glasgow, N. H. Narayanan, & B. Chandrasekaran (Eds.), *Diagrammatic reasoning* (pp. 273-301). Menlo Park, CA: AAAI Press.

Palmer, S. E. (1978). Fundamental aspects of cognitive representation. In E. Rosch & B. B. Lloyd (Eds.), *Cognition and categorization* (pp. 259-303). Hillsdale, NJ: Lawrence Erlbaum.

Peterson, D. (1996). Introduction. In D. Peterson (Ed.), *Forms of representation* (pp. 1-27). Wiltshire, GB: Cromwell Press.

Pratt, I. (1993). Map semantics. In A. U. Frank & I. Campari (Eds.), *Spatial Information Theory: A theoretical basis for GIS (Proc. COSIT'93)* (pp. 77-91). Berlin etc.: Springer.

Rothkegel, R., Wender, K. F., & Schumacher, S. (this volume). Judging spatial relations from memory.

Schlieder, C. (1996). Räumliches Schließen. In G. Strube, B. Becker, C. Freksa, U. Hahn, K. Opwis, & G. Palm (Eds.) *Wörterbuch der Kognitionswissenschaft.* (pp. 608-609). Stuttgart: Klett-Cotta.

Schweizer, K., Herrmann, T., Janzen, G., & Katz, S. (this volume). The route direction effect and its constraints.

Sloman, A. (1971). Interactions between philosophy and artificial intelligence: The role of intuition and non-logical reasoning in intelligence. *Artificial Intelligence, 2,* 209-225.

Sloman, A. (1975). Afterthoughts on analogical representations. *1st Workshop on Theoretical Issues in Natural Language Processing (TINLAP-1)* (pp. 164-168). Cambridge, MA.

Stevens, A., & Coupe, P. (1978). Distortions in judged spatial relations. *Cognitive Psychology, 10,* 422-437.

Tversky, B. (1993). Cognitive maps, cognitive collages, and spatial mental models. In A. Frank & I. Campari (Eds.), *Spatial information theory* (pp. 14-24). Berlin: Springer.

A Hierarchy of Qualitative Representations for Space *,**

Benjamin Kuipers

Computer Science Department,
University of Texas at Austin,
Austin, Texas 78712 USA

Abstract. Research in Qualitative Reasoning builds and uses discrete symbolic models of the continuous world. Inference methods such as qualitative simulation are grounded in the theory of ordinary differential equations. We argue here that cognitive mapping — building and using symbolic models of the large-scale spatial environment — is a highly appropriate domain for qualitative reasoning research.

We describe the *Spatial Semantic Hierarchy* (SSH), a set of distinct representations for space, each with its own ontology, each with its own mathematical foundation, and each abstracted from the levels below it. At the control level, the robot and its environment are modeled as a continuous dynamical system, whose stable equilibrium points are abstracted to a discrete set of "distinctive states." Trajectories linking these states can be abstracted to actions, giving a discrete causal graph level of representation for the state space. Depending on the properties of the actions, the causal graph can be deterministic or stochastic. The causal graph of states and actions can in turn be abstracted to a topological network of places and paths. Local metrical models, such as occupancy grids, of neighborhoods of places and paths can then be built on the framework of the topological network while avoiding their usual problems of global consistency.

This paper gives an overview of the SSH, describes the kinds of guarantees that the representation can support, and gives examples from two different robot implementations. We conclude with a brief discussion of the relation between the concepts of "distinctive state" and "landmark value."

---

* This reprints an article that first appeared in *Working Papers of the Tenth International Workshop on Qualitative Reasoning about Physical Systems (QR-96)*, Fallen Leaf Lake, California. AAAI Technical Report WS-96-01, AAAI Press, May 1996.

** This work has taken place in the Qualitative Reasoning Group at the Artificial Intelligence Laboratory, The University of Texas at Austin. Research of the Qualitative Reasoning Group is supported in part by NSF grants IRI-9216584 and IRI-9504138, by NASA contract NCC 2-760, and by the Texas Advanced Research Program under grant no. 003658-242.

1 The Spatial Semantic Hierarchy

Building on recent progress in robot exploration and map-building, we propose an *ontological hierarchy* of representations for knowledge of large-scale space.

An ontological hierarchy shows how multiple representations for the same kind of knowledge can coexists. Each level of the hierarchy has its own *ontology* (the set of objects and relations it uses for describing the world) and its own set of inference and problem-solving methods. The objects, relations, and assumptions required by each level are provided by those below it.

The dependencies among levels in the hierarchy help clarify which combinations of representations are coherent, and which states of incomplete knowledge are meaningful.

In this paper, we formalize the computational model of the cognitive map as developed by Kuipers and his students [13, 11, 12]. That theory was motivated by two insights from observations of human spatial reasoning skills and the characteristic stages of child development [18, 27, 8]. First, a *topological* description of the environment is central to the cognitive map, and is logically prior to the metrical description. Second, the spatial representation is grounded in the sensorimotor interaction between the agent and the environment.

The *Spatial Semantic Hierarchy* (SSH) [15] abstracts the structure of an agent's spatial knowledge in a way that is relatively independent of its sensorimotor apparatus and the environment within which it moves. The following informally describes the knowledge at the different SSH levels, which will be described more formally below.

- The *sensorimotor system* of the robot provides continuous sensors and effectors, but no direct access to the global structure of the environment, or the robot's position or orientation within it.
- At the *control level* of the hierarchy, the ontology is an egocentric sensorimotor one, without knowledge of fixed objects or places in an external environment. A *distinctive state* is defined as the local maximum found by a hill-climbing control strategy, climbing the gradient of a selected sensory feature, or *distinctiveness measure*. Trajectory-following control laws take the robot from one distinctive state to the neighborhood of the next, where hill-climbing can find a local maximum, reducing position error and preventing its accumulation.
- At the *causal level* of the hierarchy, the ontology consists of *views*, which describe the sensory images at distinctive states, and *actions*, which represent trajectories of control laws by which the robot moves from one view to another. A causal graph of associations $\langle V, A, V' \rangle$ among views, actions, and resulting views represents both declarative and imperative knowledge of routes or action procedures.
- At the *topological level* of the hierarchy, the ontology consists of *places*, *paths*, and *regions*, with connectivity and containment relations. Relations among the distinctive states and trajectories defined by the control level, and among

their summaries as views and actions at the causal level, are effectively described by the topological network. This network can be used to guide exploration of new environments and to solve new route-finding problems. Using the network representation, navigation among distinctive states is not dependent on the accuracy, or even the existence, of metrical knowledge of the environment.

- At the *metrical level* of the hierarchy, the ontology for places, paths, and sensory features is extended to include metrical properties such as distance, direction, shape, etc. Geometrical features are extracted from sensory input, and represented as annotations on the places and paths of the topological network.

Two fundamental ontological distinctions are embedded in the SSH. First, the continuous world of the sensorimotor and control levels is abstracted to the discrete symbolic representation at the causal and topological levels, to which the metrical level adds continuous properties. Second, the egocentric world of the sensorimotor, control, and causal levels is abstracted to the world-centered ontologies of the topological and metrical levels.

Formalizing the levels of the hierarchy draws on different bodies of relevant theory: the sensorimotor and control levels on control theory and dynamical systems; the causal level on logic and stochastic transition models; the topological level on logic and simple topology; the geometrical level on estimation theory and differential geometry.

The Spatial Semantic Hierarchy approach contrasts with more traditional methods, which place geometrical sensor intepretation (the most expensive and error-prone step) on the critical path prior to creation of the topological map [6, 23]. The SSH is consistent with, but more specific than, Brooks' [5] subsumption architecture, particularly levels 2 and 3.

The SSH representational framework has been implemented on several different simulated and physical robots. Figure 1 (modified from [12]) shows how the control level definition of states and trajectories grounds the topological description of places and paths, which in turn supports exploration and planning while more expensive sensor fusion methods accumulate metrical information. When metrical information is available, it can be used to optimize travel plans or to disambiguate apparently identical places, but when it is absent navigation and exploration remain possible. Figure 2 demonstrates a fragment of behavior of an RWI B12 robot, using a ring of 12 sonar sensors, as it follows control laws and identifies a distinctive place in the indoor office environment.

2 The Sensorimotor Level

The robot has an objective location in the environment, but it does not have direct access to a representation of that location in an absolute frame of reference. Assume that the environment is two-dimensional, so that the *state* of the robot has three dimensions: position (x, y) and orientation θ. The vector of state

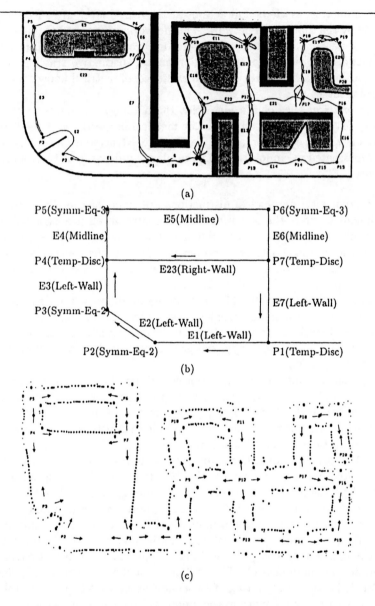

Fig. 1. Simulated NX robot applies SSH exploration and mapping strategy.

(a) The simulated NX robot uses range-sensors to explore and map an environment [12]. The exploration and control strategies identify random and systematic sensor errors, and thus provide robustness. (b) The topological map (fragment) identifies places and paths with the distinctiveness measures that define them (e.g., equidistance from nearby obstacles, discontinuous sensor changes), and represents their connectivity relations. (c) The metrical map consists of annotations on each place and path, and can be relaxed into a global 2D frame of reference.

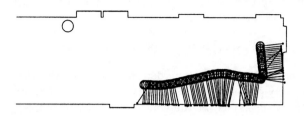

Fig. 2. Spot, a physical robot, applying SSH control strategies.

Spot moves along a right wall, identifies a distinctive place, turns left, and begins to follow the next wall. This figure plots the position of the robot and the single most relevant sonar reading, on a map of the corridor it is exploring.

variables is $\mathbf{x} = [x, y, \theta]^T$. The robot also has a memory M including symbolic descriptions of goals, beliefs, etc., which can influence the choice of control law, hence behavior.

The robot has a vector of sensors providing input $\mathbf{s} = [s_0, \ldots s_{n-1}]^T$ and a vector of motor outputs $\mathbf{u} = [u_0, \ldots u_{k-1}]^T$ by which it can change its position in the environment.

The sensor values are a function of the robot's state,

$$[s_0, \ldots, s_{n-1}]^T = \mathbf{s} = \Psi(\mathbf{x}) = \Psi(x, y, \theta). \tag{1}$$

All variables are piece-wise continuous functions of time. This model treats the environment as static, with the only changes being to the robot's position and orientation.

The "physics of the environment" (or dynamics of the robot),

$$[\dot{x}, \dot{y}, \dot{\theta}]^T = \dot{\mathbf{x}} = \Phi(\mathbf{x}, \mathbf{u}) = \Phi(x, y, \theta, u_0, \ldots u_{k-1}) \tag{2}$$

specifies how the state, and hence the sensory values, change with time as a function of the current state and the motor outputs. The robot does not have direct access to its state variables, but only to the sensory information $\mathbf{s}(t)$ provided to it as it moves through the environment.

3 The Control Level

The purpose of the SSH control level is to select and execute control laws for travel through the environment. During exploration, locally well-behaved features of the sensory input are identified and used to construct suitable control laws. During travel through a known environment, control laws are retrieved from the causal level of the cognitive map.

During a particular segment i of reactive behavior, the robot moves through the environment by setting its motor vector in response to its sensory inputs, according to a control law χ_i.

$$[u_0, \ldots, u_{k-1}] = \mathbf{u} = \chi_i(\mathbf{s}) = \chi_i(s_0, \ldots s_{n-1}) \tag{3}$$

Although χ_i is purely reactive (i.e., determined by $\mathbf{s}$), its selection depends both on the currently perceived environment, and on goals and other aspects of the robot's state not described at the control level. For example, sitting at an intersection, the robot's goals determine whether to invoke a control law for a left or right turn, or to continue straight. For a given choice of control law χ_i, equations (1), (2), and (3) define a dynamical system that describes the behavior of the robot interacting with its environment.

3.1 Distinctiveness Measures

A critical step in our approach is the identification of a discrete set of *locally distinctive states* within a continuous state-space. A locally distinctive state can be defined in terms of the behavior of a control law if we can identify a continuous *distinctiveness measure* with an isolated local maximum in the current neighborhood.

A *distinctiveness measure* (or "d-measure") d is a continuous function $d(\mathbf{s}) \to \Re$. The set $D = \{d_0, \ldots d_{m-1}\}$ of distinctiveness measures depends on the environment and sensorimotor system of the particular robot. A d-measure can be used to define a point-like distinctive state, such as the state equidistant from three obstacles and oriented midway between two of them; or a path-like trajectory, such as the midline of a corridor. Pierce and Kuipers [28] show how d-measures and control laws can be learned from unguided experience.

Each d-measure d has an *appropriateness measure* $a_d(\mathbf{s}) \to [0, 1]$ that specifies the degree to which d is useful for control. a_d need not be continuous, and it may depend on goals or other aspects of the robot's state, as well as the sensory input stream $\mathbf{s}(t)$ to the robot. It is sometimes useful to think of a d-measure d as having a prerequisite $\pi_d(\mathbf{s})$ which must be true for d to be defined. This can be subsumed by the appropriateness measure: $\pi_d(\mathbf{s}) \equiv a_d(\mathbf{s}) > \epsilon$, for some user-specified $\epsilon \geq 0$.

A *neighborhood nbd(d)* of the distinctiveness measure d is a connected subset of the set of states where $\pi_d(\mathbf{s})$ is true. That is, a given distinctiveness measure may have several disconnected neighborhoods in different parts of the environment.

3.2 Local Control Laws

Navigation at the control level is an alternation between two different types of control laws: *hill-climbing* control laws to reach a nearby local maximum of a d-measure, and *trajectory-following* control laws to move from one part of the state space to another. One way to express these is through a simple but general local control law associated with a given d-measure d, specifying a direction of change in the state space of the robot:

$$\dot{\mathbf{x}} = [\dot{x}, \dot{y}, \dot{\theta}]^T = k_1 \nabla d + k_2 N_d \tag{4}$$

where ∇d is the gradient of d in the state space, and N_d is a unit vector orthogonal to ∇d.

Since more than one d-measure $d \in D$ may be appropriate during a given trajectory, we take the weighted average in the spirit of heterogeneous control [14]:

$$\dot{\mathbf{x}} = \frac{\sum_{d \in D} a_d(t) [k_1 \nabla d + k_2 N_d]}{\sum_{d \in D} a_d(t)} \tag{5}$$

where $a_d(t)$ is an *appropriateness measure* for d. When d is not meaningful, $a_d(t) = 0$. Note that as the robot moves, the effective number of participating local control laws may change.

Other compositional approaches to control include potential field methods [2, 31] and fuzzy control [20, 10]. Appropriateness measures and other parameters of the control laws χ_i may be acquired and optimized by function-learning methods including neural nets (e.g. [29]) and memory-based learning [3, 22].

Hill-Climbing. Starting in the state where a trajectory-following control law terminates, identify the applicable hill-climbing d-measure(s). For a hill-climbing control law of the form (4), the $k_1 \nabla d$ term points the robot toward the local maximum, and $k_2 = 0$. A *distinctive state* $\langle x, y, \theta \rangle$ is the state of the robot when a hill-climbing control law terminates; i.e., when $\dot{\mathbf{x}} = \nabla d = 0$.

Trajectory-Following. Starting at a locally distinctive state, select and obey a trajectory-following control law (or a sequence of local control laws) until it terminates. For a trajectory-following control law of the form (4), the term $k_1 \nabla d$ keeps the robot on the desired trajectory, and $k_2 N_d$ moves it along the trajectory in one direction or the other depending on the sign of k_2. A trajectory-following control law terminates when $\dot{\mathbf{x}}$ changes discontinuously (or very quickly). In equation (5) this would happen when some $a_d(t)$ suddenly becomes zero while the corresponding control action $[k_1 \nabla d + k_2 N_d]$ is non-zero.

For example, a robot starts at one end of a corridor, facing "open space." It takes a trajectory consisting of open-loop motion into the corridor it faces, then following the midline to the end of the corridor. Upon reaching the end, the robot does hill-climbing to position itself equidistant from nearby obstacles.

3.3 Putting Control into Action

The local control law (5) provides a desired direction of motion $\dot{\mathbf{x}}$ in state space, which must be translated into values for the robot's motor output variables $\mathbf{u}$.

In simple cases, the dynamics of the robot (equation (2)) will have a pseudo-inverse Φ^{-1} so that, given $\mathbf{x}$ and a desired $\dot{\mathbf{x}}$, we can directly compute

$$\mathbf{u} = \Phi^{-1}(\mathbf{x}, \dot{\mathbf{x}}) \text{ such that } \dot{\mathbf{x}} = \Phi(\mathbf{x}, \mathbf{u}). \tag{6}$$

In general (i.e., for a robot with non-holonomic motion constraints), there may be no way to achieve a desired $\dot{\mathbf{x}}$ for a given state $\mathbf{x}$ (cf. [16]). In such a case, we specify the control goal as a net change $\Delta\mathbf{x}$ to be obtained over some period of time. Then we assume the ability to plan a sequence of continuous actions (e.g., [26]), or to retrieve a previously developed control plan:

$$p = plan(\mathbf{x}, \Delta\mathbf{x}), \text{ such that } \mathbf{u} = p(\mathbf{x}, t) \tag{7}$$

has the desired effect of reaching the state $\mathbf{x} + \Delta\mathbf{x}$. Note that, as with parallel parking, the intermediate states of the plan p may be farther from the goal than the initial or final states. Further extensions will be required to cope with pedestrians and other unexpected obstacles.

Equation (5), along with either (6) or (7), provides an instance of the control law χ_i required by equation (3). Thus, the robot's behavior *during a single hill-climbing or trajectory-following segment* consists of the state-evolution of a particular dynamical system. Higher-level symbolic reasoning intervenes at the joints between these segments to determine which dynamical system controls the behavior.

4 The Causal Level

When a sequence of control laws — trajectory-following then hill-climbing — reliably takes the robot from one distinctive state to another, we abstract the sequence of control laws to an action A, and the two distinctive states to the sensory images, or views, V and V', obtained there. Their association is represented by the schema $\langle V, A, V' \rangle$.

When this abstraction can be applied across the environment, the continuous state space in which the robot is described as following the trajectories of a dynamical system is abstracted to a discrete state space in which the robot is described as performing a sequence of discrete actions.

4.1 Views, Actions, and Schemas

A *view* is a description of the sensory input vector $\mathbf{s}(t) = [s_1(t), \ldots s_n(t)]$ obtained at a locally distinctive state, $\langle x, y, \theta \rangle$. A view could be a complete snapshot of $\mathbf{s}(t)$, or it could be a partial description, consistent with more than one value of $\mathbf{s}$.

An *action* denotes a sequence of one or more control laws which can be initiated at a locally distinctive state, and terminates after a hill-climbing control law with the robot at another distinctive state. A typical action might consist of an open-loop trajectory-following control law to escape from the current neighborhood, then a closed-loop trajectory-following control law to reach a new neighborhood, and finally a hill-climbing control law to reach a new distinctive state.

A *schema* is a tuple $\langle V, A, V' \rangle$, representing the (temporally extended) event in which the robot takes a particular action A, starting with view V and terminating with view V'.

In the following, $holds(V, s_0)$ means that the view V is observed in situation s_0; $do(A, s_0)$ means that action A is initiated in situation s_0; and $result(A, s_0)$ denotes the situation resulting after action A is initiated in situation s_0 and terminates in a new distinctive state. The schema $\langle V, A, V' \rangle$ has two meanings:

$$\text{declarative: } holds(V, s_0) \rightarrow holds(V', result(A, s_0))$$
$$\text{imperative: } holds(V, now) \Rightarrow do(A, now).$$

The declarative meaning is standard situation calculus [21]. The imperative meaning is intuitively clear, but not formalized.

Procedurally, in order for a complete schema $\langle V, A, V' \rangle$ to be created from observations during behavior, the partially filled schema $\langle V, A, nil \rangle$ must be preserved in working memory during the time required to carry out the action A to termination. In case of interruption, it may be that only the partial schema is stored in long-term memory. The partially filled schema $\langle V, A, nil \rangle$ lacks the declarative meaning of the complete schema, but has a restricted version of the imperative meaning:

$$\text{imperative: } holds(V, now) \Rightarrow do(A, now).$$

4.2 Routines

A *routine* is a set of schemas, indexed by initial view. It represents the sequence of actions and intermediate views in a behavior that moves the robot from an initial to a final distinctive state. A routine can be used either as a description of the behavior, or as a procedure for reproducing it.

Consider the alternating sequence of views and actions $V_0, A_0, V_1, A_1, V_2, \ldots, V_{n-1}, A_{n-1}, V_n$ leading from V_0 to V_n.

- A routine R is *complete* from view V_0 to V_n if R contains the schema $\langle V_i, A_i, V_{i+1} \rangle$ for each i from 0 to $n - 1$.
- A routine R is *adequate* from V_0 to V_n if R contains either $\langle V_i, A_i, V_{i+1} \rangle$ or $\langle V_i, A_i, nil \rangle$ for each i from 0 to $n - 1$.

An adequate routine supports "situated action": physical travel from state V_0 to V_n within the environment [1]. It also generalizes naturally to causal graphs such as *universal plans* [30], which are sets of rules specifying the actions to take at *each* state in a state-space to move toward a given goal. In addition to situated action, a complete routine supports cognitive operations such as mental review or verbal description of the route in the absence of the environment.

5 The Topological Level

The topological map describes the environment as a collection of places, paths, and regions, linked by topological relations such as connectivity, order, containment, boundary, and abstraction. Places, paths, and boundary regions are created from experience represented as a sequence of views and actions. They are created by *abduction*, positing the minimal additional set of places, paths, and regions required to explain the sequence of observed views and actions.

- A *place* describes part of the robot's environment as a zero-dimensional point. A place may lie on zero or more paths. A place may also be defined as the abstraction of a region.
- A *path* describes part of the robot's environment, for example a street in a city, as a one-dimensional subspace. It may describe an order relation on the places it contains, and it may serve as a boundary for one or more regions. The two directions along a path are $dir = +1$ and $dir = -1$.
- A *region* represents a two-dimensional subset of the robot's environment. The set of places in a region share a common property. A region may be defined by one or more boundaries, by a common frame of reference, or by its use in an abstraction relation.

5.1 Co-occurrence Implies Topological Connections

The "current context" or "You-Are-Here pointer" describes the current state of the explorer. The topological level adds the current place, path, and 1-D direction to the current context. Simultaneous presence of several descriptions in the current context implies a topological connection.

$current\_place(p) \wedge current\_view(v) \rightarrow at(v, p)$
$current\_path(p) \wedge current\_direction(d) \wedge current\_view(v) \rightarrow along(v, p, d)$
$current\_place(p) \wedge current\_path(path) \rightarrow on\_path(p, path).$

The 1D topological order of places along the current path is inferred from a *Travel* action and the current direction. We can also infer, or abduce, topological boundary and containment relations among regions, paths, and places.

5.2 Abduction to Places and Paths from Views and Actions

The definition of topological places is coupled with a categorization of actions into those that change the current place, called *Travel* actions, and those that leave the current place the same, called *Turn* actions. An action description includes a term representing the observed magnitude of the corresponding control laws, from internal effort sensors such as odometry. Since an action must begin and end at a locally distinctive state, not every magnitude of *Turn* or *Travel* is a meaningful action.

- $\langle V, (\mathtt{Turn}\ a), V' \rangle$ means that $place(V) = place(V')$ and a is monotonically related to the magnitude of the turn from $\theta(V)$ to $\theta(V')$. ($place(V)$ denotes the place where the robot observed V. Since places can have identical views, it is not necessarily a function only of V.)
- $\langle V, (\mathtt{Travel}\ d), V' \rangle$ means that $place(V) \neq place(V')$ if $d \neq 0$, and d is monotonically related to the distance traveled from $place(V)$ to $place(V')$.

We use the following observations as the basis for abduction of the connectivity properties of places and paths, given sequences of views and actions. Since different places could provide the same sensory image, sophisticated inference and even physical travel may occasionally be required to identify the current place from the current view [11, 7].

- Every view is observed at a place.

$$\forall view \; \exists place \; at(view, place)$$

- A *Turn* leaves the traveller at the same place.

$$\langle V, (\text{Turn } a), V' \rangle \rightarrow \exists place \; [at(V, place) \land at(V', place)] \qquad (8)$$

- A *Travel* leaves the traveller on the same path, facing the same direction. If the distance traveled is non-zero, the starting and ending places are different.

$$\langle V, (\text{Travel } d), V' \rangle \land d \neq 0 \rightarrow \exists p_1, p_2 \; [p_1 \neq p_2 \land at(V, p_1) \land at(V', p_2)]$$

$$\langle V, (\text{Travel } d), V' \rangle \rightarrow \exists path, dir \; [along(V, path, dir) \land along(V', path, dir)]$$

The topological level supports an array of problem-solving methods, augmenting graph search with heuristics based on the boundary and containment relations (not described here) that regions add to the topological map.

We have implemented a system that takes alternating sequences of views and actions from tours of a simulated urban environment and builds causal, topological, and local 1-D metrical descriptions of the environment.

6 The Metrical Level

6.1 Local 1-D Geometry

Observations of the magnitudes of actions provide information about the local geometry of places and paths. $\langle V, (\text{Travel } d), V' \rangle$ provides evidence about the distance between two places on the current path. $\langle V, (\text{Turn } a), V' \rangle$ provides evidence about the angle between obstacles and/or paths at the current place. This information can be represented as 1-D (linear or circular) metrical properties of the individual places and paths in the topological map. These properties are accumulated incrementally by the same abductive process that builds the topological map.

6.2 Local 2-D Geometry

If we take into account the fact that the topological map is embedded in a 2-D space, we can incrementally accumulate local descriptions of place neighborhoods and path segments as 2-D manifolds. Occupancy grids [24, 9], sonar target maps [17, 19], and generalized cylinders [25, 4] are three representations for 2-D manifold descriptions of local place neighborhoods and path segments.

6.3 Global 2-D Geometry

Once the topological and local metrical descriptions are sufficiently rich and reliable, these descriptions can be relaxed into a global 2-D frame of reference (figure 1(c)). However, this representational transformation is never on the critical path for exploration, map-learning, route-planning, or navigation.

7 Guarantees

A state s is *localizable* if, starting from s, there is a reliable method for traveling to a distinctive state, and thus being localized within the topological map. The localizable states are defined in terms of (a) the selection criteria for control laws, as embodied in the appropriateness measures $a_d(\mathbf{s})$, and (b) the basins of attraction defined by those control laws, considered as dynamical systems. A state s is *reachable* if, starting at a localizable state, there is a reliable method for the robot to travel to s. Using the framework defined above, we can analyze which states in the physical environment are localizable and/or reachable, giving various levels of knowledge in the cognitive map.

8 Discussion

The concept of "distinctive state" as used here for cognitive mapping appears to generalize certain aspects of the concept of "landmark value" as used for qualitative simulation.

Landmark values in QSIM corresponding to sign changes, operating region transitions, or extreme points in the behavior all represent individual real numbers. Ordinal relations with these landmarks support straight-forward qualitative hill-climbing, so they are distinctive within their quantity spaces. The only landmark values with a different character are those corresponding to initial values, which represent universally quantified variables ranging over a set defined by pre-existing landmark values.

Both cognitive mapping and qualitative simulation rely on abstracting a continuous underlying space to a discrete set of objects with symbolic names and symbolic relationships. In spite of the differences between the domains, I believe that the common structure will prove to be important.

References

1. P. E. Agre and D. Chapman. Pengi: An implementation of a theory of activity. In *Proc. 6th National Conf. on Artificial Intelligence (AAAI-87)*. Morgan Kaufmann, 1987.
2. R. C. Arkin. Motor schema-base mobile robot navigation. *International Journal of Robotics Research*, 8(4):92–112, 1989.
3. C. G. Atkeson, A. W. Moore, and S. Schaal. Locally weighted learning. *Artificial Intelligence Review*, 1996. in press.

4. R. A. Brooks. Symbolic reasoning among 3D models and 2D images. *Artificial Intelligence*, 17:285–348, 1981.
5. R. A. Brooks. A robust layered control system for a mobile robot. *IEEE Trans. on Robotics and Automation*, RA-2(1):14–23, 1986.
6. Raja Chatila and Jean-Paul Laumond. Position referencing and consistent world modeling for mobile robots. In *IEEE International Conference on Robotics and Automation*, pages 138–170, 1985.
7. Gregory Dudek, Michael Jenkin, Evangelos Milios, and David Wilkes. Robotic exploration as graph construction. *IEEE Trans. on Robotics and Automation*, 7(6):859–865, 1991.
8. R. A. Hart and G. T. Moore. The development of spatial cognition: A review. In R. M. Downs and D. Stea, editors, *Image and Environment*. Aldine Publishing Company, Chicago, 1973.
9. K. Konolige. A refined method for occupancy grid interpretation. unpublished draft manuscript, 1995.
10. Bart Kosko. *Neural Networks and Fuzzy Systems*. Prentice-Hall, Englewood Cliffs, NJ, 1992.
11. B. Kuipers and Y. T. Byun. A robust qualitative method for spatial learning in unknown environments. In *Proc. 7th National Conf. on Artificial Intelligence (AAAI-88)*, Los Altos, CA, 1988. Morgan Kaufmann.
12. B. Kuipers and Y.-T. Byun. A robot exploration and mapping strategy based on a semantic hierarchy of spatial representations. *Journal of Robotics and Autonomous Systems*, 8:47–63, 1991.
13. B. J. Kuipers. Modeling spatial knowledge. *Cognitive Science*, 2:129–153, 1978. Reprinted in *Advances in Spatial Reasoning, Volume 2*, Su-Shing Chen (Ed.), Norwood NJ: Ablex Publishing, 1990.
14. B. J. Kuipers and K. Åström. The composition and validation of heterogeneous control laws. *Automatica*, 30(2):233–249, 1994.
15. B. J. Kuipers and Tod Levitt. Navigation and mapping in large scale space. *AI Magazine*, 9(2):25–43, 1988. Reprinted in Advances in Spatial Reasoning, Volume 2, Su-shing Chen (Ed.), Norwood NJ: Ablex Publishing, 1990.
16. Jean-Claude Latombe. *Robot Motion Planning*. Kluwer Academic Publishers, Boston, 1991.
17. John J. Leonard and Hugh F. Durrant-Whyte. *Directed Sonar Sensing for Mobile Robot Navigation*. Kluwer Academic Publishers, Boston, 1992.
18. Kevin Lynch. *The Image of the City*. MIT Press, Cambridge, MA, 1960.
19. Paul MacKenzie and Gregory Dudek. Precise positioning using model-based maps. In *IEEE International Conference on Robotics and Automation*, volume 2, pages 1615–1621, Los Alamitos, CA, 1994. IEEE Computer Society Press.
20. E. H. Mamdani. Applications of fuzzy algorithms for control of a simple dynamic plant. *Proc. IEE*, 121:1585–1588, 1974.
21. J. McCarthy and P. J. Hayes. Some philosophical problems from the standpoint of artificial intelligence. In B. Meltzer and D. Michie, editors, *Machine Intelligence 4*, pages 463–502. Edinburgh University Press, Edinburgh, 1969.
22. A. W. Moore, C. G. Atkeson, and S. Schaal. Memory-based learning for control. *Artificial Intelligence Review*, 1996. to appear.
23. H. Moravec and A. Elfes. High resolution maps from wide angle sonar. In *IEEE International Conference on Robotics and Automation*, pages 116–121, 1985.
24. Hans P. Moravec. Sensor fusion in certainty grids for mobile robots. *AI Magazine*, Summer 1988.

25. R. Nevatia and T. O. Binford. Description and recognition of curved objects. *Artificial Intelligence*, 8(1):77–98, 1977.

26. J. S. Penberthy and D. S. Weld. Temporal planning with continuous change. In *Proc. 12th National Conf. on Artificial Intelligence (AAAI-94)*, pages 1010–1015, Cambridge, MA, 1994. AAAI Press/The MIT Press.

27. Jean Piaget and Baerbel Inhelder. *The Child's Conception of Space*. Norton, New York, 1967. First published in French, 1948.

28. D. Pierce and B. Kuipers. Learning to explore and build maps. In *Proc. 12th National Conf. on Artificial Intelligence (AAAI-94)*. AAAI/MIT Press, 1994.

29. Dean A. Pomerleau. Knowledge-based training of artificial neural networks for autonomous robot driving. In J. H. Connell and S. Mahadevan, editors, *Robot Learning*. Kluwer Academic Publishers, 1993.

30. Marcel J. Schoppers. Universal plans for reactive robots in unpredictable environments. In *Proc. 10th Int. Joint Conf. on Artificial Intelligence (IJCAI-87)*, San Mateo, CA, 1987. Morgan Kaufmann.

31. Marc G. Slack. Navigation templates: mediating qualitative guidance and quantitative control in mobile robots. *IEEE Trans. on Systems, Man and Cybernetics*, 23(2):452–466, 1993.

Spatial Reasoning with Topological Information

Jochen Renz and Bernhard Nebel

Institut für Informatik, Albert-Ludwigs-Universität
Am Flughafen 17, D-79110 Freiburg, Germany
`renz,nebel@informatik.uni-freiburg.de`
`www.informatik.uni-freiburg.de/~sppraum`

Abstract. This chapter summarizes our ongoing research on topological spatial reasoning using the Region Connection Calculus. We are addressing different questions and problems that arise when using this calculus. This includes representational issues, e.g., how can regions be represented and what is the required dimension of the applied space. Further, it includes computational issues, e.g., how hard is it to reason with the calculus and are there efficient algorithms. Finally, we also address cognitive issues, i.e., is the calculus cognitively adequate.

1   Introduction

When describing a spatial configuration or when reasoning about such a configuration, often it is not possible or desirable to obtain precise, quantitative data. In these cases, qualitative reasoning about spatial configurations may be used.

Different aspects of space can be treated in a qualitative way. Among others there are approaches considering orientation, distance, shape, topology, and combinations of these. A summary of work on these and other aspects of qualitative spatial reasoning can be found in [Coh97].

One particular approach in this context has been developed by Randell, Cui, and Cohn [RCC92], the so-called *Region Connection Calculus* (RCC), which is based on binary topological relations. One variant of this calculus, RCC-8, uses eight mutually exhaustive and pairwise disjoint relations, called base relations, to describe the topological relationship between two spatial regions. A similar calculus was developed by Egenhofer [Ege91], who defined relations by comparing the intersection of the interior, the exterior, and the boundary of different planar regions and identified the same base relations.

In this chapter we are addressing different aspects of using RCC-8. Among these are cognitive aspects of RCC-8, namely, whether a formally defined topological calculus like RCC-8 can also be regarded as cognitively adequate. We will report about an empirical investigation on that topic [KRR97] that resulted from a cooperation with the project MEMOSPACE (see their chapter in this volume [KRSS98]).

One aspect is concerned with representational properties. As spatial regions used by RCC-8 are arbitrary regular subsets of the topological space, it is unclear how these regions should be represented. We will present a canonical model that allows a simple representation where regions are reduced to their important points and information about the neighborhood of these points [Ren98].

Most applications of spatial reasoning deal with two- or three-dimensional space and not with arbitrary topological space, where dimension is not considered. Therefore there might be consistent sets of RCC-8 relations which are not realizable in the desired dimension. Using the canonical model, we can prove that any consistent set is always realizable in any dimension $d \geq 1$ if arbitrary regions are used and in any dimension $d \geq 3$ if regions must be internally connected [Ren98].

Another aspect is concerned with computational issues of reasoning with RCC-8. We will prove that reasoning with RCC-8 is NP-hard in general and identify a large maximal tractable subset of RCC-8 which can be used to make reasoning much more efficient even in the general NP-hard case [RN97].

This chapter is organized as follows. In the second section we introduce RCC-8, Section 3 summarizes our empirical investigation on cognitive validity of RCC-8. In Section 4 we introduce the modal encoding of RCC-8 and identify the canonical model. In Section 5 this model will be interpreted topologically, which allows a simple representation of regions and also predications about the dimension of regions. Section 6 summarizes our results on computational properties of RCC-8.

2 Qualitative Spatial Reasoning with RCC

RCC is a topological approach to qualitative spatial representation and reasoning where *spatial regions* are regular subsets of a topological space $\mathcal{U}$ [RCC92]. $\mathcal{U}$ is called the *universe*, i.e., the whole space. Relationships between spatial regions are defined in terms of the relation $C(r,s)$ which is true if and only if the closure of region r is connected to the closure of region s, i.e. if their closures share a common point. We consider only regular closed regions, i.e., regions that are equivalent to the closure of their interior. This is no restriction, as with the above definition of C it cannot be distinguished between open, semi-open, and closed regions. Regions themselves do not have to be internally connected, i.e., a region may consist of different disconnected parts. The domain of *spatial variables* (denoted as X, Y, Z) is the whole topological space.

In this work we will focus on RCC-8, but most of our results can easily be applied to RCC-5, a subset of RCC-8 [Ben94]. RCC-8 uses a set of eight pairwise disjoint and mutually exhaustive binary relations, called *base relations*, denoted as DC, EC, PO, EQ, TPP, NTPP, TPP^{-1}, and NTPP^{-1}, with the meaning *DisConnected, Externally Connected, Partial Overlap, EQual, Tangential Proper Part, Non-Tangential Proper Part*, and their converses. Examples for these relations are shown in Figure 1. In RCC-5 the boundary of a region is not taken into account, i.e., one does not distinguish between DC and EC and between TPP and NTPP. These relations are combined to the RCC-5 base relations DR for *DiscRete* and PP for *Proper Part*, respectively.

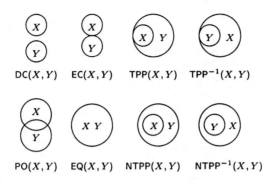

Fig. 1. Two-dimensional examples for the eight base relations of RCC-8

Sometimes it is not known which of the eight base relations holds between two regions, but it is possible to exclude some of them. In order to represent this, unions of base relations can be used. Since base relations are pairwise disjoint, this results in 2^8 different relations, including the union of all base relations, which is called *universal relation*. In the following we will write sets of base relations to denote these unions. Using this notation, the RCC-5 base relation DR = DC ∪ EC, e.g., is identical to {DC, EC}. *Spatial formulas* are written as XRY, where R is a spatial relation. A *spatial configuration* can be described by a set Θ of spatial formulas.

Apart from union (∪), other operations are defined, namely, converse (⌣), intersection (∩), and composition (∘) of relations. The formal definitions of these operations are:

$$\forall X, Y : X(R \cup S)Y \leftrightarrow XRY \lor XSY,$$
$$\forall X, Y : X(R \cap S)Y \leftrightarrow XRY \land XSY,$$
$$\forall X, Y : \quad XR^\smile Y \leftrightarrow YRX,$$
$$\forall X, Y : X(R \circ S)Y \leftrightarrow \exists Z : (XRZ \land ZSY).$$

The compositions of the eight base relations are shown in Table 1. Every entry in the composition table specifies the relation obtained by composing the base relation of the corresponding row with the base relation of the corresponding column. Composition of two arbitrary RCC-8 relations can be obtained by computing the union of the composition of the base relations.

Given a particular subset S of RCC-8, the *closure* of S under composition, intersection, and converse contains all relations that can be obtained by applying these operations to the relations of S. The closure of S is denoted $\widehat{S}$. The closure of the set of RCC-8 base relations $\mathcal{B}$, e.g., contains among other relations all relations in the composition table, as they can be obtained by composing the base relations.

One important computational problem is deciding *consistency* of a set Θ of spatial formulas. Θ is consistent, if it is possible to find a *realization* of Θ, i.e., an instantiation of every spatial variable with a spatial region such that all relations

∘	DC	EC	PO	TPP	NTPP	TPP⁻¹	NTPP⁻¹	EQ
DC	*	DC,EC PO,TPP NTPP	DC,EC PO,TPP NTPP	DC,EC PO,TPP NTPP	DC,EC PO,TPP NTPP	DC	DC	DC
EC	DC,EC PO,TPP⁻¹ NTPP⁻¹	DC,EC PO,TPP TPP⁻¹,EQ	DC,EC PO,TPP NTPP	EC,PO TPP NTPP	PO TPP NTPP	DC,EC	DC	EC
PO	DC,EC PO,TPP⁻¹ NTPP⁻¹	DC,EC PO,TPP⁻¹ NTPP⁻¹	*	PO TPP NTPP	PO TPP NTPP	DC,EC PO,TPP⁻¹ NTPP⁻¹	DC,EC PO,TPP⁻¹ NTPP⁻¹	PO
TPP	DC	DC,EC	DC,EC PO,TPP NTPP	TPP NTPP	NTPP	DC,EC PO,TPP TPP⁻¹,EQ	DC,EC PO,TPP⁻¹ NTPP⁻¹	TPP
NTPP	DC	DC	DC,EC PO,TPP NTPP	NTPP	NTPP	DC,EC PO,TPP NTPP	*	NTPP
TPP⁻¹	DC,EC PO,TPP⁻¹ NTPP⁻¹	EC,PO TPP⁻¹ NTPP⁻¹	PO TPP⁻¹ NTPP⁻¹	PO,EQ TPP TPP⁻¹	PO TPP NTPP	TPP⁻¹ NTPP⁻¹	NTPP⁻¹	TPP⁻¹
NTPP⁻¹	DC,EC PO,TPP⁻¹ NTPP⁻¹	PO TPP⁻¹ NTPP⁻¹	PO TPP⁻¹ NTPP⁻¹	PO TPP⁻¹ NTPP⁻¹	PO,TPP⁻¹ TPP,NTPP NTPP⁻¹,EQ	NTPP⁻¹	NTPP⁻¹	NTPP⁻¹
EQ	DC	EC	PO	TPP	NTPP	TPP⁻¹	NTPP⁻¹	EQ

Table 1. Composition table for the eight base relations of RCC-8, where * specifies the universal relation.

hold between the regions. We call this problem RSAT. For example consider the set $\Theta = \{X\{\mathsf{NTPP}\}Y, Y\{\mathsf{TPP}\}Z, Z\{\mathsf{TPP}, \mathsf{NTPP}\}X\}$. Θ is inconsistent as it follows from Table 1 that NTPP composed with TPP is NTPP, so in our example $X\{\mathsf{NTPP}\}Z$ should be true which contradicts $X\{\mathsf{TPP}^{-1}, \mathsf{NTPP}^{-1}\}Z \in \Theta$. This is easy to see, as it is not possible that a region r is part of a region s which is part of another region t which is part of r. When only relations of a specific set S are used in Θ, the corresponding reasoning problem is denoted RSAT(S).

A *canonical model* of RCC-8 is a model by which every consistent set of RCC-8 formulas can be interpreted. The standard canonical model for RCC-8 is the topological space, as every region can be interpreted as a subset of the topological space. A canonical model for Allen's interval calculus [All83], e.g., is the set of all convex intervals of real numbers. This model allows each interval to be represented using the two endpoints of the interval. Such a simple representation is not possible with the topological space as a canonical model for RCC-8.

3   Cognitive Plausibility of RCC-8

Qualitative temporal and spatial calculi are usually justified by application requirements and/or the introspection of the researchers developing the calculi. The cognitive significance of these calculi is usually not investigated. One exception is Allen's interval calculus, which has been analyzed from a cognitive point

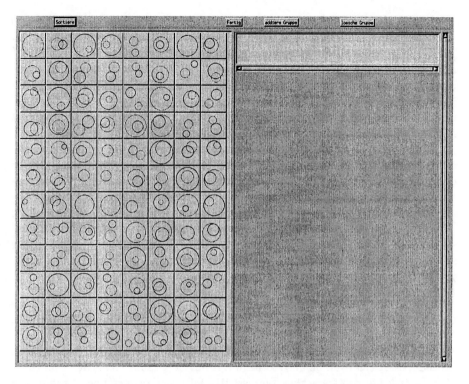

Fig. 2. Screen dump of the monitor at the beginning of the grouping task

of view by the MEMOSPACE project (see Chapter [KRSS98]). Here the authors distinguish between *conceptual cognitive adequacy* and *inferential cognitive adequacy* [KRS95].

According to Knauff et al [KRR97], a spatial calculus is inferentially cognitive adequate if "the reasoning mechanism of the calculus is structurally similar to the way people reason about space" and it is conceptually cognitive adequate if "empirical evidence supports the assumption that a system of relations is a model of people's conceptual knowledge of spatial relationships." Our main aim in assessing the cognitive plausibility of RCC-8 was to find out whether the distinctions made in RCC-8 are conceptually adequate. In particular, we were interested in finding out whether sub-calculi such as RCC-5 are more plausible than RCC-8. In cooperation with the MEMOSPACE project, we investigated these questions [KRR97] using the grouping task paradigm. 20 subjects (students of Albert-Ludwigs-Universität, Freiburg) were presented 96 items with varying configurations of one red and one blue circle. The task of the subjects was to group similar configurations together, where the number of groups was not given to the subjects (see Figure 2). After having completed the grouping task, subjects were (unexpectedly) asked to give natural language descriptions of the groups they had formed.

Applying a cluster analysis to the data obtained in this investigation revealed that after some clustering steps items for the RCC-8 relations were clustered together. After some more clustering steps items for the relations TPP and TPP^{-1} as well as items for the relations NTPP and NTPP^{-1} were clustered together, but at no level of the cluster analysis other sub-calculi of RCC-8 were detected. Clustering of TPP and TPP^{-1} as well as NTPP and NTPP^{-1} probably happened because some subjects ignored the distinction between reference object and to-be-localized object.

In the analysis of the natural language description of the groupings it became evident that in more than 95 % of all cases topological terms were used to describe the groupings. This and the above described finding led us to the conclusion that there is evidence that the RCC-8 system of relations is conceptually cognitive adequate, i.e., people use them to conceptualize spatial configurations [KRR97]. However, more investigations are necessary to confirm this. For instance, one should investigate whether the RCC-8 assumption of regions that are not internally connected is adequate. Further, it will be interesting to investigate the inferential cognitive adequacy of RCC-8.

4 Modal Encoding of RCC-8 and a Canonical Model

As RCC is defined in first-order logic, this does not lead to efficient decision procedures. It can even be derived from a result of [Grz51] that RCC is undecidable. In order to overcome this, Bennett [Ben94] used an encoding of the RCC-8 relations in propositional intuitionistic logic whereby RCC-8 is proven to be decidable. In this chapter we are using Bennett's encoding of RCC-8 in modal logic [Ben95]. After making a brief introduction to modal logic, we are describing the modal encoding and based on this identify a canonical model of RCC-8.

4.1 Propositional Modal Logic and Kripke Semantics

Propositional modal logic [Fit93,Che80] extends classical propositional logic by additional unary *modal operators* $\Box_i$. A common semantic interpretation of modal formulas is the *Kripke semantics* which is based on a set W of so-called *worlds* and a set $\mathcal{R}$ of *accessibility relations* between these worlds, where $R \subseteq W \times W$ for every accessibility relation $R \in \mathcal{R}$. Worlds are entities in which modal formulas can be interpreted as either true or false. In different worlds modal formulas are usually interpreted differently. A different accessibility relation $R_{\Box_i}$ is assigned to every modal operator $\Box_i$. For example if $u, v \in W$ are worlds, $R_{\Box_i} \in \mathcal{R}$, and $uR_{\Box_i}v$ holds, then the world v is *accessible* from u with $R_{\Box_i}$. v is also called $R_{\Box_i}$-*successor* of w.

A *Kripke model* $\mathcal{M} = \langle W, \mathcal{R}, \pi \rangle$ uses an additional valuation π that assigns each propositional atom in each world a truth value $\{true, false\}$. Using a Kripke model, a modal formula can be interpreted with respect to the set of worlds, the accessibility relations, and the valuation. For example, a propositional atom a is true in a world w of the Kripke model $\mathcal{M}$ (written as $\mathcal{M}, w \Vdash a$) if and only

Relation	Model Constraints	Entailment Constraints
$DC(X,Y)$	$\neg(X \wedge Y)$	$\neg X, \neg Y$
$EC(X,Y)$	$\neg(\mathbf{I}X \wedge \mathbf{I}Y)$	$\neg(X \wedge Y), \neg X, \neg Y$
$PO(X,Y)$	—	$\neg(\mathbf{I}X \wedge \mathbf{I}Y), X \to Y, Y \to X, \neg X, \neg Y$
$TPP(X,Y)$	$X \to Y$	$X \to \mathbf{I}Y, Y \to X, \neg X, \neg Y$
$TPP^{-1}(X,Y)$	$Y \to X$	$Y \to \mathbf{I}X, X \to Y, \neg X, \neg Y$
$NTPP(X,Y)$	$X \to \mathbf{I}Y$	$Y \to X, \neg X, \neg Y$
$NTPP^{-1}(X,Y)$	$Y \to \mathbf{I}X$	$X \to Y, \neg X, \neg Y$
$EQ(X,Y)$	$X \to Y, Y \to X$	$\neg X, \neg Y$

Table 2. Modal encoding of the eight base relations [Ben95].

if $\pi(w, a) = true$. An arbitrary modal formula is interpreted according to its inductive structure. A modal formula $\Box_i \varphi$, e.g., is true in a world w of the Kripke model $\mathcal{M}$, i.e., $\mathcal{M}, w \Vdash \Box_i \varphi$, if and only if φ is true in *all* worlds accessible from w with $R_{\Box_i}$. $\mathcal{M}, w \Vdash \neg\Box_i \varphi$ if and only if there is a world accessible from w with $R_{\Box_i}$ where φ is false. The operators $\neg, \wedge$ and $\vee$ are interpreted in the same way as in classical propositional logic.

Different modal operators can be distinguished according to their different accessibility relations. In this chapter we are using so-called S4-operators and S5-operators. The accessibility relation of an S4-operator must be reflexive and transitive, the accessibility relation of an S5-operator must be reflexive, transitive, and euclidean. With the accessibility relation R of a *strong* S5-operator all worlds are accessible from each other, i.e., $R = W \times W$. The use of Kripke models should become more clear in Section 4.3 and Section 5, where worlds and accessibility relations are displayed (see Figure 3 and Figure 4) .

4.2 Modal Encoding of RCC-8

The modal encoding of RCC-8 was introduced by Bennett [Ben95] and extended in [RN97]. In both cases the encoding is restricted to regular closed regions, i.e., regions which are equivalent to the closure of their interior. The modal encoding is based on a set of *model* and *entailment constraints* for each base relation, where model constraints must be true and entailment constraints must not be true. Bennett encoded these constraints in modal logic by introducing an S4-operator $\mathbf{I}$ which he interpreted as an interior operator [Ben95]. Table 2 displays these constraints for the eight base relations. Every spatial variable corresponds to a propositional atom, so the modal formula $X \wedge Y$ corresponds to the intersection of the spatial regions X and Y, $X \vee Y$ to the union of X and Y, $\neg X$ to the complement of X, and $\mathbf{I}X$ to the interior of X. If a modal formula φ must be true in all worlds, then the spatial region corresponding to φ is equal to the universe. The model constraint for the relation $EC(X,Y)$, e.g., states that the complement of the intersection of the interior of region X with the interior of region Y is equal to the universe. This constraint guarantees that regions X and Y have no common interior. The entailment constraints of $EC(X,Y)$ state that the complement of the intersection of region X and region Y is not equal to the universe. Also the complements of both X and Y are not equal to the universe. These constraints guarantee that regions X and Y have points in common and that both regions are not empty.

In order to combine the model and entailment constraints to a single modal formula, Bennett introduced a strong S5-operator $\square$, where $\square\varphi$ is written for every model constraint φ and $\neg\square\psi$ for every entailment constraint ψ [Ben95]. $\square\varphi$ can be interpreted as *the spatial region φ is equal to the universe* and $\neg\square\varphi$ as *the spatial region φ is not equal to the universe*. All constraints of a single base relation are then combined conjunctively to a single modal formula. In order to represent unions of base relations, the modal formulas of the corresponding base relations are combined disjunctively. In this way every spatial formula XRY can be transformed to a modal formula $m_1(XRY)$. Two additional constraints $m_2(X)$ are necessary to guarantee that only regular closed regions X are used [RN97]: every region X must be equivalent to the closure of its interior and the complement of a region must be an open region.[1]

$$m_2(X) = \square(X \leftrightarrow \neg\mathbf{I}\neg\mathbf{I}X) \wedge \square(\neg X \leftrightarrow \mathbf{I}\neg X).$$

So any set of spatial formulas Θ can be written as a single modal formula $m(\Theta)$ where $Reg(\Theta)$ is the set of spatial variables of Θ:

$$m(\Theta) = \left(\bigwedge_{XRY \in \Theta} m_1(XRY) \right) \wedge \left(\bigwedge_{X \in Reg(\Theta)} m_2(X) \right).$$

As follows from the work by Bennett [Ben95], Θ is consistent if and only if $m(\Theta)$ is satisfiable.

4.3 A Canonical Model of RCC-8

A canonical model of a calculus is a structure that allows to model any consistent formula of the calculus. An obvious canonical model of RCC-8 is the topological space, as every spatial region can be modeled by a subset of the topological space. As described above, the modal encoding of RCC-8 can be interpreted by Kripke models. As the modal encoding of RCC-8 is equivalent to a set of RCC-8 formulas, a canonical model of RCC-8 is a structure that allows a Kripke model for any modal formula obtained by the modal encoding of RCC-8. In order to obtain a canonical model we distinguish different levels of worlds. A *world of level 0* is a world which cannot be accessed from any other world with $R_{\mathbf{I}}$, the accessibility relation corresponding to the **I**-operator. A *world of level l* is a world which can be accessed with $R_{\mathbf{I}}$ from a world of level $l - 1$ but not from other worlds with a higher level than $l - 1$.

Definition 1. *An RCC-8-structure $\mathcal{S}_{RCC8} = \langle W, \{R_\square, R_{\mathbf{I}}\}, \pi \rangle$ has the following properties (see Figure 3):*

---

[1] It can be easily verified that $\neg\mathbf{I}\neg\varphi$ corresponds to the closure of φ.

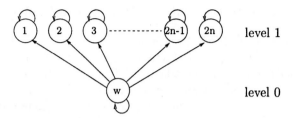

Fig. 3. A world w of level 0 together with its $2n$ R_I-successors as used in an RCC-8-structure. Worlds are drawn as circles, the arrows indicate the accessibility of worlds with the relation R_I

1. There are only worlds of level 0 and 1.
2. For every world u of level 0 there are exactly $2n$ worlds v of level 1 with uR_Iv.
3. For every world u of level 1 there is exactly one world w of level 0 with wR_Iu.
4. For all worlds $w, v \in W$: wR_Iw and $wR_\square v$.

S_{RCC8} contains worlds with all possible instantiations with respect to $R_\square$ and R_I. An RCC-8-model $\mathcal{M}$ of $m(\Theta)$ is a finite subset of S_{RCC8}. In a polynomial RCC-8-model the number of worlds is polynomially bounded by the number of regions.

Every world of level 0 together with its $2n$ R_I-successors forms an independent cluster (see Figure 3). From the definition of "level" and Definition 1 it follows that R_I is reflexive and transitive, so it is guaranteed that **I** is an S4-operator. As the number of regions is countable, the number of worlds of W is also countable.

Lemma 1. If $m(\Theta)$ is satisfiable, then there is a polynomial RCC-8-model $\mathcal{M}$ with $\mathcal{M}, w \Vdash m(\Theta)$ with at most $3n^2$ worlds of level 0.

Therefore the RCC-8-structure is a canonical model of the modal encoding of any set of spatial formulas. The number of required worlds of level 0 results from the number of different entailment constraints.

5 Representational Properties of RCC-8

It was shown in the previous section that the RCC-8-structure is a canonical model of RCC-8. This model was obtained from the modal encoding of topological relations, so the model depends mainly on the modal encoding but not on topology. In order to use this model for representational purposes, we have to find a way to interpret it topologically. Then the model can also be used for dealing with other properties of regions, e.g., dimension. A more detailed description of representational issues of RCC-8 can be found in [Ren98].

5.1 Topological Interpretation of the RCC-8 Model

The modal encoding of RCC-8 was obtained by introducing a modal operator $\mathbf{I}$ corresponding to the topological interior operator and transferring the topological properties and axioms to modal logic. Using the intended interpretation of $\mathbf{I}$ as an interior operator, it is unclear how the RCC-8-model, especially the accessibility relations $R_{\square}$ and $R_{\mathbf{I}}$, can be topologically interpreted. In this section we present a way of topologically interpreting the RCC-8-model such that all parts of the model can be interpreted consistently. The $\mathbf{I}$-operator will not be interpreted as an interior operator, but we will prove that it satisfies the intended interpretation of an interior operator.

Because $\mathbf{I}$ is an S4-operator and because of the additional constraints m_2, exactly one of the following formulas is true for every world w of $\mathcal{M}$ and every region X.

1. $\mathcal{M}, w \Vdash \mathbf{I}X$
2. $\mathcal{M}, w \Vdash \mathbf{I}\neg X$
3. $\mathcal{M}, w \Vdash X \wedge \neg \mathbf{I}X$

Consider a particular world w. Then the set of all spatial variables can be divided into three disjoint sets according to which of the three possible formulas is true in w. Let $\mathcal{X}_w$ be the set of spatial variables where the first formula is true in w, $\mathcal{Y}_w$ be the set where the second formula is true in w, and $\mathcal{Z}_w$ be the set where the third formula is true in w, i.e., $\mathcal{M}, w \Vdash \mathbf{I}X_i \wedge \mathbf{I}\neg Y_j \wedge (Z_k \wedge \neg \mathbf{I}Z_k)$ for all $X_i \in \mathcal{X}_w$, $Y_j \in \mathcal{Y}_w$, and $Z_k \in \mathcal{Z}_w$.

Some relations between these spatial variables cannot hold as they contradict the modal and entailment constraints of these relations. In the following table the excluded relations and their topological consequences are shown for two regions X and Y. $i(.)$ denotes the interior, $e(.)$ the exterior, and $b(.)$ the boundary of a region.

Set of X	Set of Y	Impossible relations	Consequences
$\mathcal{X}_w$	$\mathcal{X}_w$	DC, EC	$i(X) \cap i(Y) \neq \emptyset$
$\mathcal{X}_w$	$\mathcal{Y}_w$	TPP, NTPP, EQ	$i(X) \cap e(Y) \neq \emptyset$
$\mathcal{X}_w$	$\mathcal{Z}_w$	DC, EC, TPP, NTPP, EQ	$i(X) \cap b(Y) \neq \emptyset$
$\mathcal{Y}_w$	$\mathcal{Y}_w$	–	–
$\mathcal{Y}_w$	$\mathcal{Z}_w$	TPP^{-1}, NTPP^{-1}, EQ	$e(X) \cap b(Y) \neq \emptyset$
$\mathcal{Z}_w$	$\mathcal{Z}_w$	DC, NTPP, NTPP^{-1}	$b(X) \cap b(Y) \neq \emptyset^2$

It can be seen, e.g., that when $\mathbf{I}X$ and $\mathbf{I}Y$ is true for a world w then the two regions X and Y have a common interior.

Considering points in the topological space, we can distinguish three different ways how a point p can be related to a region X:

---

[2] Actually this is not necessarily the case for PO(X, Y) if X or Y are not internally connected, but assuming this does not contradict any constraint since RCC-8 is not expressive enough to distinguish different kinds of partial overlap.

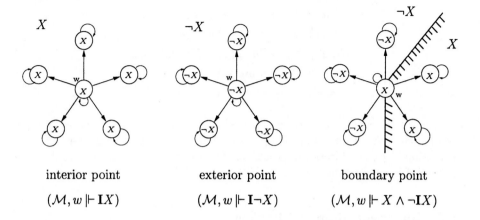

interior point

$(\mathcal{M}, w \Vdash \mathbf{I}X)$

exterior point

$(\mathcal{M}, w \Vdash \mathbf{I}\neg X)$

boundary point

$(\mathcal{M}, w \Vdash X \wedge \neg \mathbf{I}X)$

Fig. 4. Three different topological interpretations of a world w. The solid line is the boundary of X where the hatched region indicates the interior of X.

1. p interior point of X: there is a neighborhood N of p such that all points of N are contained in X
2. p is exterior point of X: there is a neighborhood N of p such that no point of N is contained in X
3. p is boundary point of X: every neighborhood N of p contains points inside of X and points outside of X

Comparing this to the three modal formulas described above, it can be seen that there is a connection between the modal formula which is true in a world w and the topological properties of a point p. It can be proven that there are functions $p : W \mapsto \mathcal{U}$ and $N : W \mapsto 2^{\mathcal{U}}$ that map every world w to a point $p(w)$ in the topological space and to a neighborhood $N(w)$ of $p(w)$ such that

$$p(w) \in X \text{ if } \pi(w, X) = \text{true},$$
$$p(w) \notin X \text{ if } \pi(w, X) = \text{false},$$
$$p(u) \in N(w) \text{ if } wR_{\mathbf{I}}u.$$

For this proof we assume that $p(w)$ is in the interior of all regions X_i, in the exterior of all regions Y_j, and on the boundary of all regions Z_k simultaneously. As there is no contradiction to this neither from the topological constraints nor from the modal constraints, it can be safely assumed. With this assumption the proof is immediate. Figure 4 shows the three different kinds of interpretations of worlds as points.

Modal formulas can now be transformed stepwise to topological formulas as follows:

$$\mathcal{M}, w \Vdash \Box\varphi \mapsto \forall u : p(u) \in \mathcal{U}.\mathcal{M}, u \Vdash \varphi$$

$$\mathcal{M}, w \Vdash/ \Box\varphi \mapsto \exists u : p(u) \in \mathcal{U}.\mathcal{M}, u \Vdash/ \varphi$$
$$\mathcal{M}, w \Vdash \mathbf{I}\varphi \mapsto \forall u : p(u) \in N(w).\mathcal{M}, u \Vdash \varphi$$
$$\mathcal{M}, w \Vdash/ \mathbf{I}\varphi \mapsto \exists u : p(u) \in N(w).\mathcal{M}, u \Vdash/ \varphi$$
$$\mathcal{M}, w \Vdash X \mapsto p(w) \in X$$
$$\mathcal{M}, w \Vdash/ X \mapsto p(w) \notin X$$

Therefore $\mathcal{M}, w \Vdash \mathbf{I}X$ can be interpreted as "there is a neighborhood $N(w)$ of $p(w)$ such that all points of $N(w)$ are in X". This satisfies the intended interpretation of $\mathbf{I}$ as an interior operator, as $\mathcal{M}, w \Vdash X$ means that $p(w)$ is in X and $\mathcal{M}, w \Vdash \mathbf{I}X$ means that $p(w)$ is in the interior of X.

5.2 Dimension of Spatial Regions

The topological space we have been using so far does not have any particular dimension. This means that a consistent set of spatial relations is realizable in some dimension, but not necessarily in the dimension an application requires, e.g., two- or three-dimensional space. In the following we examine what dimension a space requires in order to realize the canonical model. Suppose that all $R_{\mathbf{I}}$ successors of a world w are mapped to points on the boundary of an n-dimensional sphere with $p(w)$ in the center. Then the neighborhoods of Figure 4, e.g., can as shown in the figure be mapped to a two-dimensional plane where all regions are also two-dimensional. This is possible because the mappings of the $R_{\mathbf{I}}$-successors of the rightmost level 0 world can be separated by two line-segments belonging to the boundary of X. If the worlds cannot be separated by two line-segments for a region, we have to find a permutation of the $R_{\mathbf{I}}$-successors such that a separation is possible. A separation is necessary only for those neighborhoods that contain boundary points of a region, as for the other neighborhoods all points are the same. By analyzing which points are boundary points of which regions and the relationship between those regions, it turns out that a permutation can always be found such that the worlds can be separated by at most two line-segments for any region. In fact only two distinct $R_{\mathbf{I}}$-successors are necessary for each world of level 0. Therefore we obtain another canonical model for RCC-8 which allows models which are much more compact than the RCC-8-model as introduced in the previous section. The new canonical model is denoted *reduced* RCC-8-*structure* and the corresponding Kripke models *reduced* RCC-8-*models*. One world of level 0 of the reduced RCC-8-structure together with its $R_{\mathbf{I}}$-successors is shown in Figure 5a.

In order to obtain regions from the neighborhoods we have to *close* every neighborhood, i.e., for every neighborhood $N(w)$ find the closure of the part of every region which is affiliated with $N(w)$. Both sides of every neighborhood (see Figure 6a) can be treated almost independently. All regions which are affiliated with the same side of a neighborhood are either overlapping or one is part of the other, i.e., TPP or NTPP. For the closure of the neighborhoods all "part of" relations must be fulfilled, the partial overlap relation is not important.

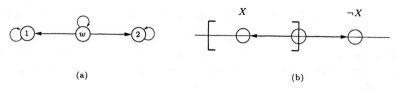

(a)   (b)

Fig. 5. (a) shows a world w of level 0 of the reduced RCC-8-structure together with its two R_1-successors. In (b) it is shown how the neighborhoods can be placed in one-dimensional space. The two brackets indicate a possible one-dimensional region X where the neighborhood defines a boundary point of X.

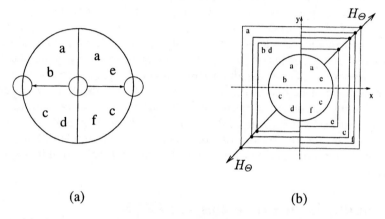

(a)   (b)

Fig. 6. (a) shows the two-dimensional neighborhood of a boundary point which is divided by the boundary. In (b) the neighborhood is closed with respect to the hierarchy H_Θ of the affiliated regions.

In order to fulfill the "part of" relations, we have to find a *hierarchy* H_Θ of the regions such that such that a "part of" b if and only if $H_\Theta(a) < H_\Theta(b)$. The parts of all regions affiliated with a neighborhood can then be closed as rectangles according to the hierarchy H_Θ, i.e., regions of the same level are equal (for a particular neighborhood) and are part of all regions of a higher level (see Figure 6b). A neighborhood can be closed in any higher dimension d. The hierarchy of regions is then measured along the diagonal of the d-dimensional hypercube. In Figure 5b it can be seen that using the reduced RCC-8-model it is also possible to place the neighborhood in a one-dimensional space where regions are disconnected intervals.

Theorem 1. *If a set of spatial formulas Θ is consistent, the RCC-8-model can be realized in any dimension $d \geq 1$.*

Starting from a two-dimensional model of possibly non connected regions, it is possible to construct a three-dimensional model of connected regions.

Theorem 2. *If a set of spatial formulas Θ is consistent, the* RCC-8-*model can be realized in any dimension $d \geq 3$ using only connected regions.*

The new canonical model is much better suited for representational purposes than the RCC-8-model, but, as we will see in Section 6, it has some computational drawbacks.

5.3   Representing Regions with the Canonical Model

The RCC-8-models give us a possibility to represent topological regions. With the topological interpretation of the model it becomes clear that regions can be reduced to points and information about their neighborhood. The points that are needed within the model represent the important features of the regions with respect to a set of relations.

Using the canonical model we can give algorithms to generate a realization of Θ in the desired dimension. This can be done by simply placing the level 0 worlds together with their neighborhoods in the desired space and close the neighborhoods according to the hierarchy H_Θ. In this realization every region consists of many disconnected parts (at most $3n^2$ pieces, as there are at most that many distinct worlds of level 0, i.e., neighborhoods). A realization using only internally connected regions can be generated in any dimension $d \geq 3$ by connecting all parts of a region of the $d - 1$ dimensional realization in a specific way [Ren98].

6   Computational Properties of RCC-8

In order to get a deeper insight into a problem and to find efficient algorithms, an analysis of the computational properties is helpful. First results on computational properties of RCC-8 were obtained by Nebel, who considered sets of base relations [Neb95]. It was shown that the consistency problem RSAT($\mathcal{B}$) (where $\mathcal{B}$ is the set of RCC-8 base relations) is polynomial and that the path-consistency method (see also Section 6.3), a popular $O(n^3)$ approximation algorithm, is sufficient for deciding consistency. Based on these results we are interested in the complexity of the general consistency problem of RCC-8, where all 256 relations are allowed. In this section we will show that RSAT is NP-hard, i.e., that every algorithm is expected to take time super-polynomial in the number of spatial regions, provided P $\neq$ NP. As we now have intractability of the general consistency problem of RCC-8 and tractability of a subset of RCC-8, we are interested in the boundary between tractability and intractability. Therefore we identify a maximal tractable subset of RCC-8 and prove that the path-consistency method is sufficient for deciding consistency of this set. A more detailed description of the computational properties of RCC-8 can be found in [RN97]

6.1   Complexity of RCC-8

All of the following NP-hardness proofs use a reduction of a propositional satisfiability problem to RSAT($\mathcal{S}$) by constructing a set of spatial formulas Θ for every instance $\mathcal{I}$ of some propositional problem, such that Θ is consistent if and only if $\mathcal{I}$ is a positive instance. These satisfiability problems include 3SAT where all clauses have exactly 3 literals, NOT-ALL-EQUAL-3SAT where every clause has at least one true and one false literal, and ONE-IN-THREE-3SAT where exactly one literal in every clause must be true [GJ79].

The reductions have in common that every literal as well as every literal occurrence L is reduced to two spatial variables X_L and Y_L and a relation $R = R_t \cup R_f$, where $R_t \cap R_f = \emptyset$ and $X_L R Y_L$ holds. L is true if and only if $X_L R_t Y_L$ holds and false if and only if $X_L R_f Y_L$ holds. Additional "polarity" constraints have to be introduced to assure that for the spatial variables $X_{\neg L}$ and $Y_{\neg L}$, corresponding to the negation of L, $X_{\neg L} R_t Y_{\neg L}$ holds if and only if $X_L R_f Y_L$ holds, and *vice versa*. Using these polarity constraints, spatial variables of negative literal occurrences are connected to the spatial variables of the corresponding positive literal, and likewise for positive literal occurrences and negative literals. Further, "clause" constraints have to be added to assure that the clause requirements of the specific propositional problem are satisfied in the reduction. We will first prove that the consistency problem for RCC-5 is NP-hard.

Theorem 3. RSAT(RCC-5) *is NP-hard.*

Proof Sketch. Transformation of NOT-ALL-EQUAL-3SAT to RSAT(RCC-5) (see also [GPP95]). $R_t = \{PP\}$ and $R_f = \{PP^{-1}\}$. Polarity constraints:

$$X_L\{PP, PP^{-1}\}X_{\neg L}, Y_L\{PP, PP^{-1}\}Y_{\neg L},$$
$$X_L\{PO\}Y_{\neg L}, Y_L\{PO\}X_{\neg L}.$$

Clause constraints for every clause $c = \{i, j, k\}$:

$$X_i\{PP, PP^{-1}\}X_j, X_j\{PP, PP^{-1}\}X_k, X_k\{PP, PP^{-1}\}X_i,$$
$$X_i\{PO\}Y_k, X_j\{PO\}Y_i, X_k\{PO\}Y_j. \qquad \blacksquare$$

Since RCC-5 is a subset of RCC-8, this result can be easily applied to RCC-8.

Corollary 1. RSAT(RCC-8) *is NP-hard.*

In the above NP-hardness proof only the relations $\{PO\}$, $\{PP, PP^{-1}\}$, and the universal relation were used, so this set of three relations is already NP-hard. The same or similar proofs can be carried out when we use one of the RCC-8 relations $\{TPP, NTPP^{-1}\}$, $\{TPP, TPP^{-1}\}$, $\{NTPP, NTPP^{-1}\}$, $\{NTPP, TPP^{-1}\}$ or $\{TPP, NTPP, TPP^{-1}, NTPP^{-1}\}$ instead of $\{PP, PP^{-1}\}$, so these sets are also NP-hard. The number of intractable subsets can be increased by using an additional property [NB95].

Theorem 4. RSAT($\widehat{S}$) *can be polynomially reduced to* RSAT(S)

Corollary 2. *Let S be a subset of* RCC-8.

1. RSAT($\widehat{S}$) $\in$ P *if and only if* RSAT(S) $\in$ P.
2. RSAT(S) *is NP-hard if and only if* RSAT($\widehat{S}$) *is NP-hard.*

With this property, all sets of RCC-8 relations whose closure contains one of the five relations mentioned above are also intractable. By computing the closure of all sets containing all base relations plus one additional relation, it turned out that for 72 relations deciding consistency is NP-hard when one of them is added to the base relations.

Lemma 2. RSAT$(\mathcal{S})$ *is* NP-*hard for any subset* $\mathcal{S}$ *of* RCC-8 *containing all base relations together with one of the 72 relations of the following sets:*

$$\mathcal{N}_1 = \{R \mid \{\text{PO}\} \not\subseteq R \text{ and } (\{\text{TPP}, \text{TPP}^{-1}\} \subseteq R \text{ or } \{\text{NTPP}, \text{NTPP}^{-1}\} \subseteq R)\},$$
$$\mathcal{N}_2 = \{R \mid \{\text{PO}\} \not\subseteq R \text{ and } (\{\text{TPP}, \text{NTPP}^{-1}\} \subseteq R \text{ or } \{\text{TPP}^{-1}, \text{NTPP}\} \subseteq R)\}.$$

6.2 Tractable Subsets

In order to identify a set of RCC-8 relations as tractable, one either has to specify a particular algorithm for deciding consistency of this set, or find another tractable decision problem to which the consistency problem of the particular set can be reduced. We have chosen HORNSAT, the tractable satisfiability problem of propositional Horn formulas, i.e., those propositional formulas where each clause contains at most one positive literal. For this reduction we first reduce RSAT to SAT, the propositional satisfiability problem, and then identify the relations which are reduced to Horn formulas.

For reducing RSAT to SAT, we specify a transformation by which every instance of RSAT, i.e., every set of spatial formulas Θ, is transformed to a propositional formula. For this we will start from the modal encoding $m(\Theta)$ and the corresponding RCC-8-model $\mathcal{M}$. Every world w of level 0 of $\mathcal{M}$ together with every spatial region X results in a propositional atom X_w. In order to preserve the structure of the RCC-8-model in the propositional formula, the $2n$ worlds of level 1 of every level 0 world w are transformed to propositional atoms X_w^i for $i = 1, .., 2n$. Using these atoms, every model and every entailment constraint can be transformed to a propositional formula. Additionally, the properties of the I-operator, i.e., reflexivity and transitivity and the m_2-formulas, also have to be transformed to a propositional formula. It turns out that all these formulas can be written as Horn formulas. As some of the model constraints can be transformed to indefinite Horn formulas, i.e., formulas where all clauses contain only negative literals, disjunctions of these constraints with any other constraint can also be transformed to Horn formulas. Thus every relation that can be written as a conjunction of constraints and Horn transformable disjunctions of constraints can be transformed to a Horn formula. For the set of these relations deciding consistency is thereby tractable. This set consists of 64 different relations and is denoted $\mathcal{H}_8$. Because of Corollary 2, the closure $\widehat{\mathcal{H}}_8$ of $\mathcal{H}_8$ is also tractable.

Lemma 3. RSAT$(\widehat{\mathcal{H}}_8)$ *can be polynomially reduced to* HORNSAT.

The reduction to HORNSAT is not possible for the reduced RCC-8-model, as the transformation of the first part of m_2 does not result in a Horn formula.

Theorem 5. $\widehat{\mathcal{H}}_8$ *contains the following* 148 *relations:*

$$\widehat{\mathcal{H}}_8 = \text{RCC-8} \setminus (\mathcal{N}_1 \cup \mathcal{N}_2 \cup \mathcal{N}_3)$$

with $\mathcal{N}_1$ and $\mathcal{N}_2$ as defined in Lemma 2 and

$$\mathcal{N}_3 = \{R | \{\text{EQ}\} \subseteq R \text{ and } ((\{\text{NTPP}\} \subseteq R, \{\text{TPP}\} \not\subseteq R) \\ \text{or } (\{\text{NTPP}^{-1}\} \subseteq R, \{\text{TPP}^{-1}\} \not\subseteq R))\}.$$

For proving that $\widehat{\mathcal{H}}_8$ is a maximal tractable subset of RCC-8, we have to show that no relation of $\mathcal{N}_3$ can be added to $\widehat{\mathcal{H}}_8$ without making RSAT intractable. For relations of the sets $\mathcal{N}_1$ and $\mathcal{N}_2$ this is already known (see Lemma 2). The following Lemma can be proven by a computer assisted case-analysis.

Lemma 4. *The closure of every set containing $\widehat{\mathcal{H}}_8$ and one relation of $\mathcal{N}_3$ contains the relation $\{\text{EQ}, \text{NTPP}\}$.*

Therefore it is sufficient to prove NP-hardness of $\text{RSAT}(\widehat{\mathcal{H}}_8 \cup \{\text{EQ}, \text{NTPP}\})$ for showing that $\widehat{\mathcal{H}}_8$ is a maximal tractable subset of RCC-8.

Lemma 5. $\text{RSAT}(\widehat{\mathcal{H}}_8 \cup \{\text{EQ}, \text{NTPP}\})$ *is NP-hard.*

Proof Sketch. Transformation of 3SAT to $\text{RSAT}(\widehat{\mathcal{H}}_8 \cup \{\text{EQ}, \text{NTPP}\})$. $R_t = \{\text{NTPP}\}$ and $R_f = \{\text{EQ}\}$. Polarity constraints:

$$X_L\{\text{EC}, \text{NTPP}\}X_{\neg L}, Y_L\{\text{TPP}\}Y_{\neg L}, \\ X_L\{\text{TPP}, \text{NTPP}\}Y_{\neg L}, Y_L\{\text{EC}, \text{TPP}\}X_{\neg L},$$

Clause constraints for each clause $c = \{i, j, k\}$:

$$Y_i\{\text{NTPP}^{-1}\}X_j, Y_j\{\text{NTPP}^{-1}\}X_k, Y_k\{\text{NTPP}^{-1}\}X_i. \qquad \blacksquare$$

Theorem 6. $\widehat{\mathcal{H}}_8$ *is a maximal tractable subset of RCC-8.*

It has to be noted that there might be other maximal tractable subsets of RCC-8 that contain all base relations, since, e.g., $\text{RSAT}(\{\text{EQ}, \text{NTPP}\} \cup \mathcal{B})$ has not been shown to be NP-hard so far.

6.3 Applicability of Path-Consistency

The path-consistency method is a very popular approximation algorithm for deciding consistency of a Constraint Satisfaction Problem (CSP). It can be applied since RSAT is a CSP where variables are nodes and relations are arcs of the constraint graph and the domain of the variables is the topological space. The path-consistency method imposes path-consistency of a CSP by successively removing relations from all edges with the following operation until a fixed point is reached:

$$\forall k : R_{ij} \leftarrow R_{ij} \cap (R_{ik} \circ R_{kj})$$

where i, j, k are nodes and R_{ij} is the relation between i and j. The resulting CSP is equivalent to the original CSP with respect to consistency. If the empty relation occurs while performing this operation, the CSP is inconsistent, otherwise the resulting CSP is path-consistent. More advanced algorithms impose path-consistency in time $O(n^3)$ where n is the total number of nodes in the graph [MF85].

It has already been mentioned that the path-consistency method is sufficient for deciding consistency of sets of base relations. It can be shown that it is also sufficient for deciding consistency of sets of $\widehat{\mathcal{H}}_8$ relations. This is done by showing that the path-consistency method finds an inconsistency whenever positive unit resolution (PUR) resolves the empty clause from the corresponding propositional formula. The only way to get the empty clause is resolving a positive and a negative unit clause of the same variable. Since the Horn formulas that are used contain only a few types of different clauses, there are only a few ways to resolve unit clauses using PUR which were covered by a case-analysis. As PUR is refutation-complete for Horn formulas [HW74], it follows that the path-consistency method decides $\mathsf{RSAT}(\mathcal{H}_8)$. Using the proof of Theorem 4, it is possible to express every relation of $\widehat{\mathcal{H}}_8$ as a Horn formula. Then the following theorem can be proven.

Theorem 7. *The path-consistency method decides* $\mathsf{RSAT}(\widehat{\mathcal{H}}_8)$.

6.4 Applicability of the Maximal Tractable Subset

One obvious advantage of the maximal tractable subset $\widehat{\mathcal{H}}_8$ is that the path-consistency method can now be used to decide RSAT when only relations of $\widehat{\mathcal{H}}_8$ are used and not only when base relations are used.

As in the case of temporal reasoning, where the usage of the maximal tractable subset ORD-HORN has been extensively studied [Neb97], $\widehat{\mathcal{H}}_8$ can also be used to speed up backtracking algorithms for the general NP-complete RSAT problem. Previously, every spatial formula had to be refined to a base relation before the path-consistency method could be applied to decide consistency. In the worst case this has to be done for all possible refinements. Supposing that the relations are uniformly distributed, the average branching factor, i.e. the average number of different refinements of a single relation to relations of $\mathcal{B}$ is 4.0.

Using our results it is sufficient to make refinements of all relations to relations of $\widehat{\mathcal{H}}_8$. Except for four relations, every relation not contained in $\widehat{\mathcal{H}}_8$ can be expressed as a union of two relations of $\widehat{\mathcal{H}}_8$, the four relations can only be expressed as a union of three $\widehat{\mathcal{H}}_8$ relations. This reduces the average branching factor to 1.4375. Both branching factors are of course worst-case measures because the search space can be considerably reduced when path-consistency is used as a forward checking method [LR97].

The following table shows the worst-case running time for the average branching factors given above. All running times are computed as $b^{(n^2-n)/2}$ where b is the average branching factor and n the number of spatial variables contained in Θ. We assumed that 100.000 path-consistency checks can be performed per second.

#spatial variables	$\mathcal{B}$ (4.0)	$\widehat{\mathcal{B}}$ (2.5)	$\widehat{\mathcal{H}}_8$ (1.4375)
5	10sec	95msec	3msec
7	500days	38min	20msec
10	10^{14}years	10^6years	2min

Recent experiments have shown that consistency can be decided much faster than these numbers indicate. Almost all instances up to a problem size of 100 spatial variables can be solved in less than a second. Using $\widehat{\mathcal{H}}_8$ for the backtracking search turns out to be about twice as fast in average than using $\widehat{\mathcal{B}}$. Also a significantly larger number of difficult instances can be solved in reasonable time when $\widehat{\mathcal{H}}_8$ is used.

7 Summary

In this chapter we reported about our ongoing work on the cognitive, representational, and computational aspects of the Region Connection Calculus. We made an empirical investigation of whether or not people use similar topological information as in RCC-8 when conceptualizing spatial arrangements and found that RCC-8 is a good candidate for a cognitively adequate spatial relation system and that RCC-5 and other sub calculi of RCC-8 are not cognitively adequate. We introduced a new canonical model of RCC-8 that resulted from the encoding of RCC-8 in modal logic. This model was topologically interpreted which allows a more simple representation of regions than it is possible with the topological space as a canonical model. It could also be shown that a consistent set of relations always has a realization in any dimension $d \geq 3$ when regions are internally connected and $d \geq 1$ otherwise. The consistency problem of RCC-8 was shown to be intractable in general, but a maximal tractable subset of RCC-8 was identified. For this set the path-consistency method was proven to be sufficient for deciding consistency.

Open problems and further work on the topics of this chapter includes a more detailed analysis of the canonical model with respect to models of internally connected two-dimensional regions. Another open problem is whether the maximal tractable subclass we found is the only one containing all base relations. We are planning to make further empirical investigations on the cognitive validity of RCC-8. This includes studying the inferential cognitive adequacy of RCC-8 as well as examining whether the complexity results have any cognitive meaning.

Acknowledgments

We would like to thank Markus Knauff and Reinhold Rauh for their collaboration concerning the investigation of the cognitive aspects of RCC-8, Ronny Fehling and Thilo Weigel for their assistance in developing the software, and Markus Knauff and Fritz Wysotzki for their helpful comments on earlier versions of this chapter.

This research has been supported by DFG as part of the project FAST-QUAL-SPACE, which is part of the DFG special research effort on "Spatial Cognition."

References

[All83] James F. Allen. Maintaining knowledge about temporal intervals. *Communications of the ACM*, 26(11):832–843, November 1983.

[Ben94] Brandon Bennett. Spatial reasoning with propositional logic. In J. Doyle, E. Sandewall, and P. Torasso, editors, *Principles of Knowledge Representation and Reasoning: Proceedings of the 4th International Conference*, pages 51–62, Bonn, Germany, May 1994. Morgan Kaufmann.

[Ben95] Brandon Bennett. Modal logics for qualitative spatial reasoning. *Bulletin of the IGPL*, 4(1), 1995.

[Che80] Brian F. Chellas. *Modal Logic: An Introduction*. Cambridge University Press, Cambridge, UK, 1980.

[Coh97] Anthony G. Cohn. Qualitative spatial representation and reasoning techniques. In G. Brewka, C. Habel, and B. Nebel, editors, *KI-97: Advances in Artificial Intelligence*, volume 1303 of *Lecture Notes in Computer Science*, pages 1–30, Freiburg, Germany, 1997. Springer-Verlag.

[Ege91] Max J. Egenhofer. Reasoning about binary topological relations. In O. Günther and H.-J. Schek, editors, *Proceedings of the Second Symposium on Large Spatial Databases, SSD'91*, volume 525 of *Lecture Notes in Computer Science*, pages 143–160. Springer-Verlag, Berlin, Heidelberg, New York, 1991.

[Fit93] Melvin C. Fitting. Basic modal logic. In D. M. Gabbay, C. J. Hogger, and J. A. Robinson, editors, *Handbook of Logic in Artificial Intelligence and Logic Programming – Vol. 1: Logical Foundations*, pages 365–448. Oxford, Clarendon Press, 1993.

[GJ79] Michael R. Garey and David S. Johnson. *Computers and Intractability—A Guide to the Theory of NP-Completeness*. Freeman, San Francisco, CA, 1979.

[GPP95] Michelangelo Grigni, Dimitris Papadias, and Christos Papadimitriou. Topological inference. In *Proceedings of the 14th International Joint Conference on Artificial Intelligence*, pages 901–906, Montreal, Canada, August 1995.

[Grz51] Andrzej Grzegorczyk. Undecidability of some topological theories. *Fundamenta Mathematicae*, 38:137–152, 1951.

[HW74] L. Henschen and L. Wos. Unit refutations and Horn sets. *Journal of the Association for Computing Machinery*, 21:590–605, 1974.

[KRR97] Markus Knauff, Reinhold Rauh, and Jochen Renz. A cognitive assessment of topological spatial relations: Results from an empirical investigation. In *Proceedings of the 3rd International Conference on Spatial Information Theory (COSIT'97)*, volume 1329 of *Lecture Notes in Computer Science*, pages 193–206, 1997.

[KRS95] Markus Knauff, Reinhold Rauh, and Christoph Schlieder. Preferred mental models in qualitative spatial reasoning: A cognitive assessment of Allen's calculus. In *Proceedings of the Seventeenth Annual Conference of the Cognitive Science Society*, pages 200–205, Mahwah, NJ, 1995. Lawrence Erlbaum Associates.

[KRSS98] Markus Knauff, Reinhold Rauh, Christoph Schlieder, and Gerhard Strube. Mental models in spatial reasoning. In this volume, 1998.

[LR97] Peter Ladkin and Alexander Reinefeld. Fast algebraic methods for interval constraint problems. *Annals of Mathematics and Artificial Intelligence*, 19(3,4), 1997.

[MF85] Alan K. Mackworth and Eugene C. Freuder. The complexity of some polynomial network consistency algorithms for constraint satisfaction problems. *Artificial Intelligence*, 25:65–73, 1985.

[NB95] Bernhard Nebel and Hans-Jürgen Bürckert. Reasoning about temporal relations: A maximal tractable subclass of Allen's interval algebra. *Journal of the Association for Computing Machinery*, 42(1):43–66, January 1995.

[Neb95] Bernhard Nebel. Computational properties of qualitative spatial reasoning: First results. In I. Wachsmuth, C.-R. Rollinger, and W. Brauer, editors, *KI-95: Advances in Artificial Intelligence*, volume 981 of *Lecture Notes in Computer Science*, pages 233–244, Bielefeld, Germany, 1995. Springer-Verlag.

[Neb97] Bernhard Nebel. Solving hard qualitative temporal reasoning problems: Evaluating the efficiency of using the ORD-Horn class. *CONSTRAINTS*, 3(1):175–190, 1997.

[RCC92] David A. Randell, Zhan Cui, and Anthony G. Cohn. A spatial logic based on regions and connection. In B. Nebel, W. Swartout, and C. Rich, editors, *Principles of Knowledge Representation and Reasoning: Proceedings of the 3rd International Conference*, pages 165–176, Cambridge, MA, October 1992. Morgan Kaufmann.

[Ren98] Jochen Renz. A canonical model of the Region Connection Calculus. In *Principles of Knowledge Representation and Reasoning: Proceedings of the 6th International Conference*, Trento, Italy, June 1998.

[RN97] Jochen Renz and Bernhard Nebel. On the complexity of qualitative spatial reasoning: A maximal tractable fragment of the Region Connection Calculus. In *Proceedings of the 15th International Joint Conference on Artificial Intelligence*, pages 522–527, Nagoya, Japan, August 1997. Technical Report with full proofs available at www.informatik.uni-freiburg.de/~sppraum.

A Taxonomy of Spatial Knowledge for Navigation and Its Application to the Bremen Autonomous Wheelchair

Bernd Krieg-Brückner[1], Thomas Röfer[1],
Hans-Otto Carmesin[2], and Rolf Müller[1]

[1] Bremer Institut für Sichere Systeme, TZI, FB3,
Universität Bremen, Postfach 330440, D - 28334 Bremen
{bkb,roefer,rmueller}@informatik.uni-bremen.de
[2] FB1, Universität Bremen, Postfach 330440, D - 28334 Bremen
carmesin@physik.uni-bremen.de

Abstract. A new taxonomy is proposed that relates different navigational behaviors in a hierarchical and compositional way. Elementary navigation tactics are combined to tactical navigation in routes; landmarks in space are contrasted to routemarks in networks of passages. Survey knowledge comes in at the level of strategic navigation. The Bremen Autonomous Wheelchair is then presented as a vehicle for experimentation in robotics, both to model biologically plausible navigational behaviors and to develop efficient navigational mechanisms for a technical application. The implementation on the autonomous system is based on the use of basic behaviors and the identification of routemarks. The actual recognition of artificial routemarks is described and early results of the current work on the identification of natural 3-D marks are presented.

1 Introduction

Perception, Spatial Knowledge and Navigational Behavior. Spatial knowledge, its representation and acquisition are intimately related with perceptual information on the one hand, and intended navigational behavior on the other. We propose a ternary relation between sensory performance, behavioral performance and spatial knowledge, cf. the "Spatial Cognition Triangle" in Fig. 1. When analyzing a specific animal or human behavior, sensory and behavioral performance are given and one can try to model the mechanisms for spatial cognition. When designing an autonomous robot, the task may be to develop navigation techniques for given sensory equipment and desired behavioral performance; alternatively, one can try to develop the best sensor to achieve optimal behavior, or ask, for given techniques and desired behavior, what minimal sensory equipment might do.

Objectives of the Taxonomy. The taxonomy proposed here relates navigational behaviors, i.e. observable behavior associated to conjectured spatial

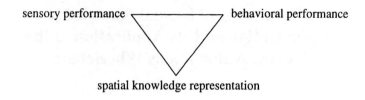

Fig. 1. The Spatial Cognition Triangle.

knowledge representations, where particular sensory equipment is given. While primarily intended to describe navigational behavior, it might well be used to structure our thinking about spatial knowledge in general. It aspires to cover, to a large extent, animal and human behavior, as well as biologically inspired robotics. Differently from Trullier *et al.* (1997) and similarly to Kuipers (1998), the taxonomy structures the field of navigation as a hierarchy.

In the context of the interdisciplinary DFG priority program on Spatial Cognition, it is intended to serve several purposes: a common framework for

1. theoretical modeling and reasoning unifying concepts and terminology of biology, psychology, artificial intelligence and robotics,
2. empirical investigations, e.g. about navigation performance and conjectured mental representations in animals or humans, and
3. experimentation with robots.

Although animal, human and robot navigation differ substantially in many respects, two recent workshops at Göttingen (Werner *et al.*, 1997) and Berlin (Krieg-Brückner, 1998) have revealed that the essential navigational issues are the same. Psychologists and biologists are concerned with understanding the mechanisms that enable humans and other animals to navigate, while the eventual objective in robotics research is to develop navigational skills for technical applications. Empirical evaluation focuses on efficiency and robustness of implemented mechanisms whereas empirical data is used in biological and psychological research to infer the underlying processes and mental representations.

Thus the synthetic and the analytic approaches complement one another: biological systems can be used as an inspiration for developing robots to perform similar tasks; moreover, synthetic approaches can be used to model biologically plausible navigational behaviors, isolating specific aspects and test hypotheses generated in biological and psychological research. On the other hand, robotics research may generate questions to be empirically investigated in the complex environment of biological systems. Synthetic approaches make the technical problems, representations and algorithms explicit that are associated with particular navigational behaviors; thus psychological theories can be restricted to computationally and biologically plausible models.

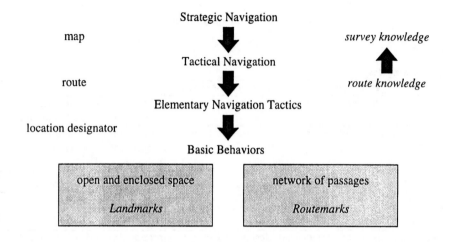

Fig. 2. Hierarchy of navigation behaviors.

The Bremen Autonomous Wheelchair. After a presentation of the taxonomy, it is applied to robot navigation. We present the Bremen Autonomous Wheelchair as a vehicle for experimentation, both to model biologically plausible navigational behaviors and to develop efficient navigational mechanisms for a technical application.

2 A Taxonomy of Spatial Knowledge for Navigation

2.1 Overview

Spatial knowledge about an environment can be separated into location, route and survey knowledge. A *location* is often characterized by a view of the surroundings from this position (Schweizer and Janzen, 1996; Schölkopf and Mallot, 1995; Franz *et al.*, 1997a,b), also sometimes called a "snapshot" (Cartwright and Collet, 1983). These views are described as consisting of a constellation of landmarks (Collet and Kelber, 1988) or cues (Poucet, 1993) that identify the location. A *route* is often identified with a sequence of locations or views (Schweizer and Janzen, 1996). *Survey knowledge* is obtained as an abstraction and integration of specific routes. The mental representation of survey knowledge is referred to as "cognitive map" (Tolman, 1948; Scholl, 1987; Poucet, 1993); e.g. in insect research, there has been a long discussion whether insects such as honeybees have a cognitive map (Gould, 1986) or not (Wehner and Menzel, 1990; Dyer, 1991).

The taxonomy keeps essentially the same levels in a hierarchy of navigational behaviors (cf. Fig. 2): starting from the top, *strategic navigation*, employing survey knowledge, is mapped to *tactical navigation* with routes; *elementary tactics* use *basic behaviors*. The classification leads to a fine-grain distinction of navigational behaviors; it should be emphasized that any actual agent—such as an animal—is likely to use a combination in order to increase robustness.

Table 1. Basic navigation behaviors.

approaching target

Title	Percept	Representation	Action
reaching target	remaining distance	distance	when remaining distance near Zero, trigger new behaviour
heading for target	view (of target)	designated view	when view corresponds to designated view, trigger new behaviour
stopping			stop (at target)
basic behaviour in (enclosed or open) space			
course following	course; orientation	course (direction)	adjust orientation to course, steer clear of obstacles
docking at target	position, orientation relative to target	designated target position, orientation	manoeuvre into target position and orientation
basic behaviour in network of passages (tunnels, corridors, roads, trails)			
passage following	walls, obstacles		follow passage centred between walls, avoiding obstacles
wall following {left | right}	wall, corners, obstacles	{left | right}	follow {left | right} wall (around corners), avoiding obstacles
turning into a passage	junction with n branches	branch designator	turn into designated passage

Landmarks and routemarks are distinguished; the actual objects in view may be the same, but their tactical use is different. *Landmarks* are used to determine the position of an agent in enclosed or open space. From one distant landmark, e.g. a church squire, an agent can determine its rotation but it needs at least three landmarks to exactly determine its position in 2-D space. *Routemarks* are used to determine an agent's position along a route. As routes are only one-dimensional, a single routemark might sufficiently characterize the location of the agent.

2.2 Basic Behaviors

At the level of basic behaviors and elementary navigation tactics (cf. Tab. 1), two settings should be distinguished: *space*, either open or enclosed, and *networks of passages*.

For example, basic behaviors in a passage, e.g. a tunnel, a corridor, a road, a trail, may be *wall-following, passage-following*, i.e. centered between walls while avoiding obstacles (Nehmzow, 1995), *turning* into a designated passage at a junction, etc. Basic behaviors are atomic tactics, the simplest to be investigated—or realized in a robot. They are used in more complex elementary tactics.

Table 2. Elementary navigation tactics in space.

searching

Title	Percept	Representation	Action
spiralling	locality in space	designated view	spiral outwards, heading for target
meandering	locality in space	designated view	meander, heading for target
vectorial navigation			
directional navigation	(compass) direction; orientation, elapsed time, speed	target vector	compute course, distance to target; follow course, approach target
dead reckoning	homing vector; self-movement	target vector	compute course, distance to target; follow course, approach target
positional navigation			
landmark navigation	view of landmarks, orientation	target position in view of landmarks	triangulate vector to target, navigate vectorially
celestial navigation	view of moving celestial bodies, orientation	target position in view or coordinate system	triangulate vector to target, navigate vectorially

2.3 Elementary Navigation Tactics

Elementary navigation tactics in space (cf. Tab. 2) include *directional navigation* guided by a compass (Wiltschko and Wiltschko, 1995; Collet and Baron, 1994), or *dead reckoning* (Gallistel, 1990; Müller and Wehner, 1988; Mittelstaedt and Mittelstaedt, 1982) using a homing vector accumulated from self-movement. The major knowledge representation for such navigation tactics in space is a *target vector* (direction and distance) that may have been derived from a map, i.e. survey knowledge, learned from a previous experience or communicated from other agents, e.g. for honeybees via the bee language (v. Frisch, 1967), transposed from a vertical 2-D notation in the hive to a horizontal situation in space. During actual navigation, the course is then computed relative to a compass direction, e.g. a magnetic compass, the polarization of light (Rossel and Wehner, 1986) in a hazy sky, or the direction of one particular landmark such as the sun.

Alternatively, the representation may be a particular configuration of several *global landmarks*, i.e. a *view*, used to determine a relative target position by triangulation; thus, a *view* from a particular location corresponds to a particular configuration of landmarks. Honeybees compare the original, learned view to the present view in order to determine the vector for "homing in" (cf. Fig. 7), both from a defined direction (Cartwright and Collet, 1982, 1983, 1987; Collet et al., 1986). In *landmark navigation*, these landmarks are fixed, e.g. prominent buildings, trees, lighthouses. In *celestial navigation* (Wehner, 1983), landmarks

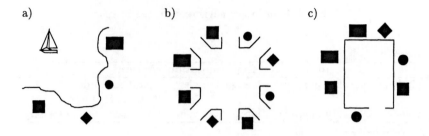

Fig. 3. Landmarks in open and enclosed space: a) Ocean. b) Town square. c) Room.

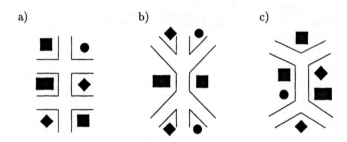

Fig. 4. Routemarks in networks of passages. a) Grid. b) Graph. c) "Hexatown" (Gillner and Mallot, 1997).

are moving over the day, e.g. the sun, and over the year, e.g. the stars, and these movements have to be compensated for by an internal clock (circadian rhythm).

It is interesting to note that navigation tactics in space seem to be equally applicable to open or enclosed space, e.g. an ocean, a town square, a room (cf. Fig. 3).

Elementary navigation tactics in passages (cf. Tab. 3) can take advantage of the limited alternatives within a passage, e.g. *binary* or *n-way branching*. A junction constitutes a *decision point* for the target passage and the switch to a new behavior. Characteristic *views*, depending on the direction from which the junction was approached, may be associated with the decision of selecting a particular target passage (Schölkopf and Mallot, 1995; Franz *et al.*, 1997a). Such a view may be characterised by *local landmarks* or *routemarks* (cf. Fig. 4) such as prominent visual objects, odors, sounds or tactile percepts, as long as they are stable, fixed and persistent.

Task Dependency. Elementary *navigation steps* are highly task dependent. This can be modeled as a pair **(tactic, location designator)**, where the **location designator** captures the spatial knowledge needed to instantiate the general tactic and determines a tactical decision. Thus the tactic represents the

Table 3. Elementary navigation tactics in networks of passages.

searching

Title	Percept	Representation	Action
quarter searching	locality in rectangular grid of passages	designated view	search each block in vicinity by spiralling or meandering

branching

Title	Percept	Representation	Action
binary branching	bifurcation	{left | right}	bear {left | right}
rectangular branching	intersection in rectangular grid	{left | straight | right}	follow passage approaching and across junction | bear {left | right}
directional branching	intersection in directional grid, direction	{N | E | S | W}	follow passage approaching and across junction | bear {left | right}
n-way branching	n-way junction	i	follow passage approaching junction; turn into i-th branch
designated branching	view of n-way junction	[view,] branch designator	follow passage approaching junction; turn into designated branch

procedural knowledge (action) and the location designator the corresponding data. The actual representation of a location designator and its reference system depends on the particular tactic. For example, for navigation in space, a vector to the target relative to the present position and orientation, or a vector in a global coordinate system, could be used; for branching, a particular view or branch designator would suffice.

2.4 Tactical Navigation along Routes

Generalized Routes. In general, a *(heterogeneous) route* consists of a sequence of arbitrary navigation steps: ⟨(**tactic, location designator**)⟩. So far, most authors regard a route just as a sequence of locations. This simplistic view only applies when the same tactic can be applied; then a sequence of location designators can be taken, where each location designator only represents the view (knowledge about a location and tactical decision) in the direction of progress along the route (cf. also Sect. 2.5). Thus a *homogeneous route* is a pair (**tactic,** ⟨**location designator**⟩).

Tactical Navigation. Different tactics may be concatenated in *heterogeneous route navigation*: consider e.g. the honeybee heading for a feeding area with a target vector communicated from a sister (v. Frisch, 1967), then foraging by searching along a route that is learned and remembered for next time, thereafter

Table 4. Tactical navigation.

route navigation

Title	Percept	Representation	Action
homoge-neous route navigation	{locations}	homogenous route: tactic, ⟨location designator⟩	apply tactic to location designators in sequence, till end of route; stop/dock at target
heteroge-neous route navigation	{locations}	heterogeneous route: ⟨⟨tactic, location designator⟩⟩	⟨apply tactic to location designator⟩ in sequence, till end of route

explorative navigation

path finding	possibly obstructed space	target designator	navigate along passages towards designated target
threading	labyrinth of passages	target view	navigate along all passages, heading for target, constructing thread (inverse route), backtracking

combined tactical navigation

positional path finding	possibly obstructed space	target designator	combine positional navigation (e.g. using GPS) and path finding
traversing	possibly obstructed space	target designator	combine vectorial/positional (e.g. landmark) navigation and explorative navigation around obstacles

returning to the vicinity of the hive with an accumulated homing vector by dead reckoning, followed by searching for the hive, and finally homing in by landmark navigation based on a view of a landmark constellation that had been acquired before leaving (Cartwright and Collet, 1982, 1983, 1987; Wittmann, 1996). Similarly, navigation in a *network of passages*, say roads, may be interspersed with landmark navigation across an enclosed *space* such as a large crowded city square.

Operations on Tactics. Some authors postulate an *inversion* operator on tactics in routes (Herrmann *et al.*, 1998). This may be appropriate for humans, but is questionable for lower animals; it is definitely a problem for robots navigating with minimalistic tactics (cf. Sect. 5.3). A better approach to allow backtracking is to overlay a kind of *threading* (cf. Tab. 4) or computation of a *homing vector* (cf. Fig. 7), as insects and other animals do, while moving forward. Moreover, different navigation tactics may be *overlaid* to achieve a tactical goal, e.g. vectorial navigation and explorative navigation such as path finding in a maze of passages. The list of combined tactics in Tab. 4 is obviously incomplete and just states examples.

Table 5. Strategic navigation.

alternative route navigation

Title	Percept	Representation	Action
making a detour	(temporarily) obstructed route	route graph	construct route to target of obstructed route via evasive location; navigate with alternative route
making a shortcut	"longer" route	route graph	construct "shorter/direct" route to target of "longer" route; navigate with alternative route

premeditated navigation

Title	Percept	Representation	Action
route map navigation	route marked on map, {locations}	route on map	construct (concrete) route from abstract route; navigate with route
map navigation	source, target marked on map, {locations}	source, target marked on map	construct route map by abstract navigation on map; navigate with route map

2.5 Strategic Navigation, Route Graphs, and Maps

Strategic navigation includes planning. The most basic form seems to be a combination of routes—in particular in networks of passages—into a net or directed graph.

Locations and Route Graphs. There are two ways in which routes may be combined by matching locations, thus regarding a set of **location designators** as equivalent: *source matching* and *target matching*. Either two target **location designators** emanate from the same *source location* but denote different targets (in space, coming from the same source direction, the target vectors may differ; in a network of passages, the (view at the) source location, coming from the passage, is the same, but the branching designator may differ), or two target **location designators** lead to the same *target location* but potentially emanate from different sources (in a network of passages, the views of the target location can be made coincidental by rotation). Thus matching leads to a notion of (abstracted) **location** as a node in a *route graph*, in which the edges are labeled with respective navigation tactics. In this modelling, a location is still represented by a position, i.e. a location does not have an extent.

It is not clear, to what extent a route graph really represents (or requires) *survey knowledge* in the sense of changing the point of reference from a field perspective to that of an observer (Herrmann *et al.*, 1998). A simple (partial) route graph, in a field perspective, may do for *alternative route navigation* to plan for detours or shortcuts (cf. Tab. 5); possibly, only minimal topological information about directions (quasi-survey knowledge) is necessary when a navigation tactic in space is overlaid (cf. Sect. 2.4).

Maps. In contrast, we see a map as a generalization of the concept of route graph where the abstraction goes along with a change of perspective. A *map* is a pair of a set of **location abstractions** as nodes and a set of **tactic abstractions** as edges, i.e. ({**tactic abstraction**}, {**location abstraction**}). Different tactical aspects of navigation may lead to different maps with different kinds of abstractions for tactic-oriented spatial knowledge contained in locations, e.g. by introducing topological or Cartesian relations (Barkowsky *et al.*, 1997), cf. also (Berendt *et al.*, 1998). This abstract information contained in a map (or an overlay of several maps) will then have to be sufficient for reconstructing routes and for planning shortcuts or detours around obstacles. A map need not be global, but may be used locally ("local chart") (Poucet, 1993); the reference system may change according to the tactic applied.

In contrast to alternative route navigation, *Premeditated navigation* requires a map as a first class object, trying to invert the previous process of abstraction before navigating in the field—or even while going along.

3   The Bremen Autonomous Wheelchair

A wheelchair is used as an experimental robot platform in Bremen. It has four wheels. The front axle drives the wheelchair while the back axle is used for steering. Therefore, the wheelchair moves like a car driving backwards. It is equipped with a Pentium 100 computer, twelve bumpers, six infrared sensors, 16 ultrasonic sensors and a camera. The infrared sensors can only detect whether there is an obstacle within a radius of approximately 15 cm; however, they cannot measure the distance. Two different kinds of ultrasonic sensors are fitted to the wheelchair: half of the sensors have an opening-angle of 80° while the other half only measure in a range of 7°. In addition, the wheelchair can measure the rotations of its front wheels. Thus, it is able to perform dead reckoning.

The sensors are assigned to four control subsystems. Three of these systems are illustrated in Fig. 5:

Collision Detection. The wheelchair uses all bumpers, all infrared sensors and the wide-angle ultrasonic sensors to detect collisions with the environment. As long as an obstacle is perceived by the infrared and ultrasonic sensors, respectively, the wheelchair is able to stop before physical contact is made.

Steering Restriction. As the wheelchair is steering with its back wheels, its rear swings out very heavily. To prevent it from colliding with obstacles on the side during driving maneuvers, the distance to the closest obstacle is measured and the steering angle is reduced as much as necessary to avoid a collision.

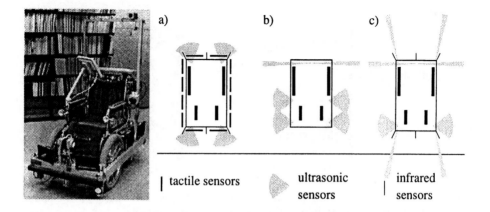

Fig. 5. The Bremen Autonomous Wheelchair (first prototype) and its sensor control subsystems. a) Collision detection. b) Steering restriction. c) Navigation.

Navigation. Six of the narrow-angle ultrasonic sensors and the six infrared sensors are employed for navigation purposes. They have been chosen because their measurements do not only reflect a certain distance to an obstacle but also determine it in a definite direction. In contrast, the wide-angle opened ultrasonic sensors would not allow a precise localization. At the moment, these sensors are supported by two wide-angle sensors to the side.

Landmark or Routemark Detection. The camera is used to scan the surroundings for landmarks or routemarks, respectively. It is mounted on a pan-tilt-head. Therefore, it can watch the environment independently from the current orientation of the wheelchair.

4 Navigation in Space

The Parti-Game Algorithm, originally developed by Moore and Atkeson (1995) was one of the first navigation algorithms adapted to the Bremen Autonomous Wheelchair (Kollmann *et al.*, 1997). Based on simple collision detection and self-localization, it allows an agent to navigate between positions between which the *direct* connection is blocked by obstacles. It divides the space of all possible states into partitions of different size. The granularity of the partitioning depends on the narrowness or spaciousness of a certain part of the state-space, i.e. the number of obstacles present in this region. On the other hand, the Parti-Game algorithm determines whether it is possible to move from one partition to another.

The Parti-Game algorithm is a very artificial navigation approach because most animals do not have a global self-localization system. Nevertheless, birds are assumed to have a global "grid-map" that may be based on the earth's

a)

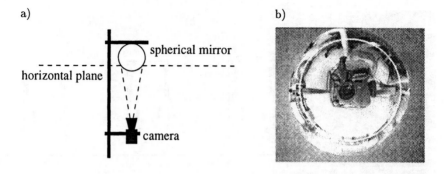

Fig. 6. Apparatus for taking panoramic images. a) Scheme. b) Image on the sphere. The circle indicates the mapping region of the horizontal plane.

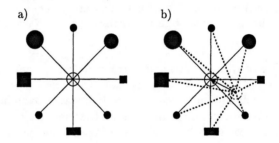

Fig. 7. Landmark navigation of honeybees. a) The learned landmark constellation. b) Calculating the homing direction from deviations.

magnetic field (Wiltschko, 1997). We hope to adopt this approach to combined navigation, e.g. traversing a room or a market place (cf. Tab. 4).

The image-based navigation approach (Röfer, 1995a) is inspired by the homing behavior of honeybees (Cartwright and Collet, 1982, 1983, 1987). It has been implemented using the simulation SimRobot (Siems *et al.*, 1994; Röfer, 1995b) and on the Bremen Autonomous Wheelchair utilizing the panoramic sensor shown in Fig. 6 (Röfer, 1997, 1998). To enable the method to be used in indoor navigation, the implementation goes beyond the bee's homing behavior in two respects: it is able to determine the rotation as well as the direction to the target position, and it learns and autonomously reproduces complete routes. The peculiarity of this approach is that no associations between perceptions and actions are learned. Instead, the actions are directly calculated from a stored view and the current one (cf. Fig. 7). Locations are characterized by panoramic views and routes are represented as sequences of these views—as described in Sect. 2.4.

The learned routes can be combined to a route graph. This enables the wheelchair to concatenate several routes to a longer one in order to drive between positions between which the direct route has not been trained. Nevertheless, the wheelchair is not able to construct new routes since it has no knowledge about the spatial relationship of the learned routes.

5 Route Navigation in Passages

An architecture with several layers has been chosen for the control system of the present Bremen Autonomous Wheelchair. In contrast to the aforementioned navigation approaches that have been inspired by biological findings or the reinforcement learning theory, the current approach follows the taxonomy presented in Sect. 2. It consists of the levels "basic behaviors" and "route knowledge".

5.1 Basic Behaviors

Several basic behaviors, e.g., wall-following and turning-into-door, form the basis of the presented navigation method. They enable the wheelchair to move in corridors, and to enter and exit rooms. They are fairly robust against changes in the environment because they hardly ever assume fixed compositions of the surroundings. Their implementation uses the sensors of the navigation subsystem (cf. Sect. 3). Ultrasonic sensors have several weaknesses (Jörg et al., 1993). The signal that has been sent out by a sensor may not return to the same sensor if it hits a smooth surface diagonally (specular reflection) or is caught by another sensor (cross-talk).

Obstacle Map. As the wheelchair's sensors with the small opening-angle do not seem to produce cross-talks at all in their current arrangement but often miss smooth objects, it is not possible to implement the basic behaviors by a straight sensor-motor-linkage. Instead, the measured distances are recorded in an obstacle map (cf. Fig. 8). This map plots the local environment around the wheelchair and represents an area of 4×4 m^2. When the wheelchair drives on, the measurements in the map are shifted in the same way as the environment passes along the moving wheelchair. Everything that is scrolled out of the map is forgotten, as are measurements that are older than 30 seconds. This allows the wheelchair to cope with dynamic obstacles. Thus the local obstacle map corresponds to a short term memory.

The information in the obstacle map is abstracted by two *virtual sensors* in the map; their measurements are used for navigation. The virtual sensors work like ultrasonic sensors: they determine the distance to the closest obstacle in their measuring range. In contrast to real sensors, the virtual ones only measure distances to objects already represented in the obstacle map. Thus, they can exploit the sensory data that has been collected in the past 30 seconds; they may even detect an obstacle that the real sensors currently overlook. The two virtual sensors scan the map from a position that, in reality, is 10 cm in front

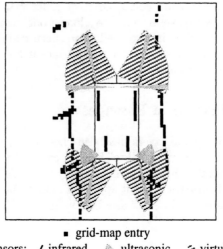

■ grid-map entry

sensors: ∕ infrared ◈ ultrasonic ⋰ virtual

Fig. 8. The local obstacle map.

of the wheelchair. One is oriented towards the left side; the other towards the right side (cf. Fig. 8).

Eight basic behaviors have been implemented:

Center between Walls. If this behavior is selected, the wheelchair will try to measure the same distance with both of the front virtual sensors. To achieve this, it always steers in the direction of the larger measurement. While performing this behavior, the wheelchair is driving forwards. As it is possible that an obstacle may have been overlooked, the collision detection subsystem perceives it and stops the wheelchair. In this case, the wheelchair performs the same behavior in backward direction for a distance of 50 cm and then returns to driving forwards. Normally, the infrared sensors have detected the missed obstacle during the collision. As their measurements have been recorded in the obstacle map, the wheelchair is aware of this obstacle during further actions. To perform the same behavior while driving backwards, the difference between the two virtual sensor readings simply has to be inverted.

Follow Left/Right Wall. Wall-following is realized by a slightly modified wall-centering behavior. The only difference is that the measurements of the virtual sensor that is opposite to the wall are limited to a maximum distance. Therefore, the wheelchair assumes that there is another wall within this distance and centers itself between this virtual wall on one side and the real one on the other side. If the real corridor is narrower than this virtual one, there is no change to the wall-centering behavior.

Turn into Left/Right Door. These two behaviors enable the wheelchair to turn into a door that is either in the left or the right wall. They are quite similar to the wall-following behaviors. The only difference is that on the side of the door, no virtual sensor is used. Instead, the measurement of the real narrow-angle ultrasonic sensor is used that is oriented toward this side. In this way, the hole between the door-jambs can be determined as soon as possible. If the wheelchair has turned more than 60°, it automatically switches to the wall-centering behavior.

Follow a Direction while Driving Forward/Backward. This behavior controls the wheelchair in a certain direction. As it is not equipped with a compass, the odometry is used as a reference system for the specification of the direction. The behavior can be performed by driving forward or by driving backward. It is based—as all behaviors that have been presented so far—on the centering between walls. As a result, the wheelchair is still able to avoid obstacles while it is driving in a certain direction. Similarly to wall-following, a virtual corridor is constructed from the given direction; the wheelchair centers itself in this virtual corridor. If the real corridor is narrower than this virtual one, there is no difference to the normal wall-centering behavior.

Stop. As it is always the goal of the wheelchair to reach a certain position, it has to stop when it has arrived at the target position.

5.2 Routemark Detection

The basic behaviors enable the wheelchair to *move* in an office environment. In order to also allow it to *navigate*, it must be able to localize itself in the environment. Therefore, it has to be capable of recognizing reference points in the surroundings, e.g. to switch to the next behavior. In the navigation approach that is presented in this article, these reference points are called routemarks because they are used to locate the wheelchair's position along a certain route. The long-term goal of the authors is to use some features of the environment's 3-D structure as routemarks. This will be described in Sect. 6. At the moment, only artificial 2-D marks are employed, determined by an image processing algorithm. These marks consist of a black circle on a white background. In the circle, there are up to four white, horizontal stripes interpreted as a scan-code. The recognition of the routemarks is illustrated in Fig. 9.

Routemark Map. As the sizes of the artificial routemarks are known, the distances to them—as with routemarks detected by 3-D vision—can be determined. Therefore, the position of a mark relative to the wheelchair can be calculated from its distance and its bearing. This allows the creation of another map: the routemark map. This map is similar to the obstacle map that has already been presented. When the wheelchair moves, the positions of the marks in the map

a)   b)   c)

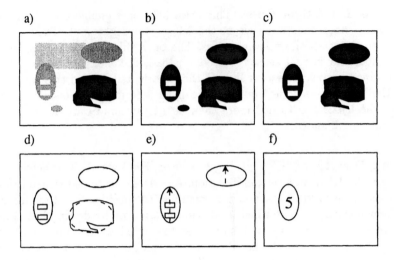

d)   e)   f)

Fig. 9. The recognition of artificial routemarks. a) Original image. b) Extracting dark areas. c) Ignoring small areas. d) Comparing outlines. e) Extracting scan-codes. f) Resulting routemark with identification number.

are adjusted on the basis of its dead reckoning system. As the obstacle map, the routemark map is local to the wheelchair, i.e. it only maps a region of a radius of 5 m around the vehicle. All routemarks that move out of this radius are forgotten. Through the routemark map, the wheelchair knows the positions of all marks in its vicinity, even if its camera does not currently see them. Hence, it represents a *view*, i.e. the current routemark constellation of a certain place.

5.3   Route Knowledge

In the second layer of the control hierarchy, the basic behaviors and the routemark recognition are combined. On this tactical level, the environment is represented as a set of routes. A route is a static way from a starting location to a target location. The wheelchair can drive along such a route by a concatenation of different basic behaviors while routemarks are used to trigger the starting and changing of these behaviors (cf. Fig. 10). They are the reference marks along the route. Therefore, a route is represented as a sequence of basic behaviors and the routemarks that trigger these behaviors. This sequence can be learned by the wheelchair, e.g. when a teacher controls the vehicle along the route by switching between the available behaviors. Meanwhile, the camera scans the surroundings for routemarks and inserts them into the routemark map. When the teacher alters the wheelchair's behavior, it stores the view of the routemark map as the trigger for the new operation.

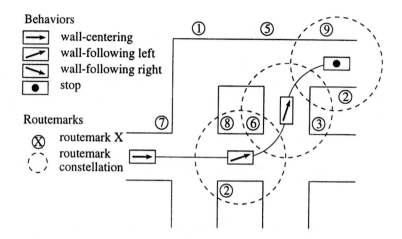

Fig. 10. Teaching of routes.

Multiple Routemarks. If the wheelchair drives along the learned route autonomously, it executes each recorded behavior until it recognizes a part of the routemark constellation that it has stored during the training to trigger the change to the next behavior. As the wheelchair will never reach exactly the same positions during the training and the autonomous repetition, it cannot compare the stored routemark constellation with the current one directly. Instead, the intersection of the stored routemark constellation and the current state of the routemark map is determined. If at least one common routemark is found, it is checked whether the wheelchair has passed all these common routemarks. It has "passed" a routemark, if the current bearing of a mark is further back than the bearing of the stored mark. As soon as all common routemarks have been passed, the wheelchair switches to the next behavior of the route. This solution is robust against the absence of single marks. Nevertheless, it tries to reproduce the moment of switching the behaviors as precise as possible by using multiple marks.

As a single recognized routemark is sufficient for switching from one behavior to the next, behaviors might be switched too early if all routemarks of a certain place are stored because it is possible that the same marks have already been present earlier in this route segment. Since the decision when to switch to the next behavior is based on the bearings of the marks, it is necessary to use only those marks that never had a bearing further back during the route segment than at its end.

Errors. In spite of the use of multiple routemarks for the switching of behaviors, errors during the autonomous drive of the wheelchair cannot be ruled out. Two types of errors are possible:

1. All routemarks for switching a behavior have been missed (cf. Fig. 11b). This can happen, e.g. if a routemark is hidden by a person.

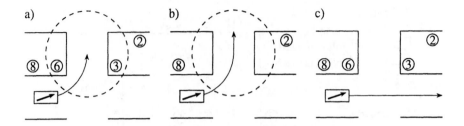

Fig. 11. Possible errors. a) Learned route segment. b) All routemarks missing. c) Erroneous carrying out of behavior.

2. During the autonomous drive, a behavior is performed that is different from the version that has been trained (cf. Fig. 11c). If, e.g., the wheelchair should follow the left wall and misses a turn-off because a person blocks the passage, the vehicle drives straight on where it should have turned left.

In both cases, two possibilities exist to terminate the erroneous behavior:

1. The execution of a behavior lasts more than 50% longer than during the training. To be able to determine this fact, the wheelchair measures the duration of behaviors. If such an overflow has been detected, it is assumed that there was an error in the reproduction of the learned route.
2. The wheelchair finds different routemarks of the same type and switches to the next behavior. In this case, the erroneous execution is not detected immediately but it can be expected that one of the subsequent behaviors cannot be completed and therefore will fail.

So, an error can only be determined if a behavior lasts too long, but the cause of this error can also be a defective execution of a previous behavior. Thus, a strategy for compensating errors must be able to correct errors in previous behaviors. If the wheelchair has missed some routemarks, it can track back the route segment and search again. If it has made an error in carrying out a behavior, it has no other possibility than to drive back to a position before the error and to try again. As it does not know where the error occurred, it drives back the last route segment and searches for the missing routemarks; then it repeats this segment. If this again results in an error, it drives back the last two segments and repeats them etc. If it has repeated the complete previous route without being successful, the execution of the route fails. Such failures can only be recovered from on the level of survey knowledge, e.g. by also back-tracking the previous route or by using a different one.

Backtracking. A straightforward approach for the implementation of backtracking along a route segment would be the use of "inverse behaviors", i.e. the same behavior but performed in backward direction. The problem of such an

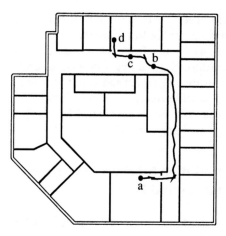

Fig. 12. A measured trajectory that has been generated by the combination of four basic behaviors. a) Follow left wall. b) Follow right wall. c) Turn into right door. d) Stop.

approach is that a behavior can be carried out incorrectly in forward direction but its inverse behavior may be executed accurately. In this case, the inverse behavior would not drive the wheelchair back to its starting-point and therefore, it would not compensate the forward behavior. To really cancel a behavior, the wheelchair uses a threading tactic: it records the positions of is dead reckoning system during the execution of the behaviors and then uses the "follow a direction while driving backward"-behavior to drive back the segment position by position. This allows it to cancel any behavior, even if the behavior was performed incorrectly.

As soon as the recognition of routemarks is not only seen as a binary decision but instead as a process with a particular uncertainty, the representation of knowledge becomes more complex. To develop a solution for this problem, several psychological findings could be employed, e.g. expected routemarks should be detected with a higher probability than unexpected ones. Marks in the same constellation could support each other in the recognition process.

5.4   Results

Although the wheelchair is 72 cm wide and 134 cm long, the basic behaviors have enabled it to drive through 94 cm wide door-frames and to turn into doors in a 176 cm wide corridor. Figure 12 shows a trajectory that has been obtained from a combination of four different basic behaviors. The trajectory has been recorded by the wheelchair's onboard odometry. As the odometers are not precise enough to dead reckon over 30 m, the recorded trajectory has been adjusted manually to allow the visualization in the floor-plan. Some obstacles are not shown in the plan, e.g. there is a coat-rack in the corridor that the wheelchair has bypassed.

The wheelchair has been able to learn and to repeat a route consisting of four basic behaviors. It has been able to recover from both types of possible errors, i.e. from missing routemarks and from the erroneous performance of behaviors. Both errors have been forced by hiding routemarks and blocking passages. An important result for future work with natural 3-D routemarks is that it is necessary to always have a wide selection of routemarks because the worst thing that can happen is that the behavior should be changed and no appropriate routemark to trigger this change is available.

6 Structure from Motion for the Visual Detection of Natural Landmarks and Routemarks

The aforementioned navigation approach is based on the recognition of routemarks. As an intermediate step, artificial routemarks are used that are easy to detect. Nevertheless, the long-term goal of the authors is to use natural features as routemarks. To this end it is desirable that the system can use local structures typically present in its surrounding as natural routemarks or landmarks. Such natural landmarks or routemarks should be recognized in a robust manner, independent of orientation, illumination or surface variations in dynamic environments. It is intended to use significant 3-D structures such as corners, doors or boxes. In the following such small 3-D structures used as landmarks or routemarks are denoted as semi-local 3-D marks, or 3-D marks for short.

The visual estimation of depth in a small (semi-local) image region is a fundamental prerequisite for the detection of such semi-local 3-D marks. The visual estimation of depth is usually based on at least two images. These are either taken simultaneously in stereo vision or subsequently in so-called "structure from motion" approaches (Barron, 1984; Aggarwal and Nandhakumar, 1988; Heeger and Jepson, 1992). Here, the second approach is used because utilizing a single camera is cheaper and takes less calibration effort, and structure from motion is computationally very efficient and thus advantageous for real time applications.

6.1 Method

The approach consists of two major steps: the estimation of the so-called *focus of expansion* which is the intersection of the heading direction vector and the image plane, and the subsequent estimation of *depth*.

The focus of expansion is estimated by a method that was developed in the authors' group. It was described in detail by Herwig (1996). From two consecutive camera images, the intensity derivatives with respect to time t and spatial coordinates x and y are estimated. Assuming short-time constancy of brightness of a point in the observed scene, the normal optical flow field (v. Helmholtz, 1909) is estimated from those derivatives. The focus of expansion is estimated from the normal optical flow field by minimizing a cost function. This method has been shown to be robust with respect to observed scenes and noise, moreover it is computationally efficient and has been used in real time robotics applications.

In the second step, the ratio of local image depth to the observer's speed is estimated from the normal optical flow field and from the position of the focus of expansion, as Herwig (1996) proposed.

6.2 Computer Simulations

Currently, this method of estimating distance from image sequences is investigated with computer simulations using SimRobot (Siems *et al.*, 1994).

From these images, the depth is estimated in the way described above. Presently, the influence of the algorithm's parameters and of the experimental settings on the estimated depth values is analyzed. The most essential parameters of the algorithm are upper and lower thresholds for the intensity derivatives. Low values of intensity derivatives are likely to enhance the effect of noise. High values of intensity derivatives maybe due to quantization errors; in such cases they cause inadequate flow field estimations. Moreover, the norm of the intensity gradient vector is required to be high enough in order to provide sufficient correspondence of the optical flow vector and the velocity vector of an image point (Verri and Poggio, 1987; Aloimonos *et al.*, 1993). Image points are not processed if their intensity derivatives do not match these criteria. Experimental settings include the walls' texture, the observer's speed and the resolution and noise intensity of the camera image.

6.3 Early Results

The results of typical simulations are shown in Fig. 13 and Fig. 14.

To judge the algorithm's performance for depth estimation, the camera approaches a single wall perpendicularly. In this setup, depth values are equal for each image point. The results of such an experiment are shown in Fig. 13. The trial started at a distance of 80 arbitrary units. Depth was estimated from consecutive pairs of images. Diamonds indicate the mean value of estimated depth of all image points. The error bars indicate the standard deviation. As this is the result of a simulation, the true distance is known. It is marked by the dashed line.

In the experiment shown in Fig. 14, the camera approaches walls that are arranged as a step in distance. This simulation's setup is sketched in part (a) of the figure. Part (b) shows the camera's view. The walls' texture consists of a quadratic grid of isotropic patches. The patches' intensity is maximal in their center, dropping towards the periphery according to a cosine function. Part (c) of the figure shows the estimated depth as grayscale values. The crosshair marks the estimated position of the focus of expansion. One may easily identify a depth-discontinuity from Fig. 14c.

6.4 Future Perspective

As mentioned before, the long-term goal is to use three-dimensional structures as landmarks or routemarks for the navigation of the Bremen Autonomous

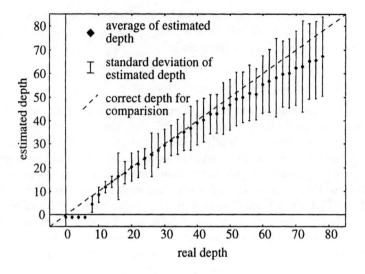

Fig. 13. Estimated depth vs. real depth.

Wheelchair. To achieve this, the distance images will have to be segmented into areas of nearly equal relative motion. In each such area, discontinuities of depth, and discontinuities and extrema of depth derivatives should be recognized. Significant depth discontinuities or extrema should be automatically selected and utilized as 3-D marks.

7 Conclusion

The taxonomy relates different navigational behaviors in a hierarchical and compositional way. It remains to be seen whether the distinction between navigation in space, using landmarks, and in networks of passages, using routemarks, is useful in psychology and artificial intelligence or robotics alike. The abstraction to survey knowledge, strategic navigation, dealing with exceptions etc. still require investigation. We hope for cooperation and feedback from research in psychology as to a separation of overlaid navigation tactics (conjectured in Sect. 2.4), e.g. vectorial navigation and route navigation in a city, such that the former allows for correction and recovery when the latter fails, e.g., due to a blocked passage.

The Bremen Autonomous Wheelchair has been a useful vehicle for experimentation, not only to model biologically plausible navigational behaviors, but also, apparently, to develop efficient navigational mechanisms for a realistic technical application; this will still have to be shown in applications with users. A robust implementation of some basic behaviors and the recognition of artificial routemarks enable the wheelchair to autonomously follow trained routes. We expect still better results with the second prototype, presently developed in cooperation with Meyra, the market leader for wheelchairs in Germany.

a)   b)   c)

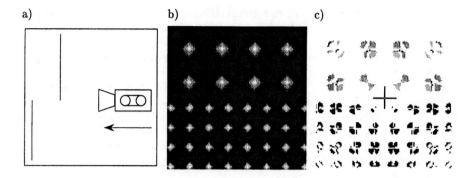

Fig. 14. A simulated camera approaching a distance step. a) Scheme of setup. b) Camera view. c) Estimated depth coded as brightness. Brighter areas correspond to farther surfaces.

In the future, the identification of the artificial routemarks will be replaced by the recognition of natural 3-D landmarks or routemarks. Early results in this field indicate that the presented algorithm is capable of estimating depth in real time, with high spatial resolution and robustly with respect to illumination, orientation, and noise.

References

Aggarwal, J. K. and Nandhakumar, N. (1988). On the computation of motion from sequences of images – a review. In *Proc. of the IEEE*, volume 76, pages 917–935.

Aloimonos, Y., Rivlin, E., and Huang, L. (1993). Designing visual systems: Purposive navigation. In Y. Aloimonos, editor, *Active Perception*. Lawrence Erlbaum.

Barkowsky, T., Freksa, C., Berendt, B., and Kelter, S. (1997). Aspektkarten - Integriert räumlich-symbolische Repräsentationsstrukturen. In C. Umbach, M. Grabski, and R. Hörnig, editors, *Perspektive in Sprache und Raum*, Wiesbaden. Deutscher Universitätsverlag. To appear.

Barron, J. (1984). A survey of approaches for determining optic flow, environmental layout and egomotion. In *Tech. Report RBCV-TR-84-5*, University of Toronto. Dept. of Computer Science.

Berendt, B., Barkowsky, T., Freksa, C., and Kelter, S. (1998). Spatial representation with aspect maps. This volume.

Cartwright, B. A. and Collet, T. S. (1982). How honey bees use landmarks to guide their return to a food source. In *Nature*, volume 295, pages 560–564.

Cartwright, B. A. and Collet, T. S. (1983). Landmark learning in bees. *Journal of Comparative Physiology A*, **151**, 521–543.

Cartwright, B. A. and Collet, T. S. (1987). Landmark maps for honeybees. In *Biological Cybernetics*, volume 57, pages 85–93.

Collet, T. S. and Baron, J. (1994). Biological compasses and the coordinate frame of landmark memories in honeybees. In *Nature*, volume 368, pages 137–140.

Collet, T. S. and Kelber, A. (1988). The retrieval of visuo-spatial memories in honeybees. *Journal of Comparative Physiology A*, **163**, 145–150.

Collet, T. S., Cartwright, B. A., and Smith, B. A. (1986). Landmarks learning and visuo-spatial memories in gerbils. *Journal of Comparative Physiology A*, **158**, 835–851.

Dyer, F. C. (1991). Bees acquire route-based memories but not cognitive maps in familiar landscape. In *Animal Behavior*, volume 41, pages 239–246.

Franz, M. O., Schölkopf, B., and Bülthoff, H. H. (1997a). Homing by parameterized scene matching. In *Proc. 4th Europ. Conf. on Artificial Life*. To appear.

Franz, M. O., Schölkopf, B., Georg, P., Mallot, H. A., and Bülthoff, H. H. (1997b). Learning view graphs for robot navigation. In W. L. Johnson, editor, *Proc. 1st Int. Conf. on Autonomous Agents*, pages 138–147, New York. ACM Press.

Gallistel, C. R. (1990). *The organization of learning*. MIT Press, Cambridge, Massachusetts.

Gillner, S. and Mallot, H. A. (1997). Navigation and acquisition of spatial knowledge in a virtual maze. Technical Report 45, Max-Planck-Institut für biologische Kybernetik, Tübingen.

Gould, J. L. (1986). The locale map of honeybees: Do insects have a cognitive map? In *Science*, volume 234, pages 861–863.

Heeger, D. J. and Jepson, A. J. (1992). Subspace methods for recovering rigid motion i: Algorithm and implementation. In *Int. J. of Computer Vision*, volume 7:2, pages 95–117.

Herrmann, T., Schweizer, K., Janzen, G., and Katz, S. (1998). The effect of route direction and its constraints. This volume.

Herwig, C. (1996). Visual motion processing for active observers. Ph.D. thesis. In B. Krieg-Brückner, G. Roth, and H. Schwegler, editors, *ZKW-Bericht*, number 1/96 in ISSN 0947-0204, Universität Bremen. Zentrum für Kognitionswissenschaften.

Jörg, K.-W., v. Puttkamer, E., and Richstein, H.-J. (1993). Integration und Fusion heterogener Multisensorinformation zur geometrischen Weltmodellierung für einen Autonomen Mobilen Roboter. In G. Schmidt, editor, *Autonome Mobile Systeme*, volume 9, pages 287–298, Technische Universität München.

Kollmann, J., Lankenau, A., Bühlmeier, A., Krieg-Brückner, B., and Röfer, T. (1997). Navigation of a kinematically restricted wheelchair by the parti-game algorithm. In *Spatial Reasoning in Mobile Robots and Animals*, pages 35–44, Manchester University. AISB-97 Workshop.

Krieg-Brückner, B. (1998). A taxonomie of spatial knowledge for navigation. In U. Schmid and F. Wysotzki, editors, *Qualitative and Quantitative Approaches to Spatial Inference and the Analysis of Movements*, number 98-2 in Technical Report, Technische Universität Berlin. Computer Science Department.

Kuipers, B. (1998). A hierarchy of qualitative representations for space. This volume.

Mittelstaedt, H. and Mittelstaedt, M.-L. (1982). Homing by path integration. In F. Papi and H. G. Wallraff, editors, *Avian Navigation*, pages 290–297.

Moore, A. and Atkeson, C. (1995). The parti-game algorithm for variable resolution reinforcement learning in multidimensional state-spaces. In *Machine Learning*, volume 21, pages 199–233, Boston. Kluwer Academic Publishers.

Müller, M. and Wehner, R. (1988). Path integration in desert ants, cataglyphis fortis. In *Proc. Natl. Acad. Sci. USA*, volume 85, pages 5287–5290.

Nehmzow, U. (1995). Animal and robot navigation. In *Robotics and Autonomous Systems*, volume 15:1-2, pages 71–81.

Poucet, B. (1993). Spatial cognitive maps in animals: New hypotheses on their structure and neural mechanisms. In *Psychological Review*, volume 100, pages 163–182.

Röfer, T. (1995a). Image based homing using a self-organizing feature map. In F. Fogelman-Soulie and P. Gallinari, editors, *Proc. Int. Conf. Artificial Neural Networks*, volume 1, pages 475–480. EC2 & Cie.

Röfer, T. (1995b). Navigation mit eindimensionalen 360° Bildern. In R. Dillmann, U. Rembold, and T. Lüth, editors, *Autonome Mobile Systeme*, volume 10 of *Informatik aktuell*, pages 193–202, Berlin, Heidelberg, New York. Springer.

Röfer, T. (1997). Controlling a wheelchair with image-based homing. In *Spatial Reasoning in Mobile Robots and Animals*, pages 66–75, Manchester University. AISB-97 Workshop.

Röfer, T. (1998). *Panoramic Image Processing and Route Navigation*. Ph.D. thesis, Universität Bremen. To appear.

Rossel, S. and Wehner, R. (1986). Polarization vision in bees. In *Nature*, volume 323, pages 128–131.

Schölkopf, B. and Mallot, H. A. (1995). View-based cognitive mapping and planning. In *Adaptive Behavior*, volume 3, pages 311–348.

Scholl, M. J. (1987). Cognitive maps as orienting schemata. *Journal of Experimental Psychology*, **13:4**, 615–628.

Schweizer, K. and Janzen, G. (1996). Zum Einfluß der Erwerbssituation auf die Raumkognition: Mentale Repräsentation der Blickpunktsequenz bei räumlichen Anordnungen. In *Sprache und Kognition*, volume 15:4, pages 217–233.

Siems, U., Herwig, C., and Röfer, T. (1994). Simrobot - Ein Programm zur Simulation sensorbestückter Agenten in einer dreidimensionalen Umwelt. In B. Krieg-Brückner, G. Roth, and H. Schwegler, editors, *ZKW-Bericht*, number 1/94 in ISSN 0947-0204, Universität Bremen. Zentrum für Kognitionswissenschaften. Also: http://www.tzi.org/~simrobot.

Tolman, E. C. (1948). Cognitive maps in rats and men. In *Psychological Review*, volume 55, pages 189–208.

Trullier, O., Wiener, S. I., Bertholz, A., and Meyer, J.-A. (1997). Biogically based artificial navigation systems: Review and prospects. In *Progress in Neurobiology*, volume 51, pages 483–544. Pergamon.

v. Frisch, K. (1967). *The dance language and orientation of bees*. Oxford University Press, London.

v. Helmholtz, H. (1909). *Handbuch der Physiologischen Optik. 3rd edn.* Verlag von Leopold Voss, Hamburg.

Verri, A. and Poggio, T. (1987). Against quantitative optical flow. In *Proc. 1st IEEE Int. Conf. Computer Vision*, pages 171–180.

Wehner, R. (1983). Celestial and terrestial navigation: Human strategies – insect strategies. In F. Huber and H. Markl, editors, *Neuroethology and Behavioural Physiology*, pages 366–381, Berlin, Heidelberg, New York. Springer.

Wehner, R. and Menzel, R. (1990). Do insects have cognitive maps? In *Ann. Rev. Neurosci.*, volume 13, pages 403–414.

Werner, S., Krieg-Brückner, B., Mallot, H. A., Schweizer, K., and Freksa, C. (1997). Spatial cognition: The role of landmark, route, and survey knowledge in human and robot navigation. In *Informatik '97 - Informatik als Innovationsmotor*, Informatik aktuell, Berlin, Heidelberg, New York. Springer.

Wiltschko, R. (1997). The navigation system of birds. In *Spatial Reasoning in Mobile Robots and Animals*, pages 5–14, Manchester University. AISB-97 Workshop.

Wiltschko, R. and Wiltschko, W. (1995). *Magnetic Orientation in Animals*. Springer, Berlin, Heidelberg, New York.

Wittmann, T. (1996). *Insektennavigation: Modelle und Simulationen*. Ph.D. thesis, Universität Bremen. Berichte aus der Biologie. Shaker, Aachen.

Human Place Learning in a Computer Generated Arena*

L. Nadel[1], K.G.F. Thomas[1], H.E. Laurance[1], R. Skelton[3],

T. Tal[4], and W.J. Jacobs[1,2]

[1]Department of Psychology, University of Arizona
Tucson, AZ, USA
[2]Department of Psychology, University of Arizona, Sierra Vista
Sierra Vista, AZ, USA
[3]Department of Psychology, University of Victoria
Victoria, British Columbia, Canada
[4]Dartmouth College, Hanover
NH, USA

Abstract. We describe the development of a computer-generated arena within which one can study human place learning by asking subjects to locate an invisible target. A series of studies demonstrate that such learning is based on acquiring knowledge about the spatial relations among the distal cues presented in this arena. We show that (1) the presence of proximal cues does not prevent learning about the distal cues; (2) the removal of individual cues does not impair performance until all distal cues are removed; and (3) the re-arrangement of distal cues profoundly impairs performance. In further studies we demonstrate that learning can proceed even when the subject is placed on the target rather than having to navigate to it, and even if the subject merely watches someone else navigate to the target. Finally, we demonstrate that learning is impaired by traumatic brain injury, and that aged subjects do not perform as well as young adults. This paradigm should prove useful in investigations of spatial cognition in general, and the role of specific neural systems inparticular.

1   Introduction

The study of spatial cognition has played a prominent role in cognitive neuroscience in recent years. This rich cognitive domain encompasses a number of important

*   **Acknowledgements.**   We thank Joe Demers for his work in writing the software controlling these experiments. We also thank Thomas Brunner and the many laboratory assistants who have helped us run these experiments.   This research and our contributions to it were supported by grants from the Zumberge Foundation and the McDonnell-Pew Program in Cognitive Neuroscience to W.J.J. and by a grant from the McDonnell-Pew Eastern European Program to L.N.

behavioral capacities. Finding one's way around in the world, locating objects of interest, and remembering where particular events transpired are only some of the critical functions that depend upon spatial cognition. Much of this behavioral repertoire can be observed in rodents and primates, thereby making possible well-controlled studies of both the behavioral properties and neurobiological underpinnings of spatial cognition. Following the discovery by O'Keefe & Dostrovsky (1971) of neurons in the hippocampal formation of the rat that are selectively active in a specific location, and the development of the cognitive map theory of hippocampal function (Nadel & O'Keefe, 1974; O'Keefe & Nadel, 1978), in the past 25 years a number of studies have explored the bases of spatial cognition, using a variety of new behavioral and physiological techniques (see Nadel, 1991, 1994).

Most recently, the emergence of non-invasive brain imaging techniques has made it possible to study the neural bases of normal human spatial cognition. Such study would benefit greatly from the development of spatial tasks for humans that are as similar as possible to the tasks used so successfully in work with rodents. We have recently worked with a computer-based version of the water-maze task developed by Morris (1981) and will report here our initial findings from this computer-generated arena (see Jacobs et al., 1997, in press, for more details).

The task Morris developed consisted of a large circular tank, filled about two-thirds with cloudy water. Within this tank, typically about 2 meters in diameter, an escape target could be hidden beneath the surface of the water, such that the rat lacked proximal information to help it locate the target. Given this, rats must use either distal information to help them navigate to the proper location, or some complex combination of such distal information with movement-feedback cues. Over a number of acquisition trials, rats learn to navigate to the target with increasing efficiency, as seen in decreasing latencies to find the hidden target. That they learn the location of the target is confirmed by their consistent circling about the target location on probe trials run without the target in place. In another maze situation demanding the use of distal cues, Suzuki et al. (1980) showed that rats use the spatial relations among these distal cues to solve the task, rather than relying on a single well-placed cue. They showed this by transposing various cues during probe trials; such transpositions cause dramatic impairments in task performance that can best be explained if one assumes that relations among multiple cues, rather than individual cues, subserve learning. This of course is what is meant by the notion of a cognitive map, as spelled out by O'Keefe & Nadel (1978).

In developing a comparable task for use with humans, we wanted first to demonstrate that our participants were solving the task in the same way, that is, using distal cues in a relational way. We also wanted to explore the role of movement cues and the possibilities of various forms of perceptive learning. We describe a series of studies exploring these and other features of a human version of the water maze.

1.1 The Task: General Methods

We modeled the water-maze task on a computer, creating a computer generated "virtual arena." This situation allowed us to explore various aspects of spatial cognition, and to do so in a situation where a wide range of people might be tested.

Participants were tested on a standard personal computer displaying a multicolored view of a circular arena. The arena was located within a square room, and the participants viewed this scene as though standing on the floor of the arena, from a perspective showing the wall of the arena on the lower part of the computer screen and the wall of the room on the upper part of the screen. If a participant stood with his or her back to a wall, the opposite arena wall would occupy the lower half of the screen, and the room wall would occupy the upper half of the screen. As the participant moves around the arena, the portion of the screen taken up by the arena and room walls varies. The floor of the arena was gray and there were no cues within the arena or on its walls. The walls of the square room were also gray, and each wall had a distinctive pattern of windows or arches, thereby providing a means by which participants could orient within the arena (Figure 1). The arena was divided into four imaginary quadrants, and a square target (hidden or visible) could be located within one of the quadrants.

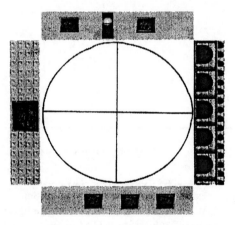

Figure 1. The shape and walls of the computer-generated arena

Participants started in a practice room where they could practice walking and looking around while using the keyboard or joystick. After becoming comfortable with this procedure, they could teleport themselves into the experimental room, close to one of the arena walls. They were told that their task was to search the arena and find a target located within it; the location of the target was fixed across trials. In some conditions the target was visible and in others it was not. Participants were instructed to stand on the target once found, to look around, and then to teleport themselves back to the practice room. In the first 3 studies acquisition trials in the arena lasted a maximum of 4 minutes; this was reduced to 3 minutes in the remaining studies. Probe trials lasted 4 minutes in the initial 3 studies, and 2 minutes in the later studies.

1.2 Study 1: Acquisition

The first study focused on the question of whether human participants could learn the location of a hidden target using only distal cues. The 15 participants in this study received 10 acquisition trials during which they were teleported to random start locations in the arena, from which they searched for the hidden target. On the 11th and final trial the target was removed and they were allowed to explore the arena for 4 minutes. This procedure closely matches the methods used by Morris in his original study with rats (1981) and in countless other studies since. Figure 2 shows the time it took participants to find the fixed target when started from random locations. The figure shows a steady decline in the time required to find the target over the 10 acquisition trials.

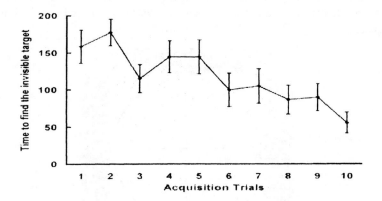

Figure 2. Learning to find the target from random start locations. An invisible target was placed in the northeast quadrant of the arena. Each acquisition trial (1-10) began from a random location next to and facing the arena wall. The figure illustrates the time in seconds (mean and SE) required by the 15 subjects in Study 1 to reach the invisible target on each of the 10 acquisition trials. A repeated-measures ANOVA detected significant changes in the time required to find the invisible target across the acquisition trials, $F(9, 126) = 4.34$, $p < .05$. Orthogonal post-hoc contrasts detected a systematic decrease in the time required to find the invisible target across the acquisition trials

Figure 3 shows the search data from the probe trial on which the target was absent from the arena. It is clear from these data that the participants learned the location of the target, as their search concentrated on the correct quadrant. Thus, in terms of both latency to find the target and search focus, human performance in the arena with a hidden target closely resembles that seen in rats in the water maze. Participants were asked to report on their performance after the probe trial, and those who knew the target location (13/15) reported using distal cues, and the relations among them, to find their way. This study shows that humans, much like rats, can engage in place learning based on the use of distal cues and their relations, and that they can do so in a computer-generated arena that does not require movement through tangible 3-dimensional space. We return to this point later, as it differentiates the human task from the rat studies.

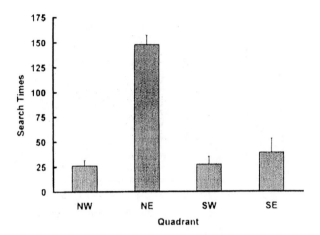

Figure 3. Searching a place for the invisible target. The invisible target was removed from the northeast quadrant on the last trial (Trial 11). The figure illustrates the time in seconds (mean and SE) the 15 subjects in Study 1 searched each quadrant during the 4-min probe trial. As can be seen, the subjects searched the target quadrant (NE) about four times more than they searched the others. A repeated-measures ANOVA detected significant differences in the time spent searching the quadrants, $F(3, 42) = 27.82$, $p < .05$. Orthogonal post-hoc contrasts detected no differences in mean time spent searching the SE, SW, and NW quadrants, but a significant difference between these taken together and the time spent searching the NE quadrant

1.3 Study 2: Proximal vs. Distal Cues

The second study considered an interaction between distal and proximal cues in the control of spatial learning in the computer-generated arena. Participants in one group (N = 15) were tested with distal cues and a hidden target (a replication of Study 1), and participants in a second group (N = 15) were tested with a visible target that provided proximal cues in addition to the distal cues. This condition was accomplished by providing a visible representation of the target on the floor of the arena. We were interested in seeing how this proximal cue condition influenced learning rates, and how the participants would perform on the probe trial when the visible target was removed. Participants in both groups started each trial from a random initial position. On the probe trial participants in the visible target group were tested without a representation of the target on the floor of the arena Figure 4 shows latency data for both groups over the 10 acquisition trials. The group with a hidden target (and only distal cues) displayed a learning curve quite similar to the participants in Study 1. The group tested with a visible target showed a much more rapid decrease in latency, as expected.

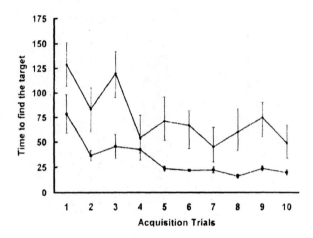

Figure 4. Learning to find a visible or invisible target from random start locations. A target was placed in the northwest quadrant of the arena. Each acquisition trial (1-10) began from a random location next to and facing the arena wall. The figure illustrates the time (mean and SE) the two groups of subjects in Study 2 required to find the target on each of the 10 acquisition trials. The squares represent the data obtained from the Visible Target group and the rhombs represent data obtained from the Invisible Target group. As can be seen, those in the Visible Target group found the target more quickly than did those in the Invisible Target group. A repeated-measures split-plot ANOVA conducted on the mean latency to find the target detected a significant difference between the Visible and the Invisible Target groups, $F(1, 28) = 13.41$, $p < .05$, significant differences across acquisition trials, $F(9, 252) = 5.69$, $p < .05$, but no significant interaction, $F < 1$. Separate within-subjects ANOVAs conducted on the mean time required to find the target detected significant differences across trials for both the Invisible Target group, $F(9, 126) = 2.67$, $p < .05$, and the Visible Target group, $F(9, 126) = 6.64$, $p < .05$. Orthogonal post-hoc contrasts conducted on the means obtained from the Invisible Target group showed the time required to find the target systematically decreased across the acquisition trials. In contrast, orthogonal post-hoc contrasts conducted on the means obtained from the Visible Target group showed the time required to find the target on the first trial differed significantly from the time required to find the target on the remaining nine acquisition trials, which did not differ from one another

Figure 5 shows the data from the probe trial, and indicates that participants in both groups searched in the target quadrant more than in any of the other quadrants. This result shows that when the proximal cues are eliminated, the participants can orient their search with reference to the distal cues. This in turn demonstrates that even though the participants in this group had access to proximal cues during initial acquisition, they nonetheless acquired information about the distal cues. In other words, we did not observe "overshadowing" of the distal cues by the proximal cue.

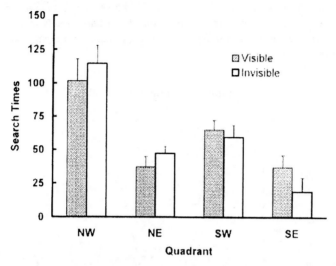

Figure 5. Searching a place for a target. The target was removed from the northwest quadrant on the last trial (Trial 11). The figure illustrates the time in seconds (mean and SE) the subjects from the two groups (Visible and Invisible Target groups) in Study 2 searched each quadrant during the 4-min probe trial. As can be seen, the subjects in the Invisible and Visible Target groups searched the target quadrant (NW) about equally and more then they searched the others. A repeated-measures split-plot ANOVA and subsequent orthogonal post-hoc contrasts confirmed this impression. As suggested by R. G. M. Morris (personal communication, 1 April 1997) the between-groups effect is meaningless and was therefore ignored in the present analysis. The group-by-quadrant interaction term obtained from the split-plot ANOVA indicated that the distribution of search time across quadrants between groups was not significantly different, $F(3, 81) < 1$, $p > .05$. The same analysis detected significant differences in the time spent searching the various quadrants, $F(3, 81) = 17.83$, $p < .05$. Post-hoc contrasts indicated that subjects searched the NW quadrant more than the NE, SW, or SE quadrants, which did not differ from one another.

1.4 Study 3: Generalizing From Familiar to Novel Start Locations

In this study, participants searched for an invisible target in the absence of proximal cues. In contrast to Studies 1 and 2, participants in this study ($N = 17$) started from the same location on each of the 10 acquisition trials. This phase was followed by 3 transfer trials on which participants were started from novel locations. The next, and last, trial was a probe trial on which the target was removed from the arena. Figure 6 shows the acquisition data for our participants. Some participants behaved rather differently in this study than did those in the first two studies. Panel A shows the

6/17 participants who looked similar to the participants in the earlier studies - their acquisition function was gradual. Panel B shows the 9/17 participants who apparently acquired the task in a single trial. Panel C shows the 2/17 participants whose acquisition function was quite erratic. Sixteen of the participants reported they knew the location of the target, and the performance of these participants on the transfer trials supports these claims.

Panel A: Gradual Acquisition

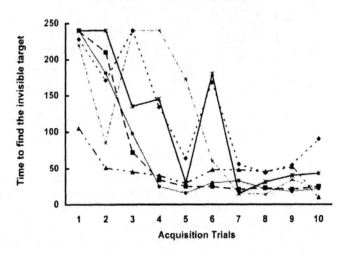

Panel B: One-Trial Acquisition

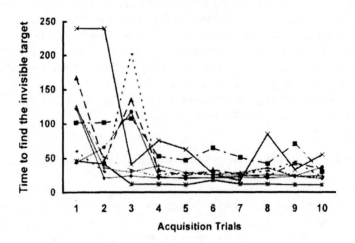

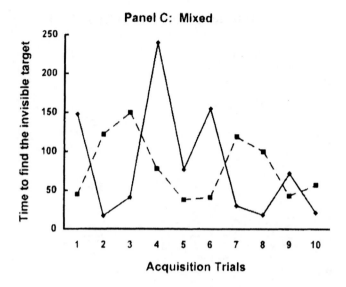

Panel C: Mixed

Figure 6. Learning to find an invisible target from a fixed start location (Study 3). Panel A illustrates the individual acquisition curves obtained from those six subjects classified as acquiring gradually. Panel B illustrates the individual acquisition curves obtained from those nine subjects classified as acquiring suddenly. Panel C illustrates the individual acquisition curves obtained from those two subjects classified as exhibiting mixed acquisition

As seen in Figure 7, participants who knew the location of the target headed directly for the appropriate quadrant and rapidly found the target when started from novel locations (P4,5,9,1). The lone participant (P3) who reported not knowing the location of the target performed quite differently on these transfer trials. Performance on the final probe trial confirmed these impressions.

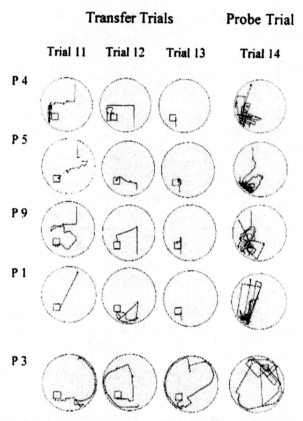

Transfer Trials **Probe Trial**

Trial 11 Trial 12 Trial 13 Trial 14

P 4

P 5

P 9

P 1

P 3

Figures 7. Representative search paths on transfer trials and probe trials. After 10 acquisition trials from a fixed start location, participants received three transfer trials from new start locations followed by a probe trial. The first four rows of Figure 7 illustrate representative individual search paths for 4 of the 16 individuals who reported knowing the location of the invisible target on the transfer trials (the first three figures in the row) and the probe trial (the last figure in the row). The last row of Figure 7 illustrates the search paths of the individual who reported not knowing the location of the target. The first column represents the individual search paths obtained from the first transfer trial (Trial 11), the second represents the individual paths obtained from the second transfer trial (Trial 12), and the third represents the individual paths obtained from the third transfer trial (Trial 13). The fourth column represents individual data obtained on the probe trial (Trial 14)

The participants who knew where the target was searched predominantly in the correct quadrant (Figure 8). The participant who did not know where the target was searched in all of the quadrants. These results indicate that people can generalize proper search behavior to new start locations, indicating that they have not simply route learned by rote.

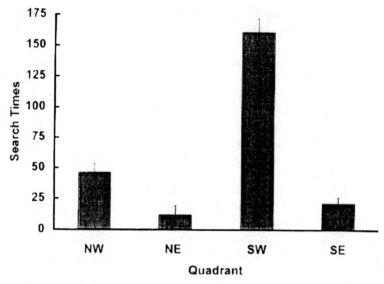

Figure 8. Searching a place for a target. The target was removed from the southwest quadrant on the last trial (Trial 14). The figure illustrates the time in seconds (mean and SE) the 17 subjects in Study 3 searched each quadrant during the 4-min probe trial. As can be seen, the subjects searched the target quadrant (SW) about three times more than they searched the others. A repeated-measures ANOVA detected significant differences in the time spent searching the quadrants, $F(3. 48) = 57.93$, $p < .05$. Orthogonal post-hoc contrasts detected the following: (a) no significant difference in mean time searching the NE and SE quadrants, (b) a significant difference in search time between the NE and SE quadrants taken together and compared against search time in the NW quadrant, and (c) a significant difference in the mean search time of the NW, NE, and SE quadrants taken together and compared against search time in the SW quadrant

1.5 Study 4: The Effects of Cue Removal

Two rather different accounts of how rats place learn in the water maze are available. According to one view, organisms use local views, or snapshots, to organize their search (eg., McNaughton, 1989), and movement through the maze is guided by the position of individual cues. This view predicts that removing subsets of the available distal cues should have devastating effects on place performance. A second view, the cognitive map theory (O'Keefe & Nadel, 1978), suggests that organisms use an integrated representation of relations among distal cues to organize their search behavior. By this view, removal of subsets of cues should have little or no impact on place performance, so long as enough cues remain to orient by. Fenton et al. (1994) showed that rats trained to find the location of a hidden platform on the basis of four distal cues could perform quite well when any two of the four cues were removed. In Study 4 we sought to examine this issue in humans exploring the computer-generated arena. After learning, our participants were tested on trials in which all cues were removed from randomly chosen distal walls. In Study 4a cues were removed from one wall at a time, while in Study 4b cues were removed from one, two, three or all four of the walls.

In these studies, participants began with 2 trials on which the target was visible. This was followed by 8 acquisition trials, and then by four cycles of test trials on which the target was hidden. Each of the four cycles consisted of two trials - the first involved the removal of cues from one or more walls, the second was a standard trial similar to that used in acquisition. The hidden target was present in the arena on all these trials. Following the test trials the participants received a 2-minute probe trial on which the target was not available. Finally, participants received one more visible-target trial to conclude the study.

Figure 9 displays the acquisition performance of the participants (N = 20) in Study 4a, and their performance on the test trials when distal cues were removed from one wall at a time. As the Figure shows, removal of the cues from any one wall has no effect on search performance, the participants continuing to find the hidden target quite efficiently. Thus, humans solve this task in a way that is not dependent upon specific solitary cues.

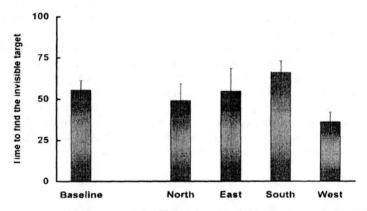

Figure 9. The time in seconds (mean and SE) the 20 subjects in Study 4a required to find the invisible target on an asymptotic baseline (the mean of acquisition trials 5-8) and the trials when distal stimuli were removed from the North, East, South, or West wall. As can be seen, no obvious differences appear among these means. A repeated measures ANOVA detected no significant differences among the mean times required to find the invisible target on acquisition trials 5-8, and trials during which the distal stimuli on either the N, E, S, or W walls were removed, F < 1

Figure 10 displays the aymptotic performance of the participants (N = 24) in Study 4b; in this study acquisition was very rapid in most of our participants. Figure 10 also displays the performance of participants on the various test trials. Removal of cues from one, two or even three walls did not disrupt performance. Removal of the cues from all four walls did have a negative effect on performance, although this manipulation did not completely abolish performance. These results suggest that although human participants utilize distal cues in solving the task, they are not completely dependent upon these cues.

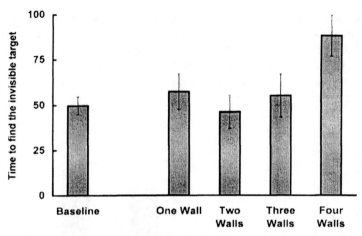

Figure 10. The time in seconds (mean and SE) the 24 subjects in Study 4b required to find the invisible target on an asymptotic baseline (the mean of acquisition trials 5-8) and the trials when distal stimuli were removed from one, two, three, or four walls. As can be seen, the mean time required to find the invisible target does not appear to change from baseline when one, two, or even three sets of distal stimuli are removed from the walls. Removing all the distal stimuli disrupted place performance. A repeated-measures ANOVA conducted on the baseline and four test trials detected overall significant differences among the mean time required to find the invisible target on baseline and the trials during which the distal stimuli on one, two, three, or four walls were absent, $F(7, 161) = 2.24$, $p < .05$. Orthogonal post-hoc contrasts detected no significant differences among the mean search times on the baseline or the trials on which the distal stimuli on one, two, or three walls were eliminated. The analysis detected a significant difference among the means from these conditions taken together and compared against the mean search time obtained when distal stimuli from all four walls were eliminated, $F(1, 23) = 9.92$, $p < .05$. This set of statistical decisions implies the following order of means: Acquisition trials $5 = 6 = 7 = 8 = $ One $=$ Two $=$ Three $<$ Four

1.6 Study 5: The Role of Spatial Relations Among Distal Cues

We have argued that the use of distal cues in solving the hidden target task depends upon the spatial relations among the cues, and not just the cues themselves. In work with rats on an eight-arm radial maze, Suzuki et al. (1980) showed that transposing various distal cues interfered with performance, while simply rotating all the cues, leaving their relations intact, had no such deleterious effect. In Study 5 we investigated the effect of cue transposition on human performance.

In this study, 2 trials with a visible target were followed by the acquisition phase of 8 trials and then 3 cycles of test trials with a hidden target. Each of the three cycles consisted of two trials, the first involving a transposition of some of the walls, the second returning the walls to the initial training condition. Two different transpositions were used, in random order for each participant: one of these involved swapping two of the walls (east and west), while the other involved relocating three walls (west, east and south). Following these transposition tests participants (N = 12) were given a single 2 minute probe trial on which the target was not in the arena. Finally, a last training trial with a visible target was administered. Figure 11 shows the time taken to locate the target on the transposition trials, broken down by whether two or three walls were relocated.

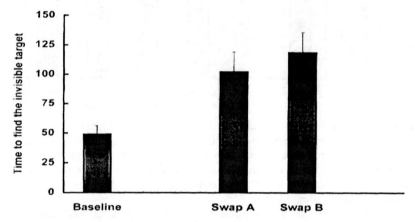

Figure 11. The time in seconds (mean and SE) the 12 subjects in Study 5 required to find the invisible target on an asymptotic baseline (the mean of acquisition trials 5-8) and the trials when the distal stimuli were transposed. During the test trials labeled Swap A, walls were rearranged in a clockwise order of North, West, South, and East. During the test trials labeled Swap B, walls were rearranged in a clockwise order of North, West, East, and South. As can be seen, the mean time required to find the invisible target increased dramatically on the transposition trials. A repeated-measures ANOVA detected significant differences among the mean time required to find the target on the baseline and the two transposition trials, $F(5, 55) = 6.60$, $p < .05$. Orthogonal post-hoc contrasts detected no differences among the baseline trials (trials 5-8), $F < 1$, or between the transposition trials (Swap A and Swap B), $F < 1$. A significant difference was detected when these baseline trials were taken together and compared against the two transposition trials taken together, $F(1, 11) = 18.35$, $p < .05$. This set of statistical decisions implies the following order of means: Acquisition trial $5 = 6 = 7 = 8 < $ Swap A $=$ Swap B

Dramatic increases in latency were observed. Figure 12 shows the paths taken by some of the participants on these two types of test trial; transposition of the distal cues causes clear disruptions in the search behavior of these participants.

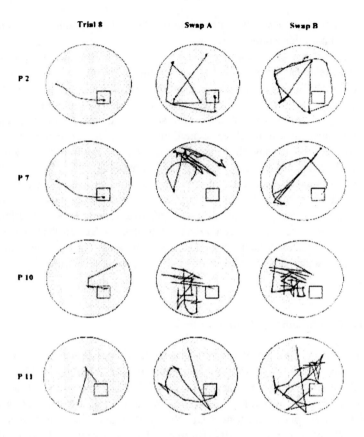

Figure 12. An aerial view of the individual search paths taken by four of the subjects (P 2, P 7, P 10, and P 11) during the last baseline trial (Trial 8) and the two test trials (Swap A and Swap B) of Study 5

1.7 Overview of the Initial Studies

This first group of studies accomplished a number of objectives. First, they demonstrated that it is possible to create a computer-based arena task for humans that can be learned relatively quickly by most participants. Second, the learning in this task seems to depend upon the same kinds of mechanisms observed in rats learning the Morris water-maze task. That is (1) the participants predominantly depend upon distal cues in learning the task; (2) when both proximal and distal cues are available they learn about both and can continue to perform efficiently when the proximal cues are absent (eg., there is no overshadowing of distal by proximal cues); (3) the participants learn about the relations among all of the cues, as shown by the fact that their performance is not disrupted by the removal of subsets of the cues, but is considerably impaired by transposing cues. The former manipulation leaves spatial relations intact, while the latter manipulation does not.

The data suggest that place learning in the computer-generated arena replicates in many ways the features of real-world place learning. While the task lacks the kinesthetic information associated with real-world spatial learning, it does include multiple-perspective information. The fact that we obtain results similar to those seen in animals moving around in the world shows that place learning, and the acquisition of sophisticated internal representations of space, do not require kinesthetic feedback. This is not to say that such information is unimportant when it is available, only that it is not necessary for this kind of learning.

Having demonstrated that place learning in the computer-generated arena replicates many of the features of place learning in the Morris water maze task for rats, we turned our attention to using it to test a number of assumptions about place learning that derive from cognitive map theory (O'Keefe & Nadel, 1978). In these studies we investigated the role of exploratory behavior of various kinds on place learning, the impact of brain damage, and the ability of aged participants to learn and perform in this situation.

1.8 Study 6: Placement Learning

We wanted to explore the kind of learning that would be possible if the participants were simply placed on the target for 60 seconds and allowed to look around but not to move off. These placement trials preceded a set of 7 normal acquisition trials and a probe trial with the target absent from the arena. We tested two experimental groups, one receiving a single placement trial (LL1, N=13), the other receiving 5 such trials (LL5, N=10). Control participants (N=10) received 7 acquisition trials and a probe trial.

The most critical piece of data concerns performance on the first acquisition trial. Figure 13 presents the path length data for the various groups on this trial. It seems clear from the data that the placement learning experience was of benefit to the participants, particularly those in group LL5. We asked participants in this study to engage in a "wall-assignment" probe after the placement trials but before the acquisition trials. Participants were given a model of the arena and asked to assign tokens of the walls to their appropriate locations. Five of the thirteen participants in Group LL1 and seven of the ten participants in Group LL5 scored perfectly on this task; that is, they placed all four walls in correct spatial relation to one another. These participants clearly knew the relations among the distal cues at this point in the study.

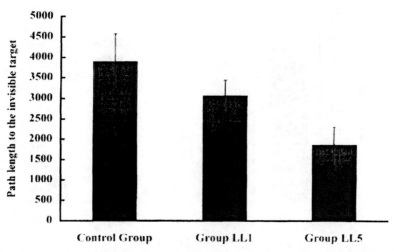

Figure 13. The path length in units (mean and SE) the three groups of subjects in Study 6 required to find the invisible target on the first acquisition trial. As can be seen, subjects in both Placement Learning groups appear to take a more direct path to the target than did those in the control group. A one-way between-groups ANOVA detected a significant difference among the mean path lengths produced by the subjects in the Control group, Group LL1, and Group LL5 on the first acquisition trial, $F(2, 30) = 3.79$, $p < .05$. A set of a prior contrasts conducted on these data detected (a) a significant difference between the mean path length of subjects in the Control group and that of subjects in Group LL5, $t(30) = 2.73$ and (b) no significant difference between the mean path length of subjects in the Control group and that of Group LL5 taken together and compared against that of Group LL1, $t(30) < 1$. This pattern of results implies the following order of means on the first acquisition trial: Control Group > Group LL1 > Group LL5

Figure 14 shows the results of the probe trial. All of the groups searched the target quadrant more than the other quadrants. In the post-training report, all but three of the participants (2 controls, 1 in group LL5) reported using relations among the distal cues to find the target.

The results of this study indicate that human participants can learn about the spatial relations among distal cues even though they do not move around within the computer-generated arena.

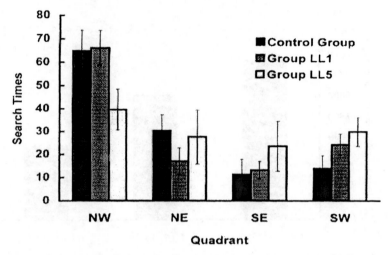

Figure 14. Searching a place for a target. The invisible target was removed from the northwest quadrant on a trial immediately following the acquisition trials. The figure illustrates the time in seconds (mean and SE) the subjects from the three groups in Study 6 searched each quadrant during the 2-min probe trial. A split-plot ANOVA detected a significant Quadrant effect $F(2,90)$ = 13.64, p < .05, but no significant Group x Quadrant effect, $F(6,90)$ = 1.71, p > .05. As above, we ignored the ANOVA's between-groups effect. Orthogonal post-hoc contrasts detected the following: (a) no significant difference in the mean time searching the NE and SW quadrants, $F(1,30)$ < 1, (b) no significant difference in mean search time between the NE and SW quadrants taken together and compared against the SE quadrant, $F(1,30)$ = 2.25, p > .05, and (c) a significant difference in mean search time between the NE, SW, and SE quadrants taken together and compared against the NW quadrant, $F(1,30)$ = 30.58, p < .05

1.9 Study 7: Observational Learning

Given that participants can learn about the arena without moving through it, we next explored whether place learning could occur simply by observing someone else perform in the computer-generated arena. We ran a control group (N=18) as usual, and an observational learning group (N=18) whose participants observed an experimenter receive four acquisition trials administered in the standard way. The experimenter used the computer keyboard rather than the joystick to move around in the arena, and experimental participants were not permitted to observe the experimenter's hands during the manipulation of the keyboard. The experimenter moved to the target as quickly and efficiently as possible on each trial, and after locating it rotated once before ending the trial by teleporting back to the waiting room. Following these observation trials, the participants were given a "wall-assignment" probe, then each experimental participant received 7 acquisition trials, with the target hidden in the same location as that used for the observation trials. This sequence was followed by a probe trial with the target absent from the arena. Figure 15 shows path length data for the initial acquisition trial from this study. Once again, the opportunity for learning, in this case by merely observing someone else perform, resulted in a marked facilitation of place perfomance on the initial trial.

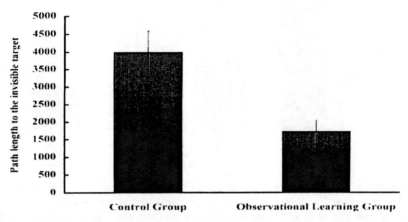

Figure 15. The path length in units (mean and SE) the two groups of subjects in Study 7 required to find the invisible target on the first acquisition trial. Subjects in the Observational Learning group appeared to take a more direct path to the target than did those in the Control group. A one-way between-groups ANOVA detected a significant difference between the mean path length of the two groups on the first acquisition trial, $F(1,34) = 10.60$, $p < .05$

Figure 16 shows the search times on the probe trial for the two groups. Both groups searched the target quadrant more intensely than the other quadrants. In the pre-acquisition wall placement task 12 of 16 participants in the observational learning group scored perfectly on the task. In their reports after the learning trials, 17/18 participants in the control group reported using distal cues to solve the task, and 16/18 participants in the observational learning group reported such distal cue use.

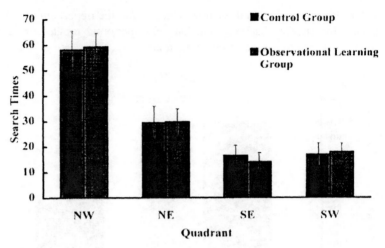

Figure 16. Searching a place for a target. The invisible target was removed from the northwest quadrant on a trial immediately following the acquisition trials. The figure illustrates the time in seconds (mean and SE) the subjects from the two groups in Study 7 searched each quadrant during the 2-min probe trial. The subjects in the Control and Observational Learning groups searched the target quadrant (NW) about equally and more carefully than they searched the other quadrants. A split-plot detected a significant Quadrant effect, $F(2,102) = 22.78$, $p < .05$, but no significant Group x Quadrant interaction, $F < 1$. Orthogonal post-hoc contrasts detected the following: (a) no significant difference in mean time spent searching the SE and SW quadrants, $F(1,34) < 1$, (b) a significant difference in mean search time between he SE and SW quadrants taken together and compared against the NE quadrant, $F(1,34) = 6.40$, $p < .05$, and (c) a significant difference in mean search time between the SE, SW and NE quadrants taken together and compared against the NW quadrant, $F(1,34) = 39.17$, $p < .05$

1.10 Study 8: Latent Learning by Movement

In the final study of this series we explored latent spatial learning by testing participants who were allowed initially to move around the arena in the absence of any target. These participants had an opportunity to learn the array of distal cues, but not the location of the target. The control group in this study (N=16) received 10 standard acquisition trials followed by a probe and a final trial with a visible target. The experimental group (N = 16) received two latent learning trials during which they were teleported into the arena without the platform. Both groups had received two practice trials at the start of the study, on which they were allowed to find a visible target. On the latent learning trials the participants were not told that the target was absent; they were simply allowed to move about for 60 seconds, after which the trial ended with the participant being returned to the waiting room. These latent learning trials were followed by 10 standard acquisition trials, a probe trial and a final trial with a visible target.

Figure 17 shows the path lengths for the two groups in this study. There were only minor differences between the groups in this study; careful analysis of the individual data indicated that some, but not all, of the participants in the experimental group showed a latent learning effect.

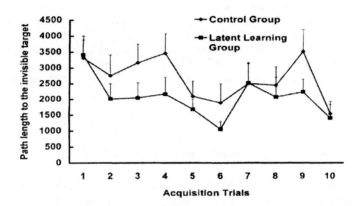

Figure 17. The path length in units (mean and SE) the two groups of subjects in Study 8 required to find the invisible target on each of the 10 acquisition trials

Figure 18 displays the data from the probe trial, indicating that participants in both groups searched extensively in the correct quadrant. In their post-testing reports, 14/16 controls and 15/16 experimentals indicated they used relations among distal cues to guide their search for the target.

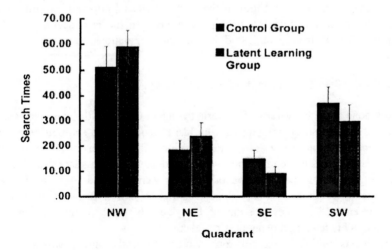

Figure 18. Searching a place for a target. The invisible target was removed from the northwest quadrant on a trial immediately following the acquisition trials. The figure illustrates the time in seconds (mean and SE) the subjects from the two groups in Study 8 searched each quadrant during the 2-min probe trial. The subjects in each group searched the target quadrant (NW) about equally and more carefully than they searched the other quadrants. A split-plot ANOVA detected a significant Quadrant effect, $F(2,90) = 15.80$, $p < .05$, but no significant Group x Quadrant interactio, $F(2,90) < 1$. Orthogonal post-hoc contrasts detected the following: (a) no significant difference in mean time spent searching the NE and SW quadrants, $F(1,30) = 2.91$, $p > .05$, (b) a significant different in mean search time between the NE and SW quadrants taken together and compared against the SE quadrant, $F(1,30) = 49.41$, $p < .05$, and (c) a significant difference in mean search time between the NE, SW, and SE quadrants taken together and compared against the NW quadrant, $F(1,30) = 22.18$, $p < .05$

1.11 Overview of the Second Series of Studies

In this second set of studies we demonstrated that participants could acquire important information about the arena in circumstances other than normal training. Thus, two forms of latent learning were found to be effective: participants could learn from being placed on the target directly, and from being allowed to observe another person navigate through the arena.

This set of results, along with those of the initial series of studies, demonstrates that the computer-generated arena task is a close analog of the Morris water-maze task. It further shows that the kind of learning going on within the arena has the properties attributed to "place learning by cognitive maps" in the work of O'Keefe & Nadel (1978). Therefore we assume that this task engages the human cognitive mapping system, and hence is a useful tool in the study of special populations with presumed damage to, or dysfunction in, the hippocampus. In the next section we describe preliminary results from several studies with such populations.

1.12 Study 9:   The Impact of Brain Damage

In work done at the University of Victoria by one of us (R.S.), the computer-generated arena was used to study participants who had experienced and subsequently recovered from traumatic brain injury (TBI). People with TBI often have trouble wayfinding. We predicted that such participants would also have trouble place learning in the computer-generated arena. Twelve participants were recruited from the Head-Injury Rehabilitation Program in Gorge Road Hospital, Victoria, British Columbia. The study concentrated on the question of how quickly these participants learned the location of a hidden target using only distal cues.

Participants initially received four acquisition trials during which they were teleported to random start locations in the arena and had to search for a target that was visible on the floor, but at different locations across trials. Following this, participants were given 10 acquisition trials, starting from random locations and searching for a fixed, invisible, target. On the next trial a probe was given with no target available, and this was followed by the final trial, with a visible target on the arena floor. Figure 19 presents the data for the 4 visible-target trials, and the 10 invisible-target trials. These data indicate that all participants were able to rapidly navigate to the visible target. As the figure shows, matched controls learned to rapidly find the invisible target, while experimental participants did not.

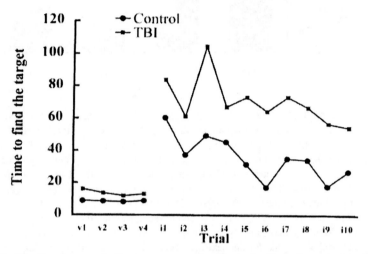

Figures 19. The time in seconds the two groups of subjects in Study 9 required to find the visible target on four trials (v1-v4) and the invisible target on 10 trials (i1-i10). Subjects in the TBI group required more time to find the visible target than did subjects in the Control group. A repeated-measures ANOVA confirmed that the difference in latency between the two groups across visible target trials was significant, $F(1,22) = 7.78$, $p < .05$. Subjects in the TBI group required more time to find the invisible target than did subjects in the Control group. A repeated-measures ANOVA confirmed that the difference in latency between the two groups across invisible target trials was significant, $F(1,22) = 7.77$, $p < .05$

Figure 20 shows that control participants concentrated their search in the correct quadrant on the probe trial with the target absent from the arena, but experimental participants failed to do so. On the post-training wall-assignment probe only 5 of the 12 participants with TBI correctly positioned the walls, and placed the target accurately. These data indicate that participants with moderate to severe TBI are deficient at learning the spatial layout of the computer-generated arena.

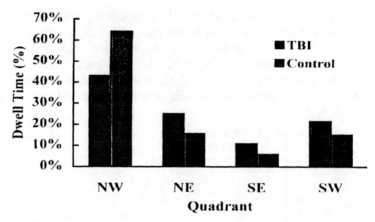

Figure 20. Searching a place for a target. The invisible target was removed from the northwest quadrant on a trial immediately following the invisible target trials. The figure illustrates the percentage of time the subjects from the two groups in Study 9 searched each quadrant during the 2-min probe trial. Subjects in the TBI group spent significantly less time searching the target quadrant (NW) than did subjects in the Control group, $t(22) = 2.20$, $p < .05$, one-tailed, and though their preference for the target quadrant was significantly above chance (43% versus 25%, $t(11) = 2.25$, $p < .05$), it was not significantly greater than the next-most preferred quadrant, $t(11) = 1.09$, $p > .05$. In contrast, the Control group subjects' preference was well above chance (64%, $t(11) = 7.77$, $p < .05$) and well above that for the next-most preferred quadrant, $t(11) = 6.25$, $p < .05$

1.13 Study 10: Changes With Aging

It is widely assumed that one of the concomitants of the cognitive deficits observed in aging is hippocampal dysfunction. Thus, we would expect aged participants to have difficulties with the computer-generated arena. In collaboration with S. Luczak at the University of Southern California we tested 8 women between the ages of 71 and 84 on the computer-generated arena. Participants were given 10 acquisition trials with a hidden target. Two of the 8 aged participants failed to complete the task, expressing frustration at their inability to locate the target. The remaining participants failed to demonstrate much place learning, as shown in Figure 21. On a final probe trial with the target absent, controls concentrated on the correct quad- rant while aged participants did not (Figure 22). Five of the 6 aged participants who completed the task reported afterwards that they did not know the location of the target.

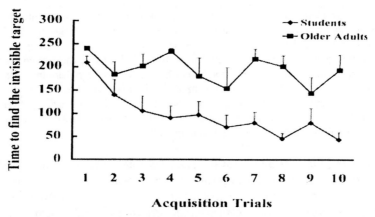

Figure 21. The time in seconds (mean and SE) the two groups of subjects in Study 10 required to find the invisible target on each of the 10 acquisition trials. A split-plot ANOVA detected a significant Group effect, $F(1,12) = 19.77$, $p < .05$, a significant Trials effect, $F(9,108) = 4.12$, $p < .05$, and a significant Group x Trials interaction, $F(9,108) = 2.03$, $p < .05$. These results imply that, across acquisition trials, (a) there was a significant difference between the mean time each group required to find the target, (b) the mean time the subjects required to find the target changed significantly across trials, and (c) the rate of this change in the Older Adult group differed from that in the Student group. It appears that across trials subjects in the Student group took progressively more direct paths to the invisible target, but participants in the Older Adult group did not; they either did not find or took circuitous paths to the target. Separate within-subjects ANOVAs conducted on the mean path length showed a significant Trials effect in the Student group, $F(5,15) = 5.39$, $p < .05$, but no such effect in the Older Adult group, $F(5,15) < 1$. Orthogonal post-hoc comparisons showed that the path length subjects in the Student group required to find the target systematically decreased across the acquisition trials

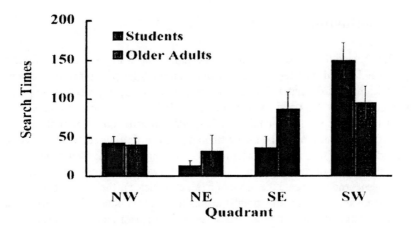

Figure 22. Searching a place for a target. The invisible target was removed from the southwest quadrant on a trial immediately following the invisible target trials. The figure illustrates the time in seconds (mean and SE) the subjects from the two groups in Study 10 searched each quadrant during the 4-min probe trial. Subjects in each group searched the target quadrant (SW) about equally and somewhat more carefully than they searched the other quadrants. In a split-plot ANOVA the differences between time spent searching the target quadrant and that spent searching the other quadrants did not reach significance at the .05 level. The analysis detected no significant Quadrant effect, $F(3,18) = 1.11$, $p > .05$, and no significant Group x Quadrant interaction, $F(3,18) < 1$

To rule out the possibility that the place learning difficulty demonstrated by the aged participants resulted from difficulties using the computers, we tested an additional four aged participants on trials where the target was visible, followed by trials on which it was invisible. These participants performed well on the visible target trials, but poorly on the invisible target trials; this result rules out any simple deficiency in computer use. After training concluded these four participants were asked to reconstruct the experimental room using puzzle pieces. Each participant received four rectangular pieces representing the walls, seven smaller pieces representing objects contained on the walls (eg., door, window), and a blue rectangular piece representing the target. They were instructed to first lay out the walls appropriately, then add the objects to the correct walls, and finally to situate the target at its appropriate location. Surprisingly, these 4 aged participants did rather well on this task. Two performed perfectly on the wall placement task, and two performed perfectly on the object placement task. The controls were only marginally better. However, the aged participants consistently failed to put the target token in the correct location, indicating that although they formed a representation of the experimental space, this representation did not include an accurate sense of where the target was within that space.

2. Conclusions

Complex behavior in space appears to be guided by various kinds of information acquired and stored for later use. To facilitate study of this phenomenon, we developed a place-learning task that helps to isolate the component processes important to spatial cognition. This task, which we have labeled the "computer-generated arena," permits investigation of the acquisition of spatial information under highly controlled and convenient conditions. The major finding from initial studies with the computer-generated arena is that human performance in a computer-generated space is strictly comparable to that of rodent performance in an equivalent 3-dimensional space.

Humans learned to locate the place of a target hidden on the floor of the computer-generated arena as seen by the gradual acquisition curves obtained over trials, intense searching of that location on probe trials, self report, performance on a task requiring a reconstruction of the experimental room, and the strong positive correlation between spatial competence as measured by self report and performance in the computer-generated arena. Taken together, the results indicate that young to middle-aged humans do not require specific distal cues, vestibular or motor information, proximal cues, previous familiarity with the environment, or taxis (response-based) strategies to learn to locate themselves and places in computer-generated space.

The designs used in Studies 4a and 4b showed that participants need not find the invisible target using specific distal cues; random elimination of any subset of distal cues did not disrupt task performance. The designs used in Studies 6, 7 and 8 showed that participants need not use vestibular or motor information to find an invisible target; both sources of information were eliminated during the acquisition phase of these studies. The designs used in Studies 1 and 3 showed that participants did not find the invisible target using proximal cues; none were programmed in this computer-generated world. The designs used in Studies 4a and 4b showed that, although participants may use taxis strategies in the computer-generated arena, this strategy does not, in and of itself, account for place learning in the arena. When we eliminated all distal cues and all of the proximal cues except the arena wall, performance in the arena was disrupted. In the absence of proximal and distal cues, experienced participants appeared to use the arena wall as a cue, searching in a circle at an appropriate distance from it. This taxis strategy allowed them to find the invisible target more quickly than experimentally naive individuals, but less quickly than participants provided with a rich set of distal cues.

As predicted by spatial mapping theory (O'Keefe & Nadel, 1978), relations among available distal cues, as seen in the effects of cue transpositions and self report, control navigation in the computer-generated arena (Study 5). In addition, the presence of proximal cues does not appear to interfere with the rapid acquisition of representations of distal cues and relations among them (Study 2).

The studies of perceptive learning demonstrated that humans learn the location of an invisible target in computer-generated space by observing an environment from the target location (Study 6), or by watching a demonstrator locate the target (Study 7). Simply exploring the environment without learning about places within it facilitates subsequent learning about a target's location in at least some participants (Study 8). The implication of these studies is that humans may learn the layout of an environment through simple observation. Perceptive learning of this type indicates that map-like information may be acquired quickly, accurately retained, and subsequently retrieved to guide motor strategies within the environment.

These findings, taken together with the data showing that trained participants can generate direct routes to hidden targets from novel start locations (Study 2), show that humans can generate novel motor strategies within the computer-generated environment. All of our participants found it quite easy to adopt appropriate finger, hand and wrist movements in operating the joystick or keyboard, for movement in the computer-generated arena. Thus, the novel motor strategies employed in this study go beyond the kinds of movements participants make in 3-dimensional space. It would be of some interest to explore if, and how, participants might translate what they have learned about this computer-generated space into movements in actual space.

The study of humans recovering from brain damage (Study 9) shows that diffuse brain injury effects performance in the computer-generated arena in much the same way that various forms of brain injury in rodents effects performance in the Morris water maze. Although not all of the recovering participants showed deficits in the spatial portion of the task, the majority did not consistently find the invisible target in the arena. In contrast, these participants easily and quickly found the same target on trials when it was visible. These data, taken together with data showing strong positive correlations between performance on the spatial acquisition trials and room reconstruction ($r^2 = 0.73$, p = .001), spatial acquisition trials and wayfinding questions on a questionnaire ($r^2 = 0.49$, p = .02), and room reconstruction and wayfinding questions on a questionnaire ($r^2 = 0.41$, p = .05) suggest that performance in the computer-generated arena provides a useful measure of a set of constructs we might label place learning, place memory, place performance, and spatial navigation.

Finally, our preliminary studies show that aging effects the performance of humans in the computer-generated arena in much the same way it effects the performance of rats in the Morris water maze (Study 10). None of our aged participants consistently found the invisible target hidden on the floor of the computer-generated arena. In contrast, these participants easily and rapidly found the same target on the trials when it was made visible. Similar data in work with rodents has led some to suggest that aged participants fail to learn the layout of the test context. The room-reconstruction data we obtained from our elderly participants, however, indicate that they learned the test context layout at least as well as young and middle aged controls. When asked to reconstruct the room from puzzle pieces their performance was accurate. What distinguished young from old participants was the fact that the latter participants did not place the target in the correct location on this reconstruction task. After accurately configuring the walls and distal cues, our elderly participants either placed the target in a wildly incorrect location or reported that they could not place it at all. Although controls were not always perfect in this task, their performance was considerably better than that of the aged participants.

These data indicate that learning the stable layout of an environment may be dissociable from locating oneself (or a target) within that environment. The failure of aged rats to perform in a water maze could, in this view, reflect problems in the latter, rather than the former as has generally been assumed. This finding is consistent with the suggestion that representing an environment may be mediated by different neural structures than navigating efficiently within the environment (Bohbot et al., 1997; Malkova & Mishkin, 1997; Nadel et al., 1997). This computer-generated task might be quite useful in helping tease apart these apparently separate spatial cognitive capacities.

In summary, our data suggest that, in this computer-generated space, representations of distal cues and spatial relations among them are learned and retained as a cognitive map of the computer-generated environment (O'Keefe & Nadel, 1978). Upon re-exposure to that computer-generated environment participants retrieve the appropriate map (or fragment of a map). They then use this information to establish their location in the environment, and to establish the location of the invisible target. Using these two locations as a reference, motor movements are planned and executed that produce action leading from the current to the target location.

Each step in such a process requires an inference-like leap. Insuring that an appropriate map fragment is retrieved, that appropriate locations on that map have been noted, and that appropriate motor plans are executed require complex and as yet poorly understood processes. Continued analysis of place learning within the computer-generated arena will, we hope, make a detailed empirical examination of these and related processes easier. Place learning in computer-generated spaces may also prove useful as a convenient assay of hippocampal function, and we are currently planning to use it with children, aged participants, and populations with brain damage of various kinds.

References

1. Bohbot, V.D., Kalina, M., Stepankova, K., Spackova, N., Petrides, M. & Nadel, L. (1997) Lesions to the right hippocampus cause spatial memory deficits in humans. Society for Neuroscience Abstracts.
2. Fenton, A.A., Arolfo, M.P., Nerad, L. & Bures, J. (1994) Place navigation in the Morris Water Maze under minimum and redundant extra-maze cue conditions. Behavioral and Neural Biology, 62, 178-189.
3. Jacobs, W.J., Laurance, H.E. & Thomas, K.G.F. (1997) Place learning in virtual space I: Acquisition, overshadowing and transfer. Learning and Motivation, 28, 521-541.
4. Jacobs, W.J., Thomas, K.G.F., Laurance, H.E. & Nadel, L. (submitted) Place learning in virtual space II: Topographical relations as one dimension of stimulus control.
5. Malkova, L. & Mishkin, M. (1997) Memory for the location of objects after separate lesions of the hippocampus and parahippocampal cortex in rhesus monkeys. Society for Neuroscience, 23, 12.
6. McNaughton, B. L. (1989) Neuronal mechanisms for spatial computation and information storage. In L. Nadel, L.A. Cooper, P. Culicover & R.M. Harnish (Eds.), Neural connections, mental computation, pp. 285-350, Cambridge, MA: MIT Press.
7. Morris, R.G.M. (1981) Spatial localization does not require the presence of local cues. Learning and Motivation, 12, 239-260.
8. Nadel, L. (1991) The hippocampus and space revisited. Hippocampus, 1, 221-229.
9. Nadel, L. (1994) Multiple memory systems: What and Why. An Update. In D.
10. Schacter and E. Tulving (Eds.) Memory Systems 1994, pp. 39-63, Cambridge, MA, MIT Press.
11. Nadel, L. and O'Keefe, J. (1974) The hippocampus in pieces and patches: an essay on modes of explanation in physiological psychology. In R. Bellairs and E.G. Gray (Eds.) Essays on the nervous system. A Festschrift for J.Z. Young. Oxford: The Clarendon Press, 1974.
12. Nadel, L., Allen, J.J.B., Bohbot, V.D., Kalina, M. & Stepankova, K. (1997) Hippocampal and parahippocampal role in processing spatial configurations in humans: ERP and lesion data. Society for Neuroscience Abstracts.
13. O'Keefe, J. & Dostrovsky, J. (1971) The hippocampus as a spatial map. Preliminary evidence from unit activity in the freely-moving rat. Brain Research, 34, 171-175.
14. O'Keefe, J. & Nadel, L. (1978) The hippocampus as a cognitive map. Oxford: The Clarendon Press.
15. Suzuki, S., Augerinos, G. & Black, A.H. (1980) Stimulus control of spatial behavior on the eight-arm radial maze. Learning and Motivation, 11, 1-8.

Spatial Orientation and Spatial Memory Within a 'Locomotor Maze' for Humans

Bernd Leplow[1], Doris Höll[1], Lingju Zeng[1] & Maximilian Mehdorn[2]

[1] Department of Psychology, University of Kiel, Olshausenstr. 62, D-24098, Kiel,
Germany
{leplow, dhoell, lzeng}@psychologie.uni-kiel.de
[2] Clinic of Neurosurgery, University of Kiel, Weimarer Str. 8, D-24106 Kiel, Germany
mehdorn@nch.uni-kiel.de

Abstract. Spatial behavior was investigated using a locomotor maze for
humans which incorporates basic features of widely used animal paradigms.
Experiments are based on the 'cognitive map' theory originally put forward
by O`Keefe & Nadel [22] and allowed the assessment of place learning, and
spatial working and spatial reference memory errors. In our procedure,
subjects and patients have to learn and remember five out of twenty
locations within a 4 x 5 m area with completely controlled intra- and
extramaze cue conditions. Usually, participants learned to reach the
criterion. A probe trial from an opposite starting position with transposed
intramaze cues followed. Results showed that it is possible to assess cue-
dependent orientation, to dissociate spatial working memory and spatial
reference memory and to identify 'place-behavior' using specific parameters
derived from inertial navigation theory [16]. This will be demonstrated in
selected cases with circumscribed cerebral lesions and in unimpaired
subjects.

1 Introduction

One of the most prominent theories of spatial behavior is that of O'Keefe & Nadel
[22]. Based on the findings and assumptions of Tolman [36, 37] they distinguished
three basic types of spatial learning: learning of places, of routes and of responses [20,
21]. Place learning is characterized by the formation of an observer-independent
representation of the external world, i.e. a so-called 'cognitive map'. This
representation is assumed to be initially established by simultaneously encoding distal
stimuli and their mutual interconnections. In contrast, 'route learning' depends on the
acquisition of single 'landmarks' spatially related to the goals, whereas 'response
learning' solely depends on the processing of proprioceptive, kinesthetic and vestibular
cues.

Despite the extensive literature about spatial behavior and its determinants in animals [e.g. 11], the experimental work in humans is only fragmentary. Apart from newer developments using virtual reality [e.g. 17], Howard and Templeton [9] summarized the older human real space maze-literature and concluded that orientation from a contraligned position depends on the subject's ability to verbalize and imagine spatial concepts. Thorndyke and Hayes-Roth [35] found subjects to be superior in orientation tasks from new perspectives if they had had the opportunity to actively move around in space instead of learning the same environment from a map. Presson and Hazelrigg [29] also demonstrated that 'alignment errors' were observed if subjects had to learn a path from a map from one perspective and then to judge directions from a different one. But if active exploration or direct visual scanning of the path had been possible, no alignment effects were detectable. Presson, deLange, and Hazelrigg [30] showed that alignment effects were minimal when subjects were able to obtain multiple orientations during learning. Presson, deLange and Hazelrigg [31] varied the sizes their arrays and the maps of these arrays and found orientation-specific behavior in small arrays and with small maps and orientation-free behavior in larger arrays and with large maps. They concluded that orientation-specific spatial behavior is primarily egocentric and very precise under aligned conditions. However, in the case of larger environments they assume that the subjects regard themselves as being in an object-based frame of reference which can flexibly be used under contraligned conditions.

Though an orientation-free, observer-independent reference system provides an individual with an allocentric frame of reference and thus allows spatial behavior of high flexibility even when response requirements or environmental conditions are changed, the exact nature of the stimulus conditions hindering or facilitating this flexibility of spatial behavior has still to be investigated. Such an examination requires an experimental setup which allows

- assessment of spatial abilities within locomotor space,
- complete control of intra- and extramaze cues
- strict definition of behavioral response requirements
- automatic recording of inter-response intervals (IRI)
- detection of problem solving behavior (excluding algorithmic strategies)
- dissociation of place-, landmark-, and response-strategies and
- identification of spatial reference and spatial working memory errors.

In animals, learning behavior is usually assessed by means of the Morris Water Maze [18] in which a rat has to swim towards a hidden platform. Place learning behavior is induced by means of rich environmental cues located in distal space. On the contrary, cue learning is induced using an elevated platform or presenting a single landmark which is well visible above the platform's location. Numerous experiments have shown that successful navigation within these two environmental conditions depends on distinct neuronal circuits, i.e. the hippocampal formation in the case of place learning and the basal ganglia, especially the striatum, in the case of the acquisition of S-R based strategies [e.g. 19]. While allocentric and egocentric based search behavior can easily be assessed in the Water Maze [e.g. 2, 3], the different types of spatial memory errors are mostly investigated using the Radial Maze [26]. In this task, rodents are placed in the center of an eight, twelve, or sixteen arm maze and allowed to explore the endpoints of the arms, which are baited with food. Sometimes, only a subset of the arms is baited. Using this setup, three types of spatial memory errors can be obtained. Firstly, within the same trial an animal can revisit an already visited arm. In such a case it has not developed a 'win-shift' rule and thus a 'working memory' error is recorded [28]. Secondly, if the animal visits an arm which has never been baited, it was unable to build up a rule which stays valid across trials and has thus violated a 'win-stay' rule. In such a case a 'reference memory' error is recorded. Thirdly, if the animal revisits an arm which has never been baited within the same trial, a combined 'reference-working memory' error can be identified. This distinction from the other two types of errors was shown to be necessary by Colombo, Davis, and Volpe [1] and Okaichi and Oshima [23], who demonstrated different psychological mechanisms underlying reference and reference-working memory processes in the case of brain damage.

Only a few attempts have been undertaken to develop experimental setups for humans which are equivalent with respect to the task characteristics used in the non-human maze literature. Foreman was one of the first who tried to overcome the specific difficulties one is confronted with if purely 'spatial' behavior is to be obtained in children. In an experimental chamber like a radial arm maze two- and four-year-olds had to find hidden chocolates from eight identically labeled positions in an unfamiliar room [5]. Results showed that working memory errors were far more frequent in younger children, and that above chance performance was controlled by distal cue configurations. This result has been extended in further work with four- and six-year-olds showing that performance was associated with choice autonomy and active locomotor behavior [6]. In experiments with children from eighteen months to five years who were subjected to a maze with subsets of baited arms, reference memory was assumed to develop earlier than working memory [7]. Moreover, differences between groups of six-year-olds who were either actively or passively moving around and who either had or had not freedom of choice were best reflected by the reference memory component of the spatial task [8].

While Foreman and coworkers emphasized the distinction of reference and working memory errors across age groups in infants, Overman, Pate, Moore & Peuster [27] explicitly tried to assess place learning in children and adults by means of an eight arm radial maze similar to that of Foreman et al. [6, 7, 8], and a Morris Water Maze adaptation. Working memory seemed to be fully developed in children above the age of five, but unfortunately, algorithmic strategies were found in about 50% of the older children and ceiling effects were observed in adults. The 'water' Maze was constructed in such a way that a large cardboard 'pool' was filled with plastic packing chips. Children were requested to find a hidden 'treasure chest' located at the bottom of the pool. In the absence of any sex differences it was shown that place representations could be established from age five onwards and that the presence of proximal cues improved performance.

Since, up to now, no experimental procedure has been available for humans which fits the requirements listed above, a maze-like open field analogon (i.e., a 'locomotor maze') was constructed which incorporates the basic features both of the Radial Maze and the Morris Water Maze. This apparatus was constructed for adults [12] and adapted for children [10]. The experiments were designed so that the full range of abilities can be tested without ceiling or bottom effects. Preliminary results in healthy adults have shown that acquisition in a place condition is superior to the landmark condition which in turn is superior to condition of egocentric encoding. Analysis of a subsequent probe trial revealed that subjects of the egocentric groups were only inferior to landmark- and place-individuals if the task required the updating of the subject's orientation or if verbal or spatial material was interpolated. If task characteristics remained stable, the three orienting conditions did not affect recall of the spatial representation. Egocentric learning errors were mostly of the working memory type, whereas place learning yielded errors which were almost exclusively of the reference memory type [13]. As in the Overman et al. [27] study, no sex differences were found. These results were replicated and extended with brain impaired patients [14].

In the following, the basic principles of the locomotor maze and testing of healthy subjects and brain impaired patients will be outlined. Especially in selected cases, it will be shown, how cue dependent behavior can be assessed in contrast to orientation free behavior, how spatial working memory errors and spatial reference memory errors can be obtained and how 'place orientation' can be assessed using specific parameters derived from inertial navigation theory of McNaughton, Chen & Markus [16].

2 METHOD

2.1 Apparatus

Subjects were exposed to a dimly lit chamber with a 'circular platform' 3.60 m in diameter. The platform was covered with a black carpet and surrounded with black cloth leaving only 'gates' of about 1.5 m within each corner (**Fig.** 1). These gates served as starting points for the acquisition and probe trials (see below). Extramaze cues were controlled completely by means of eight distinct fluorescent symbols of about 30 x 30 cm in size each. Two of these symbols were attached deep in the corner of the four gates, respectively. The gates were enclosed by the same black cloth enclosing the platform. The chamber was completely painted black and was prepared in such a way that the subjects were prevented from orienting themselves according to acoustic stimuli from outside the experimental room.

The circular platform consisted of a wooden floor of about 20 cm height Twenty magnetic capacity detectors were fixed to this floor in a semi-irregularly fashion (Fig. 1). These detectors registered the presence or absence of a human limb and are thus a mean of assessing the track of spatial behavior within an experimental chamber. This arrangement was supposed to resemble the 'hidden platform' paradigm by Morris [18]. The detectors were connected individually to a microcomputer in a neighboring room. The location of each detector was marked on the carpet by identical light points provided by very thin glass fiber cables inserted into the wooden floor next to the capacity detectors. Such a light point could only be seen when the subject positioned himself/herself about 30 cm away from a capacity detector. Since the brightness of the diodes could be adapted according to the subject's height, no array of light points could be scanned and only two to three light points could be seen simultaneously. Thus, the subjects were prevented from employing simple geometric encoding strategies. Subjects had to move towards these light points and to step on them. Five out of twenty were designated 'correct'. This was signaled by a 40-Hertz tone whose source could not be located. A second step on one of these five 'correct' detectors did not yield another 40-Hertz tone. Thus, within one experimental trial, only the first stepping on a correct detector was characterized as a 'correct response'. The other fifteen detectors were labeled 'incorrect locations'. Stepping onto an incorrect location did not yield a 40-Hertz tone. IRIs as well as incorrect and correct responses were recorded automatically.

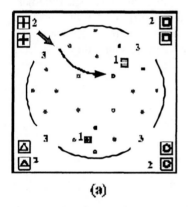 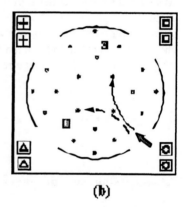

(a)   (b)

Fig. 1. Standard layout of the 'Locomotor Maze' (a) Acquisition phase; (b) probe trial. Probe trial manipulation is characterized by rotation of the starting point and rotation of the proximal cues by 180°, respectively. a: 1 = proximal cues; 2 = distal cues; 3 = gates; b: bold arrow = correct path (place orientation); dotted arrow = incorrect path (cue-/ egocentric orientation); a & b: black dot = correct location; grey dot = incorrect location (not visible for Ss)

2.2 General Procedure

Prior to exposure to the experimental chamber subjects heard and read elaborate instructions to convince them about the nature of the task. After informed consent had been obtained, subjects were then guided to the experimental chamber and given the following instructions (i) to explore the chamber, to visit each location, to step onto each detector, to remember the correct locations and (ii) to try to visit correct locations only once within each trial.

Each subject was then guided by the experimenter to his/her initial starting position and was again given the instructions for the exploration phase of the experiment ("please visit all of the twenty light points, step on them and try to remember the 'correct locations'. After having visited all twenty locations the subject was guided back to his/her starting point.

Before the first learning trial began, the subject again listened to the learning instruction ("now, please try to visit only the 'correct locations' - i.e., "those with the tone" - and "try to visit these 'correct locations' only once"). The subject then began to visit the correct locations while trying to avoid reference memory errors, working memory errors and reference-working memory errors. During each trial the experimenter herself moved to different positions within the experimental chamber. In order to spare the subjects the additional memory load of memorizing the number of successfully found correct locations, the experimenter counted aloud the number of correctly identified locations from different positions in the chamber. When the five 'correct locations' had been visited, this was signaled by a double tone and the subject was guided back to his/her starting point by means of a meander walk. Then the next acquisition trial began. The acquisition phase was always performed from the same starting point but the subjects were free to move around and to make their own choices. Subjects had to learn the location of the correct detectors until they completed two subsequent trials without errors.

3 Experiments

3.1 Assessing cue dependent orientation

Background. Within the framework of spatial navigation theory the distinction between an observer independent, 'orientation-free' and an observer-centered, 'orientation-specific' type of navigation is of major importance [31]. Farah, Brunn, Wong, Wallace and Carpenter [4] designated this dichotomy as 'environment centered' and 'viewer centered'. These dichotomies correspond to O'Keefe & Nadels [22] 'locale' and 'taxon' systems. Within the locale system memories are formed in a spatial-temporal context whereas the taxon system operates by means of the rules of category inclusion [21]. The locale system is driven by novelty and determined by distal cue configurations thus enabling the observer to encode relations between stimuli instead of single landmarks. On the contrary the operation of the taxon system depends on a distinct above-threshold stimulus. Repeated presentation of such a stimulus enhances response probabilities whereas the locale system will cease to operate if the same stimulus is presented repeatedly. This can be investigated experimentally by presenting a set of stimuli attached in the distal space while transposing stimuli attached in the proximal space. If the subject is bound to a viewer-centered taxon system he or she will use these proximal cues for navigation even if this is no longer adaptive. If, on the contrary, the locale system is activated and an orientation-free, environment-centered perspectivecan be obtained, spatial behavior has to rely on the set of distal cues. For this investigation no dissociation of error types will be undertaken.

Procedure, Recordings and Participants. After the subjects had reached the learning criterion of two successive error-free trials, a break of about two minutes was filled by informal conversation. This was done to prevent the subject from developing rehearsal strategies. Then a subsequent probe trial was scheduled. For this purpose the subject was guided by the experimenter to the new starting position by a meander-walk. The new starting position was rotated 180° with respect to the initial starting point. Moreover the proximal cues were also rotated by 180°. This manipulation leads to a viewer's perspective which is equivalent to that obtained during acquisition (Fig. 1b). Thus, this probe trial can only be mastered if the distal cue configuration is taken into account. If the subject relies on a cue or a response strategy (Fig. 1b, dotted lines) he or she will not be able to complete the task successfully. If, on the contrary, the set of distal stimuli are taken into account - i.e. 'place'-learning is obtained - the subject will orient towards the correct locations, irrespective of proximal cue distribution.

For each trial in the acquisition phase the total number of errors was calculated. In order to detect response stereotypes the path of the last acquisition trial is analyzed by means of a graph and compared to that of the probe trial. This is demonstrated in two experimental subjects with no known history of CNS-disorders. Subject S1, was male 20 years of age and subject S2 was female, 24 years of age.

Results. Fig. 2 shows that the performance of the two subjects with respect to the sum of errors across trials is largely comparable if the course of errors across trials is inspected. Though the S1 subject needed six trials to fulfill the learning criterion of two consecutive error-free trials whereas the S2 subject showed error-free performance immediately after the exploration phase, the total number of errors displayed within acquisition trials was quite low even in subject S1. Moreover, the exploration trials were performed comparably with nine and ten errors, respectively. Even the probe trial yielded comparable results for both subjects with seven errors in S1 and five errors in S2.

A distinct pattern of results emerges when the paths taken by the subjects are analyzed (Fig. 3).

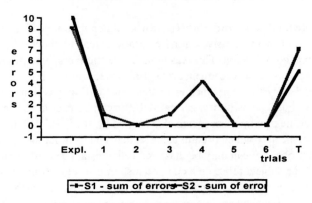

S1, male, 20 years old S2, female, 24 years old

Fig. 2. Sum of errors across acquisition trials and in the final probe-trial of two experimental subjects

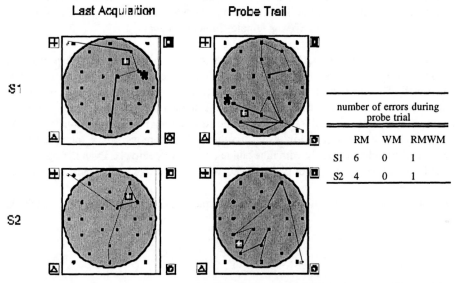

	number of errors during probe trial		
	RM	WM	RMWM
S1	6	0	1
S2	4	0	1

Fig. 3. Performance in the last acquisition trial and the probe trial of subjects S1 and S2, respectively. For simplicity only one proximal cue is shown; * denotes an example for two locations which are equivalent with respect to the S1 viewer's perspective (because the proximal cue and the subject's starting position were rotated). (RM = reference memory errors; WM = working memory errors; RMWM = reference-/ working memory errors)

Though subject S1 was comparably good at acquiring the locations within the spatial layout, his first probe trial move was directed towards a location which was equivalent to step 2 within his last acquisition phase (see * in Fig. 3, upper row). His search behavior in the probe trial was then bound to the 'southern' part of the environmental space which was identical to the 'northern' part of the environment during acquisition. The search seems to be guided by the proximal cue located in the 'northern' part during acquisition and in the 'southern' part during probe trial testing. On the contrary, probe trial manipulation did not much affect subject S2 because she initially moved towards the 'northern' part of the experimental chamber.

Discussion. Obviously the two experimental subjects oriented in different ways. The S2 subject showed an errorless performance immediately after the exploration phase. According to O'Keefe and Nadel [22] this 'one trial learning' is indicative for the locale system. But overall, performance across acquisition trials did not differ remarkably between subjects (Fig. 2). On the contrary, probe trial behavior showed that while the S1 subject seemed to be bound to a cue strategy, the S2 subject was initially able to maintain an orientation-free behavior. Thus, it has to be assumed that she oriented with respect to the distal cues. Using the locale and not the cue (or taxon) system she was enabled to update her position in the very difficult probe trial condition. Since the viewer's perspective of the experimental chamber looked alike irrespective of whether the subject's position was in the 'North-West' or in the 'South-East', the task could only be solved if the subjects relied on distal cue information.

3.2 Dissociating spatial reference and working memory errors

Background. As outlined in the introduction the dissociation of spatial reference memory errors and spatial working memory errors has been shown to be of major importance in animal research investigating the effects of different cerebral lesions. With respect to this matter the basal ganglia, a subcortical structure of the telencephalon, are of specific importance. Lesions within these ganglia have been shown to lead to deficiencies in sequencing motor acts and switching behavior from one mode of response to another. In general, behavioral and cognitive responses deteriorate if the algorithms necessary to solve these tasks have to be generated in the absence of external cues or if these algorithms have to be adapted rapidly to changing task characteristics and varying response demands. The concept of reference memory errors is an example of such an algorithm. Since the position of the correct and incorrect locations remains stable across trials it incorporates the development of a 'win-stay' rule [28]. On the contrary, working memory depends on the updating of one's ongoing behavior and thus remains valid only for the current trial. If patients with lesions of the basal ganglia are exposed to a locomotor maze which enables the investigator to dissociate reference memory errors and working memory errors, these patients should display persistent reference memory errors.

Procedure, Recordings, and Participants. Within the working memory paradigm, subjects had to learn and remember the five correct locations out of twenty locations and not to return to any of these correct locations during the same trial. Revisiting a previously visited correct location was called a 'working memory' error and was considered functionally equivalent to visiting a previously visited baited arm in the radial arm maze paradigm of Olton et al. [26]. Stepping on a detector of an incorrect location was called a 'reference memory' error and was considered functionally equivalent to visiting an arm without food in a radial maze. Pressing an incorrect location in one trial more than once was considered a 'reference-working' memory error. These three types of spatial memory errors were recorded for each acquisition trial. Two patients with Parkinson's disease, a degenerative disorder with known lesions within the basal ganglia, and two age and sex-matched controls were taken from an ongoing study [15]. Both patients had verified diagnoses but differed with respect to age and duration of disease (Table 1).

Results. Fig. 4 shows that the PD1 patient not only displayed the largest number of errors (see Table 1) but that the distribution of error types markedly differs between participants.

Table 1. Characteristics of Parkinson patients (PD) and controls (PC)

	PD1	PC1	PD2	PC2
age	66	66	35	35
sex	male	male	female	female
diagnosis of disease	7 yrs.	-	2 weeks	-
number of trials	6	7	9	5
criterion	no	yes	yes	yes
sum of errors	157	10	40	16
mean of errors	10.7	1.3	3.2	3.0
error type				
reference memory	64	9	29	12
mean reference	10.7	1.3	3.2	3.0
working memory	27	0	8	3
mean working	4.5	0	0.9.	0.8
reference-working memory	66	1	3	1
mean reference-working	11	0.1	1.0	0.3

Thus, the PD1 patients showed a constant rate of about eleven reference memory errors without any sign of improvement. On the contrary, the rate of working memory errors decreased across trials 1-4 to close to zero and then increased to a score near twenty in trial 5. Such a dissociation of error types was absent in the two control subjects who displayed hardly any working memory errors and constantly improved with respect to reference memory errors. Moreover, the recently diagnosed, younger PD2 patient showed a pattern of results similar to that of the healthy subjects. Especially, the slope of reference memory errors decreased to quite a similar degree to that of her age-matched control. Though the two unimpaired subjects differed by 30 years in age the distribution of their error types was roughly similar.

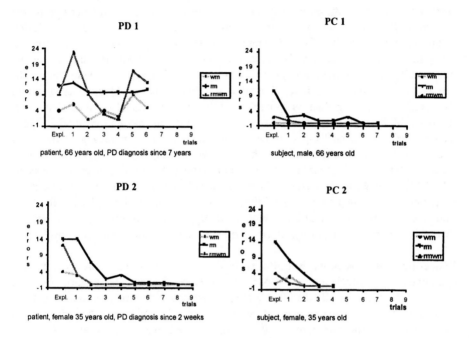

Fig. 4. Types of spatial memory errors across acquisition trials in two Parkinsonian patients (PD) and their controls (PC) (wm: working memory error; rm: reference memory error; rmwm: reference-/ working memory error)

Discussion. This presentation underlines the use of dissociating error types because persisting reference memory errors are exactly what can be predicted from a basal ganglia related disease. Moreover, this typical result of a PD patient corresponds to animal research with experimental lesions within the striatum, an important part of the basal ganglia. These lesions usually lead to severe deficits in procedural learning. Thus, the reference memory component may be an indicator for specific alterations in spatial behavior in advanced Parkinson's disease which are not reflected by the total sum of errors.

3.3 Identifying place orientation

Background. 'Place' strategies [25] have been shown to depend on distal cue configurations. If the spatial task is manipulated in such a way that distal and proximal cues are dissociated, it can be shown that spatial orientation behavior in younger children [10] and in brain impaired patients [14] is controlled by proximal cues. Thus, it can be concluded that place orientation has not occurred. In order to investigate the nature of place orientation in more detail, the actual behavior of the participants in the experiment has to be quantified. For this task, at least two aspects seem to be important: the participant's rotational behavior and the distances he or she is moving. Traveling through our experimental chamber necessarily induces a rotation around the vertical body axis. This angle is of theoretical and practical importance. Since 'head direction cells' have been identified in the dorsal presubiculum, a structure functionally related to the hippocampus, it is known that at least in rodents these cells fire with respect to their 'preferred direction' in extrapersonal space and irrespective of the actual place an organism is moving to [32, 33, 34]. Based on these and other findings, McNaughton et al. [16] proposed an inertial navigation model which incorporates a so-called 'H'-part recording angular size (i.e. by means of the 'head direction cells') and an 'H''-part which computerizes angular velocity and which is located within the vestibular and sensorimotor systems. Some cells of both H-system-parts are presumed to converge to a so-called H-H'-system. The whole system is supported to be started by means of the hippocampal 'local view'-'place cells' [24]. Thus it can be expected that lesions within the hippocampal formation should lead to severe disturbances in acquiring information between local views and directions [16].

Procedure, Recordings, and Participants. For our purposes we used the 'H'-component from McNaughton's model and included distance information [3]. We calculated the mean of all rotational moves a participant performed within one trial. This angular 'A'-component was compared with a second angular measure derived from the distance a participant moved after having turned his/her body axis. For this calculation each angle was divided by the distance moved. The mean of all divisions within one trial served as the second measure, denoted henceforth as relative angular or 'Ar'-component. If a participant shows a large 'A'-component and a comparably smaller 'Ar'-component, he or she has traveled long distances with respect to the angular turns. If, in addition, the error rates are low, it can be concluded that he or she was able to move well throughout space, obtaining a large number of perspectives and approaching the goals only. If, on the contrary, a participant shows an 'Ar'-component which is larger than the 'A'-component, it means that he or she has traveled rather short distances with respect to angular turns. This pattern of results would be typical for inefficient, stereotyped behavior.

The following examples serve to demonstrate the relationship between the 'A'- and 'Ar'-components of the navigational system and error rates in patients with circumscribed cerebral lesions and unimpaired subjects. For this purpose four patients with cerebral tumors and four age and sex-matched controls were taken from another ongoing study [14]. The patient characteristics are shown in table 2. Results are presented by showing the paths taken by each participant within the exploration trial, the composed paths of acquisition trials 1 to 3, the last acquisition trial and the probe trial, which was performed twenty minutes later from a starting point rotated by 90°. In this investigation, only distal stimuli were used. The position of these stimuli remained constant throughout the experiments.

Table 2. Characteristics of patients with cerebral tumors (CT)

CT1	large right frontal meningeoma, affecting the corpus callosum
CT2	circumscribed small right hippocampal astrocytoma
CT3	large left frontal, anterior temporal meningeoma
CT4	medium-sized, right temporal, parahippocampal glioblastoma

Results. Fig. 5 shows the paths of the participants and the courses of the 'A'- and the 'Ar'-measures, respectively. The CT1 patient was obviously not able to solve the task and after the sixth acquisition trial the experiment had to be aborted. As can be seen from Fig. 5, this patient was not only unable to acquire a strategy, but she also displayed the expected 'A-Ar' discrepancy. This is indicated by the relatively large 'Ar'-component compared to the 'A'-part of the 'A-Ar'-system. Contrary to this 'A < Ar'-behavior her control (TC1) not only reached the learning criterion after eight trials but also showed a corresponding 'A > Ar'-behavior, as indicated by the last graph in the upper row of Fig. 5. A completely different picture is displayed by the CT2 patient and his TC2 counterpart. Both participants displayed 'one trial learning' and showed the optimal 'A > Ar' pattern but the CT2 patient performed very poorly in the delayed probe trial condition (Fig. 5).

A largely stereotyped behavior is revealed both by the paths and the 'A-Ar'-system of the CT3 patient (Fig. 5). Despite a relatively small overall number of errors he showed a circular exploration behavior which was largely reproduced within his last acquisition trial. Furthermore and in correspondence with the theoretical assumptions, this patient displayed an 'A < Ar'-behavior whereas his control subject showed the 'A > Ar'-behavior. Since the CT3 patient did not reach the learning criterion and was obviously unable to display an efficient spatial strategy to solve the task, he showed extremely poor performance within the subsequent probe trial.

The CT4 patient showed paths which are largely similar to those of his TC4 counterpart. But although he made nearly twice the number of errors within acquisition trials 1 to 3 and though he tried unsuccessfully for more than twice the number of trials to reach the criterion the behavioral pattern as shown by the paths does not seem to be very different from that of his control (Fig. 5). Again, and more importantly, the CT4 patient displayed an 'A < Ar'-behavior throughout the experiment, while his counterpart was successful in developing an 'A > Ar'-behavior from acquisition trial 3 on. This may, at least partly, account for CT4's poor performance in the delayed probe trial.

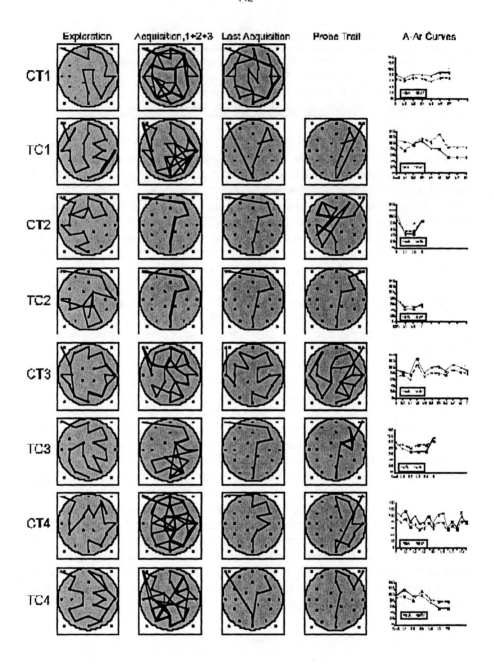

Fig. 5. Paths of patients with circumscribed brain lesions (CT) and their controls (TC). Last column: 'A-Ar'-system (further explanation see text). Ar - component: bold lines; A - component: dotted lines

Table 3. Results from tumor patients (CT) and their controls (TC)

	sum expl.	no. trials	mean err.	mean 1-3	crit.	probe
CT1	9	6	20.0	19.0	no	-
TC1	16	8	5.6	12.0	yes	0
CT2	12	2	0.0	0.0	yes	16
TC2	11	2	0.0	0.0	yes	0
CT3	22	9	16.1	9.0	no	22
TC3	12	4	2.3	3.0	yes	7
CT4	11	14	10.0	20.0	no	4
TC4	13	6	6.7	3.3	yes	2

Note: sum expl. = sum of errors during the exploration phase; no. trials = number of trials necessary to reach the learning criterion; mean err. = mean sum of errors during acquisition; mean 1-3 = mean sum of errors during acquisition trials 1-3; crit. = learning criterion; probe = probe trial.

Discussion. The results of this investigation show that it is possible to identify different types of spatial problem solving strategies and to relate the breakdown of these strategies to different cerebral lesions. For example, the CT1 patient with a large right frontal tumor including the corpus callosum was obviously unable to develop orientation-free behavior, as indicated both by her paths in the experimental chamber and by her 'A-Ar'-pattern. On the contrary, the 'A > Ar'-behavior of her control may indicate the activation of a viewer-independent, 'place'-orientation. From the results of the CT2 patient, however, it can be concluded that the deterioration in probe trial performance did not result from poor strategy development as in the previous CT1 patient. This is in accordance with the cerebral lesion of the CT2 patient which was restricted to the hippocampus. Thus, it can be concluded that in this patient the major deficit may be attributed to a consolidation deficit of a spatial layout which was successfully acquired.

A different type of spatial problem solving deficiency is revealed by the CT3 recording. This patient showed a highly stereotyped behavior which is often seen in experimental animals. Though the hippocampus was spared in the CT3 patient with a left frontal/anterior temporal tumor, his circling behavior at the outer border of the maze strikingly resembles that of rodents with hippocampal lesions. Again, the deficiencies were reflected both by the paths obtained and the 'A < Ar'-pattern. The CT4 patient with a right temporal tumor including the parahippocampal region showed a strategy deficit which seems to be less severe than in the previous patients. Since he had a temporal lobe tumor and his paths resembled that of the TC4 participant, the patient's probe trial performance may at least partly be due to a temporal lobe-specific memory deficit.

444

4 General Discussion

The aim of the present investigation was to study spatial behavior within a maze-like environment for humans. The following criteria had to be fulfilled: (1) Spatial orientation and spatial memory should include gross motor behavior, (2) assessment of spatial behavior should be performed under completely controlled cue conditions, (3) investigation of spatial behavior should be conceptually equivalent to animal studies, (4) the spatial task should allow the investigation of spatial reference and spatial working memory, and (5) basic assumptions of one of the most prominent theories of spatial behavior should be testable. For this purpose, a locomotor maze was developed incorporating basic features of the Morris Water Maze [18] and the Radial Maze of Olton et al.[26].

In the present report we documented the principles of the experimental setup, data recording, methods of data analysis and its interpretation. In order to underscore the benefit of this approach selected patients and unimpaired subjects were examined. Group studies are now required to examine interindividual differences and their determinants. What we have shown is that it is possible to comply with the requirements mentioned above and in the introduction. Moreover, it is of theoretical and practical interest that we have outlined a procedure for the assessment of cue dependent behavior in contrast to place orientation, for the dissociation of spatial working memory and spatial reference memory errors, and to identify place behavior, which we have defined as a behavior characterized by large distance with respect to angular turn (i.e. 'A > Ar'). Obviously, this type of response pattern is highly associated with low error rates, whereas the opposite behavior characterized by short distance with respect to angular turns is associated with high error rates.

This last level of analysis has to be extended with respect to the velocity component. Up to now we have only included the angular transposition of a participant's body and hence only the so-called 'H'-part of the HH'-system of McNaughton et al. [16]. Though we integrated the distances of the moves between locations which are not part of McNaughtons's system, velocity information (i.e. distances and angles by time, respectively) has to be taken into account too. This angular velocity corresponds to the H'-part of the HH'-system.

Acknowledgment

The authors are indebted to Dipl.-Ing. Arne Herzog, an engineer who intensively supported us by working on our hardware and data recording techniques. In addition, we wish to thank cand. phil. Franka Weber, cand. phil. Roy Murphy, and cand. phil. Thorsten Schütze who worked in this project as student research assistants.

References

1. Colombo, P.J., Davis, H.P. & Volpe, B.T. (1989). Allocentric spatial and tactile memory impairments in rats with dorsal caudate lesions are affected by preoperative behavioral training. Behavioral Neuroscience, 103, 1242-1250.
2. Eichenbaum, H., Stewart, C. & Morris, R.G.M. (1990a). Hippocampal representation in spatial learning. Journal of Neuroscience, 10, 331-339.
3. Eichenbaum, H., Stewart, C. & Morris, R.G.M. (1990b). Hippocampal representation in place learning. Journal of Neuroscience, 10, 3531-3542.
4. Farah, M.J., Brunn, J.L., Wong, A.B., Wallace, M.A. & Carpenter, P.A. (1990). Frames of reference for allocating attention to space: Evidence from the neglect syndrome. Neuropsychologia, 28, 335-347.
5. Foreman, N., Arber, M. & Savage, J (1984). Spatial memory in preschool infants. Developmental Psychobiology, 17, 129-137.
6. Foreman, N., Foreman, D., Cummings, A. & Owens, S. (1990). Locomotion, active choice, and spatial memory in children. The J. of General Psychology, 117, 215-232.
7. Foreman, N., Warry, R. & Murray, P. (1990). Development of reference and working spatial memory in preschool children. The Journal of General Psychology, 117 (3), 267-276.
8. Foreman, N,. Gillet, R. & Jones, S. (1994). Choice autonomy and memory for spatial locations in six-year-old children. British Journal of Psychology, 85, 17-27.
9. Howard, I.P. & Templeton, W.B. (1966). Human spatial orientation. London: Wiley. Kirk, R.E. (1982). Experimental designs: Procedures for the behavioral sciences. Belmont: Brooks & Cole.
10. Lehnung, M., Leplow, B., Friege, L., Herzog, A., Mehdorn, M. & Ferstl, R. (1997). Development of spatial memory and spatial orientation in preschoolers and primary school children. Accepted for publication by "British Journal of Psychology".
11. Leonard, B.J. & McNaughton (1990). Spatial representation in the rat: Conceptual, behavioral and neurophysiological perspectives. In R.P. Kesner and D.S. Olton (Eds.) Neurobiology of comparative cognition, 363-422. Hillsdale: Lawrence Erlbaum.
12. Leplow, B. (1994). Diesseits von Zeit und Raum: Zur Neuropsychologie der räumlichen Orientierung (Neuropsychology of spatial orientation). Habilitation thesis University of Kiel.
13. Leplow, B. (1997). Experimentelle Analyse räumlicher Orientierungs- und Gedächtnisleistungen (experimental analysis of spatial orientation and spatial memory performance). Submitted for publication to "Zeitschrift für Experimentelle Psychologie".
14. Leplow, B., Höll, D., Zeng, L., Behrens, Chr. & Mehdorn, M. Spatial behavior within a Locomotor Maze for patients with focal brain lesions and healthy subjects. Submitted to "Neuropsychologia".
15. Leplow, B., Höll, D., Behrens, K., Zeng, L., Deuschl, G. & Mehdorn, M. Deficits of spatial orientation in patients with Parkinson's Disease. In preparation for submission to "Neurology".
16. McNaughton, B.L., Chen, L.L. & Markus, E.J. (1991). 'Dead reckoning,' landmark learning, and the sense of direction: A neurophysiological and computational hypothesis. Journal of Cognitive Neuroscience, 3, 190-202.
17. May, M., Péruch, P. & Savoyant, A. (1995). Navigating in a virtual environment with map -acquired knowledge: Encoding and alignment effects. Ecological Psychology, 7, 21-36.

18. Morris, R.G.M. (1981). Spatial localization does not require the presence of local cues. Learning and Motivation, 12, 239-260.

19. Morris, R.G.M., Garrud, P., Rawlins, J.N.P. & O'Keefe, J. (1982). Place navigation impaired in rats with hippocampal lesions. Nature, 297, 681-683.

20. Nadel, L. (1990). Varieties of spatial cognition. Psychological considerations. Annals of the New York Academy of Sciences, 163-636.

21. Nadel, L. (1991). The hippocampus and space revisited. Hippocampus, 1, 221-229.

22. O'Keefe, J. & Nadel, L. (1978). The hippocampus as a cognitive map. Oxford: Clarendon Press.

23. Okaichi, H. & Oshima, Y. (1990). Choice behavior of hippocampectomized rats in the radial arm maze. Psychobiology, 18, 416-421.

24. O'Keefe, J., Dostrovsky, J. (1971). The hippocampus as a cognitive map. Preliminary evidence from unit activity in freely moving rat. Brain Research, 34, 171-175.

25. O'Keefe, J. & Nadel, L. (1978). The hippocampus as a cognitive map. Oxford: Clarendon Press.

26. Olton, D.S., Becker, J.T. & Handelmann, G.E. (1979). Hippocampus, space and memory. The Behavioral and Brain Sciences, 2, 313-365.

27. Overman, W.H., Pate, B.J., Moore, K. & Peuster, A. (1996). Ontogenecy of place learning in children as measured in the radial arm maze, Morris search task, and open field task. Behavioral Neuroscience, 110, 1205-1228.

28. Packard, M.G., Hirsh, R. & White, N.M. (1989). Differential effects of fornix and caudate nucleus lesions on two radial maze tasks: Evidence for multiple memory systems. The Journal of Neuroscience, 9, 1465-1472.

29. Presson, C.C. & Hazelrigg, M.D. (1984). Building spatial representations through primary and secondary learning. Journal of Experimental Psychology: Learning, Memory and Cognition, 10, 716-722.

30. Presson, C.C., deLange, N. & Hazelrigg, M.D. (1987). Orientation-specificity in kinesthetic spatial learning: The role of multiple orientation. Memory and Cognition, 15, 225-229.

31. Presson, C.C., deLange, N. & Hazelrigg, M.D. (1989). Orientation specificity in spatial memory: What makes a path different from a map of the path? Journal of Experimental Psychology, Learning, Memory, and Cognition, 15, 887-897.

32. Ranck, J.B. (1973). Studies on single neurons in dorsal hippocampal formation and septum in unrestrained rats. I. Behavioral correlates and firing repertoires. Experimental Neurology, 41, 461-531.

33. Taube, J. S., Muller, R. U. & Ranck, J. B. (1990a). Head-direction cells recorded from the postsubiculum in freely moving rats. I. Description and quantitative analysis. Journal of Neuroscience, 10, 420-435.

34. Taube, J. S., Muller, R. U. & Ranck, J. B. (1990b). Head-direction cells recorded from the postsubiculum in freely moving rats. II. Effects of environmental manipulations. Journal of Neuroscience, 10, 436-447.

35. Thorndyke, P.W. & Hyes-Roth, B. (1982). Differences in spatial knowledge acquired from maps and navigation, Cogntivie Psychology, 14, 560-589.

36. Tolman, E.C. (1932). Purposive behavior in animals and men. Appleton-Century-Crofts.

37. Tolman, E.C. (1949). There is more than one kind of learning. Psychological Review, 56, 144-155.

38. Zeng, L., Leplow, B., Höll, D. & Mehdorn, M. A computerized model for inertial navigation behavior within a locomotor maze. Prepared for submission.

Behavioral Experiments in Spatial Cognition Using Virtual Reality

Hanspeter A. Mallot, Sabine Gillner, Hendrik A.H.C. van Veen,
and Heinrich H. Bülthoff

Max–Planck–Institut für biologische Kybernetik
Spemannstr. 38, D-72076 Tübingen, Germany
http://www.kyb.tuebingen.mpg.de

Abstract. Virtual reality is used as a novel tool for behavioral experiments on humans. Two environments, Hexatown and Virtual Tübingen, are presented. Experiments on cognitive maps carried out in the Hexatown environment are reported in this paper. Results indicate that subjects are able to acquire configuration knowledge of the virtual town even in the absence of physical movement. Simpler mechanisms such as associations of views with movements are also present. We discuss the results in relation to a graph–theoretic approach to cognitive maps.

1 General Introduction

1.1 Mechanisms and competences

In animal navigation, at least three basic mechanisms of spatial memory have been identified which we will refer to as path integration, guidance, and direction. Guidance and direction use local position information, i.e., sensory input characteristic of a given place, but use this information in different ways. Path integration uses egomotion data and can function even in the absence of local position information. In more detail, the characteristics of the three mechanisms are as follows:

- *Path integration* or dead reackoning is the continuous update of the egocentric coordinates of the starting position based on instantaneous displacement and rotation data (see Maurer & Séguinot 1995 for review). Odometry data are often taken from optic flow but other modalities such as proprioception (e.g., counting steps) may be involved as well. Since error accumulation is a problem, the use of global orientation information ("compasses", e.g., distant landmarks or the polarization pattern of the skylight) is advantageous. Path integration involves some kind of working memory in which only the current "home–vector" (coordinates of the starting point) is represented, not the entire path.

- *Guidance* is a mechanism in which the navigator keeps a fixed or otherwise well–defined orientation with respect to a landmark or a group of landmarks (see O'Keefe & Nadel, 1987, p. 82). A well studied guidance mechanism in insect navigation is snapshot–based homing, i.e. the approach of a place whose local view matches a stored "snapshot" (Cartwrigth & Collett 1982). This mechanism requires long–term storage of the view or snapshot visible at that point. From a comparison of the stored view with the current view, an approach direction can be derived. Moving in this direction will lead to a closer match between the two views (Cartwright & Collett 1982, Mallot et al. 1997). More generally, "view" and "snapshot" may be replaced by local position information from any sensory modality.

- *Direction* is the association of a recognized view (local position information) to a movement. As for guidance, long–term memory of the local position information (view) is required. In addition to that, a movement direction is stored, i.e., the recognized view acts as a pointer to some other goal. The existence of such associations has been shown in bees (Collett & Baron 1995) and humans (Mallot & Gillner 1997). The direction mechanism can be generalized to associate views with more complex behaviors such as wall following or passing through a door (Kuipers & Byun, 1988).

For a more comprehensive discussion of these and other mechanisms, see O'Keefe & Nadel (1978; page 80ff) and Trullier et al. (1997).

Distinguishing between these mechanisms leads to a related distinction between different types of landmarks. We use the term "global landmarks" for distant markers serving as a compass system in path integration. In contrast, the terms "local landmark" or "view" refer to local position information as used for guidance and direction. Note that the same object can have both roles in a given scene.

Using these basic mechanisms, different levels of complexity of spatial knowledge and behavior can be formulated. Concatenating individual steps of either guidance or direction results in routes. These routes will be stereotyped and could be learnt in a reinforcement scheme. More biologically plausible, however, is instrumental learning, i.e., the learning of associations of actions with their expected results. This can be done step–by–step without pursuing a particular goal (latent learning). Instrumental learning entails an important extension of the two view–based mechanisms in that the respective consequences of each of a number of possible choices (either movements or snapshots to home to) are learnt. This offers the possibility of dealing with bifurcations and choosing among alternative actions. Thus, the routes or chains of steps can be extended to actual graphs which are a more complete representation of space, or cognitive map (Schölkopf & Mallot 1995, Mallot et al. 1997). The overall behavior is no longer stereotyped but can be planned and adapted to different goals.

1.2   The view–graph approach to cognitive maps

We discuss the representation of spatial configurations in graph structures for the direction mechanism. Similar schemes can be developed for the other mechanisms or combinations thereof. As the basic element, let us consider the association of

Landmark plus Vector	Route	Configuration

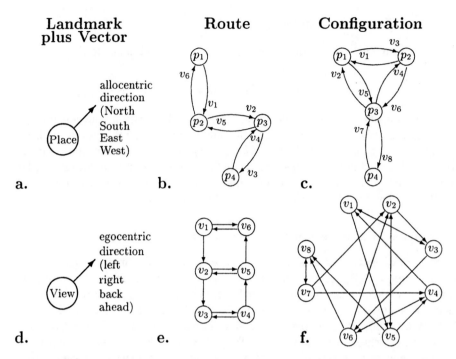

Fig. 1. The graph approach to space representation. Top row (**a.–c.**): Place–graphs. The nodes are places recognized irrespective of body orientation, the links (arrows) between them carry allocentric direction information. Bottom row (**d.–f.**): View–graphs. The nodes are recognizable views or other positional information (i.e., depend on the observer's viewing direction) and the arrows carry directional information relative to gaze. Each view v_i in Figs. **e.** and **f.** corresponds to a directed connection in Fig. **b.** and **c.** From left to right, increasingly complicated spatial layouts are shown.

a recognized view with a movement direction. Additionally, we store the view expected when performing this movement:

$$(\text{current view}, (\text{movement direction}, \text{expected next view})). \qquad (1)$$

This association is illustrated in Fig. 1d. When going from one view to the next, navigation can initially follow the movement direction associated with the present view. The additional information on what view to expect next can be used to improve view recognition at the next step. If direction and snapshot–based homing are to be combined, the expected next view is the one to home to.

Chains of such association structures implement a route memory. If different routes are to be learned that share some common section, the decision at the crossroads requires more complicated memory. One way to think of this memory is to store all possible connections

$$(\text{current view}, (\text{movement direction } 1, \text{expected next view}), \qquad (2)$$
$$\vdots$$
$$(\text{movement direction } n, \text{expected next view})),$$

and have a separate planning device select one of the possible movements. The resulting memory structure is a graph of views and movement decisions as shown in Fig. 1e,f. A neural network theory for storing the required information in the form of a labelled graph has been presented by Schölkopf and Mallot (1995). For related approaches including hippocampal modelling, see Touretzky and Redish (1996), Muller, Stead & Pach (1996), and Prescott (1996).

1.3 Behavioral experiments in virtual reality

In the work reported in this paper, we chose interactive computer graphics, or virtual reality (VR), as our experimental method. Previous studies using virtual reality have focussed on the transfer of knowledge between different media used for acquisition and testing. May, Péruch & Savoyant (1995) and Tlauka & Wilson (1996), for example, have tested map–acquired knowledge in a pointing task performed in virtual reality. Tong, Marlin & Frost (1995), using a VR bicycle, showed that active exploration leads to better spatial knowledge than passive stimulus presentation. Sketch maps produced after exploration of various virtual environments have been studied by Billinghurst & Weghorst (1995). Design principles for constructing easy–to–navigate virtual environments have been studied by Darken & Sibert (1996). Ruddle et al. (1997), using a VR–setup similar to the one used in this study, showed that navigation performance in a simulated indoor environment is essentially as good as found in experiments carried out in real buildings. In the present paper, we use virtual reality to isolate the various cues used for the build–up of spatial knowledge and to study the underlying mechanisms. The advantages of virtual reality for this application are (i) the high controllability of computer graphics stimuli, and (ii) the easy access to behavioral data, such as the subject's movement decisions.

Stimulus control. When investigating the information sources used in navigation, it is advantageous to be aware of the exact movement trajectories of the subjects and the visual information available along these trajectories. This can easily be achieved with interactive computer graphic (see Section "Methods"). The various parameters of the sensory input can be easily separated. For instance, in our experiments, we varied the number of buildings visible simultaneously in one view without changing the illumination etc. In real world experiments, such separate stimulus conditions are much harder to realize. Another interesting experimental paradigm is the modification or exchange of various features of the environment after learning. Aginsky, Harris, Rensink & Beusmans (1996) exchanged landmarks after training in a route–learning task. The effects of landmark exchange on navigation have been addressed by Mallot & Gillner (1997).

In principle, the method also allows complete control over vestibular and proprioceptive feedback. In our experiments, for instance, both were completely absent allowing the effects of visual input to be studied in isolation.

Measuring behavior. Navigation performance can be accessed most directly by the paths or trajectories that the subjects take during the exploration. In virtual reality experiments, egomotion is very simple to record, since it is equivalent to the course of the "simulated observer" used for rendering the computer graphics. In this paper, we present a number of novel techniques for data evaluation that are particularly suited for the virtual reality experiments described.

2 Virtual environments

The experiments which we report in this paper have all been conducted in a specially designed virtual village called "Hexatown". This section describes the layout and construction of this village and explains how subjects interact with it. We also list and motivate the different stimulus conditions that are used in the experiments. A valid question that one can ask is whether results obtained using this kind of virtual environments can be transferred to navigation in the natural environment. The second part of this section describes another virtual environment that we are developing and that hopefully will help to address this difficult question.

The virtual environments were constructed using Medit 3D–modelling software and animated with a framerate of 36 Hz on a SGI Onyx RealityEngine^2 using IRIX Performer software.

2.1 Hexatown

Geometry. A schematic map of Hexatown appears in Fig. 2a. It is built on a hexagonal raster with a distance between two places of 100 meters. At each junction, one object, normally a building, was located in each of the 120 degree angles between the streets; thus each place consisted of three objects. In the places with less than three incoming streets, dead ends were added instead, ending with a barrier at about 50 meters. The whole town was surrounded by a distant circular mountain ridge showing no salient features. The mountains were constructed from a small model which was repeated periodically every 20 degrees.

An aerial view of Hexatown is shown in Fig. 3[1]. It gives an impression of the objects used. The spacing and position of the trees corresponds to viewing condition 1 and 2 (see below).

The hexagonal layout was chosen to make all junctions look alike, such that no information is contained in the geometry of the junctions themselves. In rats

---

[1] Color versions of Figures 3 and 4 are available from the world wide web, http://www.kyb.tuebingen.mpg.de/links/hexatown.html.

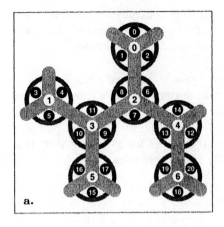 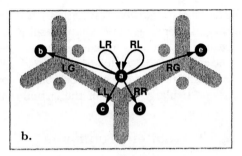

Fig. 2. a. Street map of the virtual maze with 7 places numbered 0 – 6 and 21 views numbered 0 – 20. The ring around each place indicates the hedges used in viewing conditions 1 – 3 to occlude distant places. **b.** Possible movement decisions when facing the view marked a. L: turn left 60 degrees. R: turn right 60 degrees. G: go ahead to next place.

navigating simple cross–like mazes with non–orthogonal arms, it has been shown that geometry may be more important than landmarks presented as cues. When a conflict between local landmarks placed within the arms and geometry information was introduced, rats behaved as if they would ignore the landmarks and followed the geometry information (Cheng 1986). Since in this study, we want to limit the available information to local landmarks ("views"), the hexagonal layout seems appropriate. An additional problem arising in Cartesian grids (city–block raster) is the fact that long corridors are visible at all times and the possible decisions at a junction are highly unequal: going straight to a visible target or turning to something not presently visible.

Simulated movements. Subjects could move about the town using a computer mouse. In order to have controlled visual input and not to distract subject's attention too much, movements were restricted in the following way. Subjects could move along the street on an invisible rail right in the middle of each street. This movement was initiated by hitting the middle mouse button and was then carried out with a predefined velocity profile without further possibilities for the subject to interact. The translation took 8.4 seconds with a fast acceleration to the maximum speed of 17 meters per second and a slow deceleration. The movement ended at the next junction, in front of the object facing the incoming street. 60 degree turns could be performed similarly by pressing the left or right mouse button. Again, the simulated movement was "ballistic", i.e., following a predefined velocity profile. Turns took 1.7 seconds with a maximum speed of 70 degrees per second and symmetric acceleration and deceleration.

Fig. 3. Aerial view of Hexatown. Orientation as in Fig. 2a. The white rectangle in the left foreground is view 15, used as "home"–position in our experiments. The aerial view was not available to the subjects. Object models are courtesy of Silicon Graphics, Inc., and Prof. F. Leberl, Graz.

Fig. 2 shows the movement decisions that subjects could choose from. Each transition between two views is mediated by two movement decisions. When facing an object (e.g., the one marked "a" in Fig. 2), 60 degrees turns left or right (marked "L", "R") can be performed which will lead to a view down a street. If this is not a dead end, three decisions are possible: the middle mouse button triggers a translation down the street (marked "G" for go), while the left and right buttons lead to 60 degrees turns. If the street is a dead end, turns are the only possible decision. In any case, the second movement will end in front of another object.

Viewing conditions. We used four stimulus conditions with varying degrees of visibility of the environment. In condition 1, navigation is view–based in the strictest sense, i.e. subjects have to rely on local views only. The other conditions contain increasingly more non–local information. For example views occuring in the four conditions, see Fig. 4.

In condition 1, a circular hedge or row of trees was placed around each junction with an opening for each of the three streets (or dead ends) connected to that junction. This hedge looked the same for all junctions and prevented subjects from seeing the objects at more distant junctions. The objects were placed 22 meters away from the center of the junction. The arrangement was such that when entering the circular hedge, the buildings to the left and right were already outside the observer's field of view (60 degrees). Thus, the three

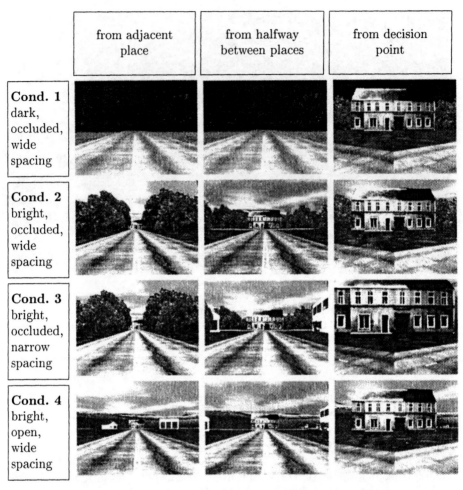

	from adjacent place	from halfway between places	from decision point
Cond. 1 dark, occluded, wide spacing			
Cond. 2 bright, occluded, wide spacing			
Cond. 3 bright, occluded, narrow spacing			
Cond. 4 bright, open, wide spacing			

Fig. 4. Viewing conditions used in the experiments. For each condition, three views are shown. The views in the left column occur when looking from a place (No. 5) into a street. The views in the middle column can be seen during the motion along a street, in this example from place 5 to place 3. The views in the right column show an object (view 11) as seen from the corresponding junction (place 3).

buildings at one junction could never be seen together in these conditions. The sky was dark, as if navigating the town at night, and illumination was as with a torch or the headlights of a car reaching about 60 meters. Thus, the building at the far end of a street was not visible in this condition.

Condition 2 was the same as condition 1, except that illumination now came form the bright sky. While this changed the overall impression considerably, the only additional information provided in condition 2 was the view at the far end of the street, which was now visible from the adjacent place.

Fig. 5. View of a preliminary version of the Virtual Tübingen model. The buildings appearing in white are geometry models without textures being added. The other buildings show complete geometry and texture mapping.

In condition 3, the building were placed closer to the junction point, at a distance of 15 meters. This had the effect that all three buildings of a place could now be seen at a glance when entering the circular hedge. Condition 3 thus provided place information, rather than mere local views.

Condition 4 was the same as condition 2, but now the hedges were removed. Therefore, subjects could not only see all buildings from the place being currently approached, but also more distant places. In this condition, landmark information is no longer locally restricted.

2.2 Virtual Tübingen

An artificial virtual environment can be designed to fit the requirements of the experiments that will be conducted in it. In the case of Hexatown, for instance, the hexagonal grid of roadways was explicitly constructed to force movement decisions at every junction. If, however, one would like to verify the applicability of experimental results obtained with virtual environments like Hexatown for navigation in the real world, one should conduct a comparative study of navigation behavior in real and virtual environments. This notion has motivated us to create a virtual copy of an existing environment: Virtual Tübingen.

We have started to construct a virtual model of the historic centre of Tübingen, a 600 by 400 m area densely packed with approximately 600 buildings. The richness and complexity of the buildings and street network in this part of the city provide a very interesting environment for navigation studies. The shapes and façades of the buildings are all different, and streets often tend to be curved and have varying width. They form an irregular network reflecting the topographical situation of Tübingen on a ridge rising between two valleys. Correspondingly, altitudes vary considerably and the resulting changes in slope add to the complexity of the navigation cues. Clearly, the construction of such a model is a long–term project and the current state of the model is far from being complete. Figure 5 shows an aerial view of an early version of Virtual Tübingen, extending from the city hall and the market place (on the left) to the north side of the "Holzmarkt" square. More information about the construction of this model can be found in Van Veen et al. (in press). Once this model has been completed, experiments on spatial behavior will be conducted both in real Tübingen and in its virtual counterpart. If the results obtained in both environments are consistent with each other, further experiments can be performed in the virtual environment taking advantage of the advanced features (such as the ability to manipulate the structure and appearance of the city).

3 Acquisition of spatial knowledge in Hexatown

3.1 Methods

Setup. Experiments were performed using a standard 19–inch SGI monitor. Subjects were seated comfortably in front of the screen and no chin–rest was used. They moved their heads in a range of about 40 to 60 cm in front of the screen which results in a viewing angle of about 35 – 50 degrees.

Procedure. In the experiment, subjects found themselves facing some view v_1. They were then presented with a target view v_2 printed out on a sheet of paper and asked to find this view in the virtual town (task $v_1 \rightarrow v_2$). When they found the view, feedback was given in the form of a little sign appearing on the screen. If they got lost in the maze, i.e., if they deviated from the shortest possible path by more than one segment, the trial was stopped and another sign announced the failure. The sequence was terminated when the shortest possible way, i.e., the way involving the minimal number of decisions (mouse clicks) was found. The whole exploration phase contained 12 such search tasks, or ways to be found. The first four ways were excursions from view no. 15, which served as a "home" position. View 15 showed a poster wall saying "Max-Planck-Institut für biologische Kybernetik". The following 8 searches were either returns to home or novel paths not touching on view 15. The return and novel path tasks were presented alternatingly in two sequence conditions: in the **returns–first condition**, the first task was a return, whereas in the **novel–first condition**, the sequence started with a novel path. In both conditions, the four excursions were performed prior to both returns and novel paths.

After the exploration phase described above, subjects were asked to rate distances in the maze and to produce a sketch map. The results from these parts of the experiments have been published elsewhere (Gillner & Mallot 1998).

Subjects. Eighty paid volunteers participated in the experiment, 40 of which were male and 40 female, aged 15 – 38. Twenty subjects (10 male, 10 female) took part in each of the four viewing conditions (Fig. 4). Within each viewing condition, the group of subjects was split equally (10 to 10) between the two sequence conditions (returns–first and novel–first), as well as the two instructions for the distance estimation ("distance" and "airline–distance").

3.2 Exploration performance

Figure 6 shows an example for the cummulative trajectories taken by a single subject in the twelve search tasks. Paths 1 – 4 are excursions, 5, 7, 9, and 11 are returns and 6, 8, 10, and 12 are novel routes. Overall, there is a tendency for lower error rates in the search tasks performed later. That is to say, there is a transfer of knowledge obtained in earlier searches to the later searches. The decrease of errors is not monotonic and is missing in some subjects.

For a quantitative analysis, errors were defined locally as movement decisions that do not reduce the distance of the goal. Each movement decision equals clicking the mouse buttons twice (cf. Fig. 2b). Distance to the goal is measured as the minimum number of decisions needed to reach it ("decision–distance[2]"). Thus, if a subject enters a street leading away from the goal, the return from that street will be counted as a correct decision even though the current position is not part of the shortest path. This is due to the fact that the return from the false street is again an approach to the goal. In cases where the correct decision is a 60 degrees turn left followed by a "go", the 120 degrees turn left would leave the decision–distance to the goal unchanged. This decision (and the mirror–symmetric case) is also counted as an error, in accordance with the above definition of local error. In cases where the subject has to perform a full 180 degrees turn, the initial 120 degrees turning step may be either left or right. Since both turns lead to a reduction in the number of steps remaining to the goal, either decision would be considered correct.

Average error rates for each path type are shown in Fig. 7. For each viewing condition (1 – 4; see Methods section), the excursions, returns, and novel paths were lumped into groups of four. As mentioned above, the excursions were performed first, while the novel and return paths were performed alternatingly, starting with a return in one group of subjects and starting with a novel route in a second group. The data show a learning effect in the sense that excursions take more errors than the later paths. They also show a clear effect of condition: higher visibility results in lower error rates. This general relation does not hold

---

[2] One could argue that false turns should not be counted as errors at all, since in a more realistic setup, they correspond to simple head turns rather than movements of the body. In fact, when accessing subjects' sense of distance, we did find in additional measurements that turns are ignored in distance estimation (Gillner 1997, Gillner & Mallot 1998). For the specification of the task, i.e., finding the shortest possible route to a goal, we think, however, that erroneous turns should be considered. Turns do take time, after all.

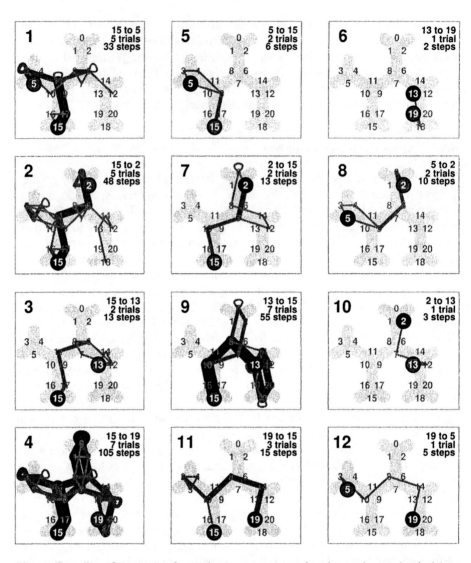

Fig. 6. Travelling frequencies for each view–transition for the twelve paths (subject GPK, viewing condition: 1; sequence condition: returns–first). Left column: excursions, middle column: returns, right column: novel paths. The overall number of errors decreases at later stages of exploration.

for the comparison of conditions 1 and 2, however, which differ in the visibility of the neighboring places.

A 3-way analysis of variance (ANOVA, 4 conditions × 3 path types × 2 genders) of error-rate as the dependent variable reveals significant effects of condition ($F(3, 72) = 17.31, p < 10^{-4}$) and path type ($F(2, 144) = 60.65, p < 10^{-4}$),

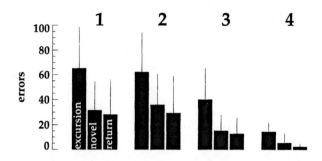

Fig. 7. Total number of wrong movements performed in the different path types (excursion, novel, return). Numbers are averaged over 20 subjects, error bars are standard deviations. 1–4: viewing conditions.

but not of gender ($F(1, 72) = 0.22$, n.s.). Additionally we found an interaction of condition and path type ($F(6, 144) = 2.66, p = 0.018$). The error rates of novel paths are slightly higher than those of the returns (see Fig. 7). This effect is not significant, however.

3.3 Knowledge Transfer across Routes

In our procedure, learning occurs on two time scales. During each of the twelve tasks, a route is learned. When switching from one route to the next, part of the knowledge acquired in the earlier routes might be transferred to the new ones. To test this, we define a transfer coefficient τ in the following way:

Let R and N denote two routes, for instance the first return and novel path, respectively. Our group of subjects is divided into two subgroups, one of which explores R first and N second, whereas the second group explores N, then R. Four such pairs of routes have been tested. We accumulate the data from these four tested pairs of returns and novel paths:

$E_{R,1}$ errors in returns in the returns–first condition
$E_{N,1}$ errors in novel paths in the novel–first condition
$E_{R,2}$ errors in returns in the novel–first condition
$E_{N,2}$ errors in novel paths in the returns–first condition

Thus, $E_{R,1}$ and $E_{N,2}$ refer to the first group of subjects (returns first condition) and $E_{N,1}$, $E_{R,2}$ to the second. If transfer occurs, the route explored first should have higher error rates in both cases. We define:

$$\tau = \frac{E_{R,1} - E_{R,2} + E_{N,1} - E_{N,2}}{E_{R,1} + E_{N,1}} \tag{3}$$

If error rates do not depend on position, τ will be zero; if everything is learned already when exploring the first route, $E_{R,2}$ and $E_{N,2}$ will be zero and τ evaluates to 1.

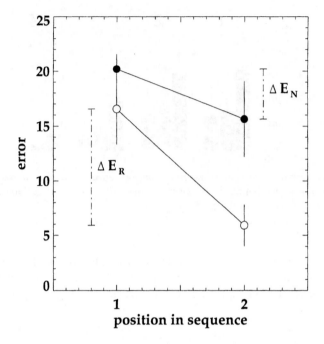

Fig. 8. Average error numbers for novel paths and returns in the novel–first and returns–first sequence conditions. All subjects from viewing conditions 1 and 2 with an overall error number below the median were included in this plot. • novel routes; ○ returns. ΔE_R: disadvantage of returns–first group in the return routes. ΔE_N: disadvantage of novel–first group in the novel routes. For both returns and novel routes, error rate drops when other routes are explored before. The transfer–coefficient (Equation 3) is $\tau = 0.4$.

For this evaluation, the subjects from viewing conditions 1 and 2 were pooled, because there were no significant differences between the respective error rates (3 way ANOVA: 2 conditions × 4 routes × 2 gender, $F(1, 36) = 0.014, p = 0.9075$). If we take the average over all 40 subjects, no significant effect of transfer is found. If, however, only the 20 subjects with the lowest overall error rate are considered, a transfer effect with $\tau = 0.4$ is found (see Fig. 8). In this case, eleven subjects were from the returns–first condition and nine subjects from the novel–first condition. The result indicates that the good navigators show significant transfer of knowledge even from one route to the next. Transfer across more steps of the exploration procedure is not visible in this evaluation, which does not mean that we exclude such a transfer.

3.4 Persistence

An inspection of Fig. 6 shows that in almost all cases where the subject started from view 15, the first movement decision was LL even though RR would have

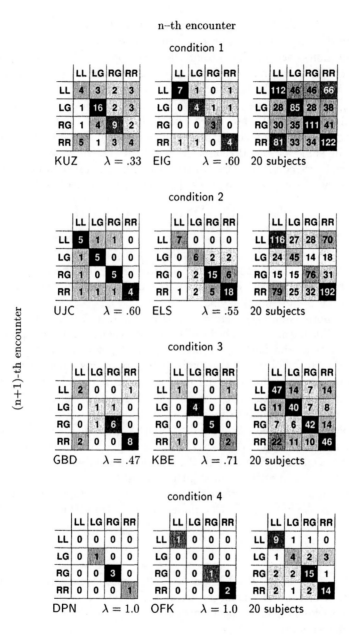

Fig. 9. Examples of return statistics for selected subjects for the four viewing conditions. In the subjects shown for the first three conditions, statistical independence of the decisions taken at the n–th and the $n - 1$–th encounter could be rejected in all cases, i.e., persistence rate λ was significantly different from zero. For condition 4, where errors were generally quite rare, statistical independence could not be rejected for any subject; still, the matrices show high diagonal entries.

been just as good. Together with similar behaviors observed from other subjects, this lead to the conjecture that at least some movement decisions reflect simple, fixed associations between the current view and some motion that is performed whenever the view occurs. In order to test this in more detail, we analysed the return statistics of the decision sequences.

Let $m_{n,v} \in \{LL, LG, LR, RL, RG, RR\}$ denote the movement decision taken at the n-th encounter of view v (see Fig. 2b for possible movement decisions). We are interested in cases where the movement chosen at the n-th encounter of view v is the same as that taken at the $n - 1$-th encounter of the same view.. More generally, we count the cases where movement j is taken at encounter $n-1$ and movement i at encounter n ($i, j \in \{LL, LG, LR, RL, RG, RR\}$):

$$F_{i,j} := \#\{(n, v)|m_{n,v} = i, m_{n-1,v} = j\}. \tag{4}$$

It is important to note two points: First, the two encounters n and $n - 1$ do not occur in subsequent time steps (unless $m_{n-1,v} \in \{RL, LR\}$). Rather, long sequences of other views may occur in between. Second, the frequency $F_{i,j}$ is accumulated over all views. Thus we are looking for an average persistence rate rather than for a view-specific one.

In the experiments, each search task is repeated until the subject finds the shortest possible path. This procedure can in itself produce repetition rates above chance if parts of the path are repeated correctly several times. To exclude this type of error, we restrict our analysis to repetitions where both decisions were false in the sense that they did not lead to an approach to the goal (local definition of errors). Finally, we dropped the cases involving the decisions LR and RL, since these are quite rare.

Example data from individual subjects are shown in Fig. 9. The numbers on the diagonal correspond to cases where the same decision was chosen in two subsequent encounters even though the decision was false in both cases. From these matrices, we can estimate average movement transition probabilities $p_{ij} := P(m_{n,\cdot} = i|m_{n-1,\cdot} = j)$; averaging is performed with respect to the different views involved. A simple statistical model for these transition probabilities is:

$$p_{ij} = \begin{cases} \lambda + p_i & \text{if } i = j \\ p_i & \text{if } i \neq j \end{cases} \tag{5}$$

where $\lambda, 0 \leq \lambda \leq 1$ and $\lambda + \sum_{i=1}^{4} p_i = 1$. It states that there is a bias λ for the repetition of the movement chosen at the previous encounter. Other than that, the decisions at subsequent encounters are independent. If $\lambda = 0$, true independence is obtained.

The analysis could be applied to data from 67 out of 80 subjects. For the remaining 13 subjects, the number of total errors was too low to fit the model. Ten of these had been tested in viewing condition 4, where the error rates were lowest. Goodness-of-fit was tested with the χ^2-Test; choosing a significance level of 5 %, the best fitting model could not be rejected in any of the 67 subjects.

In order to get an impression of the confidence intervals for λ, we repeated the analysis with fixed $\lambda = 0$ in Eq. 5. By testing goodness-of-fit for this model with the χ^2–Test, 18 cases could be rejected on the 10 %–level, 9 of which could be rejected on the 1 %–level as well.

Average persistence rate over all subjects was 0.33, indicating that about one third of the erroneous movement decisions were based on persistence. A regression analysis of persistence rate λ with the overall number of errors for each subject did not reveal a significant correlation.

4 Discussion

4.1 Navigation in virtual environments

The results show that spatial relations can be learned from exploration in a virtual environment even under rather restricted viewing conditions. Here, we briefly summarize the most important findings:

Effect of viewing condition. The four viewing conditions differ in the amount of information available to the subjects. Not surprisingly, the number of errors during the search phase decreases as more information is provided. This is in spite of the fact that the field of view was the same in all four conditions. Bringing the objects closer to the places in condition 3 and removing the occluding hedges in condition 4 is reminiscent of zooming out the whole scene with a wide–field lens. Perúch et al. (1997) showed that this zooming does not improve path–integration performance in a triangle–completion task. This discrepancy may characterize a difference between path integration and landmark navigation. Alternatively, his negative result may be due to the marked errors in perspective associated with zooming.

The comparison between conditions 1 and 2 (night and day) does not show an improvement in error rates. This is surprising since more information is available in condition 2 (objects at the far end of the streets become visible). This finding may be related to the fact that the local structure of the maze becomes more complicated in condition 2, where six objects are visible from each place.

Transfer and latent learning. The overall number of errors was smaller for the later search tasks. For the 50 % best subjects, this effect was already clearly visible for the comparison of one search task with the next (Fig. 8). If subjects simply learned a set of independent routes, e.g., by reinforcement learning, each search would be a new task and no such transfer would be expected. The knowledge being transferred from one route to the next is not just a route–memory but involves the recombination of route segments; this is to say, it is of the configuration type. Its acquisition is akin to latent learning, since knowledge obtained during one search can be employed later in other, unrelated search tasks.

As can be seen from Fig. 8, transfer was strong from the novel to the return paths, but not the other way around. One possible explanation of this finding is that the novel paths are more difficult than the returns. When considering the shortest possible paths, the novel paths involve 14 different views, 8 of which also occur in the returns. The returns involve only 9 different views, i.e., almost all of their views are already known from the novel paths. The only view not occuring in the novel paths is the final goal of the returns, view 15. The transfer assymmetry may thus be due to the fact that the novel routes contained more information about the returns than vice versa.

The occurrence of transfer from one route to another is also evidence for the presence of goal–independent memory of space, i.e., a cognitive map.

Persistence. Along with these arguments for configuration knowledge, evidence for simpler types of spatial learning was also found. The persistence rates presented in Section 3.4 indicate that at least some of the subjects based a considerable part of their movement decisions on simple associations of views with movements. This strategy is efficient for learning non–intersecting routes but will lead to errors for views at a crossroads where the correct motion decision depends on the current goal. We speculate that the persistence rate will decrease if longer training sequences are used.

Subject differences. Subjects differed strongly in terms of the number of errors made when searching a goal as well as in the amount of transfer learning. However, no clustering in different groups can be obtained from our data. In particular, no significant gender differences were found.

4.2 View–based navigation

In viewing conditions 1 and 2, subjects had to rely on local views as their only position information. Their performance and the transfer learning is therefore view–based in an obvious sense. However, this result does not exclude the possiblity that some more complicated representation of space is constructed from the local view information. Here we summarize the evidence against such a representation, i.e., evidence for a view–based mechanism of navigation.

Returns aren't easy. After having learned the four excursions, the returns to the starting point along the very same paths are almost as difficult as novel paths (Fig. 7). The advantage on the order of just one error per search task, is not significant ($F(1, 76) = 2.860$, $p = 0.095$). If the subjects acquired a place–based representation of space, it would be the same for excursions and returns, since the according place–graphs are symmetric (see Fig. 1). In this case, we would therefore expect that returns should be much easier and more reliable than novel paths. The weak difference between the number of errors occuring in returns and novel paths seems to indicate that this is not the case. It is rather more in line with a view–based mechanism, since the views occuring along the return path are as different from the original views as any other views in the maze.

Recognition and action. The average persistence rate of 33% indicates that direct associations of views to movement decisions can be learned. As was pointed out in the introduction, the association pair of view and motion decision is the basic element of a view–based memory of space.

4.3 Local information combined to a graph?

If the representation is in fact view–based, a graph structure is the only representation we can think of that would account for the transfer and planning behavior observed. Independent evidence for a graph–like representation comes mainly from the sketch maps recorded by Gillner & Mallot (1998). Maps are often locally correct but globally inconsistent. Also, places with correct local connectivity have been translocated to erroneous positions. Connectivity can be correct even though metric properties of the sketch maps, such as angles and lengths are grossly mistaken.

The distance estimates do not reflect the decision–distance, which is the graph–distance of the view–graph, but correlate better with walking distance, i.e. the graph distance of the place graph (Gillner & Mallot, 1998). It therefore appears that we cannot decide between the view–graph and the place–graph representations at this point. In ongoing work (Mallot & Gillner, 1997), this question is addressed with additional experiments.

5 Conclusion

In our view, the most important result of this study is the fact that configuration knowledge can be acquired in virtual environments. This is in spite of the fact that the subjects did not actually move, but where interacting with a computer graphics simulation. With respect to the high controllability of visual input, this result may well make virtual reality a valuable addition to more realistic field studies, where stimulus control is often a problem. More realistic simulations such as Virtual Tübingen and large displays are currently being developed and tested in our laboratory.

With respect to our starting point, i.e. view–based navigation, we think that three conclusions can be drawn:

1. *Views suffice.* Map learning is possible if only local, i.e., view–information is provided. In this sense, navigation can be view–based.

2. *Graph vs. view from above.* The representation contains local elements, i.e., a place or view with one or several movement decisions and the respective outcome associated with it. These local elements need not be globally consistent, they need not combine into a metric survey map. Rather, a graph–like representation is sufficient to account for our results.

3. *Places vs. views.* It is not clear from our data, whether the nodes of this graph are places or views. We have not found evidence that the local views are combined into a representation of space independent of the orientation of the viewer. Thus, a view–based representation seems more likely at this point.

References

1. V. Aginsky, C. Harris, R. Rensink, and J. Beusmans. Two strategies for learning a route in a driving simulator. Technical Report CBR TR 96-6, Cambridge Basic Reseach, 4 Cambridge Center, Cambridge, Massachusetts 02142 U.S.A., 1996.
2. M. Billinghurst and S. J. Weghorst. The use of sketch maps to measure cognitive maps of virtual environments. In *Proc. IEEE 1995 Virtual Reality Annual International Symposium*, pages 40 – 47, Piscataway, NJ, 1995. IEEE Press.
3. B. A. Cartwright and T. S. Collett. How honey bees use landmarks to guide their return to a food source. *Nature*, 295:560 – 564, 1982.
4. K. Cheng. A purely geometric module in the rat's spatial representation. *Cognition*, 23:149 – 178, 1986.
5. T. S. Collett and J. Baron. Learnt sensori-motor mappings in honeybees: interpolation and its possible relevance to navigation. *Journal of Comparative Physiology A*, 177:287 – 298, 1995.
6. R. P. Darken and J. L. Sibert. Navigating large virtual spaces. *International Journal of Human-Computer Interaction*, 8(1):49-71, 1996.
7. M. O. Franz, B. Schölkopf, H. A. Mallot, and H. H. Bülthoff. Learning view graphs for robot navigation. *Autonomous Robots*, in press.
8. S. Gillner. *Untersuchungen zur bildbasierten Navigationsleistung in virtuellen Welten*. Knirsch-Verlag, Kirchentellinsfurt, 1997.
9. S. Gillner and H. A. Mallot. Navigation and acquisition of spatial knowledge in a virtual maze. *Journal of Cognitive Neuroscience*, in press.
10. B. J. Kuipers and Y.-T. Byun. A robust, qualitative approach to a spatial learning mobile robot. In *SPIE Vol. 1003 Sensor Fusion: Spatial Reasoning and Scene Interpretation*. International Society for Optical Engineering (SPIE), 1988.
11. H. A. Mallot, M. Franz, B. Schölkopf, and H. H. Bülthoff. The view-graph approach to visual navigation and spatial memory. In *Artificial Neural Networks — ICANN 97*, 1997.
12. H. A. Mallot and S. Gillner. Psychophysical support for a view-based strategy in navigation. *Investigative Ophthalmology and Visual Science*, 38(Suppl.):4683, 1997. ARVO-Abstract No. 4683.
13. R. Maurer and V. Séguinot. What is modelling for? A critical review of the models of path integration. *Journal of theoretical Biology*, 175:457 – 475, 1995.
14. M. May, P. Péruch, and A. Savoyant. Navigating in a virtual environment with map-acquired knowledge: Encoding and alignment effects. *Ecological Psychology*, 7(1):21-36, 1995.
15. R. U. Muller, M. Stead, and J. Pach. The hippocampus as a cognitive graph. *Journal of General Physiology*, 107:663 – 694, 1996.
16. J. O'Keefe and L. Nadel. *The hippocampus as a cognitive map*. Clarendon, Oxford, England, 1978.
17. P. Péruch, M. May, and F. Wartenberg. Homing in virtual environments: Effects of field of view and path layout. *Perception*, 26:301 – 311, 1997.
18. T. Prescott. Spatial representation for navigation in animals. *Adaptive Behavior*, 4:85 – 123, 1996.
19. R. A. Ruddle, S. J. Payne, and D. M. Jones. Navigating buildings in "desk-top" virtual environments: Experimental investigations using extended navigational experience. *Journal of Experimental Psychology: Applied*, 3:143 – 159, 1997.
20. B. Schölkopf and H. A. Mallot. View-based cognitive mapping and path planning. *Adaptive Behavior*, 3:311 – 348, 1995.

21. M. Tlauka and P. N. Wilson. Orientation–free representations from navigation through a computer–simulated environment. *Environment and Behavior*, 28:647 – 664, 1996.

22. F. H. Tong, S. G. Marlin, and B. J. Frost. Visual–motor integration and spatial representation in a visual virtual environment. *Investigative Ophthalmology and Visual Science*, 36:1679, 1995. ARVO abstract.

23. D. S. Touretzky and A. D. Redish. Theory of rodent navigation based on interacting representations of space. *Hippocampus*, 6:247 – 270, 1996.

24. O. Trullier, S. I. Wiener, A. Berthoz, and J.-A. Meyer. Biologically based artificial navigation systems: Review and prospects. *Progress in Neurobiology*, 51:483 – 544, 1997.

25. H. A. H. C. van Veen, H. K. Distler, S. J. Braun, and H. H. Bülthoff. Using virtual reality technology to study human action and perception. *Future Generation Computer Systems*, in press. See also: http://www.kyb.tuebingen.mpg.de/ (Technical Report 57).

Spatial Orientation in Virtual Environments: Background Considerations and Experiments

Fredrik Wartenberg[1], Mark May[1], and Patrick Péruch[2]

[1] Institut für Kognitionsforschung, Universität der Bundeswehr, Postfach, D-22039 Hamburg, Germany
wartenbe@elcafe.com, mark.may@unibw-hamburg.de
[2] Centre de Recherche en Neurosciences Cognitives, CNRS, Marseille, France
peruch@lnf.cnrs-mrs.fr

Abstract. Spatial orientation strongly relies on visual and whole-body information available while moving through space. As virtual environments allow to isolate the contribution of visual information from the contribution of whole-body information, they are an attractive methodological means to investigate the role of visual information for spatial orientation. Using an elementary spatial orientation task (triangle completion) in a simple virtual environment we studied the effect of amount of simultaneously available visual information (geometric field of view) and triangle layout on the integration and uptake of directional (turn) and distance information under visual simulation conditions. While the amount of simultaneously available visual information had no effect on homing errors, triangle layout substantially affected homing errors. Further analysis of the observed homing errors by means of an Encoding Error Model revealed that subjects navigating under visual simulation conditions had problems in accurately taking up and representing directional (turn) information, an effect which was not observed in experiments reported in the literature from similar whole-body conditions. Implications and prospects for investigating spatial orientation by means of virtual environments are discussed considering the present experiments as well as other work on spatial cognition using virtual environments.

1 Introduction

Virtual environments (VEs) introduce new methodological possibilities to carry out research on spatial cognition (Cutting, 1997; Ellis, 1991; Wickens & Baker, 1995). Compared to the standard inventory of methods used in research on spatial orientation and learning, methodological advantages result from the high degree of freedom in constructing spaces and in controlling and varying environmental variables, straightforward methods of variation and control of perceptual-motor factors during navigation, and convenient methods of online-measurement of spatial behavior. Up to

---

[1] Send correspondence to Mark May.

now there is a comparatively small number of empirical studies using navigations in VEs for purposes of basic research in spatial cognition, but with the ongoing rapid development in VE-technology this picture is likely to change in the near future (Brauer, Freksa, Habel, & Wender, 1995; Durlach & Mavor, 1995).

One of the central open questions in using VEs for research on spatial cognition is whether spatial performances and achievements (e.g., building up survey knowledge) observed in virtual and real world environments are comparable. While a navigator in a VE has to almost exclusively rely on the visually simulated information supplied by the VE-system (e.g., depth cues, optical flow), a navigator in a real world environment can make use of multisensory information from the outer world (i.e., visual, auditory, tactile, etc.) as well as different internal sources of information generated by the whole-body movements through space (i.e., motor efferent commands, proprioceptive and vestibular signals). Whole-body motion information is known to be of central importance for spatial orientation and learning in animals (Gallistel, 1990a) and humans (Klatzky, Loomis, & Golledge, 1997). How the lack of whole-body information affects spatial orientation and learning performances in VEs is not known yet (Durlach & Mavor, 1995).

After giving some background information on the role of vision and whole-body motion in coding spatial information and the potentials of VEs as a tool for research on spatial cognition, this chapter will report research on spatial orientation in VEs. In the empirical part, two experiments on triangle completion (i.e., homing to the starting point after navigating two legs of a triangle) conducted in a VE (only visual information available) will be presented. The resulting error patterns will be compared to those of experiments on non-visual triangle completion (only whole-body information available) reported in the literature (Loomis, Klatzky, Golledge, Cicinelli, Pellegrino, & Fry, 1993). On the basis of this comparison, we will discuss how visual information typically present in VEs contributes to spatial orientation.

2 Actor-centered coding of spatial information

By actor-centered coding we refer to the sensory and memory mechanisms of picking up, preserving and integrating spatial information (e.g., distances and directions to objects) during locomotion through space. Vision and whole-body motion provide the two most important spatial coding mechanisms in humans and most animal species (Nadel, 1990; Rieser & Garing, 1994).

Visual coding of spatial information

Vision provides a rich source of spatial information which is assumed to dominate spatial orientation and the development of spatial knowledge in humans (Gibson, 1979/1986; Sholl, 1996; Thinus-Blanc & Gaunet, 1997). For the stationary observer visual information contains a multitude of spatial cues such as relative size of objects, density gradients, occlusion, binocular disparity, surface slants, height in the visual field to name a few. The moving observer obtains further spatial information from optical flow, motion parallax, dynamic occlusion and disocclusion of objects and surfaces, etc. A prominent view of visual spatial orientation and learning is that the navigator extracts the spatial structure of the environment ('invariant structure') from the spatial information contained in the continually changing flow of retinal information (Gibson, 1979/1986). There is no doubt that vision (among the different sensory modalities) plays a predominant role in guidance problems such as following a trail or avoiding collisions, in identifying and localizing objects in the surrounding, or in perceiving spatial relations between simultaneously available landmarks. Much more controversial is the dominant role ascribed to vision (for instance by Gibson) for more complex spatial orientation problems such as knowing about object locations momentarily occluded from sight or building up spatial knowledge of large-scale space (for a critical discussion see Strelow, 1985).

Most empirical research on the role of vision in human spatial orientation has concentrated on questions of perceiving spatial layouts or events from the perspective of a stationary observer (Sedgwick, 1982, 1986; Cutting, 1986). In the last years, research has started to examine the contribution of visual mechanisms to solving more complex problems of spatial orientation and learning. Vishton and Cutting (1995) examined the ability of subjects to correctly determine their visually defined direction of self-movement using computer simulated motion sequences through virtual 'forests' composed of multisegment lines ('trees'). Their experiments show that the displacements of identifiable objects (displacement fields) and not the continuous flow of optical information as such (velocity fields) are used as input for determining heading direction, which the authors consider to be a simple mode of wayfinding.

Alfano and Michel (1990) examined the effects of restrictions of horizontal field of view on the construction of spatial knowledge. Subjects had to explore an unfamiliar room while wearing goggles with different fields of view (9°, 14°, 22°, 60°) or no goggles (natural field of view ca. 200°). An immediate memory test using a miniature model reconstruction task showed that the accuracy of object placements decreased as a function of restrictions in field of view. Width of field of view seems to be an important factor in integrating spatial information over successively perceived parts of space. In a similar vein, Rieser, Hill, Talor, Bradfield and Rosen (1992) found decrements in distance and direction judgments between landmarks in a city for subjects with early-onset losses of visual field as compared to subjects with a normal size of the visual field, even when the latter subjects had suffered from early- or late-onset acuity losses.

Whole-body coding of spatial information

A second important source of information underlying spatial orientation and learning are internal signals generated by whole-body movements through space. Three different information sources can be distinguished: (a) motor efferent commands in control of body movements, (b) proprioceptive signals from mechanical forces exerted on joints, muscles and tendons, and (c) vestibular signals informing about translational and rotational accelerations of the body. There is a long tradition of research on the role of whole-body motion information in animal spatial cognition (O'Keefe & Nadel, 1978; Poucet, 1993). Gallistel (1990b) proposed a model for various animal species based on path integration (or dead reckoning) mechanisms which allow animals to return to the starting point of a trip even in the absence of exteroceptive (esp. visual) spatial information (for a review of path integration models see Maurer & Séguinot, 1995). In an unfamiliar terrain, path integration mechanisms function to construct a representation of the environment ('cognitive map'). Limitations in the precision of path integration result from the cumulation of errors over time- or space-extended movements, which can make intermittent correction by exteroceptive information sources (e.g., perceived distances and direction to landmarks) necessary. Animal models of spatial cognition stress the interplay of whole-body and exteroceptive information and often consider exteroceptive information to enrich whole-body information and not vice versa.

Research on the role of whole-body movement information in human spatial orientation and learning has intensified in the last years. This research shows that whole-body motion provides effective mechanisms for picking up, preserving and retrieving spatial information as well as integrating it to spatial knowledge. Loomis, DaSilva, Fujita and Fukusima (1992) found that blindfolded locomotor judgments (walking to a target location) following a short visual preview are more accurate than perceptual judgments (numerical estimates) manifesting the characteristic psychophysic distortions of visual space (Wagner 1985). Haber, Haber, Penningroth, Novak and Radgowski (1993) found that bodily performed direction judgments are more accurate than numerical or symbolical judgments. Other work shows that people reach a high degree of accuracy in dynamically updating distances and directions to objects while locomoting in the absence of vision (Loomis et al. 1992; Rieser, Guth, & Hill, 1986) and that whole-body information can exert facilitating as well as interfering effects on imaginal retrieval of object locations in the actual or a remote spatial surrounding (Rieser, 1989; Presson & Montello, 1994; May, 1996; Rieser, Garing, & Young, 1994).

Klatzky, Loomis, Golledge and coworkers have taken up the study of path integration abilities in humans using tasks of nonvisual path completion and path reproduction (Klatzky, Loomis, Golledge, Cicinelli, Doherty, & Pellegrino, 1990; Loomis et al. 1993; Klatzky et al., 1997). In a typical task blindfolded subjects are led along two legs of a triangle and have to return to the assumed starting point on a direct route ('homing'). Homing performances turn out to be quite accurate with systematic errors observed for different geometrical triangle layouts (angles and leg lengths). Fujita, Klatzky, Loomis, and Golledge (1993) have reported an Encoding Error Model for path completion tasks that accounts for these systematic errors by assuming erratic subjective distance and turn coding while moving along the path.

Up to now, comparatively little is known about the relative contribution and the mutual interplay of visual and whole-body coding mechanisms in more complex problems of human spatial orientation and learning (e.g., cognitive mapping on the basis of navigational experience). One way to approach this question is to isolate the contribution of one of the information sources, as is done in research on nonvisual path integration or research comparing spatial abilities of sighted and blind people (Golledge, Klatzky, & Loomis, 1996; Sholl, 1996). While it is relatively easy to isolate the contribution of whole-body information by blindfolding subjects and depriving them of auditory and other exteroceptive information, it is much more difficult to examine the contribution of visual information in isolation. The problem is one of isolating visual sources of spatial information without bringing the navigator into the role of a passive recipient of visually displayed information (e.g., as induced by slide shows or video displays). Here, the study of behavior in interactive VEs could open up new ways of examining the separate contribution of visual coding mechanisms to spatial orientation and learning.

3 Virtual environments as a tool for research on spatial cognition

VEs are a new type of human-computer interface which allow users to interact with an egocentrically defined graphic simulation of a 3D spatial environment (for technical descriptions cf. Ellis, 1994; Kalawsky, 1993; Pimentel & Teixeira, 1993). Actor and computer system are coupled in form of a perception-action cycle, in which actions (via hand or head movements) lead to a spatiotemporally realistic updating of the actor's viewpoint in the simulated environment (via monitor or other types of displays) forming the basis for further viewpoint-dependent actions. Techniques used to display VEs are desktop monitors, projection screens or head-mounted stereoscopic displays (HMDs); actions of the navigator are realized by recording inputs to keyboards, 2- or 3-D mice, joysticks, data-gloves, or head-trackers. These input devices constitute abstract interfaces, which do not convey the whole-body motion information accompanying navigations in real world environments (Durlach & Mavor, 1995). Head-tracking techniques are a first step to convey body locomotion information, but at the current state of VE-technology usually only allow to record head- or whole-body rotations, and have to be combined with one of the other input devices to realize translational movements through simulated space.

Whole-body interfaces such as motion platforms, treadmills or bicycles are still in the phase of prototype development and testing, but could lead to interesting research perspectives in the future (Distler & Bülthoff, 1996). The lack of ready-to-use whole-body motion interfaces suggests to employ VE-technology for examining visual modes of spatial orientation and learning. Furthermore, the lack of whole-body information requires to be careful with generalizations beyond the predominantly visual definition of spatial cognition set by the current state of the technology.[2]

---

[2] Limitations in the visual display of spatial information in current VE-technology go back to restrictions in horizontal field of view (mostly below 90°), reduced fidelity of displaying visual scenes, and time lags between user actions and display updating (esp. when using high-resolution displays with large fields of view). It is difficult to discuss implications of these limitations for research on spatial cognition as they will very much depend on the concrete context of research. Constraints set by visual display limitations are likely to relax with further technological developments in VE-technology.

VEs bring different methodological advantages to the field of spatial cognition. As researchers in the field know, experiments in real world settings are very often troublesome (e.g., problems in controlling spatial pre-knowledge of the participants, problems in manipulating environmental features, problems in experimental replications or interexperimental comparisons because of differences in experimental settings). VEs can help to avoid such problems. Methodological advantages result for instance from the higher degree of freedom in constructing spaces as well as manipulating environmental variables (e.g., one can freely define scene properties, landmark positioning or the geometry of the environmental layout). There are straight-forward methods of controlling and varying the perceptual-motor factors during navigation at hand (e.g., one can define visual parameters such as field of view or movement parameters such as speed). Furthermore, there are convenient methods of online-measurement of spatial behavior (e.g., one can record trajectories during exploratory behavior or keep track of movement decisions and latencies during wayfinding).

So far, only a small number of experimental studies have used VEs to examine questions of spatial orientation and learning. Henry and Furness (1993) reported an experiment indicating deficits in building up spatial knowledge while navigating a virtual museum as compared to navigations in the corresponding real-world museum. Subjects in the simulation conditions underestimated the spatial dimensions of the rooms navigated through and were less precise (increased variable errors) in pointing to unseen object locations in other rooms. Results revealed no significant performance differences between three different display techniques used (desk-top monitor, HMD without and HMD with head-tracking).

May, Péruch and Savoyant (1995) examined goal-oriented navigations in a VE on the basis of routes learned from topographic maps. The maps depicted routes that were either aligned or 90°- or 180° misaligned; alignment refers to the spatial correspondence of upward direction in the map and goal direction in navigation space (Levine, 1982). The greater the degree of misalignment the more time was needed to find the goal and the more deviations from the requested route were observed. An analysis of the spatial distribution of stops subjects made while traveling from start to goal indicated systematic differences in memory retrieval between the different alignment conditions.

Péruch, Vercher and Gauthier (1995) examined spatial orientation performances between conditions of active exploration of space (interactive mode; self-performed movements through a VE) and two conditions of passive exploration (video and slide presentation modes; presenting the same sequences as the active conditions either unsampled 18 frames per second or sampled 1 frame each 4 seconds). Following exploration, subjects had to find target locations obstructed from sight by using the shortest possible path. Results showed that subjects in the condition of active exploration performed significantly better (higher scores and shorter completion times); differences between the two passive conditions were not found (similar results were obtained by Tong, Marlin, & Frost, 1995; Williams, Hutchinson, & Wickens, 1996).

Gillner and Mallot (1997) used a high-fidelity simulation of a virtual town with a hexagonal layout of streets to evaluate a view-based approach to spatial navigation. The underlying model (Schölkopf & Mallot, 1995) assumes a view-graph spatial representation, where nodes representing local views are connected by edges representing movement decisions. Subjects had to perform different online- and off-line tasks reflecting the quality of the route and survey knowledge they had acquired during navigations in the virtual town. Results partially support the local representation assumptions of their model, although their data do not entirely differentiate whether the nodes in the graph represent local scenic views or local place information. An interesting variation in Gillner and Mallot's experiment concerned the use of different viewing conditions such as navigating under day or night conditions with different degrees of occlusions by trees along the routes; degree of occlusion revealed strong impacts on wayfinding performances, but no differences between the comparable night and day conditions were found.

Recently, Ruddle, Payne and Jones (1997) reported a series of experiments on the development of spatial knowledge on the basis of navigations in a complex virtual building. One experiment aimed at replicating the experimental findings of Thorndyke & Hayes-Roth (1982) obtained in a real-world building. The overall pattern of results was quite similar between the two studies with subjects learning the environments from maps showing better Euclidean knowledge of distances and directions to unseen landmarks than subjects learning the environments from navigational exploration. Although not statistically substantiated by the authors, interexperimental comparisons revealed clear decreases in accuracy of Euclidean distance and direction knowledge of the VE navigators as compared to the real-world navigators in the Thorndyke and Hayes-Roth's study. Other experiments examined the impact of landmark availability (landmarks present vs. not) and landmark familiarity (abstract paintings vs. everyday objects) on the development of route (wayfinding errors) and survey knowledge (Euclidean estimates errors) over nine consecutive navigation sessions. Results showed that landmark availability only influenced the development of route knowledge and this only when landmarks were distinguishable everyday objects. Analysis of the development of route and survey knowledge over consecutive sessions indicated that both types of knowledge improved very much in parallel from an early stage of learning (third session) on.

In summary, this short review of recent VE studies shows that very different aspects of spatial cognition can be examined using visually simulated environments. The two studies relevant to the question of effects of missing whole-body information (Henry and Furness, 1993; Ruddle et al., 1997) hint at deficits in building up spatial knowledge on the basis of navigations in virtual as compared to real world environments. Examining spatial performances in relatively complex environments (i.e., virtual buildings) makes it difficult to determine the exact causes of such deficits. Furthermore, studies comparing real and virtual environments generally reveal similarities as well as differences. In particular, the conditions of transfer of spatial information between real and virtual environments (or vice-versa) may vary from one experiment to the other (Bliss, Tidwell, & Guest, 1997; Kozack, Hancock, Arthur, & Chrysler, 1993; Wilson & Foreman, 1993).

The experiments reported below aim at investigating spatial knowledge acquisition in VEs. For this end we used simple VEs, which allowed to present and test well defined spatial parameters (distances and directions) and compare performances between real and virtual environments. The experiments can be considered as a first step in evaluating potential deficits in picking up and using spatial information in VEs.

4 Triangle completion experiments

The experiments to be reported here used VEs to examine triangle completion on the basis of purely visual information and compare the results to performances obtained by Loomis et al. (1993) under whole-body motion conditions (subjects blindfolded).[3] When navigating in a VE portions of the spatial environment are successively laid open; rotational movements of the virtual camera let elements appear on one side of the screen while elements on the opposite side vanish from sight. The portion of the environment seen during a time-extended journey is considerably larger than the portion visible in any of the still scenes. This raises the question of how observers are able to integrate the spatial information perceived during navigations in a visually simulated environment (Beer, 1993). The triangle completion task allows to examine this question with a behavior-based approach, letting the observer become a situatively embedded actor in a well-defined spatial task.

The central aim of our experiments was to examine the characteristics of spatial knowledge acquired during purely visual navigations in VEs. For this purpose we remodeled the nonvisual triangle completion experiment of Loomis et al. (1993, Experiment 1) in a simple VE with no task-relevant landmark information. The Encoding Error Model of Fujita et al. (1993) was used to systematically compare homing performances between visual simulation and whole-body conditions. The question was whether visually simulated navigations through space result in any systematic deficits in coding distance and direction information while moving along the pathway.

A second question asked in our experiments was whether the navigator's field of view would influence the accuracy of homing performances. Research in real world environments shows that the size of field of view can influence spatial orientation and learning performance, either due to the amount of spatial information available at a time or due to the participation of peripheral visual mechanisms (Alfano & Michel, 1990; Johansson & Börjesson, 1989; Rieser et al., 1992). Virtual environments allow to make a distinction between the absolute field of view (FOV) and the geometric field of view (FOVg). While FOV is defined by the visual angle subtended at the observer's eye, FOVg describes the portion of the entire 360° surrounding of a world model projected onto the screen at a time. Our experiments varied the FOVg and kept the FOV constant. By this manipulation only the amount of simultaneously available spatial information was varied. The question was whether a larger amount of

---

[3] These experiments have been reported in Péruch, May, & Wartenberg (1997).

simultaneously available spatial information during navigation leads to improved homing performances.

4.1 Experiment 1

Method

Subjects. 10 subjects (4 male, 6 female) participated in the experiment. Ages ranged from 23 to 42 years. All subjects had normal or corrected-to-normal vision.

Experimental room. The experiment took place in a completely darkened room. The subject sat in front of a projection screen (2.05 m wide, 1.65 m high). The distance to the projection screen was 2.5 m resulting in a FOV of approximately 45° horizontal by 37° vertical. The joystick was fixed on a table in front of the subject.

Virtual environments. The rendering of VEs was based on a three-dimensional world model. The VE was made up by an arena formed by 16 homogeneous white cylinders evenly spaced on a circle. In relation to the simulated eye height of the observer (1.7 m) the circular arena appeared to have a diameter of about 60 meters. The white cylinders had a radius of 0.4 m and a height of 2 m. Inside the arena two colored cylinders were placed ($r = 0.4$ m, $h = 1.0$ m) which together with the non-marked starting point defined the triangle to be navigated (see Fig. 1). The first leg (a) of the triangle was defined by the starting point (C) and a red cylinder (A). The length of the first leg always corresponded to 20 m (2 distance units). The second leg (b) was defined by the red cylinder and a second, blue cylinder (B) while the third leg (c) was defined by the blue cylinder (B) and the non-marked starting point (C; equal to origin). The location and orientation of the triangle inside the arena were randomly chosen for each trial. Triangle geometries used (length of legs a and b and α-angle) were a subset of the triangle geometries used by Loomis et al. (1993). The variation of the geometric field of view (FOVg = 40°, 60°, 80°, see Fig. 2) was realized by the rendering software.

Subjects moved in the VE by operating a self-centering joystick, providing left/right rotation, forward (but no backward) translation, combinations of both movements, and immobilization. Translational and rotational speed were linearly coupled to joystick movement, maximum values were 3.3 m/sec (translation) and 33°/sec (rotation), respectively. Movement parameters (speed in relation to the triangle dimensions) were defined to obtain triangle completion times comparable to those in the Loomis et al. (1993) experiment. For facilitating interexperimental comparisons all distances below will be specified in terms of distance units (DU).

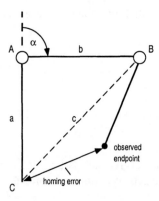

Figure 1. Triangle completion task. Starting at (C), subjects move along leg (a) until reaching the red cylinder (A). After turning an angle α they move along leg (b) until they reach the blue cylinder (B). From this point they have to return to the non-marked starting point (C). Homing error is measured as the Euclidean distance between the observed homing endpoint and the origin (equal to starting point).

Apparatus. Virtual environments were generated using a PC equipped with a 3D-vector-graphics card (Matrox SM 1281, 256 colors, 1280 x 1024 pixels). The hardware allowed for the rendering of about 10.000 Gouraud-shaded z-buffered polygons per second yielding a frame rate of about 15 frames per second. Scenes were displayed by a video-projector offering a screen resolution of 640 x 480 pixels. Joystick-data were gathered by the PC via a 12-bit analog-to-digital converter. The trajectories were recorded by the PC with a temporal resolution of approximately 10 Hz.

Procedure. Before starting the experiment the triangle completion task was explained to the subjects and they were allowed to practice on the device for about 15 minutes. The environments used for practice were not used in the experimental trials.

The experiment consisted of 27 different triangle completion tasks (3 FOVg x 9 triangle layouts) presented in random order. At the beginning of each experimental trial the subject was positioned at the starting point facing the red cylinder. Depending on FOVg-condition and the position of the triangle in the arena, a variable number of white cylinders could be visible from the starting point (see Fig. 2). The blue cylinder was always positioned to the right of the red cylinder. If it was not visible from the starting position (outside the field of view) subjects were instructed to just turn (without translation) to the right until the blue cylinder appeared and then turn back again so that they were facing the red cylinder before starting to proceed along the triangle. Subjects then first moved along the first leg a (length always 2 DU) to the position of the red cylinder (it temporarily disappeared when reached), turned to the right, and then moved to the position of the blue cylinder until it disappeared. Reaching this point, the s's task was to turn towards the assumed direction of the non-marked starting point and home to it as accurately as possible. No feedback with respect to homing accuracy was given. The next trial followed after a short pause. The entire experiment lasted about one hour.

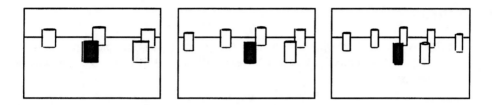

Figure 2. Geometric field of view (FOVg). Schematic drawings of spatial scenes as seen by a subject located at the starting point under different conditions of FOVg: 40° (left), 60° (center), 80° (right).

Design. The experiment made up for a 9 (triangle layout[4]) x 3 (FOVg = 40°, 60°, or 80°) within-subject design. Homing errors were defined by the Euclidean distance between observed homing endpoint and origin (see Fig. 1).

Results

Homing errors were analyzed by a repeated measures ANOVA (9 triangle layouts x 3 FOVg). A significant main effect of triangle layout was found ($p < .001$). Neither the FOVg effect ($p = .06$) nor the interaction between triangle layout and FOVg ($p = .36$) were significant. Analysis of turning angles revealed that the effect of triangle layout was mainly due to subjects generally underestimating the final turn in B (overall average = - 24° ± 1.5 (SEM)).

Discussion

The results show that triangle layout determines homing performances to a high degree. With respect to the pronounced misjudgment of the origin's direction under visual simulation conditions it could be possible that an answering bias (i.e., s's stop turning too early while making the final directional judgment) was at work. A second experiment was run in order to control for this possible answering bias.

---

[4] Three angles a (60°, 90°, or 120°) crossed with three lengths of the second leg b (1, 2, or 3 DU).

4.2 Experiment 2

Method

Subjects. A new group of 8 female and 8 male subjects participated in this experiment. Ages ranged from 20 to 26 years. All subjects had normal or corrected-to-normal vision.

Experimental room, apparatus, VEs and experimental design were the same as in Experiment 1.

Procedure. The procedure was essentially the same as in Experiment 1. The only difference concerned instructions given to s's when reaching the blue cylinder (point B). Before homing to the assumed origin they were requested to turn right until the red cylinder (point A) appeared on the right side of the projection screen. From here subjects could turn back to the left again until they assumed to be facing towards the origin and subsequently move to it. This manipulation intended to exclude any systematic misjudgment of homing direction due to an answering bias as described above.

Results

A 9 (triangle layout) x 3 (FOVg) repeated measures ANOVA on homing error revealed a significant main effect of triangle layout ($p < .001$). As in Experiment 1, neither the FOVg effect ($p = .36$) nor the interaction between triangle layout and FOVg ($p = .24$) turned out to be significant. Once again subjects exhibited a clear tendency to underestimate the angle of the final turn (overall average = $-19° \pm 1.5$ (SEM)). For testing whether errors observed in the two experiments differed systematically a 2 (Experiment 1 vs. 2) x 9 (triangle layout) x 3 (FOVg) ANOVA with goal error as dependent variable was performed. The main effect of experiment was not significant ($p = 0.35$). The interaction between experiment and triangle layout was significant ($p < 0.001$) due to the fact that for $\alpha = 120°$ deviations between the experiments were smaller than for $\alpha = 60°$ or $\alpha = 90°$.

Discussion

The larger systematic homing errors under visual simulation conditions apparently do not result from an answering bias for the final directional judgment. The comparison between Experiment 1 and 2 showed that homing errors did not differ due to s's performing the final directional judgment coming from a rightward (Experiment 1) or a leftward (Experiment 2) turn. In our experiments, triangle layout seems to be the main factor determining the observed homing behavior under visual simulation conditions.

5 Applying the Encoding-Error Model

As a first step for comparing homing performances under visual simulation and whole-body conditions the observed homing endpoints were plotted for Experiments 1 and 2 (Fig. 3, solid lines, the data were combined as no substantial differences showed up) and for whole-body conditions (Fig. 3, dashed lines, data from Loomis et al. 1993, Experiment 1), respectively. In both conditions, a systematic influence of triangle geometry on homing performance could be observed: homing errors increased with decreases in turning angle α and increases in length of leg b. However, the homing performance differed significantly between whole-body and visual simulation conditions, which is revealed by confidence ellipses overlapping only in one of the nine cases (leg b = 3 and $\alpha = 90°$).

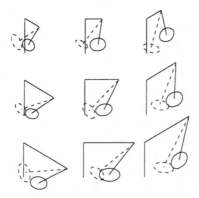

Figure 3. Two-dimensional plot of homing performances. Centroid and 95%-confidence ellipses for homing endpoints under whole-body motion conditions (dashed lines, reanalysis of data from Loomis et al. 1993) and visual simulation conditions (solid lines, data from Experiments 1 and 2 combined). Rows correspond to lengths of leg b (upper: b = 1 DU, middle: b = 2 DU, lower b = 3 DU); columns correspond to different turning angles (left: $\alpha = 120°$, middle: $\alpha = 90°$, right: $\alpha = 60°$). Non-overlapping confidence intervals indicate significant differences between experiments (Batschelet, 1981).

Systematic homing errors observed under visual simulation conditions (solid lines) were considerably larger than those observed by Loomis et al. (1993) under whole-body conditions (dotted lines): in 8 of 9 cases the homing endpoints under visual simulation conditions were farther away from the origin than under whole-body conditions (exception: leg b = 1 DU and $\alpha = 120°$). The systematic differences go back to the s's tendency to underestimate the final turning angle in the visual simulation condition, an effect which was not observed under whole-body conditions.

To determine the potential causes of the systematic differences in homing errors observed between whole-body and visual simulation conditions, we used the Encoding Error Model (EEM) formulated by Fujita et al. (1993). The EEM was developed in order to theoretically account for systematic errors in homing during path completion tasks. The model assumes four component processes to underlie pathway completion: (i) sensing the pathway, (ii) building up a route representation, (iii) computing (and representing) the desired trajectory back to the origin, and (iv) executing the homeward trajectory. According to the EEM, sensing the pathway (i.e., coding distance and turn information while moving along the pathway) is the only determinant of homing errors. While homing in a triangle, navigators have the impression of correctly completing it, as they do not notice errors arising from erratic coding of distance and turn information during locomotion. Mathematically the EEM is specified by four axioms. (1) The internal representation satisfies the Euclidean axioms. (2) The length of a straight line segment is internally coded by a function $d_{subj} = f(d_{obj})$. (3) The value of a turning angle is internally coded by a function $t_{subj} = g(t_{obj})$. (4) There is no systematic error in either computing nor executing the homeward trajectory. Calculating back from the observed homing performances, the model leads to separate distance and turn coding functions. Thus, the EEM allows to reduce the observed complex pattern of homing behavior to two psychophysical coding functions, one for distance and one for turn information.

We applied the model to the combined data from Experiment 1 and 2 and reanalyzed the raw data of Loomis et al. (1993) on the basis of the subset of the nine triangle layouts we chose from their total set of 27 triangles. In applying the model to the triangle completion data three steps can be distinguished: (1) Calculating the subjective coding of the three distances (b = 1, 2, 3 DU) and turning angles ($\alpha = 60°$, $90°$, $120°$) by fitting hypothetical coded values to the observed data using the least squares method. (2) Determining the goodness of fit (R^2) between actual and model fitted data over the total set of triangles. (3) Describing hypothetical distance and turn coding functions underlying the observed homing behavior.[5]

Step 1. The subjectively coded values for the three lengths of leg b (b=1, 2, 3 DU) and the three turning angles ($\alpha = 60°$, $90°$, $120°$) are plotted in Figures 4 (a) and (b). For the visual simulation data (Experiment 1 and 2) the subjective distance coding values obtained by least squares fitting were 1.2 DU, 1.9 DU, and 2.4 DU, while the coded turning angle values were 113°, 120°, and 132°. The reanalysis of Loomis et al. (1993) data yielded values of 1.1 DU, 1.6 DU, and 2.3 DU for distance coding and 75°, 89°, and 102° for turn coding, respectively.

---

[5] We used the EEM only as a tool to describe the observed homing performances. It was not our intention to test any of the theoretical assumptions formulated by the model.

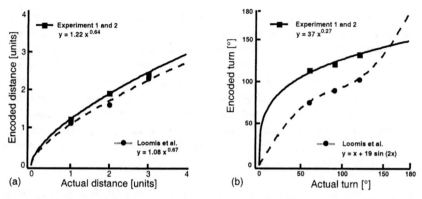

Figure 4. Hypothetical coding functions for (a) distances and (b) turns under conditions of whole-body motion (dashed lines; data from Loomis et al.) and visual simulation (solid lines; combined data from Experiment 1 and 2) (from Péruch, May, & Wartenberg, 1997).

Step 2. Figure 5 shows the homing performances as predicted by the EEM and the observed values for visual simulation and whole-body conditions. The small deviations between predicted and observed values indicate a good fit of the model-derived subjective distance and turn codings for all nine triangles. Correlations between observed and predicted homing errors reveal that the model fit in case of distance coding accounts for 89% of the variance (R^2) under visual simulation conditions and for 94% under whole-body conditions. In the case of turn coding the respective values are 93% for visual simulation conditions and 95% for whole-body conditions. These values are comparable in size to those reported by Fujita et al. (1993) on the basis of the total set of 27 triangles.

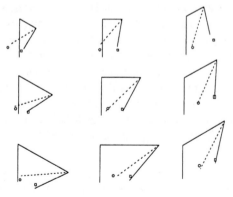

Figure 5. Observed and EEM-predicted homing endpoints. Whole-body motion endpoints (data from Loomis et al., 1993) are shown as dashed lines, model predictions as circles. Visual simulation endpoints (data from Experiment 1 and 2) are shown as solid lines, model predictions as squares.

Step 3. In order to achieve a more general description of the subjects' homing behavior under visual simulation and whole-body conditions we additionally assumed that objective nil movements (zero translations and rotations) are subjectively coded as nil movements.[6] On the basis of this additional assumption and the subjective values obtained in step 1 the coding of distance information under visual simulation and whole-body conditions can be adequately described by power-functions; i.e., relatively small distances are slightly overestimated and the larger distances become, the more they tend to be underestimated. As can be seen in Figure 4a, the coding functions for visual simulation ($y = 1.2 \, x^{0.64}$) and whole-body conditions ($y = 1.1 \, x^{0.67}$) only show marginal differences.

With respect to turn coding under whole-body conditions results reported in the literature indicate that people are quite accurate at judging the amount of whole-body turns over a wide range of angles (Klatzky et al., 1990). Assuming an additional value ($180°$ is coded as $180°$) the subjective turn codings obtained in step 1 (see Fig. 4b) can be described by a function ($y = x + 19 \sin 2x$) oscillating around the function for perfectly correct coding ($y = x$). For the visual simulation experiments the coded values are well described by a power-function ($y = 37 \, x^{0.27}$).[7]

The complex pattern of systematic homing errors found in the different experiments can be consistently accounted for by applying the EEM (Fujita et al., 1993). Coding of distance information under visual simulation as well as under whole-body conditions could be described by similar power functions indicating that differences in distance coding between both modes of navigation were negligible. The observed systematic differences in homing behavior between visual simulation and whole-body conditions seem to go back to pronounced differences in coding turns during locomotion. That these differences indeed are of a systematic nature is indicated by the EEM accounting for about 90% of the variance for whole-body as well as for visual simulation data. Whereas the whole-body movement data (Loomis et al., 1993) can be adequately described by a function oscillating around the correct values (with subjective values in the middle range distorted towards $90°$; $60°$ is coded as $72°$; $120°$ is coded as $102°$), visual simulation data appear to be best described by a power function. Participants under visual simulation conditions showed a strong overestimation of the small turns ($60°$ is coded as $112°$) which diminishes for larger turns ($120°$ is coded as $132°$) and might change to underestimations for larger angles as data from a further experiment suggest (see footnote 7).

---

[6] In contrast to Fujita et al. (1993), we assume nonlinear distance and direction coding functions. Reasons for our preference of nonlinear coding functions lie in the additional assumptions made here.

[7] Independent support for such a function comes from a third experiment not reported here. Using the same VE, subjects were requested to turn into the direction of the assumed starting point after completing the first leg (i.e., reaching the red cylinder). While the required turn was $180°$, averaging the observed turns yielded $150.4° \pm 0.9°$ (SE), corresponding to an underestimation of about $30°$; this is what is to be expected according to the described power function (see Fig. 4 b; dashed line).

6 General Discussion

The experiments reported here aimed at examining triangle completion on the basis of purely visual information and comparing the observed performances to triangle completion under whole-body conditions. It was found that errors in triangle completion were systematically larger for navigators in the VE than for subjects performing the task under whole-body motion conditions (Loomis et al., 1993). Applying the EEM (Fujita et al., 1993) to the data of the different experiments leads to a consistent picture, according to which differences in homing accuracy between whole-body and visual simulation conditions go back to systematic deficits in picking up information about turns under conditions of visually simulated navigations; differences in distance coding between both modes of navigation were negligible. This pattern of results was independent of the final turning direction before homing (Experiment 1 vs. 2) as well as independent of the FOVg (40°, 60°, 80°) realized.

That amount of simultaneously available spatial information during navigation (FOVg) had no substantial influence on the accuracy of homing behavior could be due to the fact that our VEs did not provide enough distinctive visual features (e.g., landmarks) for the variation of FOVg to become effective. A recent study of Neale (1996), in which subjects navigated in a visually richer VE (an office building) than the one used here, indicates that FOVg can indeed exert effects on perception (judging room sizes) and memory (distance estimates) of spatial environments. Research in real world environments leaves open the question whether the amount of simultaneously available spatial information or the exclusion of peripheral visual mechanisms is the determining factor causing decreases in spatial learning performances when the field of view is restricted (Alfano & Michel, 1990; Rieser et al., 1992). Virtual environments could be used for further examinations of this question as they allow for independent variation of FOV and FOVg.

The most important result of the experiments reported here are the systematic deficits found in coding directional information while moving along a path. Research on path-integration has shown that elementary errors in turn coding accumulate over multisegment paths when corrective azimuthal cues (e.g., sun) are missing (Benhamou, Sauvé, & Bovet, 1990). Errors in coding of turns of the magnitudes found in our visual simulation experiments are likely to lead to problems of disorientation and deficits in developing spatial knowledge over extended journeys in more complex environments. The turn coding functions reported here imply that navigators who can make use of whole-body movement information (real-world environment) should be at a considerable advantage compared to navigators who have to exclusively rely on visually simulated information as given in our realization.

7 Conclusions

Taken together, the experimental results reported here indicate that whole-body movement information (vestibular and proprioceptive signals, motor efferent commands) could be a decisive factor in picking up directional information correctly while navigating through space. This does not mean that spatial orientation on the basis of purely visual information is impossible, as for instance the experimental work reviewed earlier in this chapter shows. It rather points to the fact that different sources of information (visual and whole-body) contribute to spatial orientation. VEs have an interesting potential to disentangle the relative contributions of these different information sources, not only as VEs allow to completely isolate visual information from whole-body information - as was the case in our experiments - but also as they allow to vary the amount of whole-body information available by comparing spatial performances in VEs using different interfaces to body movements, as Henry & Furness (1993) for instance do when comparing spatial performance in VEs with head tracking vs. spaceball control. Furthermore, VE technology offers the possibility to investigate the contribution of the various types of visual information to spatial orientation by controlled variation of visual spatial information like FOVg - as done in our experiments - , binocular disparity, or richness of spatial structure (Liu, Tharp, French, Lai, & Stark, 1993; Ruddle et al., 1997; Gillner & Mallot, 1997). The triangle completion paradigm has proved to be well suited to analyze and disentangle the uptake of elementary spatial information about distances and directions. However, for research focusing on more complex forms of spatial knowledge, experiments with more complex pathways in visually enriched, cluttered environments will be needed. In the context of the questions discussed above such research could help to clarify how the structure of more complex spatial knowledge is affected by the interplay of whole-body and visual information.

Acknowledgments

The research reported here was supported by a grant from the French-German cooperation program PROCOPE to P. Péruch and M. May and a grant from the Deutsche Forschungsgemeinschaft (ma-1515-2/1) to M. May. We thank all authors of Loomis et al. (1993) and Fujita et al. (1993) for letting us use their data, and Constanze Wartenberg for comments and suggestions on earlier versions of the manuscript.

References

Alfano, P. L. & Michel, G. F. (1990). Restricting the field of view: Perceptual and performance effects. *Perceptual and Motor Skills, 70*, 35-45.

Batschelet, E. (1981). *Circular statistics in biology.* London, New York: Academic Press.

Beer, J. M. A. (1993). Perceiving scene layout through aperture during visually simulated self-motion. *Journal of Experimental Psychology: Human Perception and Performance, 19*, 1066-1081.

Benhamou, S., Sauvé, J. P., & Bovet, P. (1990). Spatial memory in large scale movements: Efficiency and limitations of the egocentric coding process. *Journal of Theoretical Biology, 145*, 1-12.

Bliss, J. P., Tidwell, P. D., & Guest, M. (1997). The effectiveness of virtual reality for administering spatial navigation training to firefighters. *Presence, 6(1),* 73-86.

Brauer, W., Freksa, C., Habel, C., & Wender, K. F. (1995). *Raumkognition: Repräsentation und Verarbeitung räumlichen Wissens. Vorschlag zur Einrichtung eines DFG-Schwerpunktprogrammes.*

Cutting, J. E. (1986). *Perception with an eye for motion.* Cambridge, MA: MIT Press.

Cutting, J. E. (1997). How the eye measures reality and virtual reality. *Behavior Research Methods, Instruments, & Computers, 29,* 27-36.

Distler, H. & Bülthoff, H. (1996). Psychophysical experiments and virtual environments. *Poster presented at the virtual reality world 1996.*

Durlach, N. I. & Mavor, A. S. (1995). *Virtual reality. Scientific and technological challenges.* Washington, D.C.: National Academy Press.

Ellis, S. R. (1991). Pictorial communication: Pictures and the synthetic universe. In S. R. Ellis, M. K. Kaiser, & A. C. Grunwald (Eds.), *Pictorial communication in virtual and real environments* (pp. 22-40). London: Taylor & Francis.

Ellis, S. R. (1994). What are virtual environments? *IEEE Computer Graphics & Applications,* 17-22.

Fujita, N., Klatzky, R. L., Loomis, J. M., & Golledge, R. G. (1993). The encoding-error model of pathway completion without vision. *Geographical Analysis, 25,* 295-314.

Gallistel, C. R. (1990a). Representations in animal cognition: An introduction. *Cognition, 37,* 1-22.

Gallistel, C. R. (1990b). *The organization of learning.* Cambridge, MA: MIT Press.

Gibson, J. J. (1979). *The ecological approach to visual perception.* Boston: Houghton-Mifflin.

Gillner, S. & Mallot, H. A. (1997). Navigation and acquisition of spatial knowledge in a virtual maze. *Max-Planck-Instiut für bilogische Kybernetik, Technical Report, 45.*

Golledge, R.G., Klatzky, R.L., & Loomis, J.L. (1996). Cognitive mapping and wayfinding by adults without vision. In J. Portugali (Ed.), *The construction of cognitive maps* (pp. 215-246). Dordrecht: Kluwer.

Haber, L., Haber, R. N., Penningroth, S., Novak, N., & Radgowski, H. (1993). Comparison of nine methods of indicating the direction to objects: Data from blind adults. *Perception, 22,* 35-47.

Henry, D. & Furness, T. (1993). Spatial perception in virtual environments. *Proceedings IEEE Virtual Reality Annual International Symposium (VRAIS), Seattle, WA,* 33-40.

Johansson, G. & Börjesson, E. (1989). Toward a new theory of vision studies in wide-angle space perception. *Ecological Psychology, 1,* 301-331.

Kalawsky, R. S. (1993). *The science of virtual reality and virtual environments.* New York: Addison-Wesley.

Klatzky, R. L., Loomis, J. M., & Golledge, R. G. (1997). Encoding spatial representations through nonvisually guided locomotion: Tests of human path integration. In D. Medin (Ed.), *The Psychology of Learning and Motivation* (Vol. 37, pp. 41-84). San Diego, CA: Academic Press.

Klatzky, R. L., Loomis, J. M., Golledge, R. G., Cicinelli, J. G., Doherty, S., & Pellegrino, J. W. (1990). Acquisition of route and survey knowledge in the absence of vision. *Journal of Motor Behavior, 22,* 19-43.

Kozak, J. J., Hancock, P. A., Arthur, E. J., & Chrysler, S. T. (1993). Transfer of training from virtual reality. *Ergonomics, 36,* 777-784.

Levine, M. (1982). You-are-here maps: Psychological considerations. *Environment and Behavior, 14,* 221-237.

Liu, A., Tharp, G., French, L., Lai, S., & Stark, L. (1993). Some of what one needs to know about using head-mounted displays to improve teleoperator performance. *IEEE Transactions on Robotics and Automation, 9*, 638-648.

Loomis, J. M., DaSilva, J. A., Fujita, N., & Fukusima, S. S. (1992). Visual space perception and visually directed action. *Journal of Experimental Psychology: Human Perception and Performance, 18*, 906-921.

Loomis, J. M., Klatzky, R. L., Golledge, R. G., Cicinelli, J. G., Pellegrino, J. W., & Fry, P. A. (1993). Nonvisual navigation by blind and sighted: Assessment of path integration ability. *Journal of Experimental Psychology: General, 122*, 73-91.

Maurer, R. & Séguinot, V. (1995). What is modelling for? A critical review of the models of path integration. *Journal of Theoretical Biology, 175*, 457-475.

May, M. (1996). Cognitive and embodied modes of spatial imagery. *Psychologische Beiträge, 38*, 418-434.

May, M., Péruch, P., & Savoyant, A. (1995). Navigating in a virtual environment with map-acquired knowledge: Encoding and alignment effects.. *Ecological Psychology, 7*, 21-36.

Nadel, L. (1990). Varieties of spatial cognition. Psychobiological considerations. *Annals of the New York Academy of Sciences, 608*, 613-636.

Neale, D. C. (1996). Spatial perception in desktop virtual environments. In: *Proceedings of the 40th Annual Meeting of the Human Factors and Ergonomics Society*. Santa Monica: Human Factors Society.

O'Keefe, J. & Nadel, L. (1978). *The hippocampus as a cognitive map*. Oxford: Oxford University Press.

Péruch, P., May, M., & Wartenberg, F. (1997). Homing in virtual environments: Effects of field of view and path-layout. *Perception, 26*, 301-311.

Péruch, P., Vercher, J.-L., & Gauthier, G. M. (1995). Acquisition of spatial knowledge through visual exploration of simulated environments. *Ecological Psychology, 7*, 1-20.

Pimentel, K. & Teixeira, K. (1993). *Virtual reality. Through the new looking glass*. New York: McGraw-Hill.

Poucet, B. (1993). Spatial cognitive maps in animals: New hypotheses on their structure and neural mechanisms. *Psychological Review, 100*, 163-182.

Presson, C. C. & Montello, D. R. (1994). Updating after rotational and translational body movements: Coordinate structure of perspective space. *Perception, 23*, 1447-1455.

Rieser, J. J. (1989). Access to knowledge of spatial structure at novel points of observation. *Journal of Experimental Psychology: Learning, Memory and Cognition, 15*, 1157-1165.

Rieser, J. J. & Garing, A. E. (1994). Spatial orientation. *Encyclopedia of human behavior, 4*, 287-295.

Rieser, J. J., Garing, A. E., & Young, M. F. (1994). Imagery, action and young children's spatial orientation: It's not being there that counts, it's what one has in mind. *Child Development, 65*, 1254-1270.

Rieser, J. J., Guth, D. A., & Hill, E. W. (1986). Sensitivity to perceive structure while walking without vision. *Perception, 15*, 173-188.

Rieser, J. J., Hill, E. W., Talor, C. R., Bradfield, A., & Rosen, S. (1992). Visual experience, visual field size, and the development of nonvisual sensitivity to spatial structure of outdoor neighborhoods explored by walking. *Journal of Experimental Psychology: General, 121*, 210-221.

Ruddle, R. A., Payne, S. J., & Jones, D. M. (1997). Navigating buildings in "desk-top" virtual environments: Experimental investigations using extended navigational experience. *Journal of Experimental Psychology: Applied, 3*, 143-159.

Schölkopf, B. & Mallot, H. A. (1995). View-based cognitive mapping and path planning. *Adaptive Behavior, 3,* 311-348.

Sedgwick, H. A. (1982). Visual modes of spatial orientation. In M. Potegal (Ed.), *Spatial abilities. Development and physiological foundations* (pp. 3-33). New York: Academic Press.

Sedgwick, H. A. (1986). Space perception. In K. R. Boff, L. Kaufman, & J. P. Thomas (Eds.), *Handbook of human perception and performance: Vol. 1 Sensory processes and perception* (pp. 21.1-21.57). New York: Wiley.

Sholl, M. J. (1996). From visual information to cognitive maps. In J. Portugali (Ed.), *The construction of cognitive maps* (Vol. 2, pp. 157-186). Dordrecht: Kluwer.

Strelow, E. R. (1985). What is needed for a theory of mobility: Direct perception and cognitive maps - lessons from the blind. *Psychological Review, 92,* 226-248.

Thinus-Blanc, C. & Gaunet, F. (1997). Representation of space in blind persons: Vision as a spatial sense? *Psychological Bulletin, 121,* 20-42.

Thorndyke, P. W. & Hayes-Roth, B. (1982). Differences in spatial knowledge acquired from maps and navigation. *Cognitive Psychology, 14,* 560-589.

Tong, F. H., Marlin, S. G., & Frost, B. J. (1995). Visual-motor integration and spatial representation in a visual virtual environment. *Investigative Ophthalmology and Visual Science, 36,* 1679.

Vishton, P. M. & Cutting, J. (1995). Wayfinding, displacements, and mental maps: Velocity fields are not typically used to determine one's aimpoint. *Journal of Experimental Psychology: Human Perception and Performance, 21(5),* 978-995.

Wagner, M. (1985). The metric of visual space. *Perception & Psychophysics, 38,* 483-495.

Wickens, C. D. & Baker, P. (1995). Cognitive issues in virtual reality. In W. Barfield & T. Furness (Eds.), *Virtual reality* (pp. 515-541). New York: Oxford University Press.

Williams, H. P., Hutchinson, S., & Wickens, C. D. (1996). A comparison of methods for promoting geographic knowledge in simulated aircraft navigation. *Human Factors, 38,* 50-64.

Wilson, P.N. & Foreman, N. (1993). Transfer of information from virtual to real space: Implications for people with physical disability. *Eurographics Technical Report Series, 93,* 21-25.

Author Index

Springer
and the
environment

 Springer

Lecture Notes in Artificial Intelligence (LNAI)

Lecture Notes in Computer Science

Lightning Source UK Ltd.
Milton Keynes UK
UKOW051221280313

208283UK00001B/4/A